Long Non-Coding RNAs - Function, Mechanisms, and Applications

Edited by Yusuf Tutar and Lütfi Tutar

Published in London, United Kingdom

Long Non-Coding RNAs - Function, Mechanisms, and Applications
http://dx.doi.org/10.5772/intechopen.1005933
Edited by Yusuf Tutar and Lütfi Tutar

Contributors
Abhijit Mandal, Akila Prashant, Anita Mann, Duygu Sari-Ak, Félix Recillas-Targa, Gopal V. Marviya, Gustavo Tapia-Urzúa, Hober Nelson Núñez-Martínez, Hossein Azizi, Komal G. Lakhani, Lütfi Tutar, Melika Emarati, Nida Shabbir, Nima Ghasemi, Poonam Ranga, Prashant Vishwanath, Priyanka Sharma, Rasmieh Hamid, Sanaz Kavoosi, Sarbani Giri, Silvia Scalabrin, Stefano Cagnin, Sumaiya Kausar S. Kalaigar, Swati Kumari, Sylvia Garza-Manero, Thomas Skutella, Varsha Ranga, Wahid Ali, Yusuf Tutar

First published in London, United Kingdom, 2025 by IntechOpen
IntechOpen is the global imprint of INTECHOPEN LIMITED, registered in England and Wales, registration number: 11086078, 167-169 Great Portland Street, London, W1W 5PF, United Kingdom

For EU product safety concerns: IN TECH d.o.o., Prolaz Marije Krucifikse Kozulić 3, 51000 Rijeka, Croatia, info@intechopen.com or visit our website at intechopen.com.

British Library Cataloguing-in-Publication Data
A catalogue record for this book is available from the British Library

Long Non-Coding RNAs - Function, Mechanisms, and Applications
Edited by Yusuf Tutar and Lütfi Tutar
p. cm.

This title is part of the Genetics Book Series, Volume 6
Topic: Genomics
Series Editor: Kenji Ikehara
Topic Editor: Irina Vlasova-St. Louis

Print ISBN 978-1-83634-479-7
Online ISBN 978-1-83634-478-0
eBook (PDF) ISBN 978-1-83634-480-3
ISSN 3049-7094

If disposing of this product, please recycle the paper responsibly.

IntechOpen Book Series

Genetics

Volume 6

Aims and Scope of the Series

"Genetics," which has been proud of its tradition since Mendel presented his research results in 1865, initially progressed quite slowly due to simple observational approaches of individuals and groups. However, the discovery of double-stranded DNA by Watson and Crick about 70 years ago triggered rapid progress in life sciences, including genetics, which was primarily conducted using Escherichia coli and bacteriophages infecting E. coli. Subsequently, genetics has achieved remarkable developments, such as understanding genetic disorders, including cancers, through research on the biogenesis and differentiation of plants and animals.
The two topics of this book series - Human Genetics, and Genomics - will address important areas of advancement in genetics.
Human Genetics: After fundamental genetics, initially studied with the main goal of revealing the functions of individual genes and proteins, genetics expanded from understanding the genetic system itself to understanding many infectious diseases caused by bacteria and viruses. Consequently, human beings are now overcoming infectious diseases by developing medicinal chemicals, including antibiotics and vaccines. However, genetic disorders remain challenging to cure up to now. Nevertheless, even the cure for them, including various cancers, is coming closer to reality due to the rapid progress of human genetics. In this way, the welfare of human life continues to improve, and even longevity, which was once a dream, has been achieved to some extent in recent years.
Genomics: On the other hand, the understanding of the comprehensive interrelationship of whole genes or whole proteins functioning in one organism has become possible now, as research has entered the era of genomics, owing to the rapid progress of base sequence analysis and bioinformatics. The development of genomics has further made it possible to understand the evolutionary processes of organisms through comparative studies among the genomes of many organisms.
This book series will discuss the findings obtained during the advancement of human genetics and genomics. It is also expected that this series will trigger the formation of a better world composed of human beings and all other organisms on Earth through discussions of research results obtained under the development of general genetics.

Meet the Series Editor

Kenji Ikehara graduated from the Department of Industrial Chemistry, Faculty of Engineering, Kyoto University in 1968. He received his B. Eng. (1968) and subsequently earned M. Eng. (1970) and D. Eng. (1976) degrees from Kyoto University. He began his career as a research associate in the Faculty of Science at the University of Tokyo before moving on to become an associate professor in the Faculty of Science at Nara Women's University. He was later promoted to professor and subsequently served as the dean of the Faculty of Science at Nara Women's University. Additionally, he held the position of director at the Nara Study Center of the Open University of Japan. For approximately 15 years, he focused his research on sporulation initiation of Bacillus subtilis. Later, he shifted his focus to the origins and evolutionary processes of microbial genes, the genetic code, proteins, and life. He has proposed several hypotheses, including the GC-NSF(a) hypothesis on the origin of genes, the GNC-SNS hypothesis on the genetic code, the protein 0th-order structure hypothesis on the origin of proteins, and the [GADV]-protein world hypothesis (GADV hypothesis) on the origin of life. Furthermore, he served as the local chair of the International Conference, Origin 2014, held in Nara in 2014.

Meet the Volume Editors

Dr. Yusuf Tutar currently conducts his research at the Recep Tayyip Erdoğan University, Faculty of Medicine, Division of Biochemistry. Dr. Tutar is the Dean of the Health Sciences Faculty. He obtained his M. Sc. and Ph.D. at Oregon State University and Texas Tech University, respectively. He pursued his postdoctoral studies at Rutgers University Medical School and the National Institutes of Health (NIH/NIDDK), USA. His research focuses on Biochemistry, Biophysics, Genetics, Molecular Biology, and Molecular Medicine with specialization in the fields of drug design, protein structure-function, protein folding, prion, microRNA, pseudogenes, molecular cancer, epigenetics, metabolites, proteomics, genomics, protein expression and characterization by spectroscopic and calorimetric methods. Having received 22 awards from national and international institutions, Dr. Tutar is the editor of several leading scientific journals and books.

Dr. Lütfi Tutar is currently an Assistant Professor at the Department of Molecular Biology and Genetics, Faculty of Art and Sciences, Kırşehir Ahi Evran University, Turkey. His interdisciplinary research focuses on bioinformatics analysis of high-throughput data, microRNAs, small RNAs, and Heat Shock Proteins (HSPs) in human diseases and other multicellular organisms.

Contents

Preface

Long non-coding RNAs (lncRNAs) play a multifaceted and critical role in cancer progression through diverse mechanisms affecting gene expression and cellular behavior. These RNA molecules, which do not encode proteins, regulate gene expression at multiple levels, including chromatin remodeling, transcriptional control, and post-transcriptional modulation. In cancer, lncRNAs can interact with chromatin-modifying complexes, serving as scaffolds that facilitate epigenetic changes to activate oncogenes or silence tumor suppressor genes. This contributes to tumorigenesis by altering the transcriptional landscape of cancer cells and promoting uncontrolled proliferation.

LncRNAs also function as competing endogenous RNAs (ceRNAs), sequestering tumor-suppressive microRNAs (miRNAs), thereby preventing miRNAs from binding to their mRNA targets. This modulation enhances the expression of oncogenic mRNAs, further supporting cancer cell growth, survival, and metastasis. Furthermore, lncRNAs regulate epithelial-mesenchymal transition (EMT), a vital process in cancer metastasis where epithelial cells acquire mesenchymal traits facilitating migration and invasion. By orchestrating the gene networks involved in EMT, lncRNAs contribute to tumor dissemination and resistance to anticancer therapies.

Beyond EMT, lncRNAs influence key cellular processes such as apoptosis, metabolism, and drug resistance, impacting cancer progression and therapeutic outcomes. Their aberrant expression patterns are often tumor type-specific, making them promising biomarkers for cancer diagnosis, prognosis, and stratification of patient subgroups. The complex regulation by lncRNAs extends to various signaling pathways critical for maintaining cellular homeostasis and responding to microenvironmental cues, highlighting their role as central regulators in cancer biology.

Given their diverse functions and disease associations, lncRNAs are emerging as potential targets for novel therapeutic interventions and as components of liquid biopsy platforms for non-invasive cancer detection. Ongoing research explores their utility in modulating gene expression networks to counteract tumor progression and improve treatment responses. Overall, lncRNAs represent a dynamic and versatile class of molecules that substantially influence cancer development and metastasis, underscoring their importance in both basic research and clinical applications.

This book has four sections for lncRNAs: 1. Biochemistry of Long Non-Coding RNAs, 2. Long Non-Coding RNA's in Diseases, 3. Long Non-Coding RNA's in Biotechnology, and 4. Long Non-Coding RNA's in Plants. The biochemistry section covers four chapters that provide a comprehensive review and overview of the relationships of pseudogenes with lncRNAs, X-Chromosome Dosage Compensation, and skeletal pathophysiology. The disease section covers lncRNAs as therapeutic targets for β-Thalassemia and cardiovascular disorder. The book covers CRISPR technology and stem cells over lncRNA in the Biotechnology section. The final section focuses on lncRNAs' role in stress tolerance and crop resilience. The content provides recent advances in the field to the book's audience.

Dr. Yusuf Tutar
Professor,
Faculty of Medicine,
Division of Biochemistry,
Recep Tayyip Erdogan University,
Rize, Türkiye

Dr. Lütfi Tutar
Assistant Professor,
Faculty of Science and Arts,
Department of Molecular Biology and Genetics,
Kırşehir Ahi Evran University,
Kırşehir, Türkiye

Section 1

Biochemistry of Long Non-Coding RNAs

Chapter 1

Perspective Chapter: Decoding Cancer's Silent Players – A Comprehensive Guide to LncRNAs

Abhijit Mandal and Sarbani Giri

Abstract

Long non-coding RNAs (LncRNAs) are RNAs that do not code for proteins and were thus earlier known as Junk RNAs. Recently, LncRNAs have emerged as critical regulators in the expression of coding genes and various important biological signaling pathways, thus controlling crucial biological and developmental processes. Reports of LncRNAs association with several diseases including cancer have also been implicated. LncRNAs play a crucial diverse role in regulating cancer pathways, thus influencing tumorigenesis, progression, and metastasis. They can function both as oncogenes or tumor suppressors, modulating key signaling pathways and cellular processes. Mutation or epigenetic-induced aberrant expression of LncRNAs dysregulates different essential biological pathways, leading to malignant phenotype and cancer hallmarks in different types of cancer. Tumor cells secrete specific endogenous LncRNAs into biological fluids depending on the cancer type, giving rise to stable circulating LncRNAs, thus proving to be of great potential as non-invasive or minimally invasive diagnostic biomarkers. In this chapter, we explore the multifaceted roles of LncRNAs in various cancer types, highlighting their potential as diagnostic/prognostic biomarkers and therapeutic targets. Additionally, we discuss innovative strategies for targeting LncRNAs in cancer treatment, including RNA interference and CRISPR technology. This chapter will provide a comprehensive overview of LncRNAs' implications in cancer research and personalized medicine.

Keywords: long non-coding RNAs, cancer, mechanisms, biomarkers, prognosis, therapeutic target, personalized medicine

1. Introduction

With the advent and development of advanced high-throughput sequencing technologies such as microarray and RNAseq, besides coding genes, many non-coding RNAs that do not code for proteins were identified. The most distinguished non-coding RNAs are microRNAs (miRNAs) and long non-coding RNAs (LncRNAs). LncRNAs are transcripts that are greater than 200 nucleotides in length and have no long open reading frames (>100 amino acids), regulating gene expression of target

 IntechOpen

genes at transcriptional, post-transcriptional, translational, and epigenetic levels by acting as guides, scaffolds, decoys, or signals [1–5]. LncRNAs play vital roles in important biological and developmental processes, such as genomic imprinting, X-chromosome inactivation, chromatin modification, and alternative splicing, as well as affecting various cellular activities, such as proliferation, differentiation, and survival [1, 6]. Due to their multifaceted and diverse functions, LncRNAs have been implicated in a large number of human diseases including coronary disorders, type 2 diabetes, neurological disorders, and various types of cancer [6]. Other than cardiovascular disorders, cancer is thought to be the leading cause of mortality worldwide [7]. Increasing evidence of LncRNAs association with cancer has been found in the literature [1, 8]. LncRNAs play a crucial role in regulating cancer pathways, thus influencing tumorigenesis, progression, and metastasis [1]. Literature suggests that altered expression of LncRNAs in tumors is closely linked to the manifestation of cancer hallmarks such as unregulated proliferative signaling, avoiding growth suppressors, preventing cell death (apoptosis), uncontrolled replication, triggering angiogenesis, initiating invasion and metastasis, aberrant metabolic pathways, immune evasion, genomic instability, and inflammation [5]. Although the mechanisms are poorly understood, LncRNAs play a pivotal role in tumorigenesis and progression by regulating gene expression and fine-tuning signaling pathways [1, 9]. Several LncRNAs, such as MALAT1, PCA3, HOTAIR, ANRIL, and MEG3, have been reported to be upregulated or downregulated in a variety of human cancers [8, 10–12]. Research indicates that LncRNAs, known for their high specificity and precision, hold potential as cancer biomarkers [12, 13]. Their unique expression and diverse functions across cancer types make them promising for diagnosis, prognosis, and treatment. Additionally, LncRNAs can be collected noninvasively from body fluids, tissues, and cells, serving either as independent or supplementary biomarkers to enhance diagnostic and prognostic accuracy [13]. This chapter provides an overview of the roles of different LncRNAs in the development and progression of various cancer types, emphasizing their promise as diagnostic, prognostic markers, and therapeutic targets (**Figure 1**). It also covers novel approaches to targeting LncRNAs, such as RNA interference (RNAi) and CRISPR, offering a thorough overview of their impact on cancer research and personalized medicine.

2. LncRNAs in tumorigenesis and cancer progression

LncRNAs can function both as oncogenes or tumor suppressors, regulating various aspects of cancer initiation and progression [14]. Abnormal expression of oncogenic and tumor-suppressive LncRNAs disrupts tumor suppressor gene activity by reducing their transcription and translation, while promoting the expression and translation of oncogenes. This imbalance drives processes such as cellular proliferation, differentiation obstruction, migration, invasion, metastasis, genomic instability, transformation into malignant type, tumor initiation and progression, and resistance to chemotherapy and radiotherapy [10].

2.1 LncRNAs in tumorigenesis

Several LncRNAs have been reported to be overexpressed in a variety of human cancers, for example, MALAT1 [10, 15], HOTAIR [16, 17], ANRIL [18], etc., whereas several others are downregulated, for example, MEG3 [19], PANDA [20], PCAT29

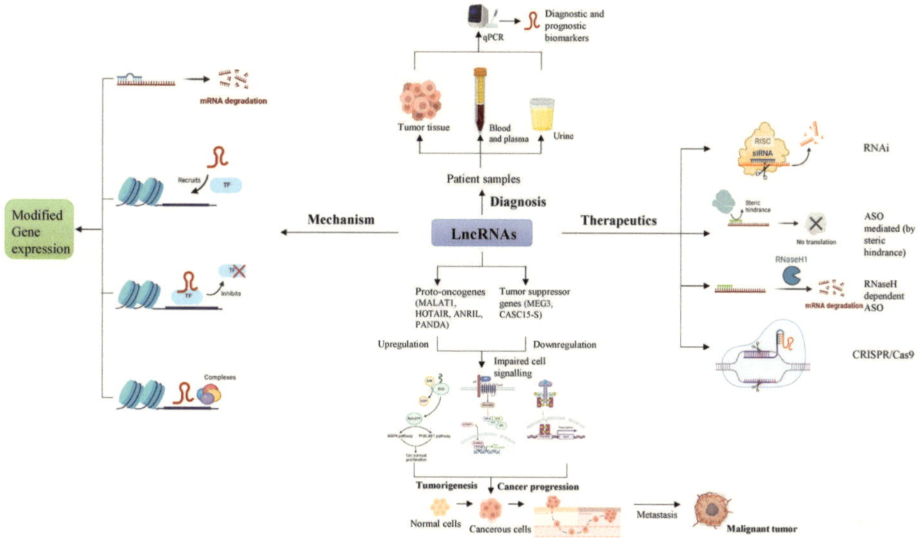

Figure 1.
Schematic representation of the influence of LncRNAs in cancer research, highlighting their role in tumorigenesis, cancer progression, diagnosis/prognosis and therapeutics.

[21], and CASC15-S [22]. Also, it has to be noted that the same LncRNA can either act as an oncogene or a tumor suppressor depending on the type of cancer; for example, MALAT1 was initially reported to be upregulated in human mantle cell lymphoma, renal cell carcinoma tissues, bladder cancer, pancreatic ductal adenocarcinoma tissues, human melanoma, ovarian cancer, lung cancer, and liver cancer [10]. On the other hand, two recent in-depth studies show that MALAT1 has tumor-suppressive effects in colorectal and breast cancers [10, 23]. Thus, demonstrating the cancer type-specific role of LncRNAs in cancer development.

MALAT1 (metastasis-associated in lung adenocarcinoma transcript 1) is one of the most researched LncRNAs in cancer because it is readily detectable and functionally investigated, and it is substantially expressed in cancer tissues [10]. Cancer tissues are thought to release MALAT1, an intergenic transcript found on human chromosome 11q13.1, into the bloodstream through exosomes since it is highly expressed in both serum and cancer tissues. As a result, its expression in blood is similar to that of the primary tumors [11]. MALAT1 was initially found to be significantly linked to non-small cell lung cancer (NSCLC) metastases [5]. Further research showed that, in contrast to normal cervical squamous cell samples, LncRNA-MALAT1 expression was elevated in cervical cancer cell lines [24]. Also, the laryngeal cancer cells' expression levels of LncRNA-MALAT1, miRNA-503-5p, and FOXK1 were investigated. The levels of miRNA-503-5p were low, while those of LncRNA-MALAT1 and FOXK1 were high. This implies that LncRNA-MALAT1 and miRNA-503-5p have a negative correlation and that miRNA-503-5p is one of its target regulatory genes [11]. Despite the fact that MALAT1 has been found to be overexpressed in the majority of human cancer tissues, its oncogenic or tumor-suppressive function in breast cancer remains up for debate [10]. MALAT1 has been found to be abnormally upregulated in human breast cancer tissues, and high expression levels of MALAT1 are associated with a poor prognosis for patients. By binding miRNA-1, miRNA-124, and miRNA-448, MALAT1 functions as a competitive endogenous RNA (ceRNA) to decrease CDC42 and increase CDK4 expression, which promotes invasion, migration, and cell cycle progression in breast cancer [10, 25]. However, recent comprehensive studies show that MALAT1 has tumor-suppressive effects in breast cancers. By binding and sequestering miRNA-17, miRNA-20a, and miRNA-106b, PTEN increases the expression of MALAT1, which in turn inhibits the migration and invasion of breast cancer cells by lowering integrin β4 (ITGB4) and the pro-metastatic Epithelial Cell Adhesion Molecule (EpCAM) [23].

LncRNA HOTAIR (HOX Transcript Antisense Intergenic RNA) dysregulation is also frequently found in several types of human cancer. Because it is overexpressed in many malignancies, HOTAIR can function as a pro-oncogene and be linked to a number of cancer hallmarks, including apoptosis inhibition, cellular proliferation, and genomic instability [5]. Overexpression of HOTAIR, an oncogene that is transcribed from the HOXC locus during normal development, can accelerate the development of gastric cancer [9]. Progesterone receptor (PGR), protocadherin 10 (PCDH10), protocadherin β5 (PCDHB5), and junctional adhesion molecule 2 (JAM2) are tumor suppressors that can be inhibited by overexpression of HOTAIR, hence increasing tumor growth [9, 26, 27]. The aberrant expression of LncRNA HOTAIR promotes the proliferation, cell cycle, and migration of gastric cancer cells, and it can suppress the expression of miRNA-217 and increase the resistance of gastric cancer cells to doxorubicin and paclitaxel [9, 28]. It has also been reported that prostate cancer (PCa) tissues and cells have high levels of HOTAIR expression which is linked to anti-apoptosis, migration, invasion, proliferation, and tumor formation [5]. LncRNA HOTAIR has also been reported to be upregulated in breast carcinoma, colorectal cancer (CRC), hepatocellular carcinoma (HCC), laryngeal squamous cell carcinoma (LSCC) tissues [5]. Peng et al. [29] in their study have shown that HOTAIR downregulates miRNA-34a to promote the development of colon cancer.

MEG3 (Maternally Expressed 3), a tumor suppressor on chromosome 14q32.2, is typically downregulated in cancer cells. In bladder cancer, MEG3 overexpression induces autophagy and promotes P53 accumulation [9]. In NSCLC, elevated Heterogeneous nuclear ribonucleoprotein A2B1 (HNRNPA2B1) expression correlates with distant metastasis, poor survival, and serves as an independent prognostic marker. HNRNPA2B1 knockdown *in vitro* and *in vivo* reduces cell proliferation and metastasis, while its overexpression has the opposite effect. Mechanistically, HNRNPA2B1 mediates m^6A modification of MEG3, and its inhibition leads to decreased MEG3 m^6A while increasing the mRNA levels. MEG3 functions as a miRNA-21-5p sponge to upregulate PTEN, inhibiting PI3K/AKT signaling and suppressing proliferation and invasion. Low MEG3 or high miRNA-21-5p expression predicts poor NSCLC survival, identifying the HNRNPA2B1/MEG3/miRNA-21-5p/PTEN axis as a potential therapeutic target [19].

LncRNA-TSLNC8 (Tumor Suppressor Long Non-Coding RNA on Chromosome 8p12) plays a critical role in various cancers, including glioma, liver, lung, breast, melanoma, and gastric cancers [30]. Acting as a tumor suppressor, TSLNC8 inhibits tumor progression and reduces chemoresistance when overexpressed. It functions as a ceRNA, sequestering specific microRNAs, and is involved in multiple signaling pathways implicated in cancer development and progression [30].

The oncofetal LncRNA H19, typically repressed postnatally, is re-expressed in various cancers, including HCC and rectal cancer [31]. Although not a classical oncogene, H19 was the first LncRNA identified as overexpressed in HCC. It influences cancer progression through pathways such as acting as a precursor to miRNA-675, whose increase promotes tumor growth, or functioning as a ceRNA that sponges miRNAs like let-7, reducing tumor suppressor activity and encouraging proliferation [9, 31]. Cancer susceptibility candidate

9 (CASC9) is a LncRNA overexpressed in oral SCC and HCC. It binds to hnRNP L to form a complex that regulates HCC cell viability by modulating the PI3K/AKT signaling pathway [32]. Small nucleolar RNA host gene 1 (SNHG1), located on chromosome 11q12.3, hosts eight small nucleolar RNAs (snoRNAs), which is another LncRNA overexpressed in various cancers [33]. In HCC, LncRNA SNHG1 levels are elevated in tissues and cell lines compared to normal counterparts. It promotes HCC progression by sponging miRNA-195, increasing AEG-1 protein expression [33, 34]. SNHG1 is also significantly upregulated in breast cancer, esophageal squamous cell carcinoma, PCa, NSCLC, colon cancer, and bladder cancer [33].

Several other LncRNAs have been reported to be involved in cancer development and progression, which has been mentioned in **Table 1.**

LncRNA	Cancer type	Oncogenic or Tumor suppressor	Expression	Mechanism	Reference
MALAT1	Colorectal	Tumor suppressor	Down-regulated	Non-canonical PTEN miRNA- MALAT1 axis	[23]
	Breast	Tumor suppressor	Down-regulated	Non-canonical PTEN miRNA- MALAT1 axis	[23]
	Bladder	Oncogenic	Upregulated	TGF-β induced MALAT 1 upregulation resulting in EMT	[15]
	Gallbladder	Oncogenic	Upregulated	Functions as a ceRNA to regulate miRNA-206	[35]
	Malignant Melanoma (Skin)	Oncogenic	Upregulated	Acting as a sponge for miRNA-22	[36]
PVT1	Lung	Oncogenic	Upregulated	Interacts with miRNAs and proteins	[37]
CASC9	HCC	Oncogenic	Upregulated	PI3K/AKT-signaling cascade	[32, 38]
PCAT29	Prostate	Tumor suppressor	Down-regulated	Directly regulated by the androgen receptor (AR), which binds to the promoter of PCAT29	[21]
loc285194	Colon	Tumor suppressor	Down-regulated	Reciprocal repression of loc285194 and miRNA-211 possibly via pathway involving RISC complex	[13]
DRAIC	Prostate	Tumor suppressor	Down-regulated	AR binds to the *DRAIC* locus and represses *DRAIC* expression	[39]
TUG1	Colorectal	Oncogenic	Up-regulated	SP1, an upstream transcription factor interacts with TUG1 and regulates the downstream miRNA-421/KDM2A/ERK axis	[40]

LncRNA	Cancer type	Oncogenic or Tumor suppressor	Expression	Mechanism	Reference
CCAT1	hepatocellular carcinoma	Oncogenic	Up-regulated	Acting as a let-7 sponge	[41]
	gallbladder cancer	Oncogenic	Up-regulated	Acting as a miRNA-218 sponge	[42]
GMDS-AS1	Colorectal	Oncogenic	Up-regulated	GMDS-AS1 targets HuR directly activating STAT3/Wnt signaling	[43]
MEG3	NSCLC	Tumor suppressor	Down-regulated	HNRNPA2B1-mediated m^6A modification of the LncRNA MEG3 facilitates tumor growth and metastasis by modulating the miRNA-21–5p/PTEN pathway.	[19]
PTTG3P	NSCLC	Oncogenic	Up-regulated	PTTG3P/ILF3/E2F1 axis	[44]
G077640	ESCC	Oncogenic	Up-regulated	Enhancing HIF1α stability and downstream glycolytic signaling pathway	[45]
GAS5	Cervical	Tumor suppressor	Down-regulated	siRNA-mediated knockdown of GAS5	[46]
HOTAIR	Pancreatic	Oncogenic	Up-regulated	Polycomb Repressive Complex 2 (PRC2) -dependent and -independent	[16]
	Gastric	Oncogenic	Up-regulated	Possibly via interaction with PRC2	[17]
ANRIL	NSCLC	Oncogenic	Up-regulated	Possibly by regulating (silencing) KLF2 and P21 transcription	[18]
TP73-AS1	Ovarian	Oncogenic	Up-regulated	Modulation of MMP2 and MMP9	[47]

Table 1.
List of some LncRNAs reported to be involved in cancer development and progression.

2.2 LncRNAs and metastasis

Metastasis, the spread of cancer to distant sites, is a leading cause of cancer-related deaths. It involves steps like epithelial–mesenchymal transition (EMT), migration, anoikis resistance, and angiogenesis [48, 49]. LncRNAs play a key role in promoting metastasis by driving EMT [5], angiogenesis, invasion, migration [32, 50, 51], organ-specific colonization, apoptosis evasion [37], and shaping the metastatic tumor microenvironment [49]. They directly regulate metastasis *in vitro* and *in vivo* across various cancers [48].

LncRNA MALAT1 was initially linked to NSCLC metastasis. Its overexpression promotes brain metastasis in NSCLC by driving EMT and enhances tumor proliferation and metastasis in osteosarcoma through the PI3K/Akt pathway [5]. MALAT1 also regulates EMT *via* this pathway in osteosarcoma progression and influences HCC through the mTOR pathway [5, 52, 53]. In ESCC, MALAT1 upregulation promotes proliferation and metastasis by dephosphorylation of the ATM-CHK2 pathway, while in gastric cancer, it accelerates tumor growth and spread [54, 55]. Clinical studies identify MALAT1 overexpression as a poor prognostic marker in pancreatic cancer [5]. In breast cancer, MALAT1 enhances angiogenesis, potentially *via* miRNA-145 regulation [56]. MALAT1's role in cancer metastasis is primarily mediated through EMT [48].

Hou et al. [57] found that LncRNA-ROR is upregulated in breast tumor tissues, promoting EMT by acting as a ceRNA for miRNA-205 in mammary epithelial cells. In triple-negative breast cancer (TNBC), ROR serves as a ceRNA sponge for miRNA-145, affecting its target ARF6, a key regulator of tumor cell invasion and metastasis. ARF6 influences E-cadherin localization and cell–cell adhesion. These findings highlight the LncRNA-ROR/miRNA-145/ARF6 pathway in regulating TNBC metastasis [48, 58]. LncRNA-TTN-AS1, an oncogene overexpressed in ESCC tissues and cell lines, promotes proliferation and metastasis. It enhances Snail1 expression by sponging miRNA-133b, triggering the EMT cascade. Additionally, LncRNA-TTN-AS1 upregulates FSCN1 expression through miRNA-133b sponging and HuR upregulation, driving ESCC invasion. miRNA-133b directly regulates FSCN1, a key protein linked to ESCC metastasis [59]. Another LncRNA SNHG1, highly upregulated in PCa and sublocalized in the nucleus, promotes PCa metastasis by activating the EMT pathway [33]. Tan et al. [60] found that SNHG1 competitively interacts with hnRNPL, disrupting CDH1 translation through the hnRNPL-CDH1 axis, thereby triggering EMT and tumor spread. LncRNA LOXL1-AS1 upregulation is linked to increased proliferation, migration, metastasis, and EMT while suppressing apoptosis in various cancers, including ovarian, cervical, endometrial, gastric, CRC, esophageal, lung, laryngeal, liver, breast, and PCa [61]. Low GAS5 expression is associated with increased apoptosis in TNBC and ER-positive breast cancer, correlating with lymph node metastasis, advanced clinical stage, and poor survival [5, 62]. In CRC, reduced GAS5 levels are linked to tumor size, advanced TNM stage, lymph node metastasis, low histological grade, poor survival, distant metastasis, and higher recurrence rates [5]. However, overexpression of GAS5 has been linked to early-stage CRC liver metastases [63]. PCAT-1 upregulation is strongly linked to TNM stage and metastasis in HCC and osteosarcoma, as well as tumor invasion and lymph node metastasis in gastric and esophageal cancers [5]. Zhang et al. [64] revealed that PCAT-1 acts as a ceRNA against miRNA-122, and PCAT-1 silencing reduced Wnt/b-catenin signaling *via* miRNA-122 suppression and WNT1 expression, which in turn prevented the advancement of ESCC. The overexpression of HOTAIR was found to be associated with tumor growth, metastasis, migration, cell proliferation, invasion, TNM stage, decreased survival, and a poor prognosis for patients with CRC as compared to healthy controls [5].

2.3 LncRNAs' interaction with signaling pathways

Abnormal signal transduction is a key driver of tumorigenesis and cancer progression, with LncRNAs acting as critical regulators by modulating various signaling pathways [1]. Several key signaling pathways regulated by LncRNA to promote cancer development and progression have been discussed in brief below.

Wnt/β-catenin pathway: The Wnt/β-catenin pathway is often dysregulated in tumorigenesis, influencing proliferation, invasion, metastasis, and apoptosis. LncRNAs regulate this pathway, acting as tumor suppressors or promoters [1]. For example, HOTAIR promotes breast cancer invasion and metastasis by activating the Wnt/β-catenin pathway [65].

STAT3 signaling pathway: The STAT3 signaling pathway, particularly JAK/STAT, is a key oncogenic driver in various cancers. For example, some miRNAs and LncRNAs act as STAT3 negative regulators, binding to its mRNA or modulating its expression to inhibit CRC progression. Conversely, certain LncRNAs, like HOTAIR, promote CRC development by enhancing STAT3 activity, functioning as oncogenes [66].

MAPK pathway: The MAPK pathway is involved in regulating cell proliferation, differentiation, transformation, and apoptosis, contributing to inflammation and tumor development by phosphorylating key cellular components [67]. Many LncRNAs modulate this pathway. For instance, LncRNA XIST targets miRNA-194-5p, reducing its expression and regulating MAPK1, while its silencing inhibits hepatoma cell proliferation and invasion [67]. Similarly, MALAT1 knockdown suppresses MEK/ERK/MAPK/JNK phosphorylation leading to inhibition of ERK/MAPK pathway, reducing gallbladder carcinoma metastasis and invasiveness [68].

PI3K/AKT signaling pathway: The PI3K/AKT signaling pathway is associated with the regulation of cell proliferation, survival, and migration and is abnormally activated in cancers like breast, colorectal, ovarian, pancreatic, and endometrial cancer [50]. Interaction of phosphatidylinositol-3,4,5-trisphosphate (PIP3) and pleckstrin homology (PH) domain drives AKT phosphorylation at Ser473 and Thr308 by PDK1 and mTOR resulting in AKT activation, which then regulates tumor growth and metastasis [50, 69, 70]. LINK-A enhances interactions between PtdIns(3,4,5)P3 and the AKT PH domain, driving AKT hyperphosphorylation, resistance to AKT inhibitors, and promoting tumorigenesis and metastasis [71].

Notch signaling pathway: The Notch signaling pathway regulates cell differentiation, proliferation, and apoptosis, playing diverse roles in tumorigenesis and progression by influencing genes like MYC, cyclin-D1, and p21 [1, 72]. LncRNA LINC01152 acts as an oncogene in glioblastoma multiforme by activating Notch signaling through miRNA-466 interaction and upregulating MAML2, a transcriptional co-activator 2, thereby promoting tumor progression [73].

TGF-β signaling pathway: The TGF-β signaling pathway maintains homeostasis in normal tissues, regulating cell proliferation, motility, and differentiation. However, in tumors, it promotes malignancy. LncRNAs significantly influence TGF-β pathway activity, impacting cancer progression [1]. For instance, downregulation of LncRNA-ANCR promotes breast cancer metastasis *via* association with TGF-β signaling pathway [74].

Unsurprisingly, multiple signaling pathways may be regulated by a single LncRNA. MALAT1 acts as an oncogene by activating the Wnt/β-catenin pathway *via* β-catenin upregulation and GSK3β downregulation. It also activates the Notch signaling pathway by increasing JAG1 (Jagged1) expression through miRNA-124 inhibition [1, 75]. Additionally, LncRNAs can link multiple pathways; for example, BCAR4 in breast cancer connects the Hippo and HH pathways. Hippo effector YAP promotes BCAR4 expression, activating HH signaling to drive glycolysis by upregulating HK2 and PFKFB3, reprogramming glucose metabolism [1, 76].

3. Diagnostic and prognostic potential

Cancer's high mortality rate is partly attributed to inadequate early detection methods and unreliable diagnostic tools like certain protein biomarkers. Most current biomarkers are protein-based, such as glycoproteins, detected through tissue biopsy and immunohistochemistry (IHC) to identify cancer subtypes. However, protein markers often produce false-positive or false-negative results. Conventional serological markers like CA153, CA125, CA27.29, and CEA are also criticized for low specificity and sensitivity. These limitations arise from antibody-based detection, where antibodies may lack specificity or cross-react with other tissues. Furthermore, traditional histology relies on invasive biopsies, discouraging some patients from undergoing diagnostics [77]. Therefore, non-invasive, non-protein biomarkers are urgently needed. Liquid biopsy enables early cancer detection through minimally invasive, serial testing of body fluids, allowing real-time tumor progression monitoring [78]. LncRNAs, detectable in blood, plasma, serum, and urine *via* real-time PCR, are stable in circulation and resistant to nuclease degradation. Their abundance and accessibility make circulating LncRNAs highly promising diagnostic biomarkers [78, 79]. Several LncRNAs, including PCA3, HOTAIR, HULC, MALAT1, and H19, are promising minimally invasive diagnostic and prognostic cancer biomarkers found in body fluids (**Table 2**). Some of these LncRNAs have already been proven effective for diagnostic and prognostic use in clinical settings [12, 79]. PCA3 has been approved by the US FDA as a urine biomarker for PCa, offering better sensitivity and specificity than the Prostate-Specific Antigen (PSA) blood test due to its higher expression in PCa patients [12]. It is more reliable than other circulating nucleic acids due to its high stability in the bloodstream and resistance to nuclease degradation [77]. Genome-wide studies have identified thousands of LncRNAs with differential expression between normal tissues and tumors, highlighting their potential as cancer-specific biomarkers. LncRNA expression profiles can also help identify cancer subtypes, providing insights into tumor behavior and prognosis [8]. MALAT1 is a promising diagnostic biomarker for lung cancer, detectable in blood, and is also elevated in the

LncRNA	Cancer type	Bioavailability	Prognosis	Reference
PCA3	Prostate	Urine	Poor prognosis	[80]
HOTAIR	OSCC	Saliva	Risk of metastasis	[81]
MALAT1	Lung	Blood	Increased risk of metastasis	[82]
H19	Gastric	Plasma, gastric juice	Poor prognosis	[83]
HULC	Pancreatic	Serum	Associated with tumor size, T staging, M staging, vascular invasion and overall survival	[84]
MEG3	Cervical	Tissue	Associated with tumor size, lymph node metastasis and overall survival	[85]
LncRNA-ATB	Breast	Serum	Associated with advanced TNM stage, large tumor size, high M stage and positive lymph node metastasis	[86]

Table 2.
List of some LncRNAs with potential diagnostic and prognostic value.

plasma and urine of PCa patients [12]. LncRNA HOTAIR, which is found to be highly expressed in the saliva samples of oral squamous cell carcinoma (OSCC) patients, is a strong candidate for diagnosing metastatic oral cancer, as its levels are higher in metastatic cases [12, 81].

Some LncRNAs are better suited as complementary biomarkers rather than standalone cancer diagnostic tools [12]. For instance, a five-LncRNA signature (AK001094, AK024171, AK093735, BC003519, and NR_003573) shows strong diagnostic and prognostic potential for gastric cancer (GC), with a combined AUC of 0.95 ± 0.025 in ROC curve analysis [87]. Similarly, three LncRNAs (lnc-MB21D1-3:5, lnc-PSCA-4:2, and lnc-ABCC5-2:1) were significantly dysregulated in GC and showed promising diagnostic performance with an AUC of 0.902 when combined [79]. For CRC, a panel of four LncRNAs (ZFAS1, SNHG11, LINC00909, and LINC00654) showed high diagnostic performance, particularly for early-stage disease, with an AUC of 0.937. Notably, SNHG11 showed the greatest potential for detecting precancerous lesions and early-stage tumors, indicating it could be a promising biomarker for CRC detection and a potential therapeutic target [88].

4. Therapeutic targeting of LncRNAs

LncRNAs have become promising pharmacological targets for treating complex malignancies due to their versatility in gene regulation and tissue-specific expression [89]. They offer advantages such as higher specificity and sensitivity, making them valuable for predicting therapeutic responses [90]. Unlike conventional treatments like chemotherapy, radiotherapy, and surgery, which often cause significant physical and mental stress and may lead to relapse, LncRNA-based approaches provide a more targeted and potentially effective alternative [91]. Furthermore, there is a growing need for personalized cancer therapy which relies on understanding genetic and molecular variations among patients that influence treatment responses [92]. A detailed analysis of LncRNA interactions with proteins, chromatin, and other RNAs can enable their use as precise biomarkers for diagnosis, prognosis, and targeted therapies in personalized medicine [93].

Abnormal levels of LncRNA in bodily fluids and tissues are consistent indicators of cancer and may be the focus of a potential treatment. Indeed, it has been demonstrated that cellular abnormalities linked to cancer can be successfully normalized by interfering with dysregulated LncRNA levels both *in vitro* and *in vivo* [91]. LncRNAs can be targeted through various approaches: (i) degrading pathogenic RNAs *via* siRNAs or chemically modified ASOs using dicer-, AGO-, or RNase H-dependent pathways; (ii) modulating LncRNA genes through promoter blockade or genome editing; and (iii) inhibiting RNA-protein interactions or formation of secondary structure by using ASOs or RNA-binding small molecules [94].

RNA interference (RNAi) is a well-established post-transcriptional gene silencing mechanism that is initiated by double-stranded RNA matching the target gene's sequence. It is a highly effective method for suppressing genes involved in specific biological or pathological processes by employing siRNAs that degrade mRNA during translation [95]. Traditional siRNA approaches have successfully targeted several LncRNAs, such as MALAT1, in human PCa cells, where they inhibited growth, invasion, and migration while inducing cell-cycle arrest [94, 96]. Similarly, siRNA-mediated knockdown of HOTAIR reduced matrix invasion in human breast cancer cells [27].

Another promising therapeutic tool for targeting LncRNAs is antisense oligonucleotides (ASOs), which interact with RNA through Watson–Crick base pairing. By binding to their target RNA, ASOs can modulate gene expression through mechanisms such as steric hindrance, splicing modification, or triggering RNA degradation [94]. In a luminal B breast cancer mouse model (MMTV–PyMT), subcutaneous delivery of MALAT 1-specific ASOs led to primary tumor differentiation and reduced metastasis by nearly 80% compared to nonspecific ASO controls. Additionally, ASOs targeting MALAT 1 decreased branching morphogenesis in 3D organoid models derived from MMTV–PyMT tumors and HER2-amplified mouse mammary tumors [97]. Beyond breast cancer, MALAT1-targeting ASOs have demonstrated strong antimetastatic effects in lung cancer xenograft models [94, 98].

Advances in genome editing technologies, such as CRISPR/Cas9, have enabled the transcriptional silencing of LncRNA-expressing genes through a technique known as CRISPR interference (CRISPRi). This method involves fusing a catalytically inactive Cas9 (dead-Cas9) with transcriptional repressors, which are guided to specific gene promoters by guide RNAs to inhibit transcription [94]. CRISPRi has been effectively utilized to selectively deactivate LncRNA genes across seven human cell lines, including six cancer types and one line of induced pluripotent stem cells (iPSCs) [99].

Nonetheless, there are several challenges associated with studying LncRNAs in the context of cancer. Many LncRNAs are expressed at low levels, raising concerns about their functional relevance in clinical oncology [100]. Differential expression patterns, alternative splicing, and tumor heterogeneity further complicate accurate analysis [101]. Techniques like RNAseq, single-cell RNAseq, and fluorescent RNA *in situ* hybridization (FISH) may address these issues but are time-consuming and resource-intensive [101, 102]. LncRNA-targeted therapies face delivery challenges, including poor membrane permeability of ASOs and siRNAs and difficulty targeting sub-nuclear LncRNAs [100, 101]. Additionally, toxicity and off-target effects remain significant hurdles, although bioinformatics and RNA-capture sequencing can help mitigate these problems. Functional validation of therapeutics *in vivo* is complicated by poor LncRNA conservation across species. Engineered models using larger human genomic segments or proteins may offer solutions. High-throughput methods and CRISPR-Cas9 technology could improve functional screening [101]. Profiling LncRNAs in body fluids or single cells is challenging due to low RNA abundance, fragmentation, and technical limitations in RNA sequencing. Advances like total RNA sequencing, RNA-capture sequencing, and specialized protocols such as SMART-seq are helping overcome these barriers [103]. Furthermore, conventional methods like RT-qPCR and microarray hybridization are time-consuming, while polyA-selection in RNAseq often misses non-polyadenylated LncRNAs [102, 103]. Emerging techniques, including total RNA sequencing, isothermal amplification and nanotechnology-based delivery systems, may improve clinical application feasibility [102, 103]. Addressing these challenges with innovative methods and technologies could unlock the full potential of LncRNAs in cancer diagnostics and therapeutics.

5. Conclusion and future perspectives

LncRNAs have been found to play a critical role in regulating cancer pathways and influencing tumorigenesis, progression, and metastasis. Aberrant expression of LncRNAs dysregulates essential biological pathways, leading to cancer hallmarks and malignant phenotype. Cancer-specific endogenous LncRNAs are secreted from tumor cells into biological fluids, giving rise to stable circulating LncRNAs, thus proving to be of great potential as minimally invasive diagnostic and prognostic biomarkers. Also, LncRNAs have emerged as promising pharmacological targets for treating different types of cancer due to their higher specificity and sensitivity, diverse functional role in gene regulation, and tissue-specific expression. However, there are various challenges that are encountered while studying LncRNAs in the context of cancer such as low and differential expression patterns, off-target effects, profiling in single cells, delivery challenges, and technical limitations. Emerging techniques such as total RNA sequencing, isothermal amplification, and nanotechnology-based delivery systems, may help in addressing these challenges so that the full potential of LncRNAs in cancer diagnostics and therapeutics can be unlocked. Nanotechnology-based delivery systems, such as lipid or polymer nanoparticles carrying ASOs or siRNAs, represent a promising avenue for targeting oncogenic LncRNAs in cancers like TNBC [89, 100]. Furthermore, LncRNAs show significant promise for personalized medicine, particularly when combined with coding genes and SNPs. This approach could soon bring precision treatments closer to clinical application, especially for complex diseases like cancer, where there are genomic, epigenomic, and transcriptomic variations [93, 104].

The advancement of LncRNA applications in the future depends on technology that can recognize and validate their functions, structures, and mechanisms. Attempts to annotate LncRNAs like the formation of GENCODE and databases such as LncRNAtor, LNCipedia, and NONCODE are rapidly expanding knowledge of LncRNA loci [93]. Cutting-edge methodologies, including genome-wide DNA binding analysis such as ChiRP, CHART, RAP, and RNA-protein interaction mapping such as CLIP-seq and PAR-CLIP-seq [92], are paving the way for functional insights. Altogether, it can be concluded that there exists a multifaceted role of LncRNAs in various cancer types, highlighting their potential as diagnostic, prognostic biomarkers and therapeutic targets. Additionally, innovative strategies like RNAi and CRISPR technology may help in improving LncRNA-based cancer therapies as well as refine targeted approaches in personalized medicine.

Acknowledgements

Abhijit Mandal sincerely acknowledges the Council of Scientific and Industrial Research (CSIR), Govt. of India, for Senior Research Fellowship (Sanction letter no. 09/747(0014)/2018-EMR-I). Figure 1 was created using BioRender.com.

Conflict of interest

The authors declare no conflict of interest.

Author details

Abhijit Mandal and Sarbani Giri*
Laboratory of Molecular and Cell Biology, Department of Life Science and Bioinformatics, Assam University, Silchar, India

*Address all correspondence to: girisarbani@gmail.com

References

[1] Zhao S et al. Long noncoding RNAs: Fine-tuners hidden in the cancer signaling network. Cell Death Discovery. 2021;**7**(1):283

[2] Yang M et al. lncRNAfunc: A knowledgebase of lncRNA function in human cancer. Nucleic Acids Research. 2022;**50**(D1):D1295-D1306

[3] Tian H et al. Current understanding of functional peptides encoded by lncRNA in cancer. Cancer Cell International. 2024;**24**(1):252

[4] Sideris N et al. LncRNAs in breast cancer: A link to future approaches. Cancer Gene Therapy. 2022;**29**(12):1866-1877

[5] Ahadi A. Functional roles of lncRNAs in the pathogenesis and progression of cancer. Genes and Diseases. 2021;**8**(4):424-437

[6] Yang X et al. A network based method for analysis of lncRNA-disease associations and prediction of lncRNAs implicated in diseases. PLoS One. 2014;**9**(1):e87797

[7] Mandal A et al. Areca nut and smokeless tobacco exposure induces micronucleus, other nuclear abnormalities and cytotoxicity in early chick embryo. Birth Defects Research. 2023;**115**(10):967-979

[8] Schmitt AM, Chang HY. Long noncoding RNAs in cancer pathways. Cancer Cell. 2016;**29**(4):452-463

[9] Wu S et al. Emerging roles of noncoding RNAs in human cancers. Discover Oncology. 2023;**14**(1):128

[10] Chen Q, Zhu C, Jin Y. The oncogenic and tumor suppressive functions of the long noncoding RNA MALAT1: An emerging controversy. Frontiers in Genetics. 2020;**11**:93

[11] Hao L et al. LncRNA-MALAT1: A key participant in the occurrence and development of cancer. Molecules. 2023;**28**(5):2126

[12] Bolha L, Ravnik-Glavač M, Glavač D. Long noncoding RNAs as biomarkers in cancer. Disease Markers. 2017;**2017**(1):7243968

[13] Liu Q et al. LncRNA loc285194 is a p53-regulated tumor suppressor. Nucleic Acids Research. 2013;**41**(9):4976-4987

[14] Aldayyeni H et al. Functions, mechanisms, and clinical applications of lncRNA LINC00857 in cancer pathogenesis. Human Cell. 2023;**36**(5):1656-1671

[15] Fan Y et al. TGF-β–induced upregulation of malat1 promotes bladder cancer metastasis by associating with suz12. Clinical Cancer Research. 2014;**20**(6):1531-1541

[16] Kim K et al. HOTAIR is a negative prognostic factor and exhibits pro-oncogenic activity in pancreatic cancer. Oncogene. 2013;**32**(13):1616-1625

[17] Endo H et al. Enhanced expression of long non-coding RNA HOTAIR is associated with the development of gastric cancer. PLoS One. 2013;**8**(10):e77070

[18] Nie F-Q et al. Long noncoding RNA ANRIL promotes non–small cell lung cancer cell proliferation and inhibits apoptosis by silencing KLF2 and P21 expression. Molecular Cancer Therapeutics. 2015;**14**(1):268-277

[19] Li K et al. HNRNPA2B1-mediated m6A modification of lncRNA MEG3 facilitates tumorigenesis and metastasis of non-small cell lung cancer by regulating miR-21-5p/PTEN axis. Journal of Translational Medicine. 2023;**21**(1):382

[20] Ghafouri-Fard S et al. Down-regulation of MEG3, PANDA and CASC2 as p53-related lncRNAs in breast cancer. Breast Disease. 2022;**41**(1):137-143

[21] Malik R et al. The lncRNA PCAT29 inhibits oncogenic phenotypes in prostate cancer. Molecular Cancer Research. 2014;**12**(8):1081-1087

[22] Russell MR et al. CASC15-S is a tumor suppressor lncRNA at the 6p22 neuroblastoma susceptibility locus. Cancer Research. 2015;**75**(15):3155-3166

[23] Kwok ZH et al. A non-canonical tumor suppressive role for the long non-coding RNA MALAT1 in colon and breast cancers. International Journal of Cancer. 2018;**143**(3):668-678

[24] Jiang Y et al. The role of MALAT1 correlates with HPV in cervical cancer. Oncology Letters. 2014;7(6):2135-2141

[25] Wang Z et al. High expression of long non-coding RNA MALAT1 in breast cancer is associated with poor relapse-free survival. Breast Cancer Research and Treatment. 2018;**171**:261-271

[26] Bhan A, Soleimani M, Mandal SS. Long noncoding RNA and cancer: A new paradigm. Cancer Research. 2017;**77**(15):3965-3981

[27] Gupta RA et al. Long non-coding RNA HOTAIR reprograms chromatin state to promote cancer metastasis. Nature. 2010;**464**(7291):1071-1076

[28] Wang H et al. HOTAIR enhanced paclitaxel and doxorubicin resistance in gastric cancer cells partly through inhibiting miR-217 expression. Journal of Cellular Biochemistry. 2018;**119**(9):7226-7234

[29] Peng C-L et al. LncRNA HOTAIR promotes colon cancer development by down-regulating miRNA-34a. European Review for Medical and Pharmacological Sciences. 2019;**23**(13):5752-5761

[30] Li X et al. Tumor suppressor LncRNA on chromosome 8p12 (TSLNC8): A concise review in human malignancies. Journal of Cancer. 2023;**14**(15):2867

[31] Gamaev L et al. The pro-oncogenic effect of the lncRNA H19 in the development of chronic inflammation-mediated hepatocellular carcinoma. Oncogene. 2021;**40**(1):127-139

[32] Bitaraf A et al. The oncogenic and tumor suppressive roles of RNA-binding proteins in human cancers. Journal of Cellular Physiology. 2021;**236**(9):6200-6224

[33] Zeng H et al. LncRNA SNHG1: Role in tumorigenesis of multiple human cancers. Cancer Cell International. 2023;**23**(1):198

[34] Zhang S, Song X. Long non-coding RNA SNHG1 promotes cell proliferation and invasion of hepatocellular carcinoma by acting as a molecular sponge to modulate miR-195. Archives of Medical Science. 2020;**16**(2):386-394

[35] Wang S-H et al. Long non-coding RNA Malat1 promotes gallbladder cancer development by acting as a molecular sponge to regulate miR-206. Oncotarget. 2016;**7**(25):37857

[36] Luan W et al. Long non-coding RNA MALAT1 acts as a competing endogenous RNA to promote malignant melanoma

growth and metastasis by sponging miR-22. Oncotarget. 2016;**7**(39):63901

[37] Hakami MA et al. PVT1 lncRNA in lung cancer: A key player in tumorigenesis and therapeutic opportunities. Pathology-Research and Practice. 2024;**253**:155019

[38] Klingenberg M et al. The long noncoding RNA cancer susceptibility 9 and RNA binding protein heterogeneous nuclear ribonucleoprotein L form a complex and coregulate genes linked to AKT signaling. Hepatology. 2018;**68**(5):1817-1832

[39] Sakurai K et al. The lncRNA DRAIC/PCAT29 locus constitutes a tumor-suppressive nexus. Molecular Cancer Research. 2015;**13**(5):828-838

[40] Liu W et al. SP1-mediated up-regulation of lncRNA TUG1 underlines an oncogenic property in colorectal cancer. Cell Death and Disease. 2022;**13**(5):433

[41] Deng L et al. Long noncoding RNA CCAT1 promotes hepatocellular carcinoma progression by functioning as let-7 sponge. Journal of Experimental and Clinical Cancer Research. 2015;**34**:1-10

[42] Mizrahi I et al. Colon cancer associated transcript-1 (CCAT1) expression in adenocarcinoma of the stomach. Journal of Cancer. 2015;**6**(2):105

[43] Ye D et al. LncGMDS-AS1 promotes the tumorigenesis of colorectal cancer through HuR-STAT3/Wnt axis. Cell Death and Disease. 2023;**14**(2):165

[44] Wang J et al. LncRNA PTTG3P promotes tumorigenesis and metastasis of NSCLC by binding with ILF3 to maintain mRNA stability and form a positive feedback loop with E2F1. International Journal of Biological Sciences. 2023;**19**(13):4291

[45] Huang X et al. Hypoxia-responsive lncRNA G077640 promotes ESCC tumorigenesis via the H2AX–HIF1α–glycolysis axis. Carcinogenesis. 2023;**44**(5):383-393

[46] Cao S et al. Decreased expression of lncRNA GAS5 predicts a poor prognosis in cervical cancer. International Journal of Clinical and Experimental Pathology. 2014;7(10):6776

[47] Wang X et al. The lncRNA TP73-AS1 promotes ovarian cancer cell proliferation and metastasis via modulation of MMP2 and MMP9. Journal of Cellular Biochemistry. 2018;**119**(9):7790-7799

[48] Li J et al. Regulation of lncRNA and its role in cancer metastasis. Oncology Research. 2016;**23**(5):205

[49] Liu SJ et al. Long noncoding RNAs in cancer metastasis. Nature Reviews Cancer. 2021;**21**(7):446-460

[50] Fu P-F et al. Role of cytoplasmic lncRNAs in regulating cancer signaling pathways. Journal of Zhejiang University-Science B. 2019;**20**(1):1-8

[51] Mahato RK et al. Targeting long non-coding RNAs in cancer therapy using CRISPR-Cas9 technology: A novel paradigm for precision oncology. Journal of Biotechnology. 2024;**379**:98-119

[52] Dong Y et al. MALAT1 promotes the proliferation and metastasis of osteosarcoma cells by activating the PI3K/Akt pathway. Tumor Biology. 2015;**36**:1477-1486

[53] Malakar P et al. Long noncoding RNA MALAT1 promotes hepatocellular

carcinoma development by SRSF1 upregulation and mTOR activation. Cancer Research. 2017;**77**(5):1155-1167

[54] Okugawa Y et al. Metastasis-associated long non-coding RNA drives gastric cancer development and promotes peritoneal metastasis. Carcinogenesis. 2014;**35**(12):2731-2739

[55] Hu L et al. Up-regulation of long noncoding RNA MALAT1 contributes to proliferation and metastasis in esophageal squamous cell carcinoma. Journal of Experimental and Clinical Cancer Research. 2015;**34**:1-13

[56] Braga EA et al. LncRNAs in ovarian cancer progression, metastasis, and main pathways: ceRNA and alternative mechanisms. International Journal of Molecular Sciences. 2020;**21**(22):8855

[57] Hou P et al. LincRNA-ROR induces epithelial-to-mesenchymal transition and contributes to breast cancer tumorigenesis and metastasis. Cell Death and Disease. 2014;**5**(6):e1287-e1287

[58] Eades G et al. lincRNA-RoR and miR-145 regulate invasion in triple-negative breast cancer via targeting ARF6. Molecular Cancer Research. 2015;**13**(2):330-338

[59] Lin C et al. Functional role of a novel long noncoding RNA TTN-AS1 in esophageal squamous cell carcinoma progression and metastasis. Clinical Cancer Research. 2018;**24**(2):486-498

[60] Tan X et al. LncRNA SNHG1 and RNA binding protein hnRNPL form a complex and coregulate CDH1 to boost the growth and metastasis of prostate cancer. Cell Death and Disease. 2021;**12**(2):138

[61] Yousefnia S. A comprehensive review on lncRNA LOXL1-AS1: Molecular mechanistic pathways of lncRNA LOXL1-AS1 in tumorigenicity of cancer cells. Frontiers in Oncology. 2024;**14**:1384342

[62] Li S et al. Long noncoding RNA GAS5 suppresses triple negative breast cancer progression through inhibition of proliferation and invasion by competitively binding miR-196a-5p. Biomedicine and Pharmacotherapy. 2018;**104**:451-457

[63] Kong H et al. Long non-coding RNAs: Novel prognostic biomarkers for liver metastases in patients with early stage colorectal cancer. Oncotarget. 2016;**7**(31):50428

[64] Zhang F et al. Long noncoding RNA PCAT1 regulates extrahepatic cholangiocarcinoma progression via the Wnt/β-catenin-signaling pathway. Biomedicine and Pharmacotherapy. 2017;**94**:55-62

[65] Thapa R et al. Unveiling the connection: Long-chain non-coding RNAs and critical signaling pathways in breast cancer. Pathology-Research and Practice. 2023;**249**:154736

[66] Rahbar Farzam O et al. Interplay of miRNAs and lncRNAs in STAT3 signaling pathway in colorectal cancer progression. Cancer Cell International. 2024;**24**(1):16

[67] Sun W et al. Interaction of long-chain non-coding RNAs and important signaling pathways on human cancers. International Journal of Oncology. 2018;**53**(6):2343-2355

[68] Wu X-S et al. MALAT1 promotes the proliferation and metastasis of gallbladder cancer cells by activating the ERK/MAPK pathway. Cancer Biology and Therapy. 2014;**15**(6):806-814

[69] Citri A, Yarden Y. EGF–ERBB signalling: Towards the systems level. Nature Reviews Molecular Cell Biology. 2006;**7**(7):505-516

[70] Manning BD, Cantley LC. AKT/PKB signaling: Navigating downstream. Cell. 2007;**129**(7):1261-1274

[71] Lin A et al. The LINK-A lncRNA interacts with PtdIns (3, 4, 5) P3 to hyperactivate AKT and confer resistance to AKT inhibitors. Nature Cell Biology. 2017;**19**(3):238-251

[72] Reicher A et al. Crosstalk between the notch signaling pathway and long non-coding RNAs. Cancer Letters. 2018;**420**:91-96

[73] Wu J et al. LINC01152 upregulates MAML2 expression to modulate the progression of glioblastoma multiforme via notch signaling pathway. Cell Death and Disease. 2021;**12**(1):115

[74] Li Z et al. LncRNA ANCR down-regulation promotes TGF-β-induced EMT and metastasis in breast cancer. Oncotarget. 2017;**8**(40):67329

[75] Guo C et al. Long non-coding RNA MALAT1 regulates ovarian cancer cell proliferation, migration and apoptosis through Wnt/β-catenin signaling pathway. European Review for Medical and Pharmacological Sciences. 2018;**22**(12):3703-3712

[76] Zheng X et al. Lnc RNA wires up hippo and hedgehog signaling to reprogramme glucose metabolism. The EMBO Journal. 2017;**36**(22):3325-3335

[77] Badowski C, He B, Garmire LX. Blood-derived lncRNAs as biomarkers for cancer diagnosis: The good, the bad and the beauty. npj Precision Oncology. 2022;**6**(1):40

[78] Beylerli O et al. Long noncoding RNAs as promising biomarkers in cancer. Non-coding RNA Research. 2022;**7**(2):66-70

[79] Yu Z et al. Aberrant non-coding RNA expressed in gastric cancer and its diagnostic value. Frontiers in Oncology. 2021;**11**:606764

[80] Bussemakers MJ et al. Dd3:: A new prostate-specific gene, highly overexpressed in prostate cancer. Cancer Research. 1999;**59**(23):5975-5979

[81] Tang H et al. Salivary lncRNA as a potential marker for oral squamous cell carcinoma diagnosis. Molecular Medicine Reports. 2013;**7**(3):761-766

[82] Guo F et al. Expression of MALAT1 in the peripheral whole blood of patients with lung cancer. Biomedical Reports. 2015;**3**(3):309-312

[83] Zhou X et al. Identification of the long non-coding RNA H19 in plasma as a novel biomarker for diagnosis of gastric cancer. Scientific Reports. 2015;**5**(1):11516

[84] Ou Z-L, Luo Z, Lu Y-B. Long non-coding RNA HULC as a diagnostic and prognostic marker of pancreatic cancer. World Journal of Gastroenterology. 2019;**25**(46):6728

[85] Zhang J et al. Downregulation of long noncoding RNA MEG3 is associated with poor prognosis and promoter hypermethylation in cervical cancer. Journal of Experimental and Clinical Cancer Research. 2017;**36**:1-9

[86] El-Ashmawy NE et al. Serum LncRNA-ATB and FAM83H-AS1 as diagnostic/prognostic non-invasive biomarkers for breast cancer. Life Sciences. 2020;**259**:118193

[87] Fan Z-Y et al. Identification of a five-lncRNA signature for the diagnosis and prognosis of gastric cancer. Tumor Biology. 2016;**37**:13265-13277

[88] Xu W et al. Circulating lncRNA SNHG11 as a novel biomarker for early diagnosis and prognosis of colorectal cancer. International Journal of Cancer. 2020;**146**(10):2901-2912

[89] Vaidya AM et al. Systemic delivery of tumor-targeting siRNA nanoparticles against an oncogenic LncRNA facilitates effective triple-negative breast cancer therapy. Bioconjugate Chemistry. 2019;**30**(3):907-919

[90] Tan H et al. Long non-coding RNAs in gastric cancer: New emerging biological functions and therapeutic implications. Theranostics. 2020;**10**(19):8880

[91] Nandwani A, Rathore S, Datta M. LncRNAs in cancer: Regulatory and therapeutic implications. Cancer Letters. 2021;**501**:162-171

[92] Huarte M. The emerging role of lncRNAs in cancer. Nature Medicine. 2015;**21**(11):1253-1261

[93] Nguyen Q, Carninci P. Expression specificity of disease-associated lncrnas: Toward personalized medicine. In: Morris K, editor. Long Non-coding RNAs in Human Disease. Current Topics in Microbiology and Immunology. Cham, Switzerland: Springer; 2015;**394**:237-258. DOI: 10.1007/82_2015_464

[94] Arun G, Diermeier SD, Spector DL. Therapeutic targeting of long non-coding RNAs in cancer. Trends in Molecular Medicine. 2018;**24**(3):257-277

[95] Ren Y-J, Zhang Y. An update on RNA interference-mediated gene silencing in cancer therapy. Expert Opinion on Biological Therapy. 2014;**14**(11):1581-1592

[96] Ren S et al. Long noncoding RNA MALAT-1 is a new potential therapeutic target for castration resistant prostate cancer. The Journal of Urology. 2013;**190**(6):2278-2287

[97] Arun G et al. Differentiation of mammary tumors and reduction in metastasis upon Malat1 lncRNA loss. Genes and Development. 2016;**30**(1):34-51

[98] Gutschner T et al. The noncoding RNA MALAT1 is a critical regulator of the metastasis phenotype of lung cancer cells. Cancer Research. 2013;**73**(3):1180-1189

[99] Liu SJ et al. CRISPRi-based genome-scale identification of functional long noncoding RNA loci in human cells. Science. 2017;**355**(6320):eaah7111

[100] Wu M et al. The potential of long noncoding RNAs for precision medicine in human cancer. Cancer Letters. 2021;**501**:12-19

[101] Fathi Dizaji B. Strategies to target long non-coding RNAs in cancer treatment: Progress and challenges. Egyptian Journal of Medical Human Genetics. 2020;**21**:1-15

[102] Salamini-Montemurri M et al. The challenges and opportunities of lncRNAs in ovarian cancer research and clinical use. Cancers. 2020;**12**(4):1020

[103] Lorenzi L et al. Long noncoding RNA expression profiling in cancer: Challenges and opportunities. Genes, Chromosomes and Cancer. 2019;**58**(4):191-199

[104] Pucci P. Combination Therapy and Noncoding RNAs: A New Era of Cancer Personalized Medicine. Epigenomics. 2022;**14**(3):117-120

Chapter 2

Uncovering the Crosstalk between Pseudogenes and Non-Coding RNAs in Tumorigenesis

Duygu Sari-Ak, Lütfi Tutar and Yusuf Tutar

Abstract

The expression of genes gets influenced by pseudogenes through their interactions with non-coding RNAs (ncRNAs) that include microRNAs (miRNAs) and long non-coding RNAs (lncRNAs) and PIWI-interacting RNAs (piRNAs). This chapter seeks to describe pseudogene-ncRNA interactions, to highlight their new roles in cancer biology, and to discuss their potential in clinics. Pseudogenes act as competing endogenous RNAs (ceRNAs) to bind miRNAs and control the expression of oncogenes and tumor suppressors, thus affecting oncogenic and tumor-suppressive pathways. Furthermore, pseudogene-derived lncRNAs and piRNAs regulate transcription and genome stability, which are crucial for cancer progression. Recent studies have shown that pseudogene-ncRNA networks can be used as biomarkers for early cancer detection, prognosis, and therapeutic intervention. However, their roles are quite different from one cancer type to another. This chapter reviews the current understanding of pseudogene-ncRNA interactions, with a focus on their significance and potential to improve cancer diagnosis and treatment.

Keywords: pseudogenes, non-coding RNAs, tumorigenesis, ceRNAs, cancer biomarkers

1. Introduction

Pseudogenes are sequences of genomic DNA that are similar to protein-coding genes but have mutations, frameshift insertions, or premature stop codons and therefore cannot produce functional proteins. Unlike other non-coding RNAs (ncRNAs), pseudogenes mainly result from retrotransposition or gene duplication events [1]. Once considered as "junk" DNA, pseudogenes are now classified as active regulators of gene expression and have been shown to influence various biological processes including cancer through interactions with microRNAs and long non-coding RNAs (lncRNAs) [2].

In cancer biology, pseudogenes regulate gene expression, thus playing a crucial role in tumorigenesis through the interaction with ncRNAs. Pseudogene-miRNA interactions are accurately predicted by computational tools, including ensemble learning-based approaches such as ELPMA, and this is significant for targeted cancer therapy [2]. For instance, in esophageal carcinoma (ESCA), pseudogenes influence cancer progression through the sequestration of miRNAs, thereby influencing transcription factors such as SOX2 and TFAP2A and cell differentiation and reprogramming pathways [3]. In breast cancer, the GBP1P1 pseudogene functions as a miRNA sponge to affect survival, self-renewal, and migration of cancer cells through the GBP1/hsa-miR-30d-5p axis [4]. These specific examples explain the molecular pathways through which pseudogenes act.

Pseudogenes also operate through RNA-independent mechanisms, including lncRNAs-induced chromatin remodeling that significantly affects oncogenic processes [5]. Furthermore, during embryonic development, pseudogenes such as PI4KAP1 and TMED10P1 interact with proteins and miRNAs, indicating their wider physiological functions beyond cancer [6]. Because pseudogene-ncRNA interactions are important in biology, it is essential to investigate them in detail. This chapter provides a review of the existing evidence on the regulatory function of pseudogenes in tumorigenesis, with an emphasis on the specifics of the molecular mechanisms of their interactions with ncRNAs. An analysis of these complex regulatory networks will help to elucidate their clinical significance and provide new ideas in cancer research, diagnosis, and treatment.

2. Pseudogenes and functional regulators

2.1 Evolutionary perspective

Pseudogenes affect embryogenesis and diseases through their evolutionarily conserved regulatory functions. Dynamic pseudogene expression, including genes such as PI4KAP1, TMED10P1, and FBXW4P1, plays important role in the maintenance of embryonic stem cell (ESC) self-renewal through modulation of transcriptional and post-transcriptional processes by protein and miRNA interactions [6, 7]. The correlated expression patterns between pseudogenes and their parent genes indicate that there are common regulatory elements that are necessary for proper embryonic development [6].

Furthermore, the conserved expression of pseudogenes across species highlights their basic function in organ identity. Pseudogenes are also related to congenital diseases, which further prove their importance in developmental biology and disease [7]. Studies on regulatory DNA mutations and promoter motifs have demonstrated that pseudogenes have evolving regulatory capabilities that are consistent with their ability to adapt to different developmental stages and environmental conditions [8]. However, to understand the function of pseudogenes in embryogenesis and their role in various diseases, more extensive studies are required.

2.2 Classification and general features of non-coding RNAs

Non-coding RNAs are functional RNA molecules that regulate gene expression without encoding proteins. Classified based on their size, structure, and biological roles, the main types of ncRNAs include long non-coding RNAs (lncRNAs), microRNAs (miRNAs), and PIWI-interacting RNAs (piRNAs) (**Figure 1**).

LncRNAs, generally longer than 200 nucleotides, regulate gene expression at chromatin, transcriptional, and post-transcriptional levels through diverse mechanisms [9, 10]. Aberrant expression of lncRNAs is associated with various diseases

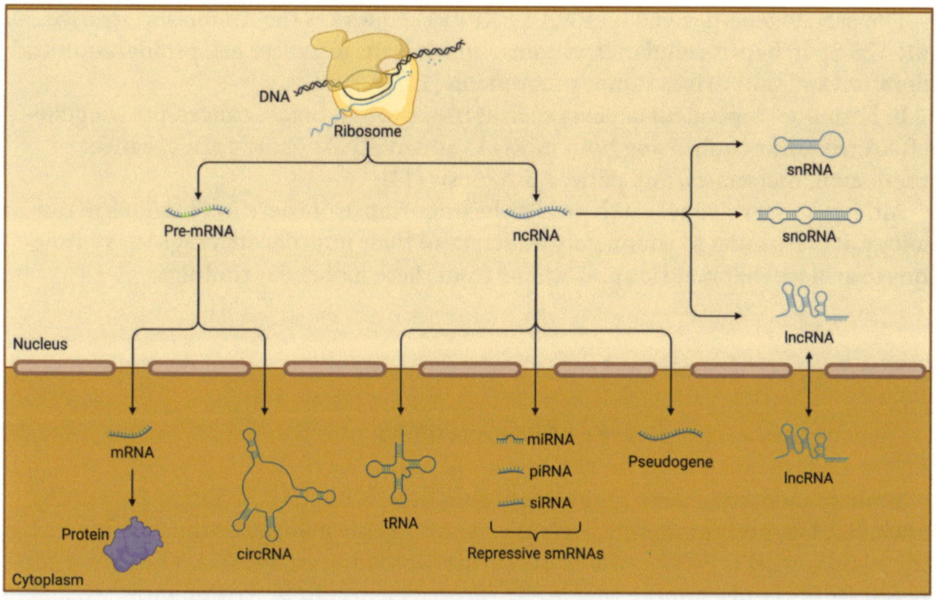

Figure 1.
The classification system for coding versus non-coding RNAs based on their structural attributes and functional relevance (created in BioRender. Tutar Y. (2025) https://BioRender.com/l65g280).

including cancer and cardiovascular disorders, making them valuable biomarkers and therapeutic targets [9, 11].

MiRNAs are short RNA molecules (~22 nucleotides) that primarily mediate post-transcriptional gene silencing by binding to complementary sequences on target messenger RNAs (mRNAs), leading to their degradation or translational inhibition. These small regulatory RNA molecules control essential cellular functions including proliferation and differentiation and apoptosis, while their altered expression causes cancer and multiple pathological states [11].

PiRNAs with lengths ranging from 24 to 31 nucleotides serve primarily to protect germline cell genomes through transposable element silencing that prevents genomic instability. The therapeutic field of piRNAs has emerged through recent studies to show promise in clinical practice especially for cancer diagnostics and therapeutics [12].

2.3 The role of pseudogene-ncRNA interactions in cancer biology

The interactions between pseudogenes and ncRNAs play a crucial role in controlling gene expression and different cellular functions in cancer biology. Primarily, pseudogenes act as competing endogenous RNAs (ceRNAs) to bind miRNAs. The ceRNA function of pseudogenes enables them to control miRNA availability that in turn regulates the expression of key oncogenes and tumor suppressors for maintaining cellular homeostasis. The disturbance of this complex equilibrium can have major implications for cancer development and its progression.

Studies in recent years have described certain ways through which these interactions occur. ESCA pseudogenes function to manage miRNAs, which later help in cancer development through their effects on cell differentiation and reprogramming processes [3]. Or, in breast cancer, GBP1 expression is to enhance with miRNA, which leads to increased cancer cell survival, clonogenic potential, and metastatic capabiliti [4].

The pseudogene-derived lncRNA SNRPFP1 sequesters the tumor-suppressive miR-126-5p in hepatocellular carcinoma, which leads to cancer cell proliferation and migration and thus drives tumor progression [13].

In hormone-dependent cancers such as prostate and breast cancer, pseudogene-ncRNA networks comprising both miRNAs and lncRNAs greatly affect cancer progression, metastasis, and patient prognosis [14].

Although extensive research proves the importance of these interactions in cancer biology, it is essential to completely understand their multifaceted regulatory functions to achieve therapeutic applications from these molecular findings.

3. Pseudogenes' functional roles and mechanisms

3.1 Pseudogenes as competing endogenous RNAs

Some pseudogenes have a regulatory function through their action as ceRNAs, which regulate protein-coding gene expression. Their main function to achieve this is by sequestering miRNAs, which otherwise would target mRNAs. The tumor suppressor PTEN is an example where PTENP1 acts as a ceRNA to bind miR-19b and miR-21 thus leading to the enhancement of PTEN expression that halts oncogenic processes [2].

The ceRNA network establishes functional crosstalk between pseudogenes, miRNAs, and mRNAs [15]. In breast cancer, for instance, GBP1P1 functions within the GBP1/hsa-miR-30d-5p axis, indirectly regulating GBP1 expression and contributing to cancer progression [4].

In addition to cancer, pseudogenes regulate miRNA activity in inflammatory diseases as well as other pathological conditions [16]. Understanding pseudogene ceRNA mechanisms in detail will generate new insights about disease progression and develop new diagnostic and therapeutic approaches [15].

3.2 Transcriptional and post-transcriptional regulation beyond ceRNA activity

Pseudogenes function beyond acting as miRNA sponges to regulate gene expression by creating endogenous siRNAs and through transcriptional interference. The described pathways emphasize the complex nature of pseudogenes in maintaining cellular balance and disease progression.

The transcription of nearby genes may be disrupted by pseudogenes. The mechanism of disruption occurs through the competition or obstruction of transcriptional machinery that binds to a pseudogene with machinery that binds to adjacent genes thus changing expression patterns [6]. In cancer, such interference can contribute to oncogenic pathways by affecting genes involved in differentiation and reprogramming [3].

Some pseudogenes produce endogenous siRNAs that serve to destroy mRNA transcripts or block their translation that adds an additional layer of post-transcriptional gene expression regulation [6]. These siRNAs have been found to regulate oncogenes and tumor suppressors, which make them important for tumorigenesis [17].

The detailed molecular mechanisms behind the regulatory functions of pseudogenes continue to be the subject of current research despite their growing recognition. The lack of complete annotation of pseudogenes together with species variations creates obstacles for thorough functional characterization [18, 19].

3.3 Dual roles of pseudogenes in cancer: Tumor suppressors and oncogenes

The biological processes of cancer are affected by pseudogenes which act as either tumor suppressors or oncogenes through their regulatory relationships with non-coding RNAs and gene expression networks.

The tumor suppressor CSPG4P12 along with other pseudogenes functions to block epithelial-mesenchymal transition (EMT) in colorectal cancer (CRC). The regulatory network leads to E-cadherin elevation and vimentin and N-cadherin and MMP9 repression, which results in reduced cancer cell proliferation and invasion as well as motility [19]. The transcriptional repressors ZEB1 and SNAIL are downregulated by CSPG4P12 that enables EMT suppression through epithelial identity restoration. The suppression of EMT serves to reduce metastatic potential and therefore functions as a tumor suppressor in CRC. PTENP1 functions as a decoy for miRNAs that target the PTEN tumor suppressor thus maintaining PTEN expression and inhibiting tumor progression [17].

In contrast to certain pseudogenes, some have oncogenic activity. OCT4-pg5 for example acts as an oncogene in bladder cancer by sponging miR-145-5p, leading to an increase in OCT4B expression and activation of the Wnt/β-catenin signaling pathway. This process increases the EMT and leads to the cancer cell dissemination and invasion [20]. In breast cancer, GBP1P1 enhances oncogenic activity through the regulation of the GBP1/hsa-miR-30d-5p axis, which leads to the upregulation of GBP1 and enhances cell viability, migration, and colony formation [4]. GBP1P1 binds to miR-30d-5p, which normally suppresses GBP1 expression. Through acting as a sponge, GBP1P1 re-established GBP1 protein levels and enhanced oncogenic signaling. This mechanism enhances the ability of cancer cells to invade and metastasize, thus highlighting its potential as a target for therapy.

Furthermore, pseudogenes can affect cancer stemness. For instance, CTB-63M22.1 has been implicated in the regulation of stemness-related pathways in hepatocellular carcinoma and may influence the prognosis and treatment of the disease [21].

These examples reveal that pseudogenes play both oncogenic and tumor suppressor roles in tumorigenesis and that they may be useful as therapeutic and diagnostic tools for different types of cancers.

4. Functional interplay between pseudogenes and long non-coding RNAs

LncRNAs are important regulators of gene expression and they do so through several mechanisms including chromatin remodeling, transcriptional regulation, and post-transcriptional regulation. They can function as scaffolds, guides, or decoys in gene expression control [9]. LncRNAs bind to chromatin modifying complexes and enzymes and determine chromatin states and transcriptional output. They also bind to mRNAs and affect their splicing, stability, and translation and thus determine the phenotype [22–24].

The aberrant expression of lncRNAs has been found to be associated with the diseases such as cardiovascular diseases and cancer. Their rapid evolutionary diversification and cell-specific functions make their classification difficult but highlight their biological significance [9, 25]. Elucidation of their structural characteristics and interaction with pseudogenes may help to elucidate their functions in disease mechanisms [22].

Pseudogenes interact with lncRNAs through various mechanisms. Some pseudogenes can function as ceRNAs, sequestrating miRNAs and thereby affecting lncRNAs levels, for instance, GBP1P1 in breast cancer which targets the GBP1/hsa-miR-30d-5p axis [4]. Others share promoter elements or exhibit antisense transcription that can hinder lncRNA transcription [17].

A notable example is the regulatory relationship between PTENP1 and the lncRNA HOTAIR, which both affects PTEN expression through the PI3K/AKT pathway. This axis is crucial for cell proliferation and metastasis, especially in cancer [26].

Furthermore, pseudogene-derived lncRNAs and associated microRNAs have been identified as potential cancer biomarkers due to their specificity and expression profiles [27]. The network of their functions affects cancer progression, and therefore, their potential in designing diagnostic and therapeutic approaches must be further investigated.

5. Small non-coding RNAs with their critical roles

5.1 Pseudogene-miRNA interactions in gene regulation

While the ceRNA function of pseudogenes is mentioned earlier, this section provides further insight into the broader category of small ncRNAs and their roles and uses. MiRNAs regulate gene expression by binding to the 3′-untranslated regions (3′-UTRs) of target mRNAs, leading to translational repression or degradation through Argonaute-containing silencing complexes. They are key regulators of proliferation, differentiation, and apoptosis. Altered miRNA expression is involved in numerous diseases, especially cancer where they can act as oncogenes or tumor suppressors [11, 28, 29]. Their specific expression patterns also render them useful as biomarkers and therapeutic targets.

Pseudogenes affect miRNA function through the mechanism of ceRNAs. As decoys, pseudogenes sequester miRNAs and thus prevent them from targeting their target mRNAs, thereby indirectly affecting gene expression. For example, PTENP1 acts as a sponge for miR-19b and miR-21 that otherwise degrade PTEN mRNA leading to PTEN expression preservation and tumor suppression [26, 30, 31]. This interaction is depicted in **Figure 2**, where PTENP1 protects PTEN mRNA from being degraded by miRNAs through competition for shared miRNAs, thereby increasing PTEN protein expression and reducing cancer progression.

In the same manner, KRASP1 binds to miR-141 and miR-185 and increases KRAS expression in colorectal and lung cancers that leads to tumor progression [32, 33].

The pseudogene network with miRNAs and mRNAs depends on the transcriptional environment together with cell tissue origin and disease condition. ELPMA and ceRNAR are analytical tools that enable researchers to study pseudogene-miRNA interactions and their regulatory functions [2].

Pseudogene networks play essential roles in both cancer biology and normal physiological developmental processes including embryonic development. Pseudogenes serve as dual regulators of developmental processes while affecting disease states, which make them useful markers for disease diagnosis and prognosis and therapeutic targets [6].

5.2 PIWI-interacting RNAs in cancer

PiRNAs bind to PIWI proteins to defend genome stability by suppressing transposable elements. PiRNAs get processed independently of Dicer and function mainly in germline cells, whereas miRNAs do not. Research now demonstrates that

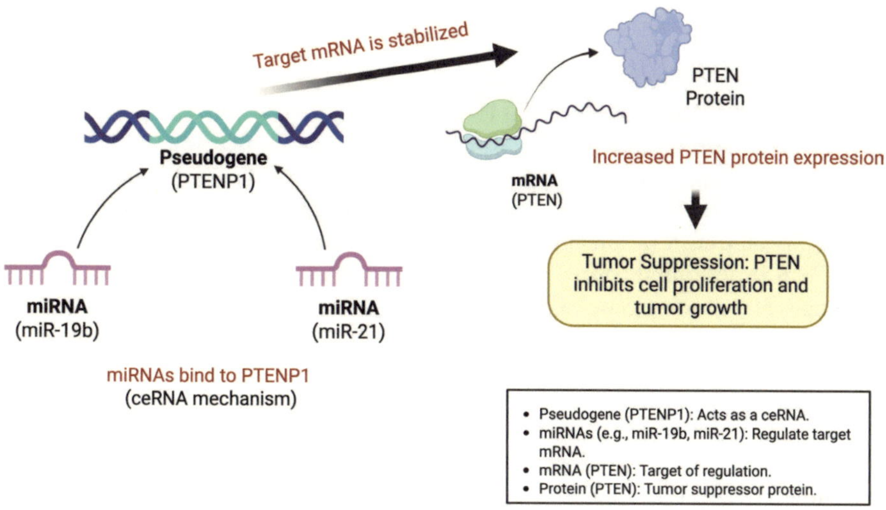

Figure 2.
Pseudogene-ncRNA interactions in gene regulation. The pseudogene, PTENP1 operates as a ceRNA by engaging with microRNAs, including miR-19b and miR-21, thus preserving the target messenger RNA, PTEN, from degradation. Consequently, this interaction facilitates an increase in PTEN protein expression, which is crucial for inhibiting tumor progression (created in BioRender. Tutar Y. (2025) https://BioRender.com/c19g442).

piRNAs play a role in cancer development through epigenetic silencing of gene expression and oncogenesis [34–36].

Abnormal piRNA expression leads to cancer onset through the inhibition of cancer-controlling genes and activation of cancer-promoting genes. The piR-823 gene accelerates multiple myeloma progression but the piR-651 gene enhances gastric cancer proliferation, while the piR-932 gene functions as an oncogene suppressor in colorectal cancer and leads to worse outcomes due to decreased expression levels [37–39].

Research has revealed that pseudogenes serve as possible sources of piRNA expression. Pseudogene-derived piRNAs control PIWI protein activity in ovarian cancer, which affect tumor progression [37]. The findings indicate that pseudogenes interact through a new regulatory mechanism with the piRNA pathway to emphasize their importance in cancer research.

Scientists investigate piR-823 among other piRNAs because of their stable expression in cancer tissues and their ability to act as cancer-specific biomarkers for early detection. Therapeutic interventions focused on oncogenic piRNA inhibition or tumor-suppressive piRNA reactivation represent possible treatment approaches [10, 39].

Current knowledge about piRNAs and their pseudogene origins does not fully explain their biological functions. Additional studies need to occur to establish the functions of these molecules within somatic tissues while evaluating their clinical value. The intricate nature of piRNA biogenesis alongside their pseudogene interactions demonstrates why extensive research must be conducted to understand their complete effects on cancer development and treatment [10, 40].

6. Cancer types and pseudogene-ncRNA networks

The development of oncogenesis through pseudogenes and ncRNAs occurs through intricate mechanisms that differ between various cancer types and their associated regulatory pathways.

6.1 Breast cancer

The interactions between pseudogenes and ncRNAs serve essential functions in the development of breast cancer and its metastatic progression. The pseudogene PTENP1 binds to miR-19b and miR-21, which results in maintaining PTEN tumor suppressor expression and reducing tumor progression [4]. According to Mohebifar et al., the pseudogene GBP1P1 functions through the GBP1/hsa-miR-30d-5p axis to drive cancer cell proliferation and migration while also enhancing colony formation [4]. HOTAIR functions as a long non-coding RNA to establish chromatin conformation changes and modify gene expression when paired with pseudogenes, which leads to metastasis development [41].

6.2 Colorectal cancer

The pseudogenes KRASP1 and BRAFP1 function as competing endogenous RNAs to bind miR-185 and miR-141 which result in KRAS oncogene activation through downstream MAPK signaling pathways [42]. MALAT1 pseudogene-lncRNA networks help CRC cells undergo EMT while also making them resistant to chemotherapy and radiotherapy. The interactions affect essential cellular processes that include proliferation, apoptosis, and differentiation. The detailed study of these complicated ceRNA interactions creates new opportunities to identify biomarkers and therapeutic approaches [11, 12, 43, 44].

6.3 Lung cancer

The process of oncogenic transformation in lung cancer relies on pseudogenes to regulate miRNAs and piRNAs. DUXAP8 acts as an miR-29a-5p sponge to enhance cell proliferation, while HSPA7 binds to miR-34a which disrupts cell cycle checkpoints [42, 45, 46]. PiRNA derived from pseudogenes acts to silence tumor suppressor genes during cancer development. The piRNAs piR-651 among others contribute to elevated cell proliferation in lung cancer cells as well as gastric cancer cells [46–48].

6.4 Prostate and ovarian cancers

PTENP1 in prostate cancer binds to oncogenic miRNAs like miR-17, which helps preserve PTEN expression to create a tumor suppressive outcome [30, 31]. PCA3 stands as a prostate-specific long non-coding RNA, which shows increased levels in prostate cancer tissues and functions as an accurate biomarker [49]. PIWI protein activity suppression by pseudogene-derived piRNAs in ovarian cancer leads to accelerated tumor progression [14].

7. Clinical implications and future directions

The detailed molecular connections involving pseudogenes and ncRNAs hold substantial importance in clinical situations, especially regarding cancer diagnosis, prognostic evaluations, and treatment plans. These regulatory frameworks have created opportunities to develop innovative personalized medicine approaches by investigating pseudogenes and ncRNAs for disease progression prediction and targeted treatment strategies.

7.1 Pseudogenes and ncRNAs as biomarkers in cancer diagnosis

Pseudogenes and ncRNAs are increasingly acknowledged as significant biomarkers for cancer diagnosis due to their consistent expression profiles and specificity. PTENP1 pseudogenes along with piRNAs and lncRNAs have proven valuable across different cancer types, which make them suitable for non-invasive diagnostic procedures. The metastatic potential and prognostic value of HOTAIR lncRNA in cancer patients has been established. These biomarkers improve diagnostic precision and allow for the early identification of tumors. **Table 1** presents a summary of pseudogenes and ncRNAs together with diagnostic tools that demonstrate effectiveness in cancer detection and prognosis. Each category presents particular examples alongside their clinical importance and the methods used to apply these biomarkers for diagnosis.

Although pseudogenes and nc-RNAs hold great promise for cancer diagnostics, several significant problems still exist regarding the standardization of these biomarkers across different populations and cancer types. To validate these biomarkers in clinical settings and incorporate them into current diagnostic protocols, additional empirical research is necessary. Moreover, the integration of multi-omics panels could enhance the accuracy of cancer risk assessment and improve the efficacy of early detection practices [54].

7.2 Therapeutic targeting of pseudogene-ncRNA networks

Pseudogene-ncRNA network modulation is new oncology strategies that disrupts oncogenic pathways. Some approaches include synthetic analogs and suppressors, antisense oligonucleotides (ASOs), RNA interference (RNAi), and CRISPR/Cas9 to reconstitute pseudogene-ceRNA networks. These techniques are currently under rigorous investigation to face the challenges of cancer therapy, particularly in the case of targeting oncogenic pseudogenes and ncRNAs.

Synthetic miRNA mimics/inhibitors modulate ceRNA networks, such as inhibiting KRASP1 to downregulate KRAS in colorectal cancer [33]. The pseudogene transcripts are targeted by ASOs, and the oncogenic functions of these transcripts are disrupted. ASOs, such as DUXAP8, suppress tumor growth in lung cancer models. Recent progress in ASO delivery systems may help to increase their usability [55]. Through RNAi, the pseudogene OCT4-pg4 is targeted, and tumor growth is reduced, and tumor suppressor activity is restored [33, 56]. CRISPR/Cas9 edits pseudogene loci or

Category	Examples	Clinical significance
Pseudogenes as biomarkers	PTENP1	Interacts with miRNAs, influencing prognosis in prostate and breast cancers. Reduced PTENP1 expression correlates with poor outcomes, making it a prognostic biomarker [50]
ncRNAs in cancer detection	piRNAs (e.g., piR-823); miRNAs; lncRNAs (e.g., HOTAIR)	piR-823 is elevated in multiple myeloma, aiding non-invasive detection; a 4-miRNA model achieves over 90% sensitivity for multiple cancers; HOTAIR is linked to metastasis in breast and CRC [50–52]
Diagnostic tools and techniques	Machine learning models, liquid biopsy	Machine learning using piRNAs provides high accuracy in CRC diagnosis; liquid biopsy techniques improve biomarker detection using reverse transcription enzymes [53]

Table 1.
Biomarkers and diagnostic tools in cancer detection.

ncRNA-binding sites to disrupt disease-causing interactions and restore normal gene expression [57]. **Figure 3** shows how pseudogenes and ncRNAs interact with each other during the progression and treatment of breast cancer.

Despite the challenges that exist, these methodologies hold promise. The use of computational techniques and artificial intelligence for the prediction and mitigation of off-target effects is critical for the advancement of small RNA therapeutics [33]. Standardized detection protocols and studies on pseudogene-ncRNA mechanisms are crucial for the therapeutic efficacy.

7.3 Challenges and opportunities in translating findings to clinical applications

The conversion of findings from pseudogene-ncRNA networks into practical clinical applications is a task that is fraught with a number of challenges as well as possibilities. The complexity of these regulatory networks, potential off-target effects, delivery system limitations, and ethical considerations are major challenges. However, progress in multi-omics integration, personalized medicine, and delivery technologies present some possible solutions. The challenges and opportunities are given in **Table 2**.

7.4 Future directions in research

Further research on the complete potential of pseudogene-ncRNA networks in oncology should be directed toward the following topics in the future. Concerning functional characterization, the roles of uncharacterized pseudogenes and non-coding RNAs in cancer pathogenesis should be determined. The therapeutic efficacy of

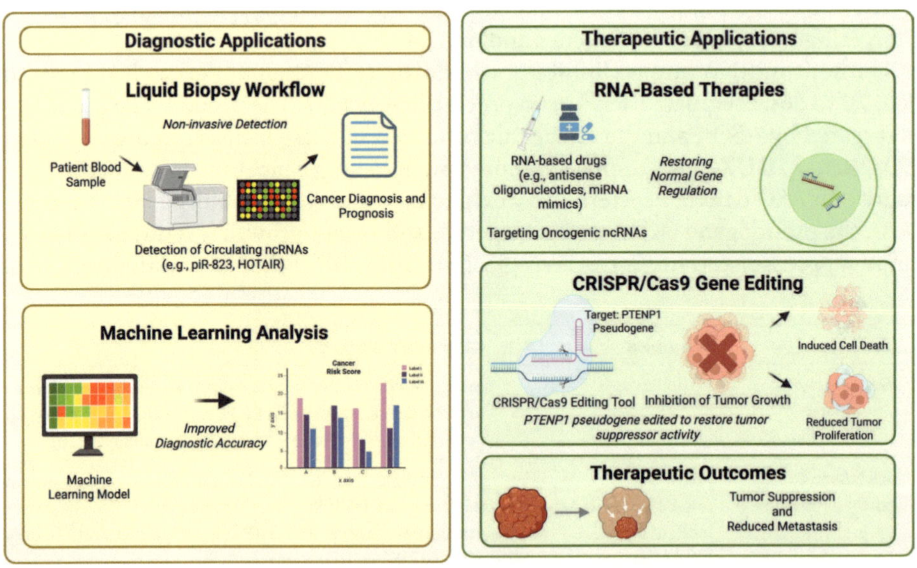

Figure 3.
Diagnostic and therapeutic applications of pseudogenes and ncRNAs in cancer. This chart shows diagnostic uses such as liquid biopsies for detecting circulating ncRNAs (e.g., piR-823, HOTAIR) and machine learning algorithms to analyze expression profiles to enhance cancer diagnosis and risk evaluation. Therapeutic applications including antisense oligonucleotides, miRNA mimetics, and CRISPR/Cas9 targeting pseudogenes (e.g., PTENP1) to restore gene regulation, reduce tumor growth, and mitigate metastasis (created in BioRender. Tutar Y. (2025) https://BioRender.com/p70j490).

Challenges	Opportunities
The ceRNA activity of pseudogenes is highly context-dependent, with variations in expression levels across tissues complicating universal therapeutic strategies [32, 58]	Combining transcriptomics, proteomics, and epigenomics helps unravel pseudogene-ncRNA networks, facilitating biomarker discovery and therapeutic targets [32]
Therapeutic techniques, such as RNAi and CRISPR/Cas9, may interfere with non-target transcripts, disrupting normal cellular functions [59]	Single-cell sequencing reveals pseudogene and ncRNA expression, while spatial transcriptomics maps pseudogene-ncRNA interactions, aiding biomarker and therapeutic target discovery [52, 60–62]
Tissue-specific targeting remains challenging despite the potential shown by nanoparticle-based delivery systems [59]	Advances in lipid nanoparticles and exosome-based systems offer promising solutions for efficient and targeted RNA therapy delivery [59]
Many pseudogenes and ncRNAs are poorly characterized, hindering their development as therapeutic targets. Comprehensive studies are needed to understand their roles in oncogenesis [1]	The specificity of pseudogene-ncRNA interactions supports tailored therapeutic approaches aligned with the personalized medicine goals [58]
Genetic material alterations present ethical and regulatory dilemmas, requiring attention to ensure patient safety and societal acceptance [59]	Panels of pseudogenes and associated ncRNAs as diagnostic tools can revolutionize cancer screening, enabling early detection and improving outcomes [63]

Table 2.
The key challenges and opportunities in pseudogene-ncRNA research and clinical applications.

the alteration of interactions between pseudogenes and ncRNAs should be validated by *in vivo* studies using preclinical models. Clinical trials should be conducted with potential candidates. To understand the relation between pseudogenes and ncRNAs and their impact on biological pathways and diseases, computational approaches and experimental practices should be integrated.

8. Conclusion

Pseudogenes, once considered evolutionary relics, are now recognized as dynamic regulatory elements that are integrated into the gene expression networks through their interactions with non-coding RNAs. Pseudogenes regulate both tumor suppressive and oncogenic pathways through mechanisms that include miRNA sponging, transcriptional interference, and piRNA generation. Their roles span across various levels of cellular regulation and are highly context-dependent, showing tissue-specific and cancer type-specific functions. The increasing knowledge of pseudogene-ncRNA interactions holds great promise for the diagnosis and treatment of cancer. These molecular interactions not only provide useful biomarkers for early detection and prognosis, but also provide potential therapeutic targets, particularly in the context of precision oncology. However, the complexity of their regulatory functions and the heterogeneity across different tumors underscore the need for further mechanistic insights and translational research. With the integration of multi-omics technologies and computational tools, pseudogenes and their ncRNA partners are ready to transform the future of cancer biology and personalized medicine.

Author details

Duygu Sari-Ak[1], Lütfi Tutar[2] and Yusuf Tutar[3]*

1 Department of Medical Biology, Hamidiye International School of Medicine, University of Health Sciences, Istanbul, Turkey

2 Department of Molecular Biology and Genetics, Faculty of Arts and Sciences, Kırsehir Ahi Evran University, Kırsehir, Turkey

3 Faculty of Medicine, Division of Biochemistry, Recep Tayyip Erdogan University, Rize, Turkey

*Address all correspondence to: yusuf.tutar@erdogan.edu.tr

References

[1] Pink RC, Wicks K, Caley DP, Punch EK, Jacobs L, Francisco Carter DR. Pseudogenes: Pseudo-functional or key regulators in health and disease? RNA. 2011;**17**:792-798. DOI: 10.1261/rna.2658311

[2] Fan C, Ding M. Inferring pseudogene–MiRNA associations based on an ensemble learning framework with similarity kernel fusion. Scientific Reports. 2023;**13**:8833. DOI: 10.1038/s41598-023-36054-y

[3] Pravallika G, Rajasekaran R. Stage II oesophageal carcinoma: Peril in disguise associated with cellular reprogramming and oncogenesis regulated by pseudogenes. BMC Genomics. 2024;**25**:135. DOI: 10.1186/s12864-024-10023-9

[4] Mohebifar H, Sabbaghian A, Farazmandfar T, Golalipour M. Construction and analysis of pseudogene-related ceRNA network in breast cancer. Scientific Reports. 2023;**13**:21874. DOI: 10.1038/s41598-023-49110-4

[5] Kafida M, Karela M, Giakountis A. RNA-independent regulatory functions of lncRNA in complex disease. Cancers (Basel). 2024;**16**:2728. DOI: 10.3390/cancers16152728

[6] Sun M, Chang L, He L, Wang L, Jiang Z, Si Y, et al. Combining single-cell profiling and functional analysis explores the role of pseudogenes in human early embryonic development. Journal of Genetics and Genomics. 2024;**51**:1173-1186. DOI: 10.1016/j.jgg.2024.07.013

[7] Qian SH, Chen L, Xiong Y-L, Chen Z-X. Evolution and function of developmentally dynamic pseudogenes in mammals. Genome Biology. 2022;**23**:235. DOI: 10.1186/s13059-022-02802-y

[8] Fuqua T, Sun Y, Wagner A. The emergence and evolution of gene expression in genome regions replete with regulatory motifs. eLife. 2024;**13**:RP98654. DOI: 10.7554/eLife.98654.1

[9] Le LTT, Nhu CXT. The role of long non-coding RNAs in cardiovascular diseases. International Journal of Molecular Sciences. 2023;**24**:13805. DOI: 10.3390/ijms241813805

[10] Wang J, He X, Corpe C. Molecular mechanisms and clinical implications of noncoding RNAs in cancer. Noncoding RNA. 2024;**10**:37. DOI: 10.3390/ncrna10040037

[11] Zhang Y, Zhan L, Jiang X, Tang X. Comprehensive review for non-coding RNAs: From mechanisms to therapeutic applications. Biochemical Pharmacology. 2024;**224**:116218. DOI: 10.1016/j.bcp.2024.116218

[12] Rinaldi S, Moroni E, Rozza R, Magistrato A. Frontiers and challenges of computing ncRNAs biogenesis, function and modulation. Journal of Chemical Theory and Computation. 2024;**20**:993-1018. DOI: 10.1021/acs.jctc.3c01239

[13] Wang N, Guo S, Hao F, Zhang Y, Chen Y, Fei X, et al. Pseudogene SNRPFP1 derived long non-coding RNA facilitates hepatocellular carcinoma progress in vitro by sponging tumor-suppressive miR-126-5p. Scientific Reports. 2022;**12**:21867. DOI: 10.1038/s41598-022-24597-5

[14] Jayarathna DK, Rentería ME, Sauret E, Batra J, Gandhi NS.

Identifying complex lncRNA/pseudogene–miRNA–mRNA crosstalk in hormone-dependent cancers. Biology (Basel). 2021;**10**:1014. DOI: 10.3390/biology10101014

[15] Zhang J, Zheng Y, Xu J. Editorial: Computational identification of ceRNA regulation. Frontiers in Molecular Biosciences. 2022;**9**:937505. DOI: 10.3389/fmolb.2022.937505

[16] Nalbant E, Akkaya-Ulum YZ. Exploring regulatory mechanisms on miRNAs and their implications in inflammation-related diseases. Clinical and Experimental Medicine. 2024;**24**:142. DOI: 10.1007/s10238-024-01334-y

[17] Nakamura-García AK, Espinal-Enríquez J. Pseudogenes in cancer: State of the art. Cancers (Basel). 2023;**15**:4024. DOI: 10.3390/cancers15164024

[18] Cohen N, Veksler-Lublinsky I. A large-scale phylogeny-guided analysis of pseudogenes in *Pseudomonas aeruginosa* bacterium. Microbiology Spectrum. 2023;**11**(4):e01704-23. DOI: 10.1128/spectrum.01704-23

[19] Song Q, Xu H, Wu H, Dong J, Ji S, Zhang X, et al. Pseudogene CSPG4P12 inhibits colorectal cancer progression by attenuating epithelial-mesenchymal transition. Brazilian Journal of Medical and Biological Research. 2024;**57**:e13645. DOI: 10.1590/1414-431x2024e13645

[20] Zhou W, Yang Y, Wang W, Yang C, Cao Z, Lin X, et al. Pseudogene OCT4-pg5 upregulates OCT4B expression to promote bladder cancer progression by competing with miR-145-5p. Cell Cycle. 2024;**23**:645-661. DOI: 10.1080/15384101.2024.2353554

[21] Ge L, Jin T, Zhang W, Zhang Z, Zhang Y, Hu X, et al. Identification of potential pseudogenes for predicting the prognosis of hepatocellular carcinoma. Journal of Cancer Research and Clinical Oncology. 2023;**149**:14255-14269. DOI: 10.1007/s00432-023-05241-9

[22] Mattick JS, Amaral PP, Carninci P, Carpenter S, Chang HY, Chen L-L, et al. Long non-coding RNAs: Definitions, functions, challenges and recommendations. Nature Reviews Molecular Cell Biology. 2023;**24**:430-447. DOI: 10.1038/s41580-022-00566-8

[23] Jones A. Long noncoding RNAs: Structure and function. Biophysical Journal. 2024;**123**:25a. DOI: 10.1016/j.bpj.2023.11.254

[24] Park Y-B, Kim J-M. Identification of lncRNA-mRNA interactions and genome-wide lncRNA annotation in animal transcriptome profiling. Journal of Animal Science and Technology. 2023;**65**(2):293-310. DOI: 10.5187/jast.2023.e17

[25] Pandey GK, Kanduri C. Long non-coding RNAs: Tools for understanding and targeting cancer pathways. Cancers (Basel). 2022;**14**:4760. DOI: 10.3390/cancers14194760

[26] Almalki WH. LncRNAs and PTEN/PI3K signaling: A symphony of regulation in cancer biology. Pathology, Research and Practice. 2023;**249**:154764. DOI: 10.1016/j.prp.2023.154764

[27] Karimi B, Mokhtari K, Rozbahani H, Peymani M, Nabavi N, Entezari M, et al. Pathological roles of miRNAs and pseudogene-derived lncRNAs in human cancers, and their comparison as prognosis/diagnosis biomarkers. Pathology, Research and Practice. 2024;**253**:155014. DOI: 10.1016/j.prp.2023.155014

[28] Ray A, Sarkar A, Banerjee S, Biswas K. Non-canonical targets of

MicroRNAs: Role in transcriptional regulation, disease pathogenesis and potential for therapeutic targets. MicroRNA. 2024;**13**:83-95. DOI: 10.2174/0122115366278651240105071533

[29] Plawgo K, Raczynska KD. Context-dependent regulation of gene expression by non-canonical small RNAs. Noncoding RNA. 2022;**8**:29. DOI: 10.3390/ncrna8030029

[30] Su S, Liu L, Fu Q, Ma M, Yang N, Pan T, et al. A black phosphorus nanosheet-based RNA delivery system for prostate cancer therapy by increasing the expression level of tumor suppressor gene PTEN via CeRNA mechanism. Journal of Nanobiotechnology. 2024;**22**:391. DOI: 10.1186/s12951-024-02659-2

[31] Ghafouri-Fard S, Khoshbakht T, Hussen BM, Taheri M, Akbari DN. A review on the role of PTENP1 in human disorders with an especial focus on tumor suppressor role of this lncRNA. Cancer Cell International. 2022;**22**:207. DOI: 10.1186/s12935-022-02625-8

[32] Ari Yuka S, Yilmaz A. Decoding dynamic miRNA:ceRNA interactions unveils therapeutic insights and targets across predominant cancer landscapes. Biodata Mining. 2024;**17**:11. DOI: 10.1186/s13040-024-00362-4

[33] Bereczki Z, Benczik B, Balogh OM, Marton S, Puhl E, Pétervári M, et al. Mitigating off-target effects of small RNAs: Conventional approaches, network theory and artificial intelligence. British Journal of Pharmacology. 2024;**182**(2):340-379. DOI: 10.1111/bph.17302

[34] Wu X, Pan Y, Fang Y, Zhang J, Xie M, Yang F, et al. The biogenesis and functions of piRNAs in human diseases. Molecular Therapy - Nucleic Acids. 2020;**21**:108-120. DOI: 10.1016/j.omtn.2020.05.023

[35] Ku H-Y, Lin H. PIWI proteins and their interactors in piRNA biogenesis, germline development and gene expression. National Science Review. 2014;**1**:205-218. DOI: 10.1093/nsr/nwu014

[36] Yamashiro H, Siomi MC. PIWI-interacting RNA in drosophila: Biogenesis, transposon regulation, and beyond. Chemical Reviews. 2018;**118**:4404-4421. DOI: 10.1021/acs.chemrev.7b00393

[37] Xu M-M, Li XZ. Enigmatic pachytene PIWI-interacting RNAs. Genome Biology and Evolution. 2024;**16**:evae162. DOI: 10.1093/gbe/evae162

[38] Ben S, Ding Z, Xin J, Li F, Cheng Y, Chen S, et al. piRNA PROPER suppresses DUSP1 translation by targeting N^6-methyladenosine-mediated RNA circularization to promote oncogenesis of prostate cancer. Advanced Science. 2024;**11**(33):e2402954. DOI: 10.1002/advs.202402954

[39] Taghizadeh M, Jafari-Koshki T, Jafarlou V, Raeisi M, Alizadeh L, Roosta Y, et al. The role of piRNAs in predicting and prognosing in cancer: A focus on piRNA-823 (a systematic review and meta-analysis). BMC Cancer. 2024;**24**:484. DOI: 10.1186/s12885-024-12180-2

[40] Lei Q, Du C, Ma Y, Shou Y, Chen L, Feng C, et al. Identifying piRNAs that regulate BaP-induced lung injuries: A bottom-up approach from toxicity pathway investigation to animal validation. Ecotoxicology and Environmental Safety. 2024;**277**:116330. DOI: 10.1016/j.ecoenv.2024.116330

[41] Lin Y-H, Yeh C-T, Chen C-Y, Lin K-H. Pseudogene: Relevant or irrelevant?

Biomedical Journal. 2024:100790. DOI: 10.1016/j.bj.2024.100790

[42] Lou W, Ding B, Fu P. Pseudogene-derived lncRNAs and their miRNA sponging mechanism in human cancer. Frontiers in Cell and Development Biology. 2020;**8**:85. DOI: 10.3389/fcell.2020.00085

[43] Li Q, Geng S, Luo H, Wang W, Mo Y-Q, Luo Q, et al. Signaling pathways involved in colorectal cancer: Pathogenesis and targeted therapy. Signal Transduction and Targeted Therapy. 2024;**9**:266. DOI: 10.1038/s41392-024-01953-7

[44] Mahajan M, Sarkar A, Mondal S. Integrative network analysis of transcriptomics data reveals potential prognostic biomarkers for colorectal cancer. Cancer Medicine. 2024;**13**(11):e7391. DOI: 10.1002/cam4.7391

[45] Xue C, Cai X, Jia J. Long non-coding RNA double homeobox A pseudogene 8: A novel oncogenic propellant in human cancer. Frontiers in Cell and Development Biology. 2021;**9**:709069. DOI: 10.3389/fcell.2021.709069

[46] Gilyazova I, Gimalova G, Nizamova A, Galimova E, Ishbulatova E, Pavlov V, et al. Non-coding RNAs as key regulators in lung cancer. International Journal of Molecular Sciences. 2023;**25**:560. DOI: 10.3390/ijms25010560

[47] Le P, Romano G, Nana-Sinkam P, Acunzo M. Non-coding RNAs in cancer diagnosis and therapy: Focus on lung cancer. Cancers (Basel). 2021;**13**:1372. DOI: 10.3390/cancers13061372

[48] Zhao M, Feng J, Tang L. Competing endogenous RNAs in lung cancer. Cancer Biology & Medicine. 2021;**18**:1-20. DOI: 10.20892/j.issn.2095-3941.2020.0203

[49] Ledesma-Bazan S, Cascardo F, Bizzotto J, Ruiz MS, Vazquez E, Gueron G, et al. Abstract 25: Competing endogenous RNA (ceRNA) networks implicated in prostate cancer progression to castration resistance identified through bioinformatics analyses. Cancer Epidemiology, Biomarkers & Prevention. 2023;**32**:25-25. DOI: 10.1158/1538-7755.ASGCR23-Abstract-25

[50] Li S, Kouznetsova VL, Kesari S, Tsigelny IF. piRNA in machine-learning-based diagnostics of colorectal cancer. Molecules. 2024;**29**:4311. DOI: 10.3390/molecules29184311

[51] Stella M, Russo GI, Leonardi R, Carcò D, Gattuso G, Falzone L, et al. Extracellular RNAs from whole urine to distinguish prostate cancer from benign prostatic hyperplasia. International Journal of Molecular Sciences. 2024;**25**:10079. DOI: 10.3390/ijms251810079

[52] Zhang J, Rui H, Hu H. Noninvasive multi-cancer detection using blood-based cell-free microRNAs. Scientific Reports. 2024;**14**:22136. DOI: 10.1038/s41598-024-73783-0

[53] Wüstmann N, Humberg V, Vieler J, Seitzer K, von Rüden S, Juratli MA, et al. Enhancing biomarker detection in cancer: A comparative analysis of preanalytical reverse transcription enzymes for liquid biopsy application. Laboratory Investigation. 2024;**104**(10):102142. DOI: 10.1016/j.labinv.2024.102142

[54] Singhal S, Gardner K, Singhal SK, Sens D. A set of molecular biomarkers as an indicator of cancer risk. Journal of Clinical Oncology.

2024;**42**:e12563-e12563. DOI: 10.1200/JCO.2024.42.16_suppl.e12563

[55] Çakan E, Lara OD, Szymanowska A, Bayraktar E, Chavez-Reyes A, Lopez-Berestein G, et al. Therapeutic antisense oligonucleotides in oncology: From bench to bedside. Cancers (Basel). 2024;**16**:2940. DOI: 10.3390/cancers16172940

[56] Chaiyawat P, Sangkhathat S, Chiangjong W, Wongtrakoongate P, Hongeng S, Pruksakorn D, et al. Targeting pediatric solid tumors in the new era of RNA therapeutics. Critical Reviews in Oncology/Hematology. 2024;**200**:104406. DOI: 10.1016/j.critrevonc.2024.104406

[57] Sighel D, Destefanis E, Quattrone A. Therapeutic strategies to target the epitranscriptomic machinery. Current Opinion in Genetics & Development. 2024;**87**:102230. DOI: 10.1016/j.gde.2024.102230

[58] Tang Z, Li X, Zheng Y, Liu J, Liu C, Li X. The role of competing endogenous RNA network in the development of hepatocellular carcinoma: Potential therapeutic targets. Frontiers in Cell and Developmental Biology. 2024;**12**:1341999. DOI: 10.3389/fcell.2024.1341999

[59] Nappi F. Non-coding RNA-targeted therapy: A state-of-the-art review. International Journal of Molecular Sciences. 2024;**25**:3630. DOI: 10.3390/ijms25073630

[60] Cao J, Chen X, Huang S, Shi W, Fan Q, Gong Y, et al. Microfluidics-based single cell analysis: From transcriptomics to spatiotemporal multi-omics. TrAC Trends in Analytical Chemistry. 2023;**158**:116868. DOI: 10.1016/j.trac.2022.116868

[61] Vandereyken K, Sifrim A, Thienpont B, Voet T. Methods and applications for single-cell and spatial multi-omics. Nature Reviews Genetics. 2023;**24**:494-515. DOI: 10.1038/s41576-023-00580-2

[62] Lim J, Park C, Kim M, Kim H, Kim J, Lee D-S. Advances in single-cell omics and multiomics for high-resolution molecular profiling. Experimental & Molecular Medicine. 2024;**56**:515-526. DOI: 10.1038/s12276-024-01186-2

[63] Qi X, Lin Y, Chen J, Shen B. The landscape of emerging ceRNA crosstalks in colorectal cancer: Systems biological perspectives and translational applications. Clinical and Translational Medicine. 2020;**10**:e153. DOI: 10.1002/ctm2.153

Chapter 3

Perspective Chapter: The Role of Long Non-Coding RNAs in X Chromosome Dosage Compensation

Sylvia Garza-Manero, Hober Nelson Núñez-Martínez, Gustavo Tapia-Urzúa and Félix Recillas-Targa

Abstract

Genetic sex determination required the specialization of an ancestral autosome pair into sexual chromosomes. This evolutionary trajectory involves recombination suppression in the sexual pair and degeneration of one chromosome. In species with a male XY heterogametic system, the Y chromosome is highly degenerated, creating a difference in the dosage of X-linked genes between females and males. To overcome this difference, one X chromosome is transcriptionally silenced in female mammals. Conversely, genes on the single-male X chromosome of lizards and the fruit fly are overexpressed, reaching the transcriptional output produced by two X chromosomes in females. Despite promoting opposite effects on transcription, these dosage compensation mechanisms share relevant features, including the modification of the chromatin landscape and three-dimensional conformation orchestrated, at least in part, by long non-coding RNAs transcribed from the target X. In this chapter, we describe the mechanisms of X chromosome dosage compensation in eutherians, marsupials, Drosophila, and a lizard. We focus on the leading role of the respective long non-coding RNA, XIST in eutherians, RSX in marsupials, roX1 and roX2 in the fruit fly, and the recently described MAYEX in lizards, in the ambitious task of equalizing the transcriptional output of X-linked genes between females and males.

Keywords: long non-coding RNAs, X chromosome, dosage compensation, X inactivation, X overexpression, sex chromosomes, genetic-sex determination

1. Introduction

In animals with sexual dimorphism, specific differentiation programs direct the development of sexual characters, including structures such as ovaries and testes. These programs can be triggered by either environmental or genetic cues (**Figure 1**). Observations along animal evolution support that genetic sex

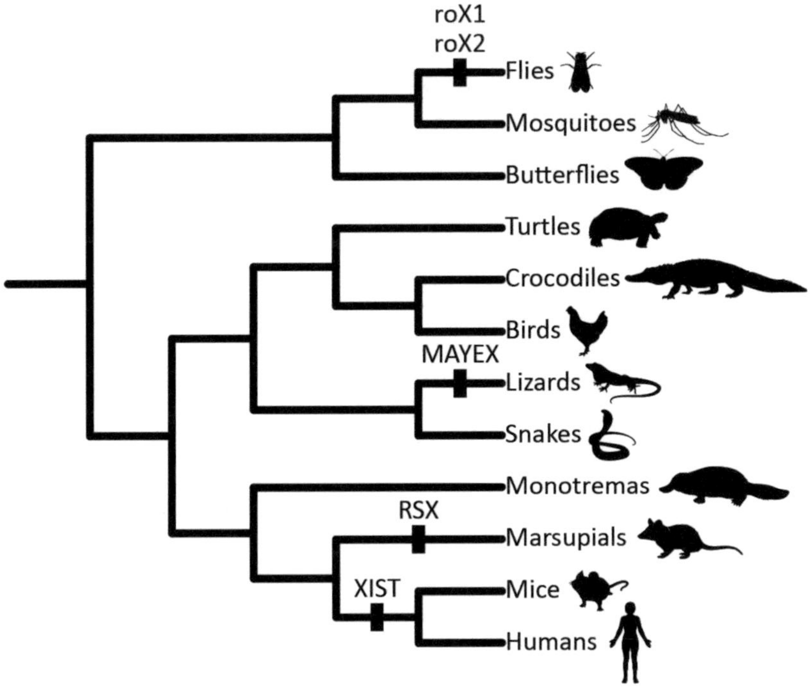

Figure 1.
LncRNAs target global dosage compensation in various XY species. *Phylogenetic associations of representative animal species with either temperature-guided: turtles and crocodiles; or genetic sex determination systems, including both the male heterogametic (XY) and the female heterogametic (ZW) sex determination systems. Groups with XY sex chromosomes include fruit flies, mosquitos, lizards, monotremes, marsupials, and eutherian mammals; while butterflies, birds, and snakes harbor ZW sex chromosomes. Global dosage compensation between sexes has been reported for eutherians, marsupials, fruit flies, and more recently, lizards, mosquitos, and butterflies [1–6]. roX, MAYEX, RSX, and XIST are lncRNAs with a leading role in global X chromosome dosage compensation between sexes, which arose in four of these different groups [7–11]. Branches indicate divergence events and do not represent a time scale. Silhouettes correspond to representative species of each group.*

determination arose after environmental sex determination, perhaps reflecting the selection pressure in response to changing environmental conditions [1]. The most recent ancestor of reptiles and mammals had a temperature-dependent sex determination that was retained in turtles and crocodiles. However, related species that transitioned further away from water, such as mammals, birds, snakes, and lizards, developed genetic sex determination systems in a similar period of time [1]. These genetic sex determination systems rely on sexual chromosomes. Although their origin is evolutionary independent, mammals and lizards developed a male heterogametic system, in which females carry two X chromosomes (XX) and males, one X and one Y chromosome (XY). In contrast, a female heterogametic system operates in birds and snakes, with males carrying two Z chromosomes and females one Z and one W chromosome. The evolution trajectory of sexual chromosomes has been recently reviewed in Ref. [12]. In this work, the authors discuss different models that intersect the main events of sexual chromosome development: the initial acquisition of the genetic cue by an ancient autosome, and the suppression of meiotic recombination between the autosomal pair, the massive loss of homologous genes in the Y or W chromosomes, and the complete degeneration of this chromosome and/or the sexual chromosome turnover [12].

The fact that the sexual chromosome pair is formed by a highly degraded Y or W and a gene-rich X or Z leads to an unbalance in the gene dosage, compared to autosomal pairs and between sexes. Counteracting this unbalance does not seem to be a rule in all studied species with heterogametic sex determination systems [13]. However, distinct mechanisms for dosage compensation have been identified and described for XY model organisms, representing eutherians [2], marsupials [3], and the fruit fly [4]; and have started to be explored in other animal species such as lizards, mosquitos, and butterflies [1, 5, 6]. Thus far, two main strategies equalize the transcriptional output of the X chromosome between females and males. The first consists in the silencing of a whole X chromosome in females, which is referred to as X chromosome inactivation (XCI) and observed in eutherians and marsupials [3, 14]. The second was first described in fruit flies [4] and, more recently, in a lizard [1], where a global twofold upregulation of the male X chromosome or an X chromosome overexpression (XCO) is observed. Importantly, this strategy also equalizes the sexual and autosomal gene expression [1]. Despite having independently evolved in the four animal groups, targeting different sexes and producing opposite transcriptional outputs, XCI and XCO share relevant mechanistic features. These include changes in chromatin organization and compaction of a complete X chromosome, mediated by pre-existent epigenetic machinery that is targeted to the X chromosome with a leading role of long non-coding RNAs (lncRNAs) (**Figure 1**).

LncRNAs are currently recognized as key regulators of gene expression and genome topology [15]. Three of the first examples describing an RNA role in this matter were the mammalian XIST (standing for X-inactive specific transcript) [7] and the Drosophila roX1 and roX2 (roX standing for RNA on the X) [8, 9], the three of them involved in the X chromosome dosage compensation. Since then, these lncRNAs have been widely studied and have become a reference point for proposing the molecular mechanisms by which lncRNAs exert their regulatory functions and for the development of RNA-based techniques [16]. In this chapter, we focus on the role of lncRNAs in X chromosome dosage compensation. First, we describe the extensively studied mechanism by which XIST as well as other lncRNAs achieves XCI in eutherian mammals. We complement the repressive role of lncRNAs on the female X chromosome by describing RSX (standing for RNA on the silent X) from marsupials, and we then contrast it with the role of roX RNAs supporting the global overexpression of the *Drosophila* male X chromosome. The last dosage compensation system has also been reported in reptiles [1], where the recently described MAYEX represents the lncRNA component [10]. Finally, we discuss the advantages of depositing relevant molecular mechanisms for the X chromosome dosage compensation on lncRNAs.

2. X chromosome inactivation in eutherian mammals

In female mammals, one of the two X chromosomes is inactivated to equalize gene dosage with males by a process-denominated X-chromosome inactivation (XCI). XCI occurs randomly in one of two X chromosomes, leading to an active (Xa) and an inactive (Xi) X chromosome [17–19]. This process occurs early in the female development (specifically during the preimplantation stage in mice and humans) to ensure the transcriptional silencing of almost the entire X chromosome across the organism's lifespan [20, 21]. XCI implies coordinated-complex layers of chromosome-wide silencing including RNA polymerase II (Pol II) eviction; deposition of repressive histone marks; and DNA methylation [14].

XCI begins within a highly conserved complex locus (around 2.3 Mb), denominated X inactivation center (XIC), harboring the necessary genetic elements to lead XCI [22–24]. These elements encode lncRNAs, including XIST, TSIX, Repa, Xite, Jpx, Ftx, and Tsx [25–28]. These elements are essential in the initiation and triggering of XCI, as demonstrated by integrating the XIC-containing transgenes into autosomal sites and recapitulating the silencing effect [29]. Efforts in characterizing the roles of these XIC-containing lncRNAs have shed light on their role in driving XCI. For instance, TSIX, Xite, and Ftx prevent XIST activation in the Xa [25, 30, 31]; whereas Jpx and RepA facilitate XIST activation [32, 33]. Thus, XIC represents an intriguing aspect of XCI due to its essential role in the choice of Xi in the same nucleus. In this section, we focus mainly on the mechanisms driving XCI by XIST on mice and humans.

2.1 XIST spreading, coating, and triggering the XCI

The major orchestrator of XCI is the X inactive-specific transcript (XIST), a monoallelic-transcribed lncRNA from the XIC of the future Xi chromosome [6, 15]. XCI initiates with the activation of XIST transcription resulting in a coordinated cascade of molecular waves, including the recruitment of chromatin regulators to induce transcriptional silencing of most of the genes located in the target X [34, 35]. Upon XIST is transcribed, it accumulates at its locus and then spreads to several nucleation sites located at gene-rich regions associated in physical proximity to the XIST locus (**Figure 2A** and **B**). Later, XIST coats the rest of the Xi, including gene-poor regions, to establish chromosome-wide silencing [36, 37]. In addition, XIST recruits numerous proteins through direct interaction with its several repeat domains (A-F) required for the multiple functions of XIST-mediated transcriptional silencing, including initiation of gene silencing (repeat-A), recruitment of Polycomb Repressive Complexes (PRC1 and PRC2) (repeat B-F), and coating across X (repeat C and E) (**Figure 2C**) [38–43]. Thus, XIST diffuses across the future Xi, exploiting its repeat domains and the chromatin structure to trigger XCI.

Histone deacetylation and H2A ubiquitination are some of the earliest events of XCI occurring across the future Xi previous to transcriptional silencing [44]. XIST, through repeat-A, directly interacts with SPEN (also known as SHARP and Msx2-interacting protein) [42]. SPEN interacts with SMRT, a co-repressor complex, to activate the histone deacetylase HDAC3 and promote Pol II eviction from chromatin [45, 46]. Later, XIST leads to PRC1 and PRC2 recruitment to induce H2AK119ub1 and H3K27me3 deposition, respectively [36, 38, 47–49]. XIST binds to heterogeneous nuclear ribonucleoprotein K (HNRNPK), through the repeats B-C, to promote PRC1 recruitment [50, 51]. JARID2, a PRC2 cofactor, interacts with the deposited H2AK119ub1 facilitating PCR2 deposition. Repeats B and F of XIST are critical for JARID2 recruitment and subsequent H3K27me3 accumulation mediated by PRC2 [52, 53]. More recently, multiple RNA G-quadruplexes (rG4s) in XIST modulate the PRC2 histone methyl transferase activity and H3K27me3 enrichment across the Xi [54]. Thus, the histone deacetylation mediated by SPEN-recruited HDACs enables early transcriptional silencing, whereas H3K27me3 deposition by PRC2 facilitates a long-term silencing effect stably maintained in subsequent cell generations. Although the interplay between XIST and PRC2 has been extensively characterized, it remains unclear how Polycomb is recruited to Xi and induces chromosome-wide silencing.

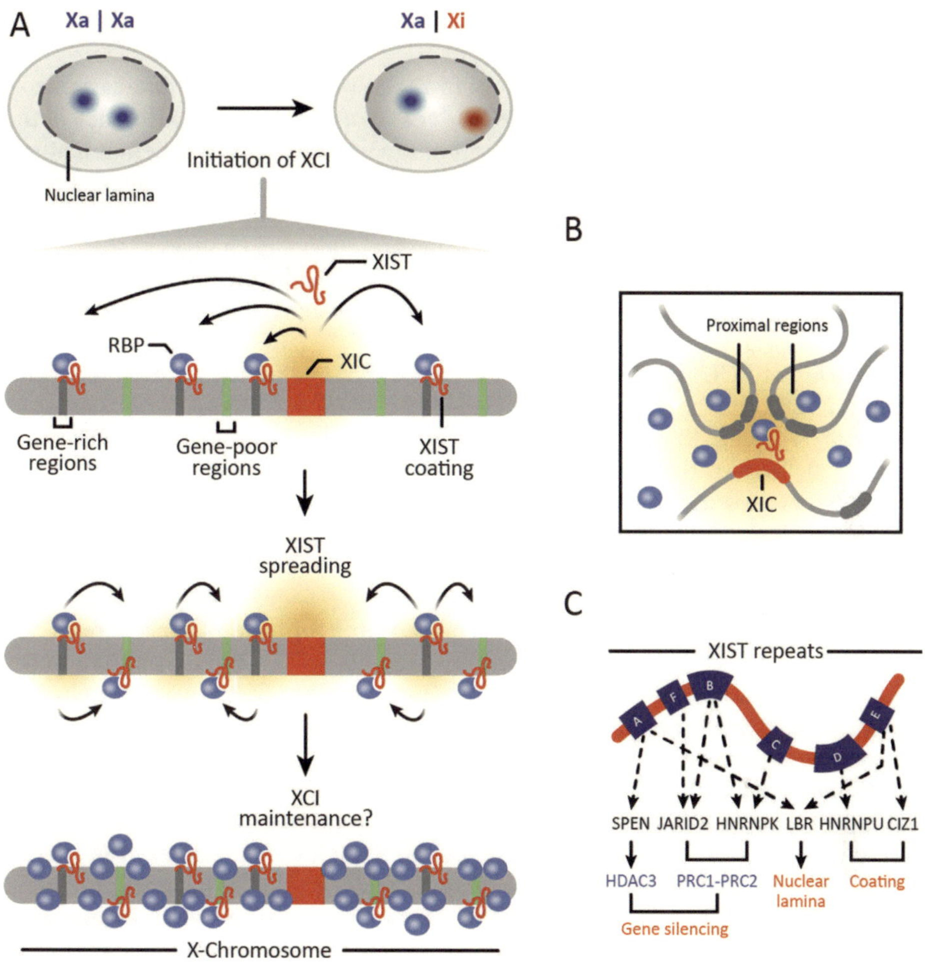

Figure 2.
Role of XIST in X-chromosome inactivation. *(A) During the early stages of development and differentiation of ESCs, one of the two active X chromosomes (Xa) becomes inactive (Xi) due to the transcriptional activation of XIST from the X-inactivation center (XIC). XIST modulates XCI through two main stages: coating and spreading. XIST coats at gene-rich genes located close to its locus (B). Subsequently, XIST spreads to gene-poor and intergenic regions to induce chromosome-wide silencing. The role of XIST at the late stages of XCI is unclear; however, its coating and spreading are essential at the early stages (initiation). (C) Repeat sequences embedded into XIST directly interact with proteins involved in gene silencing, coating, and localization of Xi to the nuclear lamina.*

Additional waves of chromatin remodeling occur later during XCI, including the deposition of H3K9me2, H3K9me3, the macroH2A histone variant, and the DNA methylation at CpG islands of X-linked genes [55–61]. These epigenetic changes may function as additional layers to maintain gene silencing and ensure XCI. Furthermore, XIST depletion analysis in somatic cells displayed limited gene reactivation, leading to the hypothesis that the central role of XIST is the initiation rather than the maintenance of XCI [62, 63]. Thus, XIST exerts a pivotal role in the initiation of XCI due to driving coordinated waves of proteins and chromatin remodelers at the future Xi, so that, the silencing effect will be maintained in subsequent cell divisions.

2.2 XIST shapes Xi topology

Xi displays a distinct nuclear structure from its counterpart Xa in the same cells. This phenomenon was first observed in adult cat neurons, where a difference in the nuclear structure was visible in females but not in males, later it was denominated "Bar body" [64]. This structure corresponds to Xi, which appears as a nuclear foci localized at the nuclear periphery [65, 66]. XIST interacts with the Lamin B receptor, leading to Xi recruitment to the nuclear lamina and inducing changes in the chromatin structure, which is essential for XIST spreading and transcriptional silencing of the future Xi [67]. However, the Xi-lamina location alone displays ineffective XCI, indicating that the localization of Xi at the nuclear lamina has a limited role in the initiation of XCI, and the additional nuclear components are required for XCI regulation [68].

In contrast to Xa, Xi appears as a compacted structure (around 1.2-fold) in somatic cells [69]. At the global level, these differences in compaction appeared insignificant but large-scale chromatin structures look more compacted in Xi, including gene-rich regions [70]. It has been proposed that these differences in the X1 and Xa result from the reorganization of higher-order structures instead of a direct chromatin compaction [71].

The development of high-throughput chromosome conformation capture technologies, such as Hi-C, has revealed the Xi topology [72]. Xa is structured into topologically associated domains (TADs), defined as units of chromatin structures characterized by high internal interaction frequency and low external interaction frequency with adjacent loci, whereas Xi is folded into two megadomains [73, 74]. This asymmetric topology between Xa and Xi could be explained by the differences in the binding patterns of architectural proteins across the chromosomes, including CTCF and cohesins. Cohesins (SMC1A and RAD21) bind the X chromosomes in an allele-specific manner, accounting for ~600 sites in Xa and ~ 20 sites in Xi. Intriguingly, the depletion of XIST in Xi results in the restoration of the cohesin-binding pattern resembling the Xa. The profiling of the XIST protein interactome revealed that XIST binds to numerous chromatin-associated proteins, including the cohesins and CTCF. The authors suggested a model in which XIST antagonizes the cohesin and CTCF binding to prevent the establishment of an Xa-like binding pattern. In the same study, the depletion of XIST and the restoration of cohesin-binding patterns in Xi lead to a transition toward an Xa-like chromatin structure, in which TAD structures appeared in Xi [41]. Thus, XIST functions as a modular platform to recruit numerous proteins and simultaneously prevents the binding of others to ensure the establishment of a Xi conformation.

2.3 TSIX, a negative regulator of XIST

In contrast to XIST, the XIST antisense non-coding transcription unit TSIX is exclusively expressed in the Xa [27]. Previous studies have pointed out that the act of TSIX transcription overlapping the *XIST* promoter is sufficient to induce chromatin remodeling, leading to the prevention of XIST upregulation in Xa [75–78]. However, recent evidence suggests that TSIX RNA by itself recruits the histone methyltransferase G9A to induce H3K9me2 and subsequent XIST silencing on the future Xa [55]. Deletion of TSIX resulted in an increased number of cells expressing XIST in both chromosomes [30]. These findings underpin the central role of TSIX in coordinating the monoallelic choice of XIST activation from both X chromosomes.

3. A parallel system of XCI in marsupials

Marsupials, as monotremes and eutherians (placentals), belong to the group of mammals, all of which evolved from a common ancestor. The sex determination system of this mammalian ancestor remains elusive; however, after its divergence into monotremes and the common ancestor of eutherians and marsupials, a genetic sex-determination system arose independently in the two phylogenetic branches, approximately 160 million of years ago (mya) [3]. Monotremes carry five pairs of sexual chromosomes, where the male is heterogametic ($X_{1-5}Y_{1-5}$) and the female homogametic ($X_{1-5}X_{1-5}$) [79]. There is no evidence of a global dosage compensation system in monotremes; rather, they equilibrate the transcriptional levels of X-linked genes by silencing female X chromosome loci in a gene-by-gene mode [79]. In contrast, eutherians and marsupials share the XY male heterogametic sex determination system and both display XCI. However, marsupial and eutherian XCI differ significantly, suggesting that at least some of the molecular actors and mechanisms evolved after the divergence of these two groups [3].

In marsupials, an apparent global upregulation of X-linked genes in both sexes accompanies the female XCI, restoring gene expression levels to those of autosomes and equalizing them between sexes [80]. Marsupial XCI is less stable and incomplete, despite the accumulation of the repressive histone modification H3K27me3 on the inactive X, which may be reminiscent of its low levels of DNA methylation in comparison with the eutherian inactive X [3]. DNA methylation seems to work differently in marsupials, since sexual and autosomal chromosomes are hypomethylated [81]. However, a flat profile of DNA methylation in the paternal germ line is associated with the predisposition of the paternal X chromosome for XCI [81]. Indeed, another great difference with eutherians is that the marsupial XCI is paternally imprinted, rather than random, meaning that the paternal X chromosome is inactive in all female cells. Finally, there is no XIST homolog in marsupials, and therefore, the leading role in XCI is played by RSX.

3.1 RNA on the silent X of female marsupials

RSX was identified by analysis of RNA-seq data of a South American marsupial, focusing on a differential male-to-female ratio in the reads mapped to the X chromosome [11]. RSX is a large (~27 Kb) lncRNA that is transcribed from and coats the inactive X chromosome in females [11]. It is silenced in the female germ line, where both X chromosomes are active, but expressed from the paternally imprinted X during embryonic development to promote XCI [11, 80]. Interestingly, integration of an *RSX* transgene in mESCs triggers gene silencing in *cis*, suggesting that RSX is a functional analog of XIST [11].

RSX and XIST share no sequence homology; however, both harbor tandem repeats form stem-loop structures critical for recruiting transcriptional repressors to the inactive X chromosome (**Table 1**) [82]. Indeed, the RSX interactome overlaps with that of XIST both in specific proteins, such as the polycomb repressive complex subunits, and in functional categories that include epigenetic transcriptional silencing [83]. Members of the last category that interact with RSX comprise the polycomb catalytic unit EZH2 that deposits H3K27me3, the histone deacetylase HDAC2, and DNA-methyl binding proteins MBD4 and MeCP2, all of which promote transcriptional silencing and chromatin compaction [83]. These findings suggest that the mechanism by which RSX facilitates a repressive chromatin landscape for gene expression is similar to that documented for XIST.

Besides the relevance of the XIST molecule, the particular three-dimensional arrangement of chromatin in the inactive X entails the *XIST* genomic locus [37].

lnRNA	Species	SSE	TOP	Size (Kb)	Repeats	Protein interactions	DNA interactions
XIST	Eutherians	F	XCI	15-17	Repeat A Repeat B Repeat C (murine) Repeat D (human) Repeat E Repeat F	hnRNPK hnRNPU RBM15 SPEN EZH2 SUZ12 CHD4	No DNA binding motif. Association between XIST spreading and LINE1 repeats.
RSX	Marsupials	F	XCI	27-30	Repeat 1 Repeat 2 Repeat 3 Repeat 4	hnRNPK SFPQ EZH2 HDAC2 MBD4 MeCP2	No DNA binding motif.
roX1 roX2	Fruit flies	M	XCO	3.8 0.6	roX-box	MSL2 MLE Polycomb?	Binding motif: MRE $(GA)_4$. Associations between satellite repeats and HAS.
MAYEX	Lizards	M	XCO	3.3		Histone 4 G1KD42 JMJD2C DNAJC2 SPT6 ZFC3H1 RANBP2 HSPA8 SASH1	Long-range interactions with $(TTA)_5$ sequences. A featured repeat sequence downstream MAYEX.

SSE: sex-specific expression. F: female. M: male. TOP: transcriptional output. XCI: X chromosome inactivation. XCO: X chromosome overexpression. Kb: Kilobases. Note that XIST repeat C is exclusive of mice, while repeat D is found in humans. RSX repeat 1 is similar to XIST repeats B, C, and D, and may serve for PRC1 indirect recruitment, while RSX repeats 2, 3 and specially 4, are similar to XIST repeats A and E, which may interact with PRC2 [82]. XIST and RSX interactomes include epigenetic repressors such as PRC components, histone deacetylases, and DNA methyl binding proteins [83]. In turn, roX and MAYEX interact with RNA-binding proteins, epigenetic activators, such as proteins holding activity or related to histone acetylases, PRC antagonists, and architectural proteins [10, 84]. X chromosome is richer in repeated sequences in comparison to autosomes, which may play a role in X chromosome recognition or spreading of XCI or XCO epigenetic mechanisms.

Table 1.
Similarities and particularities of lncRNAs targeting X chromosome dosage compensation in mammals, lizards, and the fruit fly.

Since RSX was discovered a decade ago, more studies are missing to elucidate whether this lncRNA similarly configures chromatin architecture to accomplish its function. Nevertheless, a recent paper reported an antisense RNA transcribed at the *RSX* locus in early embryos of both sexes, resembling TSIX, so the authors called it XSR [80]. The previous finding points out that *RSX* locus could represent a scaffold for a particular three-dimensional chromatin arrangement favorable for XCI.

4. X chromosome overexpression in *Drosophila*

Drosophila also harbors a XY genetic sex determination system; however, the dosage compensation strategy contrasts that of mammals, in which a whole X chromosome is inactivated in female cells. In the fruit fly, both X chromosomes are active in females and the gene dosage differences are compensated by an overexpression of the single-male X chromosome (XCO). Therefore, the transcriptional output of the single-male X chromosome equals twofold the transcriptional levels of each female X chromosome. Since *Drosophila* represents a reference animal model, and X chromosome dosage compensation is a paradigm for epigenetic regulation, vast research from decades ago has permitted to have a clear, although not yet complete, picture of XCO [4].

The *Drosophila* dosage compensation complex (DCC) is integrated by five proteins and one of the two redundant lncRNAs, roX1 and roX2, all of which were identified since its genetic perturbation leads to specific male lethality [85]. Male sex lethal 1 (MSL1) forms a dimer that serves as a scaffold for the assembly of the other protein subunits, and particularly, its binding to MSL2 is important for chromatin association [85]. MSL2 interacts with roX1/2 to recognize a GA-rich DNA motif, called the MSL-responsive element (MRE), and therefore, it targets the DCC to specific chromatin sites, referred as high affinity sites (HAS) [86]. MSL3 and MOF are relevant for the spreading of the DCC from the HAS to the rest of the X chromosome [87]. MSL3 harbors a chromodomain that reads histone methylation marks; it is possible that this subunit reads the histone mark H4K20me1 associated with active transcription to spread the complex to active genes for its upregulation [88]. H3K36me3 is a histone modification that marks transcribing gene bodies; although MSL3 does not seem to read this mark, H3K36me3 is required for DCC spreading [89]. MOF is a histone acetyl transferase (HAT) that deposits an acetyl group in the lysine 16 of histone 4 (H4K16ac) [90]; as other acetylation marks, H4K16ac is associated with chromatin relaxation and transcriptional activation, which, in this case, underline XCO. MLE is indispensable for the incorporation of roX1/2 to the DCC [91]; and the roXs play fundamental roles in the overall process of *Drosophila* dosage compensation. Noteworthy, MSL2 and roXs are the only components of DCC whose expression is restricted to males, highlighting their role in the selectivity of male XCO.

4.1 roX on the single-male X chromosome

roX1 and roX2 were first identified by two independent studies as male-specific transcripts from the X chromosome, with nuclear localization and lacking an apparent open-reading frame [8, 9]. In one of the studies, the authors were searching for sex-specific transcripts involved in sexual behavior in cDNAs libraries of female and male dorsal brains [9]. The second study aimed to identify the predicted RNA component of the DCC, and reported roX1 as an RNA molecule that specifically recognizes and coats the entire male X chromosome [8]. The roXs are expressed in the male cells from zygotic genome activation (ZGA) onward, under the control of MSL proteins [8, 92], since an MSL2 transgene activates roX transcription in females and mutations in MLE abolish roX expression in males [8]. roX1 and roX2 are functionally redundant; only a simultaneous removal of both loci results in the specific male lethality phenotype [93]. Although there are subtle spatiotemporal differences, their expression patterns mostly overlap and both lncRNAs are expressed from, localize to, and coat the entire X chromosome in male cells [93].

4.2 roX molecules

Despite of being functionally homologs, roX1 (3.7 Kb) and roX2 (0.6 Kb) show no sequence conservation, with the exception of repeated regions in tandem that are called roX-boX (**Table 1**) [84]. These tandem repeats form stem-loop structures that interact with MLE and MSL2 [84]. Importantly, the search of stem-loop underprints revealed roX orthologous across *Drosophila* genus capable of extrinsically executing their dosage compensation function in distinct species than that of origin [94]. This is reminiscent of the *RSX* transgene inserted in mESCs triggering gene silencing in *cis* [11], and to the tandem repeats forming stem loop structures that harbor RSX and XIST to recruit a similar interactome [82]. These observations suggest that the lncRNAs involved in X chromosome dosage compensation share particular features, such as the tandem repeats that form stem loops structures.

As previously mentioned, stem loops serves for the RNA-protein interactions, and therefore, for recruiting transcriptional activators and repressors. Interestingly, similar to XIST and RSX, the roXs seem to interact with polycomb. A recent study reported a novel function of the roXs in regulating autosomal gene expression [95]. Although the roXs predominantly localize to the X chromosome, some autosomal occupancy is detected [16, 95], and loss of the repressive histone mark H3K27me3 is observed upon knocking out both roXs [95]. These results can be related to the sequence similarities between the MRE, occupied by the DCC, and the polycomb responsive element (PRE), as both are GA-rich [16]. Alternatively, polycomb subunits may have affinity for the stem loop structures on the roXs. Further studies are needed to establish whether the stem loops of lncRNAs mediating X chromosome dosage compensation share any structural similarity that facilitates interaction with polycomb subunits.

4.3 roX loci encompass high-affinity sites

A major difference between the mammalian and *Drosophila* lncRNAs is that XIST and RSX act in *cis*, while roX1 and roX2 act in *trans* (**Figure 3**). When an *XIST* transgene is inserted into an autosome, XIST coats, compacts, and silences exclusively this autosome [29]. This represents an indispensable feature, as a hypothetical action in *trans* of XIST could result in the inactivation of the two female X chromosomes (**Figure 3**). In contrast, autosomal insertions of roX have local and global effects; locally, autosomal roX recruits the DCC to the site of insertion, but the spreading of the H4K16ac and transcriptional upregulation largely depends on an active and permissive chromatin environment [8, 96]. In turn, when a roX transgene is inserted into an autosome in a double rox1/2 null background, roX still targets and decorates the X chromosome (**Figure 3**), regarding the location of its genomic locus [8, 96]. These observations emphasize that, as in other examples of experiments evaluating lncRNA function, it is important to discern between the roles of the roX1 and roX2 transcripts and that of their genomic loci.

Each roX1 and roX2 genomic locus harbors an intronic DNAse I hypersensitivity site (DHS) [92], representing a predominant HAS. MSL1 and MSL2 bind hundreds of HAS on the X chromosome, but complete incorporation of the additional subunits MSL3, MOF, MLE, and either roX1 or roX2, is needed for DCC spreading, H4K16ac deposition, and gene upregulation [96]. Therefore, HAS represents DCC nucleation points, and spreading relies on a complete assembly of the DCC, together with an active transcriptional state of neighboring genes [96]. However, the sequence composition of the HAS on the X chromosome can be also found in autosomes and how the DCC discriminates between the HAS sequences on autosomes and preferentially binds to the X chromosome has been a subject of extensive research.

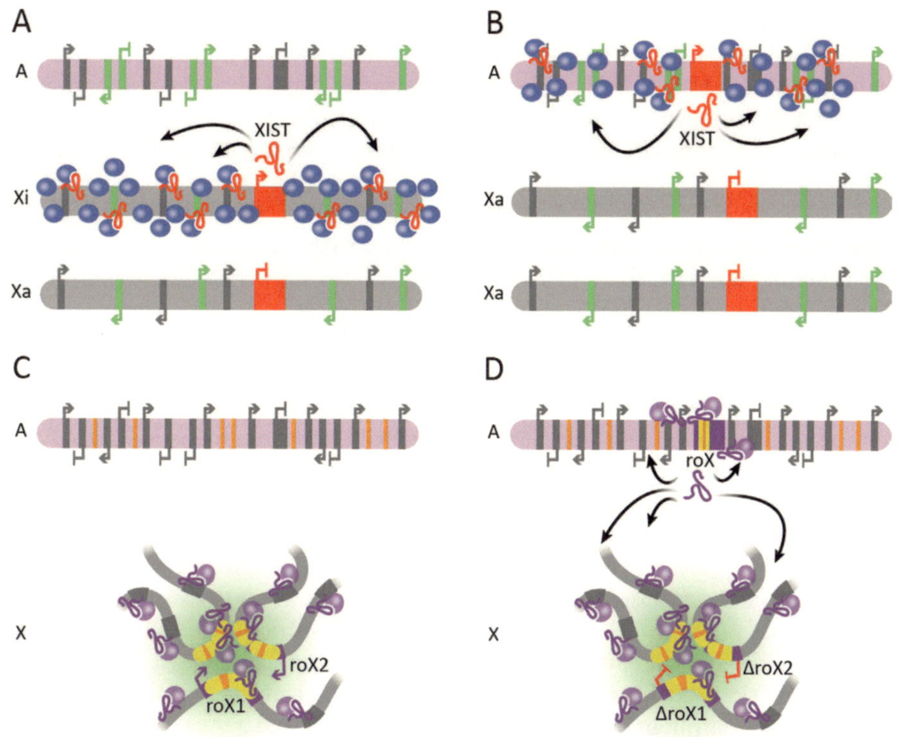

Figure 3.
XIST transcript acts in cis, while roX transcripts act in trans. *(A) XIST (in red) is transcribed from and exclusively coats the future inactive X chromosome (Xi) in mammalian females. XIST is silenced in the active X chromosome (Xa), and therefore, genes of Xa and autosomes (A) display their spatiotemporal specific expression pattern. (B) A XIST transgene is inserted into an autosome in a cell-type lacking XIST expression; for instance, naïve mouse embryonic stem cells have active their two X chromosomes. XIST is transcribed from the autosome transgene and coats exclusively this autosome, triggering its transcriptional inactivation [29]. The two X chromosomes remain active, since XIST function in cis. (C) roX 1 and roX2 comprise two predominant high affinity sites (HAS, in yellow), enclosing the MSL responsive element (MRE, in organge). HAS are in close proximity in the three-dimensional space, and a high density of clustered MRE strongly recruits MSL proteins. roX expression is upregulated and roX molecules (in purple) integrate to the dosage compensation complex (DCC, in purple), triggering its spreading to the entire single male fruit fly X chromosome and promoting global transcriptional overexpression (XCO). Autosomes harbor multiple MRE; however, roX molecules preferentially bind to the X chromosome [16]. (D) A roX transgene is inserted into an autosome of double-knockout roX1/2-/- male cells. Since roX comprises a predominant HAS [96], MSL proteins are recruited to the site of insertion, the roX transgene is upregulated, and roX molecules are integrated into the DCC. Limited DCC spreading is observed in the autosome, which relies on an active chromatin landscape of neighboring regions [96], rather than the presence of MRE. Active neighboring genes may be overexpressed. Nevertheless, most of the roX transcripts migrate to the single-male X chromosome, target the clustered HAS, spread to coat the entire X chromosome, and promote XCO.*

HAS on the X chromosome, including roX1 and roX2 genomic loci, are in close proximity in the three-dimensional space, creating a higher local density that is not observed on autosomes [97]. This particular three-dimensional conformation of the X chromosome strongly attracts the DCC and distributes it, firstly to all interacting HAS, and from there, to the entire X chromosome [97]. Indeed, a great percentage of HAS overlaps TAD boundaries on the X chromosome, in both females and males [97]. This HAS interaction network is set on prior ZGA, by DCC-independent mechanisms, which involves the chromatin-linker adaptor of MSL proteins (CLAMP) [98]. CLAMP is maternally supplied to the zygote, which recognizes GA-rich sequences on the X chromosome as well as on autosomes of females and males [99]. Only on the X chromosome, CLAMP promotes a clustering of HAS, which is targeted by MSL2 in males upon ZGA [98, 99]. The selective targeting of the X chromosome is further supported by the presence of roX1 and roX2 and possibly by satellite repeats, which are twofold more abundant on the X chromosomes in comparison with autosomes.

5. X chromosome transcriptional output restores ancestral autosome expression in a lizard

A male heterogametic sex determination system as old as the mammalian also arose in the phylogenetic branch of the reptiles [1]. In particular, the Iguania ancestor developed the XY sexual chromosome pair, after the divergence with snakes [1]. The green anole genome was the first non-avian reptile to be sequenced, and from then, this lizard has become a reference species for studying vertebrate evolution [100]. As in the other XY species, the Y chromosome of the green anole is highly degenerated, and therefore, the XY males have a reduced gene dosage in comparison with XX females and to the ancestor of origin [1, 101]. Analysis of RNA-seq data, however, displays similar expression levels of X-linked genes between sexes throughout development and in adult tissues, evidencing mechanisms of X chromosome dosage compensation [1]. Intriguingly, the green anole developed an XCO system, similar to the fruit fly and in contrast to the XCI selected in the more phylogenetic-related mammalian species [1]. XCO achieves a twofold upregulation of transcriptional output in the green anole, balancing X-linked gene expression between sexes and also restoring the predicted expression for the ancestor of origin [1]. The molecular mechanisms of the green anole XCO share additional relevant features with those in the fruit fly, including the deposition of H4K16ac along the X chromosome, and the male-biased expression of proteins relevant for HAT activity and of a lncRNA whose locus is on the X chromosome [1, 10].

5.1 A male-specific lncRNA amplifying the expression of the X

The sex-specific RNA-seq data revealed the differential expression of two lncRNAs within neighboring loci on the X chromosome [10]. The authors called FEREX, for female-specific expressed region on the X, and MAYEX, for male-specific lncRNA amplifying the expression of the X [10]. The expression of these lncRNAs is mutually exclusive, resembling the XIST/TSIX and RSX/XSR antagonism, but restricted to females and males [10]. MAYEX is a long RNA molecule of 3.3 Kb (**Table 1**), which is conserved across various lizard species [10]. In line with a role in XCO, its interactome comprehends proteins that deposit active transcriptional histone marks, while counteracting mechanisms and histone modifications related to chromatin compaction and silencing, such as those mediated by polycomb [10]. In addition, MAYEX interacts with architectural proteins that may mediate long-range contacts with distal regions on the X chromosome [10]. Similar to the HAS enclosed at roX1/2 loci, *MAYEX* locus is engaged in long-range interactions in females and males, which may be supported by a neighboring region of repetitive sequences [10]. Nevertheless, direct contacts with some distant regions are male-specific and may rely on *MAYEX* transcription or the presence of the transcript itself [10]. Upcoming studies will provide more insights into the XCO in the green anole and will uncover the role of MAYEX in this lizard.

6. Conclusions

X chromosome dosage compensation has been widely investigated in reference animal models, partly, due its relevance in the animal biology, and partly for the evolutionary contribution of its understanding. In addition, it represents an epigenetic paradigm of coordinated transcriptional regulation at the level of an entire chromosome. The main epigenetic actors and mechanisms participate in the X chromosome dosage compensation, as DNA methylation, histone modifications, chromatin remodeling complexes, the three-dimensional chromatin architecture, and therefore, all of these as well as their interplay can be studied in this model. We have learnt, from this extensive research, that the basic strategy for regulating transcription at a whole chromosome level has common features. These include the modification of the chromatin landscape and three-dimensional conformation. The cells operated its tool box to specialize some of the pre-existent epigenetic actors and mechanisms under the control of a novel component that results to be a lncRNA transcribed from the target X.

LncRNAs can act as an interphase of the chromatin-modifying machinery and the genome. Part of the sequence of the RNA molecule engages with and reads specific DNA sequences, while the other part recruits protein complexes holding enzymatic activities. Beside the lncRNA transcripts, their genomic loci serve as a scaffold of distinct combinatorial of DNA-RNA-protein interactions. In particular, the lncRNAs involved in the X chromosome dosage compensation share similar features across mammals, lizards, and the fruit fly. These are all nuclear transcripts whose genes locate in the target X chromosome [8–11, 102]. They are transcribed only in one sex, either in females for XCI or in males for XCO. Their genomic loci are relevant regulatory targets to trigger dosage compensation. Once expressed, XIST, RSX, roX1, roX2, and MAYEX coat the whole X chromosome. Due these transcripts are particularly large, repeat rich, and enclose tandem repeats capable to form stem-loop structures, they engage in RNA-protein interactions [82, 84]. In a first layer of global transcriptional regulation, the lncRNAs associate with histone acetyl transferases, lysine methylases, histone deacetylases, and demethylases to modify the X chromosome chromatin landscape [10, 83, 84]. In a second layer, the lncRNAs interact with proteins involved in long-range contacts, such as polycomb, CLAMP, and SASH1, to strengthen a chromatin three-dimensional arrangement favorable for global dosage compensation [10, 83, 95, 99].

Evidences of global dosage compensation have been observed in other insects with genetic sex determination systems, including mosquitos and butterflies [5, 6]. Further studies will elucidate whether these animals also selected lncRNAs as orchestrators of chromosome-wide transcriptional regulation to balance the expression between males and females.

Acknowledgements

We acknowledge our funding sources: Fundación Miguel Alemán A.C. and Programa de Apoyo a Proyectos de Investigación e Innovación Tecnológica (DGAPA-PAPIIT IA202723 and IN206823).

Nomenclature

CLAMP	chromatin-linker adaptor of MSL proteins
DCC	dosage compensation complex
DHA	DNase I hypersensitivity sites
HAS	high affinity sites
HAT	histone acetylase
HDAC	histone deacetylase
hnRNPK	heterogeneous nuclear ribonucleoprotein K
lncRNA	long non-coding RNA
MRE	MSL responsive element
MSL	male sex lethal
mya	millions of years ago
Pol II	RNA polymerase II
PRC	polycomb repressive complex
PRE	polycomb responsive element
rG4	RNA G-quadruplex
roX	RNA on the X
RSX	RNA on the silent X
TAD	topologically associated domain
TSIX	XIST antisense non-coding transcription unit
Xa	active X chromosome
XCI	X chromosome inactivation
XCO	X chromosome overexpression
Xi	inactive X chromosome
XIC	X inactivation center
XIST	X-inactive specific transcript
ZGA	zygotic genome activation

Author details

Sylvia Garza-Manero*, Hober Nelson Núñez-Martínez, Gustavo Tapia-Urzúa and Félix Recillas-Targa
Department of Molecular Genetics, Institute of Cellular Physiology, National Autonomous University of Mexico, Mexico City, Mexico

*Address all correspondence to: sgarza@ifc.unam.mx

References

[1] Marin R, Cortez D, Lamanna F, Pradeepa MM, Leushkin E, Julien P, et al. Convergent origination of a Drosophila-like dosage compensation mechanism in a reptile lineage. Genome Research. 2017;**27**(12):1974-1987

[2] Cecalev D, Vicoso B, Galupa R. Compensation of gene dosage on the mammalian X. Development. 2024;**151**(15):1-16

[3] Deakin JE. Marsupial X chromosome inactivation: Past, present and future %J. Australian Journal of Zoology. 2013;**61**(1):13-23

[4] Samata M, Akhtar A. Dosage compensation of the X chromosome: A complex epigenetic assignment involving chromatin regulators and long noncoding RNAs. Annual Review of Biochemistry. 2018;**87**:323-350

[5] Rose G, Krzywinska E, Kim J, Revuelta L, Ferretti L, Krzywinski J. Dosage compensation in the African malaria mosquito Anopheles gambiae. Genome Biology and Evolution. 2016;**8**(2):411-425

[6] Gu L, Reilly PF, Lewis JJ, Reed RD, Andolfatto P, Walters JR. Dichotomy of dosage compensation along the neo Z chromosome of the monarch butterfly. Current Biology. 2019;**29**(23):4071-7 e3

[7] Brown CJ, Hendrich BD, Rupert JL, Lafreniere RG, Xing Y, Lawrence J, et al. The human XIST gene: Analysis of a 17 kb inactive X-specific RNA that contains conserved repeats and is highly localized within the nucleus. Cell. 1992;**71**(3):527-542

[8] Meller VH, Wu KH, Roman G, Kuroda MI, Davis RL. roX1 RNA paints the X chromosome of male drosophila and is regulated by the dosage compensation system. Cell. 1997;**88**(4):445-457

[9] Amrein H, Axel R. Genes expressed in neurons of adult male drosophila. Cell. 1997;**88**(4):459-469

[10] Tenorio M, Cruz-Ruiz S, Encarnacion-Guevara S, Hernandez M, Corona-Gomez JA, Sheccid-Santiago F, et al. MAYEX is an old long noncoding RNA recruited for X chromosome dosage compensation in a reptile. Science. 2024;**385**(6715):1347-1354

[11] Grant J, Mahadevaiah SK, Khil P, Sangrithi MN, Royo H, Duckworth J, et al. Rsx is a metatherian RNA with Xist-like properties in X-chromosome inactivation. Nature. 2012;**487**(7406):254-258

[12] Zhu Z, Younas L, Zhou Q. Evolution and regulation of animal sex chromosomes. Nature Reviews. Genetics. 2025;**26**(1):59-74

[13] Mank JE. Sex chromosome dosage compensation: Definitely not for everyone. Trends in Genetics. 2013;**29**(12):677-683

[14] Loda A, Collombet S, Heard E. Gene regulation in time and space during X-chromosome inactivation. Nature Reviews. Molecular Cell Biology. 2022;**23**(4):231-249

[15] Henninger JE, Young RA. An RNA-centric view of transcription and genome organization. Molecular Cell. 2024;**84**(19):3627-3643

[16] Chu C, Qu K, Zhong FL, Artandi SE, Chang HY. Genomic maps

of long noncoding RNA occupancy reveal principles of RNA-chromatin interactions. Molecular Cell. 2011;**44**(4):667-678

[17] Lyon MF. Gene action in the X-chromosome of the mouse (*Mus musculus* L.). Nature. 1961;**190**:372-373

[18] Starmer J, Magnuson T. A new model for random X chromosome inactivation. Development. 2009;**136**(1):1-10

[19] Disteche CM. Dosage compensation of the sex chromosomes. Annual Review of Genetics. 2012;**46**:537-560

[20] van den Berg IM, Laven JS, Stevens M, Jonkers I, Galjaard RJ, Gribnau J, et al. X chromosome inactivation is initiated in human preimplantation embryos. American Journal of Human Genetics. 2009;**84**(6):771-779

[21] Monk M, Harper MI. Sequential X chromosome inactivation coupled with cellular differentiation in early mouse embryos. Nature. 1979;**281**(5729):311-313

[22] Augui S, Nora EP, Heard E. Regulation of X-chromosome inactivation by the X-inactivation Centre. Nature Reviews. Genetics. 2011;**12**(6):429-442

[23] Lee JT, Strauss WM, Dausman JA, Jaenisch R. A 450 kb transgene displays properties of the mammalian X-inactivation center. Cell. 1996;**86**(1):83-94

[24] Chureau C, Prissette M, Bourdet A, Barbe V, Cattolico L, Jones L, et al. Comparative sequence analysis of the X-inactivation center region in mouse, human, and bovine. Genome Research. 2002;**12**(6):894-908

[25] Chureau C, Chantalat S, Romito A, Galvani A, Duret L, Avner P, et al. Ftx is a non-coding RNA which affects Xist expression and chromatin structure within the X-inactivation center region. Human Molecular Genetics. 2011;**20**(4):705-718

[26] Lee JT, Lu N, Han Y. Genetic analysis of the mouse X inactivation center defines an 80-kb multifunction domain. Proceedings of the National Academy of Sciences of the United States of America. 1999;**96**(7):3836-3841

[27] Lee JT, Davidow LS, Warshawsky D. Tsix, a gene antisense to Xist at the X-inactivation Centre. Nature Genetics. 1999;**21**(4):400-404

[28] Simmler MC, Cunningham DB, Clerc P, Vermat T, Caudron B, Cruaud C, et al. A 94 kb genomic sequence 3′ to the murine Xist gene reveals an AT rich region containing a new testis specific gene Tsx. Human Molecular Genetics. 1996;**5**(11):1713-1726

[29] Lee JT, Jaenisch R. Long-range cis effects of ectopic X-inactivation centres on a mouse autosome. Nature. 1997;**386**(6622):275-279

[30] Lee JT. Regulation of X-chromosome counting by Tsix and Xite sequences. Science. 2005;**309**(5735):768-771

[31] Ogawa Y, Lee JT. Xite, X-inactivation intergenic transcription elements that regulate the probability of choice. Molecular Cell. 2003;**11**(3):731-743

[32] Sun S, Del Rosario BC, Szanto A, Ogawa Y, Jeon Y, Lee JT. Jpx RNA activates Xist by evicting CTCF. Cell. 2013;**153**(7):1537-1551

[33] Tian D, Sun S, Lee JT. The long noncoding RNA, Jpx, is a molecular switch for X chromosome inactivation. Cell. 2010;**143**(3):390-403

[34] Plath K, Fang J, Mlynarczyk-Evans SK, Cao R, Worringer KA, Wang H, et al. Role of histone H3 lysine 27 methylation in X inactivation. Science. 2003;**300**(5616):131-135

[35] Heard E, Rougeulle C, Arnaud D, Avner P, Allis CD, Spector DL. Methylation of histone H3 at Lys-9 is an early mark on the X chromosome during X inactivation. Cell. 2001;**107**(6):727-738

[36] Simon MD, Pinter SF, Fang R, Sarma K, Rutenberg-Schoenberg M, Bowman SK, et al. High-resolution Xist binding maps reveal two-step spreading during X-chromosome inactivation. Nature. 2013;**504**(7480):465-469

[37] Engreitz JM, Pandya-Jones A, McDonel P, Shishkin A, Sirokman K, Surka C, et al. The Xist lncRNA exploits three-dimensional genome architecture to spread across the X chromosome. Science. 2013;**341**(6147):1237973

[38] Colognori D, Sunwoo H, Kriz AJ, Wang CY, Lee JT. Xist deletional analysis reveals an interdependency between Xist RNA and polycomb complexes for spreading along the inactive X. Molecular Cell. 2019;**74**(1):101-17 e10

[39] Sunwoo H, Colognori D, Froberg JE, Jeon Y, Lee JT. Repeat E anchors Xist RNA to the inactive X chromosomal compartment through CDKN1A-interacting protein (CIZ1). Proceedings of the National Academy of Sciences of the United States of America. 2017;**114**(40):10654-10659

[40] Ridings-Figueroa R, Stewart ER, Nesterova TB, Coker H, Pintacuda G, Godwin J, et al. The nuclear matrix protein CIZ1 facilitates localization of Xist RNA to the inactive X-chromosome territory. Genes and Development. 2017;**31**(9):876-888

[41] Minajigi A, Froberg J, Wei C, Sunwoo H, Kesner B, Colognori D, et al. Chromosomes. A comprehensive Xist interactome reveals cohesin repulsion and an RNA-directed chromosome conformation. Science. 2015;**349**(6245):1-12

[42] Chu C, Zhang QC, da Rocha ST, Flynn RA, Bharadwaj M, Calabrese JM, et al. Systematic discovery of Xist RNA binding proteins. Cell. 2015;**161**(2):404-416

[43] Wutz A, Rasmussen TP, Jaenisch R. Chromosomal silencing and localization are mediated by different domains of Xist RNA. Nature Genetics. 2002;**30**(2):167-174

[44] Zylicz JJ, Bousard A, Zumer K, Dossin F, Mohammad E, da Rocha ST, et al. The implication of early chromatin changes in X chromosome inactivation. Cell. 2019;**176**(1-2):182-97 e23

[45] Collombet S, Rall I, Dugast-Darzacq C, Heckert A, Halavatyi A, Le Saux A, et al. RNA polymerase II depletion from the inactive X chromosome territory is not mediated by physical compartmentalization. Nature Structural and Molecular Biology. 2023;**30**(8):1216-1223

[46] McHugh CA, Chen CK, Chow A, Surka CF, Tran C, McDonel P, et al. The Xist lncRNA interacts directly with SHARP to silence transcription through HDAC3. Nature. 2015;**521**(7551):232-236

[47] Bousard A, Raposo AC, Zylicz JJ, Picard C, Pires VB, Qi Y, et al. The role of Xist-mediated polycomb recruitment in the initiation of X-chromosome inactivation. EMBO Reports. 2019;**20**(10):e48019

[48] Schoeftner S, Sengupta AK, Kubicek S, Mechtler K, Spahn L,

Koseki H, et al. Recruitment of PRC1 function at the initiation of X inactivation independent of PRC2 and silencing. The EMBO Journal. 2006;**25**(13):3110-3122

[49] Pinter SF, Sadreyev RI, Yildirim E, Jeon Y, Ohsumi TK, Borowsky M, et al. Spreading of X chromosome inactivation via a hierarchy of defined polycomb stations. Genome Research. 2012;**22**(10):1864-1876

[50] Pintacuda G, Wei G, Roustan C, Kirmizitas BA, Solcan N, Cerase A, et al. hnRNPK recruits PCGF3/5-PRC1 to the Xist RNA B-repeat to establish polycomb-mediated chromosomal silencing. Molecular Cell. 2017;**68**(5):955-69 e10

[51] Almeida M, Pintacuda G, Masui O, Koseki Y, Gdula M, Cerase A, et al. PCGF3/5-PRC1 initiates polycomb recruitment in X chromosome inactivation. Science. 2017;**356**(6342):1081-1084

[52] Cooper S, Grijzenhout A, Underwood E, Ancelin K, Zhang T, Nesterova TB, et al. Jarid2 binds mono-ubiquitylated H2A lysine 119 to mediate crosstalk between polycomb complexes PRC1 and PRC2. Nature Communications. 2016;**7**:13661

[53] da Rocha ST, Boeva V, Escamilla-Del-Arenal M, Ancelin K, Granier C, Matias NR, et al. Jarid2 is implicated in the initial Xist-induced targeting of PRC2 to the inactive X chromosome. Molecular Cell. 2014;**53**(2):301-316

[54] Lee YW, Weissbein U, Blum R, Lee JT. G-quadruplex folding in Xist RNA antagonizes PRC2 activity for stepwise regulation of X chromosome inactivation. Molecular Cell. 2024;**84**(10):1870-85 e9

[55] Szanto A, Aguilar R, Kesner B, Blum R, Wang D, Cifuentes-Rojas C, et al. A disproportionate impact of G9a methyltransferase deficiency on the X chromosome. Genes and Development. 2021;**35**(13-14):1035-1054

[56] Keniry A, Gearing LJ, Jansz N, Liu J, Holik AZ, Hickey PF, et al. Setdb1-mediated H3K9 methylation is enriched on the inactive X and plays a role in its epigenetic silencing. Epigenetics and Chromatin. 2016;**9**:16

[57] Escamilla-Del-Arenal M, da Rocha ST, Spruijt CG, Masui O, Renaud O, Smits AH, et al. Cdyl, a new partner of the inactive X chromosome and potential reader of H3K27me3 and H3K9me2. Molecular and Cellular Biology. 2013;**33**(24):5005-5020

[58] Gendrel AV, Apedaile A, Coker H, Termanis A, Zvetkova I, Godwin J, et al. Smchd1-dependent and -independent pathways determine developmental dynamics of CpG island methylation on the inactive X chromosome. Developmental Cell. 2012;**23**(2):265-279

[59] Sharp AJ, Stathaki E, Migliavacca E, Brahmachary M, Montgomery SB, Dupre Y, et al. DNA methylation profiles of human active and inactive X chromosomes. Genome Research. 2011;**21**(10):1592-1600

[60] Pasque V, Gillich A, Garrett N, Gurdon JB. Histone variant macroH2A confers resistance to nuclear reprogramming. The EMBO Journal. 2011;**30**(12):2373-2387

[61] Rasmussen TP, Mastrangelo MA, Eden A, Pehrson JR, Jaenisch R. Dynamic relocalization of histone MacroH2A1 from centrosomes to inactive X chromosomes during X inactivation. The Journal of Cell Biology. 2000;**150**(5):1189-1198

[62] Adrianse RL, Smith K, Gatbonton-Schwager T, Sripathy SP, Lao U, Foss EJ, et al. Perturbed maintenance of transcriptional repression on the inactive X-chromosome in the mouse brain after Xist deletion. Epigenetics & Chromatin. 2018;**11**(1):50

[63] Csankovszki G, Panning B, Bates B, Pehrson JR, Jaenisch R. Conditional deletion of Xist disrupts histone macroH2A localization but not maintenance of X inactivation. Nature Genetics. 1999;**22**(4):323-324

[64] Barr ML, Bertram EG. A morphological distinction between neurones of the male and female, and the behaviour of the nucleolar satellite during accelerated nucleoprotein synthesis. Nature. 1949;**163**(4148):676

[65] Borden J, Manuelidis L. Movement of the X chromosome in epilepsy. Science. 1988;**242**(4886):1687-1691

[66] Belmont AS, Bignone F, Ts'o PO. The relative intranuclear positions of Barr bodies in XXX non-transformed human fibroblasts. Experimental Cell Research. 1986;**165**(1):165-179

[67] Chen CK, Blanco M, Jackson C, Aznauryan E, Ollikainen N, Surka C, et al. Xist recruits the X chromosome to the nuclear lamina to enable chromosome-wide silencing. Science. 2016;**354**(6311):468-472

[68] Pollex T, Heard E. Nuclear positioning and pairing of X-chromosome inactivation centers are not primary determinants during initiation of random X-inactivation. Nature Genetics. 2019;**51**(2):285-295

[69] Teller K, Illner D, Thamm S, Casas-Delucchi CS, Versteeg R, Indemans M, et al. A top-down analysis of Xa- and Xi-territories reveals differences of higher order structure at >/= 20 Mb genomic length scales. Nucleus. 2011;**2**(5):465-477

[70] Naughton C, Sproul D, Hamilton C, Gilbert N. Analysis of active and inactive X chromosome architecture reveals the independent organization of 30 nm and large-scale chromatin structures. Molecular Cell. 2010;**40**(3):397-409

[71] Jegu T, Aeby E, Lee JT. The X chromosome in space. Nature Reviews. Genetics. 2017;**18**(6):377-389

[72] Lieberman-Aiden E, van Berkum NL, Williams L, Imakaev M, Ragoczy T, Telling A, et al. Comprehensive mapping of long-range interactions reveals folding principles of the human genome. Science. 2009;**326**(5950):289-293

[73] Giorgetti L, Lajoie BR, Carter AC, Attia M, Zhan Y, Xu J, et al. Structural organization of the inactive X chromosome in the mouse. Nature. 2016;**535**(7613):575-579

[74] Deng X, Ma W, Ramani V, Hill A, Yang F, Ay F, et al. Bipartite structure of the inactive mouse X chromosome. Genome Biology. 2015;**16**(1):152

[75] Navarro P, Chantalat S, Foglio M, Chureau C, Vigneau S, Clerc P, et al. A role for non-coding Tsix transcription in partitioning chromatin domains within the mouse X-inactivation Centre. Epigenetics and Chromatin. 2009;**2**(1):8

[76] Ohhata T, Hoki Y, Sasaki H, Sado T. Crucial role of antisense transcription across the Xist promoter in Tsix-mediated Xist chromatin modification. Development. 2008;**135**(2):227-235

[77] Navarro P, Page DR, Avner P, Rougeulle C. Tsix-mediated epigenetic switch of a CTCF-flanked region of the Xist promoter determines the Xist

transcription program. Genes and Development. 2006;**20**(20):2787-2792

[78] Sado T, Hoki Y, Sasaki H. Tsix silences Xist through modification of chromatin structure. Developmental Cell. 2005;**9**(1):159-165

[79] Whitworth DJ, Pask AJ. The X factor: X chromosome dosage compensation in the evolutionarily divergent monotremes and marsupials. Seminars in Cell and Developmental Biology. 2016;**56**:117-121

[80] Mahadevaiah SK, Sangrithi MN, Hirota T, Turner JMA. A single-cell transcriptome atlas of marsupial embryogenesis and X inactivation. Nature. 2020;**586**(7830):612-617

[81] Milton AM, Marin-Gual L, Lister NC, McIntyre KL, Grady PGS, Laird MK, et al. Imprinted X chromosome inactivation in marsupials: The paternal X arrives at the egg with a silent DNA methylation profile. Proceedings of the National Academy of Sciences of the United States of America. 2024;**121**(36):e2412185121

[82] Sprague D, Waters SA, Kirk JM, Wang JR, Samollow PB, Waters PD, et al. Nonlinear sequence similarity between the Xist and Rsx long noncoding RNAs suggests shared functions of tandem repeat domains. RNA. 2019;**25**(8):1004-1019

[83] McIntyre KL, Waters SA, Zhong L, Hart-Smith G, Raftery M, Chew ZA, et al. Identification of the RSX interactome in a marsupial shows functional coherence with the Xist interactome during X inactivation. Genome Biology. 2024;**25**(1):134

[84] Ilik IA, Quinn JJ, Georgiev P, Tavares-Cadete F, Maticzka D, Toscano S, et al. Tandem stem-loops in roX RNAs act together to mediate X chromosome dosage compensation in drosophila. Molecular Cell. 2013;**51**(2):156-173

[85] Belote JM, Lucchesi JC. Control of X chromosome transcription by the maleless gene in drosophila. Nature. 1980;**285**(5766):573-575

[86] Li F, Schiemann AH, Scott MJ. Incorporation of the noncoding roX RNAs alters the chromatin-binding specificity of the drosophila MSL1/MSL2 complex. Molecular and Cellular Biology. 2008;**28**(4):1252-1264

[87] Straub T, Zabel A, Gilfillan GD, Feller C, Becker PB. Different chromatin interfaces of the drosophila dosage compensation complex revealed by high-shear ChIP-seq. Genome Research. 2013;**23**(3):473-485

[88] Kim D, Blus BJ, Chandra V, Huang P, Rastinejad F, Khorasanizadeh S. Corecognition of DNA and a methylated histone tail by the MSL3 chromodomain. Nature Structural and Molecular Biology. 2010;**17**(8):1027-1029

[89] Larschan E, Alekseyenko AA, Gortchakov AA, Peng S, Li B, Yang P, et al. MSL complex is attracted to genes marked by H3K36 trimethylation using a sequence-independent mechanism. Molecular Cell. 2007;**28**(1):121-133

[90] Akhtar A, Becker PB. Activation of transcription through histone H4 acetylation by MOF, an acetyltransferase essential for dosage compensation in drosophila. Molecular Cell. 2000;**5**(2):367-375

[91] Meller VH, Gordadze PR, Park Y, Chu X, Stuckenholz C, Kelley RL, et al. Ordered assembly of roX RNAs into MSL complexes on the dosage-compensated X chromosome in drosophila. Current Biology. 2000;**10**(3):136-143

[92] Bai X, Alekseyenko AA, Kuroda MI. Sequence-specific targeting of MSL complex regulates transcription of the roX RNA genes. The EMBO Journal. 2004;**23**(14):2853-2861

[93] Franke A, Baker BS. The rox1 and rox2 RNAs are essential components of the compensasome, which mediates dosage compensation in drosophila. Molecular Cell. 1999;**4**(1):117-122

[94] Quinn JJ, Zhang QC, Georgiev P, Ilik IA, Akhtar A, Chang HY. Rapid evolutionary turnover underlies conserved lncRNA-genome interactions. Genes & Development. 2016;**30**(2):191-207

[95] Li J, Xu S, Liu Z, Yang L, Ming Z, Zhang R, et al. A noncanonical role of roX RNAs in autosomal epigenetic repression. Nature Communications. 2025;**16**(1):155

[96] Kelley RL, Meller VH, Gordadze PR, Roman G, Davis RL, Kuroda MI. Epigenetic spreading of the drosophila dosage compensation complex from roX RNA genes into flanking chromatin. Cell. 1999;**98**(4):513-522

[97] Ramirez F, Lingg T, Toscano S, Lam KC, Georgiev P, Chung HR, et al. High-affinity sites form an interaction network to facilitate spreading of the MSL complex across the X chromosome in drosophila. Molecular Cell. 2015;**60**(1):146-162

[98] Jordan W, Larschan E. The zinc finger protein CLAMP promotes long-range chromatin interactions that mediate dosage compensation of the drosophila male X-chromosome. Epigenetics and Chromatin. 2021;**14**(1):29

[99] Rieder LE, Jordan WT, Larschan EN. Targeting of the dosage-compensated male X-chromosome during early drosophila development. Cell Reports. 2019;**29**(13):4268-75 e2

[100] Alfoldi J, Di Palma F, Grabherr M, Williams C, Kong L, Mauceli E, et al. The genome of the green anole lizard and a comparative analysis with birds and mammals. Nature. 2011;**477**(7366):587-591

[101] Rupp SM, Webster TH, Olney KC, Hutchins ED, Kusumi K, Wilson Sayres MA. Evolution of dosage compensation in Anolis carolinensis, a reptile with XX/XY chromosomal sex determination. Genome Biology and Evolution. 2017;**9**(1):231-240

[102] Brown CJ, Ballabio A, Rupert JL, Lafreniere RG, Grompe M, Tonlorenzi R, et al. A gene from the region of the human X inactivation Centre is expressed exclusively from the inactive X chromosome. Nature. 1991;**349**(6304):38-44

Chapter 4

The Role of Long Non-Coding RNAs in Skeletal Muscle Pathophysiology: A Therapeutic Perspective

Silvia Scalabrin, Sanaz Kavoosi and Stefano Cagnin

Abstract

Skeletal muscle, one of the largest organs in mammals, plays a pivotal role in regulating body temperature, metabolism, and movement. It is a versatile tissue that can undergo degeneration due to various factors such as disease, aging, or cancer, posing challenges to cancer therapy. Conversely, it possesses the remarkable ability to regenerate. Skeletal muscle is composed of both mononucleated cells, each with distinct functions during degeneration and regeneration, and multinucleated cells called myofibers. Recently, it was demonstrated that the transcription of mammalian cells is pervasive, highlighting the importance of junk DNA. It frequently codes for non-coding RNAs that have gained increasing importance in the pathophysiology of all tissues over the past two decades. Following the advent of COVID-19 vaccines, RNA has assumed a greater significance in therapeutic applications. In this manuscript, we will explore the impact of long non-coding RNAs (lncRNAs) on muscle degeneration and regeneration and investigate their potential therapeutic applications.

Keywords: long non-coding RNAs, circular RNAs, skeletal muscle, myogenesis, metabolic disorders, muscle atrophy, therapy, metabolism

1. Introduction

The human body is composed of over 600 skeletal muscles, which constitute approximately 40% of total body weight and contain 50–75% of all body proteins. Skeletal muscle tissue is not only involved in producing movement, maintaining body posture, controlling body temperature, and stabilizing joints, but also plays a crucial role in maintaining organisms' energy homeostasis. Notably, it can be considered a biomechanical tissue with numerous interacting elements, including vasculature used for sharing and providing components essential for energy production, such as oxygen, or myokines as brain-derived neurotrophic factor, leukemia inhibitory factor IL6, IL8, and IL15, with autocrine, paracrine, or endocrine actions and involvement in inflammatory processes [1]. Additionally, it exhibits a unique interaction with motor neurons, enabling the transmission of contracting impulses through the motor neuron junction (NMJ). Furthermore, satellite cells contribute to the plasticity of this tissue, allowing it to adapt its structure in response to specific stimuli, such as unloading, exercise, or certain pathological conditions.

The interplay between electrical impulses, the number of blood vessels, and metabolism determines the differences in muscle types and metabolism. Generally, we can categorize two distinct muscle types: fast-contracting and slow-contracting. The former comprises myofibers that rely on glycolysis as their primary energy source, while the latter comprises myofibers that utilize mitochondrial respiration and, consequently, require more oxygen. Myofibers (containing multiple myofibrils) are the smallest complete contractile units of skeletal muscle, influencing its contraction velocity and metabolic rate. Bundles of myofibers form fascicles, and bundles of fascicles constitute muscle tissue. Each layer of muscle tissue is subsequently encapsulated by the extracellular matrix (ECM) [2] and supported by the cytoskeletal networks.

Mouse myofibers can be classified in four distinct types: Type I (slow twitching), Type IIa, Type IIb, and Type IIX (fast twitching). Each muscle fiber is characterized by the expression of a specific type of myosin heavy chain (MyHC), which in turn determines the muscle's contractility properties. A difference between rats and humans is that the latter do not have the MYH4 gene coding for myosin in type IIb myofibers [3].

A unique feature of myofibers is their multinucleated nature. Each myofiber is formed by the fusion of hundreds of myoblasts, resulting in the retention of each myoblast nucleus within the myofiber [4]. Interestingly, nuclei are arranged along the length of myofibers, in the peripheral compartment near the membrane, and express different genes based on their position. Nuclei located near the NMJ express genes coding for essential components of the NMJ while those in proximity to the myotendinous junction express genes for proteins crucial to that compartment [5]. This means that myonuclei can be functionally distinguishable according to their position within myofiber.

In the past, RNA was primarily regarded as an intermediary between DNA and proteins. However, advancements in DNA/RNA sequencing technologies have unveiled RNA's multifaceted role beyond its functional role as a mediator. The human genome is pervasively transcribed (75%), but only a very small fraction of these transcripts serve for protein synthesis (2%) [6]. This suggests that the information contained in 98% of transcribed DNA is used to produce non-protein-coding RNAs. The substantial energy expenditure associated with this extensive transcription cannot be attributed solely to random interactions between RNA polymerases and DNA. Rather, it is likely necessary for crucial functions that are not effectively performed by DNA and proteins alone [7]. In fact, considering the genome size and the fact that in complex eukaryotes as *Homo sapiens* and *Mus musculus* there are about 20,400 genes (information from Ensembl and Mouse Genome Informatics (MGI) database, respectively), the regulation of all genomic components (about 228,000 mapped genes in *H. sapiens* and 319,600 in *M. musculus*, according to the Ensembl and MGI database, respectively) is not possible only through the use of transcription factors (TFs). This is because TFs are relatively scarce, with an estimated count of approximately 1500 in mammals [8] compared to the number of genes they regulate (90% of the human genome is transcribed [9]). Chromatin immunoprecipitation and DNA sequencing experiments (ChIp-seq) permitted to estimate that a single TF regulate only 10 thousand genes [10]. Additionally, there are evidences that regulation of gene expression also occurs through the intervention of non-coding RNAs [11, 12].

The evidence presented underscores the active transcription and crucial functions of the non-protein-coding component of the human genome, thereby diminishing the significance of coding genes in genome regulation. Non-coding transcripts have emerged as pivotal entities in contemporary biology, particularly due to their involvement in a diverse range of regulatory processes and redefining the perception of non-coding regions as mere "junk DNA" [13]. To date, numerous types of non-coding RNAs have been identified. These can be categorized into short non-coding RNAs (sncRNAs) and long non-coding RNAs (lncRNAs), with the latter being the most recently studied. In this chapter, we present our comprehensive understanding of the most recently investigated lncRNAs that have significantly contributed to the advancement of the regulatory RNA field in skeletal muscle and have the potential to be further investigated as potential therapeutic target in muscle atrophy. This review delves into the intricacies of lncRNAs, elucidating their underlying mechanism of action in skeletal muscle. The primary objective of this manuscript is to position lncRNAs as potential therapeutic targets for pathologies of skeletal muscle.

2. Long non-coding RNAs

The term "long non-coding RNA" (lncRNA) encompasses a diverse group of genes or transcripts that presents a limited conservation of primary structure across species. Since this characteristic is important in gene or transcript classification, different methods of classification were proposed for lncRNAs, each basing on a distinct characteristic.

Only recently was used the term lncRNA. Initially, for the classification of non-coding transcripts were used their functions (rRNA, tRNA, and cRNA). However, the increase in the number of recognized non-coding RNAs and evidence of their abundance in the early 2000s [6] new classification approaches were introduced.

It is well established that mammalians [14] and plant organisms [15] use most of their energy to transcribe lncRNAs instead of coding RNAs. Nevertheless, only a few of these lnRNAs has been functionally characterized. Studies have demonstrated that these transcripts exhibit a wide range of roles and mechanisms of action, which are influenced by their subcellular localization or cell compartmentalization. Genomic location, subcellular localization or function can be used for the categorization of lncRNAs. Below is a description of each category, along with some recent examples of lncRNAs involved in muscle biology.

2.1 Genomic location

2.1.1 Genic lncRNAs (exonic or intronic)

LncRNAs can localize within genomic regions that overlap protein-coding genes. These transcripts are referred to as genic lncRNAs. Genic lncRNAs can be classified as intronic or exonic based on their position within protein-coding genes. The overlap with an intron of a protein-coding gene defines intronic lncRNAs, while the overlap with exons of protein-coding genes defines exonic lncRNAs. Consequently, considering the transcriptional orientation of lncRNAs, they can be classified as sense lncRNAs if they are transcribed in the same direction of protein-coding genes or antisense lncRNAs if they are transcribed in the opposite direction. Antisense lncRNAs are transcribed from the opposite strand of the closest protein-coding gene.

They can either overlap with a 5′ head-to-head divergent orientation (SINEUPs) [16–18] or with 3′ tail-to-tail convergent orientation (3′ overlapping lncRNAs). The Synaptopodin-2 (SYNPO2) intron sense-overlapping lncRNA (SYISL) is an example of intronic lncRNA that regulates myogenesis through the interaction with the enhancer of zeste homolog 2 (EZH2). EZH2 is a protein that participates in histone methylation and therefore in epigenetic gene expression regulation [19].

2.1.2 Long non-coding RNAs between coding genes (lincRNAs)

Long intergenic non-coding RNAs (lincRNAs) are located within genomic regions that encompass two coding genes, thereby excluding them from protein-coding loci. A substantial proportion (40%) of lincRNAs expressed in HeLa cells, human foreskin, and human lung fibroblasts is involved in chromatin structure modification suggesting their involvement in the regulation of gene expression [20]. The number of lincRNAs in the human, mouse, zebrafish, and sheep genomes is not limited. Indeed, 214 lincRNAs have been annotated in the human genome (Ensembl GRCh38), 390 in the mouse genome (Ensembl GRCm39), 41 in the zebrafish genome (Ensembl GRCz11), and 2020 in the sheep genome (Ensembl Oar_v3.1). These numbers indicate they may have important functions in various organisms. Furthermore, lincRNAs can be further categorized based on their transcription direction relative to the neighboring protein-coding gene. They can be classified as follows: (i) divergent when they are transcribed in the opposite direction of nearest protein-coding gene, (ii) convergent when they are transcribed in an end-to-end manner, which is the reverse of divergent, and (iii) transcribed in the same sense of the nearest protein-coding gene. A recently discussed lincRNA in muscle biology is lincMYH, which is involved in regulating the satellite cell pool [21].

2.1.3 Splice-based classification

The majority of lncRNAs are transcribed by RNA polymerase II (RNA pol II). However, some lncRNAs are transcribed by RNA polymerase III (RNA pol III) [22]. Following transcription, lncRNAs may undergo processing by the splicing machinery, resulting in various categories of lncRNAs: (i) macro lncRNAs, which originate from unspliced transcripts that are several kilobases in size; (ii) retained intron lncRNAs, which are alternative transcripts of coding genes that lose their coding properties after an intron is retained during the splicing process.

2.2 Subcellular localization of lncRNAs

The subcellular localization of lncRNAs is crucial for their functional roles. For instance, nuclear lncRNAs may interact with chromatin, while cytoplasmic lncRNAs may function as microRNA (miRNA) sponges.

2.2.1 LncRNAs within nucleus

Nuclear lncRNAs are exclusively located in the nucleus. Specific sequences within lncRNAs permit their retention in the nucleus [23]. The majority of nuclear lncRNAs function as transcriptional regulators or impact on mRNA processing. They can either activate or inhibit gene activity, for instance, by recruiting transcription factors or operating as decoys that hinder the interaction between transcription factors and

DNA. Notable examples of nuclear lncRNAs include Neat1, whose inhibition is beneficial for inducing muscle hypertrophy, but detrimental to injured muscle [24], and Malat1, whose expression decreased with age in mouse skeletal muscles [25].

2.2.2 LncRNAs in the cytoplasm

LncRNAs confined in the cytoplasm are classified as cytoplasmic lncRNAs. RNAs are synthesized in the nucleus, and their transport to the cytoplasm entails numerous distinct proteins located on both sides of nuclear pore. The RNA protein complex (RNP) composed of lncRNAs or mRNAs and the Tho complex, the RNA helicase UAP56 (or its paralog URH49), and Aly is formed in the nucleus. Subsequently, RNPs are transported through the nuclear pore thanks to its interaction with the nuclear export receptor composed of Nuclear RNA Export Factor 1 (Nxf1) and Nuclear Transport Factor 2 Like Export Factor 1 (Nxt1) (also known as TAP and p15). In the cytoplasm, RNPs undergo remodeling to recycle nuclear-associated exported factors and ultimately result in the new stable localization of the mRNA or lncRNA [26].

Cytoplasm contains processed RNAs, leading to the predominant involvement of cytoplasmic lncRNAs in post-transcriptional regulation. These lncRNAs can serve as alternative targets of miRNAs, thereby mitigating the severity of RNA interference. Alternatively, they can enhance the translation of target mRNAs without impacting mRNA transcription (SINEUPs) [27].

2.2.3 LncRNAs located within mitochondria and chloroplast

Mitochondria and chloroplasts are specialized organelles of eukaryotic cells for energy production or photosynthesis (plant cells). An endosymbiotic origin for both organelles is now generally accepted with the maintenance of their own genomes. During evolution, these genomes have undergone significant reduction, with most genes transferred to the nuclear genome. Mitochondrial DNA retains the capacity to synthesize 13 proteins (essential for certain parts of OxPhos complexes), 22 tRNAs, and two ribosomal RNAs (12S and 16S).

Mitochondrial DNA is a double-stranded circular molecule characterized by a guanine-rich strand (heavy strand) and a guanine-poor strand (light strand). The first strand is rich in genes while the latter codes only for one subunit of nicotinamide adenine dinucleotide dehydrogenase (ND), 8 tRNAs, and 3 lncRNAs. Notably, these lncRNAs regulate two genes in the heavy DNA strand (ND5; NADH dehydrogenase subunit 5 and CYTB; cytochrome b) and one in the light strand (ND6; NADH dehydrogenase subunit 6) throughout RNA complementary binding [28].

Nuclear genes coding for mitochondrial proteins regulates their expression in concert with mitochondrial necessity confirming the intercommunication among nucleus and mitochondria. In addition to lncRNAs synthesized from the light strand of mitochondrial DNA, other lncRNAs are synthesized and transported in the cell nucleus. For instance, a RNA composed of 16S mitochondrial RNA and an inverted repeat of 120 bp was located in nuclei of mouse and human sperm [29], further confirming the importance of intercommunication among these "organelles". Mitochondrial DNA also enables the synthesis of long intergenic noncoding RNA predicting cardiac remodeling (LIPCAR), which is associated with the risk of heart failure [30].

The genome of chloroplasts, such as mitochondrial one, also synthesizes for lncRNAs. Given the focus of this article on the function of lncRNA in skeletal muscle, a review on the function of long non-coding RNAs in plant organelle is provided in [31].

2.3 Classification of long non-coding RNAs based on their functional characteristics

Historically, rRNAs have been the earliest long non-coding transcripts described, primarily due to their abundance in cells. They are essential components of ribosomes and allow the interaction between ribosomes and mRNAs. Below is a classification of newer long non-coding RNAs based on their functions.

2.3.1 RNAs interacting with the chromatin

Chromosomal RNAs (cRNAs) were introduced in the late 1960s by James Bonner as a group of RNAs capable of binding chromatin [32]. LncRNAs interact with chromatin through various mechanisms, with the recruitment of the polycomb repressive complex (PRC) being the most prevalent. The PRC is an evolutionary conserved system of epigenetic gene silencer, and consequently, lncRNAs, that interact with them, are gene silencers that use an epigenetic mechanism. PRC1 and PRC2 are the two major PRCs. PRC1 components were initially characterized in Drosophila [33], and later homologs genes were identified in human: CBXs (polycomb homolog), PHC1, 2, and 3 (polyhomeotic homologs), Ring1a and Ring1b (dRING homologs), BMI1 (Polycomb Ring Finger Proto-Oncogene), and six minor others (posterior sex combs homologs) [34]. Functionally, DNA CpG density and methylation status regulate PRC2 binding to chromatin. In contrast, PRC1 may facilitate the positioning of PRC2 in unmethylated CXXC DNA domains, guiding H3K27 lysine 3 methylation-mediated chromatin silencing [35]. PRC2-associated proteins with DNA binding capabilities enable PRC2 binding to unmethylated DNA independently of PRC1 both in Drosophila and mammalian systems [36–38]. Notably, it was published a protocol to visualize the tissue-specific PcG distribution in the aorta, dorsal skin, and hindlimb muscles [39].

2.3.2 MicroRNA sponges

LncRNAs and miRNAs can interact to regulate protein expression. miRNAs are short non-coding RNAs (18–28 nt) [40] interacting with mRNAs to regulate protein expression by degrading mRNA or repressing translation [41]. LncRNAs that interact with miRNAs compete for miRNA regulatory function and for this reason are named competing endogenous RNAs (ceRNAs) or miRNA sponges [42]. One notable example of a ceRNA is the long intergenic non-coding RNA-Muscle Differentiation 1 (linc-MD1), one of the first discovered lincRNA in skeletal muscle. It interacts with miR-133 and -135 which modulate the expression of two important transcription factors that activate muscle-specific gene expression: Mastermind Like Transcriptional Coactivator 1 (MAML1) and Myocyte Enhancer Factor 2C (MEF2C) [43]. Pseudogenes are another significant class of lncRNAs that function as miRNA sponges. These are particularly significant in skeletal muscle. For example, EEF1A1P24 and TMSB4XP8 are two hypertrophy-related pseudogenes [44]. Another example of miRNA sponging RNAs are circular RNAs (circRNAs), which will be following discussed.

2.3.3 Enhancer RNAs

Enhancers can be transcribed to produce lncRNAs known as enhancer RNAs (eRNAs). Enhancers are short DNA regions (50–1500 bp), typically isolated from coding genes, that possess the capacity to modulate gene transcription through the binding of specific protein activators [45]. eRNAs were identified in 2010 using RNA-seq and Chip-seq techniques and can augment the expression of target genes in cis [46]. Enhancers were initially discovered in 1981 by Banerji and colleagues, who observed the enhancement of the expression of a beta-globin gene by distant SV40 DNA sequences [47]. eRNAs hold significant importance in skeletal muscle development and pathology. For a comprehensive review, refer to [48]. Recently, it was evidenced that seRNA PAM-1 is involved in the regulation of skeletal muscle satellite cell and aging through trans regulation of Timp2 expression synergistically with Ddx5 [49]. Timp2 encodes for an extracellular matrix, and increased extracellular matrix proteins expressions are essential to satellite cells activation [50]. The mechanism of transcriptional enhancement exhibited by eRNAs may vary. Some eRNAs are involved in the strength of the enhancer-promoter looping interaction, while others hinder the binding of negative elongation factors (NELFs) to the promoter, thereby reducing transcriptional repression [51]. eRNAs can be categorized into two distinct types based on their transcriptional direction: 1D (unidirectional) or 2D (bidirectional). Unidirectional eRNAs are longer (>4 kb) and polyadenylated, whereas bidirectional are shorter (0.5–2 kb) and non-polyadenylated [52, 53].

2.3.4 SINEUPs

SINEUPs, already mentioned in the paragraph entitled "Genic lncRNAs (exonic or intronic)", are antisense lncRNAs with two important sequence specificities: the presence of an inverted SINEB2 sequence and a small complementarity sequence with the targeted mRNA. Due to these characteristics, SINEUPs can contact specific mRNAs by sense-antisense binding to up-regulate the translation without affecting gene expression. SINEUPs are coded by genes that overlap with the gene encoding the target protein to allow a 5′ head-to-head binding. SINEUPs can also be artificially designed and can be used to achieve the opposite effect of miRNA/siRNA translational silencing [16].

2.3.5 Target position

Certain chromatin-interacting lncRNAs, known as cis-acting lncRNAs, function by targeting the chromatin regions from which they are transcribed. The chromatin targeted by these lncRNAs is not necessarily limited to the immediate vicinity of the lncRNA. A prominent example is the well-studied lncRNA XIST, which targets the whole X chromosome through a cis approach.

Conversely, if the lncRNA target is not located in close proximity to its gene, it is classified as a trans-acting lncRNA. An example of trans-acting lncRNA is seRNA PAM-1 previously described for its involvement in the regulation of skeletal muscle satellite cell activation and aging [49].

3. Circular RNAs

The first circular RNA (circRNA) was identified in pathogens in the late 1970s [54, 55]. Until approximately 2010, their potential impact was largely underestimated. In the late 2010s, new sequencing technologies and bioinformatic pipelines were developed leading to an explosion in circRNA research. Metazoan express thousands of distinct types of circRNAs [56–58].

CircRNAs can originate from five different RNA regions: (1) exon, (2) intron, (3) exon-intron, (4) fusion gene, and (5) read-through circular RNA formed by polymerase II. Currently, three mechanisms are recognized for the formation of circRNAs. One mechanism is based on complementary pairing of intron elements, such as reverse complementary matches, intronic complementary sequences, and Alu sequences favoring the vicinity of the 5′ splice donor site of the mRNA precursor to the 3′ splice acceptor site. A second mechanism relies on RNA binding proteins that promote cyclization by binding to flanking introns of circularized exons. Last method occurs during the splicing of mRNA precursor and exon skipping, when the formation of lasso intermediate containing an intron and an exon. The exon and intron form a circular RNA. Most circRNAs are formed from exons, while a few are formed solely from introns. The majority of circRNAs are non-coding, highly conserved, and located in the cytoplasm. However, a few circRNAs are located in the nucleus [59]. CircRNAs can play regulatory roles in transcriptional [60], post transcriptional [61] regulation, and can also be translated [62].

Translated circRNAs play crucial role in maintaining skeletal muscle mass. For example, a conserved circular RNA damage-specific DNA binding protein 1 (circDdb1) exhibits significantly elevated expression in various muscle atrophy types both *in vivo* and *in vitro*, as well as in human aging muscle. CircDdb1 encodes for the protein circDdb1-867aa which interacts with and augments the phosphorylation level of eukaryotic elongation factor 2 (eEF2) at Thr56. This interaction reduces protein translation and promotes muscle atrophy [63]. Another protein encoded by circRNA (circTmeff1) that mediates pro-atrophic effects is circTMEFF1-339aa. Overexpression of TMEFF1-339aa leads to increased expression of Atrogin-1 and MuRF-1, activation of the ubiquitin proteasome system, alkaline phosphatase, and inhibition of the AKT/FOXO3A/mTOR signaling pathway [64]. All these mechanisms induced by circTMEFF1-339aa contribute to muscle atrophy.

Most studied activities of circRNAs at post-transcriptional level are their ability to sponge miRNAs. For example, by using RNA binding protein immunoprecipitation and RNA pull-down assays it was demonstrated that circMYBPC1 binds miR-23a inducing myoblast differentiation by relieving miR-23a inhibition on myosin heavy chain (MyHC) [65]. Another interesting circRNA involved in the maintenance of satellite cell population (MuSC) is circFUT10. It suppresses MuSC proliferation and differentiation by sponging miR-365a-3p and the 3′UTR of HOXA9 mRNA [66]. Several studies have indicated that HOXA9 is implicated in cell proliferation [67, 68]. Notably, one study demonstrated that HOXA9 may promote denervated muscle atrophy [69]. CircRNAs exhibit remarkable stability compared to linear RNAs [70], making them crucial actors in prolonging RNA transfection effects in therapeutic interventions. In fact, when it is necessary to modulate the expression of a specific gene, it is possible to either act or regulate mRNA degradation by sponging miRNAs that are involved in this process, or in the case that the protein is lacking, favor protein synthesis by the maintenance of mRNA within the cell. This last aspect can also be achieved by limiting mRNA degradation by favoring its circularization. For a further description of circRNAs may be useful in skeletal muscle therapy, refer to the paragraph entitled "Circular RNAs therapeutic potential in muscle atrophy".

In addition to miRNAs, circRNAs can also interact with proteins to influence muscle development. For example, over the past 5 years, two circRNAs, circSmad4 and circNFIX, have been associated with this phenomenon. Affinity pulldown followed by mass spectrometry have revealed the association of circSamd4 with

PURA and PURB, two proteins that repress myogenesis by inhibiting the transcription of the myosin heavy chain (MHC) protein family [71]. Therefore, the interaction of circRNA with PURA and PURB impacts the ability of two proteins to interact with DNA. Another circRNA that interacts with proteins is circNfix. circNfix promotes cardiac regenerative repair and functional recovery following myocardial infarction by reinforcing the interaction between Ybx1 (Y-box binding protein 1) and Nedd4l (an E3 ubiquitin ligase). This stimulates Ybx1 degradation through its ubiquitination. Indeed, Ybx1 depletion prevents pathological cardiomyocyte growth *in vitro* and promotes cardiac function *in vivo* by regulating protein synthesis [72].

4. Micro-peptides derived from linear lncRNAs

As previously described, lncRNAs are RNAs longer than 200 nucleotides that lack the ability to code for proteins. The advent of novel sequencing techniques, such as ribosomes RNA sequencing, has demonstrated the ability of ribosomes to associate with lncRNA molecules [73]. This finding supports the possibility that lncRNAs may be translated. Although ribosome profiling provides evidence of lncRNA-ribosomes interactions, it is crucial to confirm the presence of translated peptides through mass spectrometry analysis (MS). For example, mass spectrometry experiments confirmed that only 18 out of 233 lncRNAs capable to interact with ribosomes lead to the synthesis of small peptides in human cell lines [74]. Notably, many micropeptides have been functionally characterized, highlighting their significance in muscle physiology. They regulate muscle regeneration via mTOR [75], development/myogenesis [76, 77], and calcium handling [78, 79]. Furthermore, mitochondria, which are crucial organelle for the energetic and metabolic functions of skeletal muscles, are regulated by micropeptides synthesized from lncRNA translation. Human LINC00116 encodes a highly conserved muscle- and heart-enriched micropeptide that localizes to inner mitochondrial membrane. This micropeptide impacts on mitochondrial respiration and Ca2+ buffering capacities [80] leading to its designation as Mitoregulin. Another important micropeptide derived from lncRNA translation and involved in the regulation of mitochondrial function is MOXI. It enhances fatty acid β-oxidation [81]. Skeletal muscle studies regarding the synthesis of micropeptides from lncRNAs were published prior to the time period considered in this work, but we have included them to demonstrate that lncRNAs expressed from skeletal muscle also code for micropeptides. Most recent study on micropeptide and skeletal muscle was published in 2020 from Wang and colleagues [77]. They showed that lncRNA-MyolncR4 encodes a 56-aa micro-peptide, which fosters muscle formation and regeneration. Recently, the significance of micropeptides derived from lncRNA in cardiac function has been elucidated. In 2021, two distinct research publications described the topic. One study elucidated the importance of the lncRNA LINC00961-encoded micropeptide SPAAR for cardiovascular function in adulthood [82]. The other study described a way to regulate cardiomyocyte hypertrophy by using three micropeptides that modulate the activities of oxidative phosphorylation, the calcium signaling pathway, and the mitogen-activated protein kinase (MAPK) pathway [83]. For a comprehensive review on micropeptides from lncRNAs, including bioinformatic and experimental approaches for their identification see [84].

5. Extracellular lncRNAs

Skeletal muscle is a secretory organ that communicates with other tissues by secreting myokines [1]. Exosome formation is an active way of cell-cell or tissue communication. The presence of miRNAs, lipid, or proteins in exosomes derived from skeletal muscle has been extensively studied [85–87]. In contrast, there is only one study that characterized the secretion of lncRNAs in a skeletal muscle pathological condition. In 2022, Li and colleagues demonstrated that plasma exosomes of patients with dermatomyositis (DM) in addition to miRNAs contain lncRNAs [88]. DM is an immune-mediated muscle disease characterized by skeletal muscle weakness [89]. Li and colleagues identified 452 differentially expressed lncRNAs in plasma of DM patients compared with NCs. These lncRNAs were likely involved in autophagy, IFN-β production, IL-12 production, and fatty acid transmembrane transport [88]. This result suggests that lncRNAs can also be utilized from skeletal muscle for communication and that lncRNAs can also serve as pathological markers.

6. Long non-coding RNAs in muscle physiology

6.1 Myogenesis

Long non-coding RNAs (lncRNAs) have been identified in different cells playing pivotal roles in both normal muscle physiology and diseases [90]. As previously discussed, lncRNAs can either stimulate or suppress gene expression by controlling mRNA translation and stability, pre-mRNA splicing, miRNA sponging, and gene transcription [91]. Advancements in high-throughput sequencing have facilitated the identification of tens of thousands of lncRNAs in skeletal muscle tissue. However, only a selected few of these lncRNAs have been demonstrated to contribute to the development and function of skeletal muscles [92]. During embryonic development, the proliferation of mononuclear myoblasts is critical in determining the size of future skeletal muscle. As development advances, myoblasts stop dividing and differentiate into multinuclear myotubes, a phase known as myogenesis [93].

A recent study identified 385 differentially expressed lncRNAs during C2C12 differentiation, elucidating transcriptional regulation and regulatory roles of lncRNAs in skeletal muscle cell differentiation [92, 94]. Gm10561 is one of these differentially expressed lncRNAs. However, its specific role in myogenesis remains unclear. In a study by Wang et al., it was revealed that Gm10561 exhibits higher expression levels in skeletal muscle compared to other tissues and undergoes an increase during myogenic differentiation. Functional study further demonstrated that Gm10561 positively regulates the proliferation and differentiation of C2C12 myoblasts. Mechanistically, Gm10561 functions as a molecular sponge of miR-432, thereby enhancing the expression of MEF2C and E2F3. This action accelerates myoblast proliferation and differentiation. Furthermore, Gm10561 was identified as a novel positive regulator of myogenesis, potentially contributing to the elucidation of the lncRNA–miRNA–target-gene regulatory network associated with myogenesis [92]. Mechanism of Gm10561 is shown in **Figure 1**.

LncRNA Irm is another lncRNA differentially expressed during skeletal muscle morphogenesis in embryonic development, suggesting its potential involvement in myogenesis. It was reported that lrm expression is higher in the tongue and skeletal muscles during embryonic development [96]. Notably, the expression of lrm

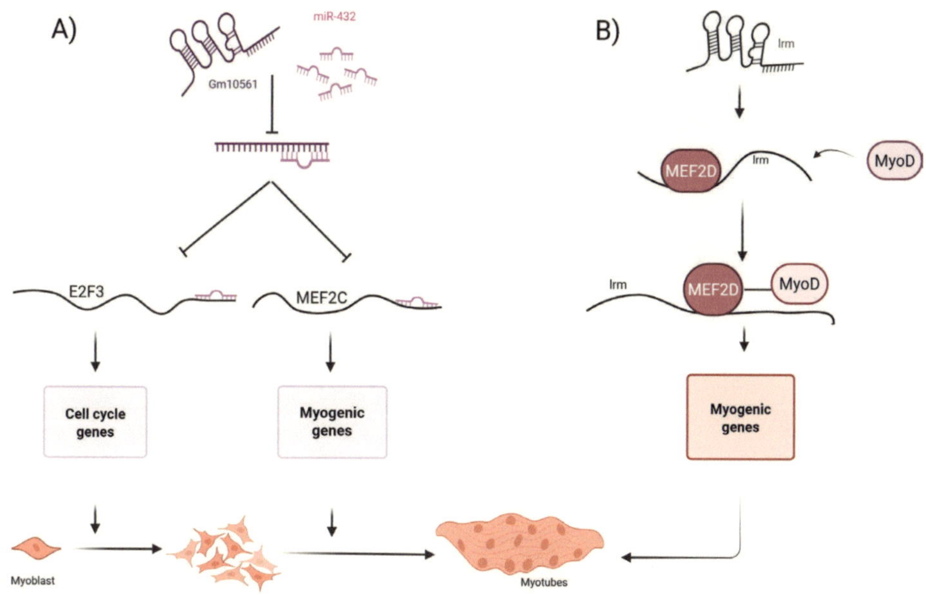

Figure 1.
(A) Gm10561 acts as a molecular sponge of miR-432 to promote myogenesis. In fact, miR-432 directly binds to the 3′UTR of E2F3 and MEF2C, which inhibits their expression. Gm10561 sponges mir-432 can increase E2F3 expression through decreasing the inhibition of miR-432 on E2F3, which leads to an increase in myoblast proliferation. Moreover, Gm10561 sponges miR-432 to enhance MEF2C expression, reduce the suppression of miR-432 on MEF2C, and thereby encourage myoblast differentiation upon differentiation [92]. (B) The lncRNA lrm plays a crucial role in regulating the expression of myogenic genes by directly binding to MyoD. This interaction enhances the assembly of MyoD/MEF2D complexes on the regulatory elements of target genes [95].

significantly increases in differentiated muscle cells, indicating its connection to active myogenesis [95]. Furthermore, using gain and loss of function experiments, Sui and colleagues demonstrated that Irm significantly promotes C2C12 differentiation. Additionally, it was shown that the absence of lrm resulted in a significant delay in the *in vivo* regeneration of injured-induced muscle. To activate its known target genes, lrm recruits MyoD/MEF2D by directly binding to MEF2D and the DNA of the promoter region. These findings revealed that lrm is a novel member of the MEF2D-regulating network that promotes myogenesis and differentiation in C2C12 cells. Furthermore, it functions as a lncRNA enhancer of MyoD/MEF2D and cooperatively controls the transcription of myogenic differentiation genes [95]. The mechanism of action of Irm is shown in **Figure 1**.

The lncRNA Has2os, situated within the coding Hyaluronan synthase 2 (Has2) gene, is another important lncRNA associated with myogenesis. Interestingly, Has2 is a gene exclusively expressed in skeletal muscles and is believed to be essential for skeletal muscle development, patterning, chondrocyte maturation, and the formation of synovial joints in growing limbs. Furthermore, myogenic differentiation requires endogenous hyaluronan synthesis, and limbal stem cell differentiation is regulated by a hyaluronan-rich microenvironment within the limbal stem cell niche. The function of lncRNAs appears to be closely linked to that of their neighboring genes through cis-acting processes [90, 97–99]. As elucidated by Cheng and colleagues, Has2os undergoes continuous accumulation during skeletal muscle differentiation. Conversely, its expression inhibition leads to the inhibition of skeletal muscle differentiation, suggesting that Has2os acts as a differentiating promoter. Additionally, Has2os plays a role in the early stages of skeletal muscle regeneration. It has been demonstrated that Has2os contributes to these processes by controlling the JNK/MAPK signaling pathway [90]. The mechanism of action of Has2os is shown in **Figure 2**.

The lncRNA 4930581F22Rik exhibits an expression pattern comparable to lrm and Has2os, suggesting a potential role in skeletal muscle development and regeneration. During myogenesis, the expression of lncRNA 4930581F22Rik progressively increases, and its knockdown significantly impedes differentiation and myogenic

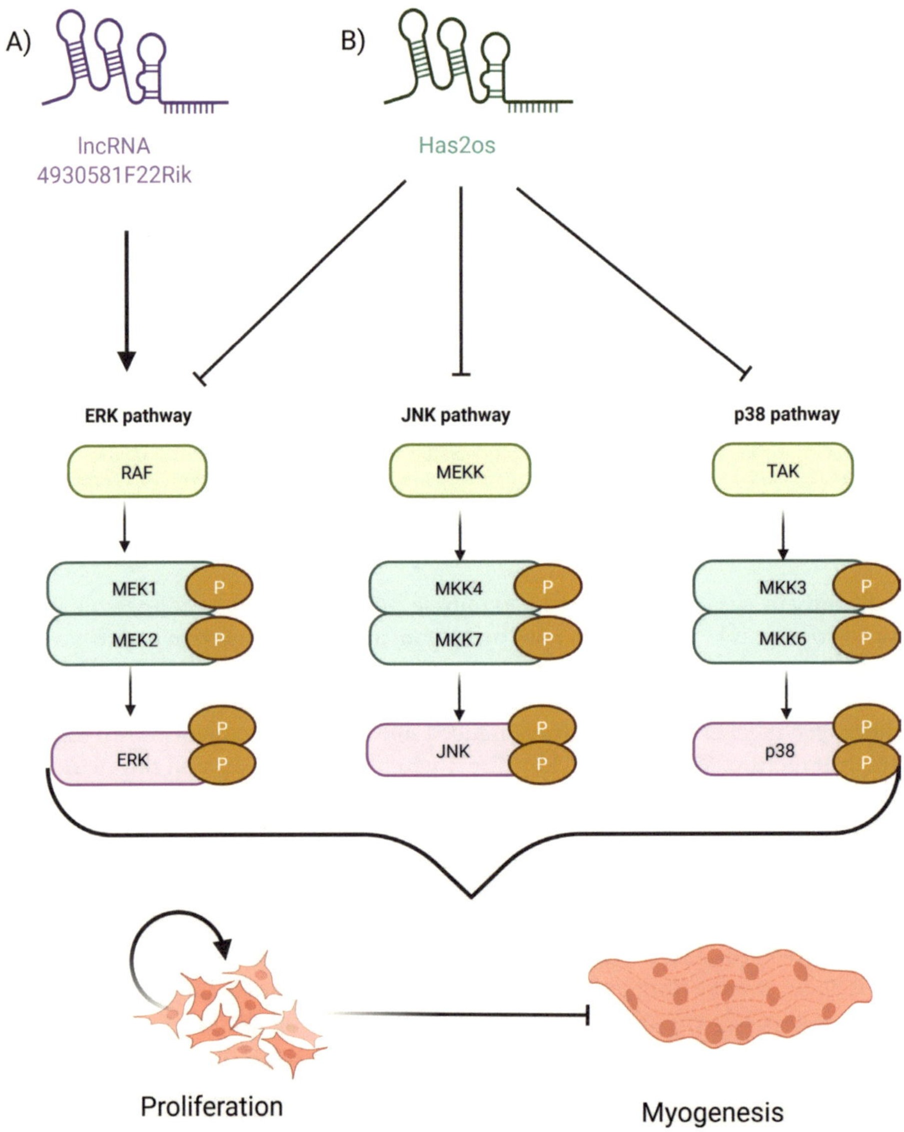

Figure 2.
(A) Expression level of lncRNA 4930581F22Rik gradually increases during skeletal muscle differentiation. LncRNA 4930581F22Rik regulates the differentiation program and the expression of myogenic markers through modulating ERK/MAPK signaling pathway [100]. (B) LncRNA Has2os is highly expressed in skeletal muscle tissue and upregulated during myogenesis. Has2os modulates myogenesis and regeneration through inhibiting the JNK/MAPK signaling pathway with three different MAPK branching signaling pathway, including JNK/MAPK, p38/MAPK, and ERK1/2/MAPK [90].

marker expression [101]. According to Chen's study, lncRNA 4930581F22Rik modulates this process by regulating the ERK/MAPK signaling pathway, which can be effectively neutralized by the ERK-specific inhibitor PD0325901. Furthermore, it has demonstrated that lncRNA 4930581F22Rik is emerging as a viable participant in the initial stages of muscle regeneration [100]. The mechanism of action of lncRNA 4930581F22Rik is shown in **Figure 2**.

To date, the physiological function of lncRNAs in muscles has been the primary focus of numerous studies [91, 102, 103]. However, there is a limited understanding of the precise mechanisms they modulate during muscle atrophy [91]. Several other lncRNAs were associated with skeletal muscle differentiation, but a limited mechanistic description was furnished. These lncRNAs are mentioned in **Table 1**.

LncRNAs	Description	Expression in myogenesis	Mechanism	Ref
Mir22hg	Mir22host gene	Increase	*Mir22hg* matures into miR-22-3p to inhibit *HDAC4* expression. This activates the downstream *MEF2C* pathway, thereby promoting skeletal muscle differentiation and regeneration.	[104]
Lnc-GD2H	*Gm13398* was named the lncRNA of growth and differentiation to histogenesis (*lnc-GD2H*).	Increase	It facilitates the expression of proliferating marker genes, including c-Myc, CDK4, CDK2, and CDK6. It establishes a feedback loop with c-Myc during myoblast proliferation. Additionally, *lnc-GD2* interacts with NACA to alleviate the inhibitory effect of NACA on Myog, thereby facilitating Myog expression and promoting differentiation.	[105]
LncMGPF	lncRNA muscle growth promoting factor (*lncMGPF*).	Increase	*lncMGPF* sponge miR-135a-5p promoting *MEF2C* gene expression. Moreover, *lncMGPF* regulates human antigen R (HuR; also known as ELAV-like RNA binding protein 1)-mediated mRNA stabilization of MyoD and MyoG genes.	[106]
Myoparr	Promoter-associated lncRNA, *Myoparr*, derived from the upstream region of the *myogenin* gene.	Increase	*Myoparr* binds Ddx17, a transcriptional coactivator of MyoD, and promotes the interaction between Ddx17 and histone acetyltransferase PCAF. *Myoparr* exerts its functions by interacting with transcriptional activator during myogenesis.	[107]
PAM	*Pax7* associated muscle lncRNA (*PAM*)	Increase	*PAM* regulates satellite cells proliferation through the interaction with Ddx5, thereby facilitating the chromatin contact between PAM SE (super enhancer) and target loci (Timp2 and Vim) during SC proliferation.	[49]

LncRNAs	Description	Expression in myogenesis	Mechanism	Ref
LncRNA-MEG3	LncRNA maternally expressed gene 3 (*lncRNA-MEG3*)	Increase	*LncRNA-MEG3* regulates muscle regeneration by sponging miR-133a-3p to regulate the expression level of proline-rich transmembrane protein 2 (PRRT2).	[108]
1700113A16RIK	A novel lncRNA, *1700113A16Rik*, that regulates muscle regeneration.	Increase	*1700113A16RIK* functions by pairing to the 3′UTR of *Mef2d* to facilitate Mef2d translation. It may help recruit ribosomes to the 3′UTR of *Mef2d* to promote translation.	[109]
Linc-MYH	*Linc-MYH* is the evolutionary conserved lncRNA, which is co-expressed with the largest myosin cluster in mammals.	Increase	*Linc-MYH* regulates the composition of the INO80 chromatin remodeler complex by directly interacting with its functional domains. Linc-MYH inhibits the INO80 complex's ability to interact with the scaffolding protein WDR5 and the transcription factor YY1. In fact, link-MYH acts as a selective molecular switch for INO80, enabling the regulation of distinct INO80 activities in MuSCs at specific stage.	[21]
2310043L19Rik	Named lnc-231	Increase	*Lnc-231* inhibits myoblasts differentiation interacting with miR-125a-5p. Inhibition activity of miR-125a-5p allows the translation of the target gene E2F3. Resulting protein promotes the cell cycle progression.	[110]
Linc-RAM	Activator of myogenesis (*Linc-RAM)*	Increase	Cytoplasmic link-RNA directly binds to glycogen phosphorylase (PYGM) and regulates its enzymatic activity, which is vital for muscle cell to differentiate.	[111]
Lnc-SMaRT	Skeletal muscle regulator of translation (*SMaRT*)	Increase	*Lnc-SmaRT* can inhibit Spire1 translation by base-pairing with its G-quadruples-forming sequence. The modulation of Spire1 expression regulates proper skeletal muscle differentiation. Furthermore, it is demonstrated that the mRNAs in the complex facilitate the indirect interaction between DHX36 and *lnc-SMaRT*.	[112]
LncMREF	LncRNA muscle regeneration enhancement factor (*lncMREF*)	Increase	*LncMREF* interacts with Smarcas5 and p300/CBP to enhance chromatin accessibility and the expression of myogenic genes. This promotes the differentiation of muscle satellite cells and muscle regeneration.	[113]

LncRNAs	Description	Expression in myogenesis	Mechanism	Ref
TUG1	Taurine-upregulated gene 1 *(TUG1)*	Increase	*TUG1* is upregulated in human skeletal muscle after exercise. Mechanistically, *TUG1* regulates transcriptional networks linked to mitochondrial calcium handling, muscle differentiation, and myogenesis.	[114]
Lnc-Malat1	Metastasis-associated lung adenocarcinoma transcript1 (*Malat1*)	Increase	Overexpression of lnc-*Malat1* competes for the binding site of miR-129-5p in the Mef2a transcription factor, thereby promoting myoblast differentiation and alleviating inhibitory effect of miR-129-5p on the same process.	[115]

Table 1.
Summary of other lncRNAs that are associated with skeletal muscle differentiation.

7. Long non-coding RNAs in muscle pathology

Dysregulated expression of lncRNAs has been implicated in various pathologies, including cancer, metabolic, and cardiovascular disorders [116, 117]. Previous studies have identified several lncRNAs, such as Linc-YY1 [118], Linc-MD1 [43, 119], Linc-RAM [120], lncRNA-Six1 [121], H19 [122], SYISL [123], lnc-mg [102], and lnc-MyoD [121], as potential regulators of myogenesis. However, with the exception of lncRNAs associated with Duchenne muscular dystrophy [DMD] [124, 125] and facioscapulohumeral muscular dystrophy [FSHD] [125, 126], the mechanistic role of lncRNAs in skeletal muscle diseases remains largely unexplored. It has been determined that a limited number of lncRNAs, including lnclRS1, lncRNA Atrolnc-1, and lncMUMA, contribute to particular forms of muscular atrophy [127]. In agreement with the expression of skeletal muscle differentiation-related lncRNAs, such as DRR [128], DUM1 [129], linc-MD1 [43], linc-YY1 [118], LncMyod [130], Neat1 [131], Myoparr [107], Malat1 [132], and SRA [133], muscle atrophy can be categorized into two subgroups: disuse-mediated atrophy, such as denervation, casting, and tail suspension, and systemic wasting atrophy, including glucocorticoid administration, cancer cachexia, and fasting. The latter subgroup is associated with genomic imprinting-related lncRNAs, such as Gtl2, H19, and IG-DMR [134]. However, it is yet unknown whether muscle atrophy in many disease models may be mediated by a common regulatory lncRNA [127].

7.1 Metabolic disease

Skeletal muscle exhibits plasticity or metabolic flexibility, enabling it to regulate fuel oxidation in response to fuel availability. Skeletal muscle metabolism has been recognized for its significant role in both health and development as well as metabolic disorders such as obesity and muscular atrophy, which are closely associated with reduced flexibility in skeletal muscle [135]. Increased protein degradation is a defining characteristic of muscular atrophy, characterized by a reduction in muscle mass and fiber size [136]. It is more probable that lncRNAs that have typically disrupted expression patterns in a variety of muscle atrophy disorders contribute to the role in onset or progression of muscle atrophy. However, there is limited research investigating lncRNAs that are commonly altered in various muscular atrophy diseases [134] and may represent a switcher important for therapeutic purposes.

7.1.1 LncRNAs and muscle insulin resistance

A pivotal indicator of type 2 diabetes (T2D) is insulin resistance, characterized by the pancreatic β-cells' inability to release insulin effectively and the diminished responsiveness of target tissues to normal insulin levels circulating in the bloodstream. LncRNAs may have a role in the development of insulin resistance in a number of organs [137]. Notably, lncRNA H19, an important imprinted gene expressed during embryogenesis, exhibits a significant decrease in its expression after birth in most tissues except skeletal muscle. It is worth to note that H19 expression is decreased in the skeletal muscle of patients with type 2 diabetes mellitus (T2DM) and in obese mice induced by high fat diet (HFD). Subsequently, downregulation of H19 disrupts glucose metabolism [138].

A study conducted by Alipoor and colleagues on insulin-resistant humans and rodent models revealed that H19 down-regulation increased let-7 expression (because H19 functions as sponge against let-7). This, in turn, led to the downregulation of let-7 target DUSP27, a cofactor of AMPK. Consequently, insulin signaling resistance in the muscle was observed. Furthermore, H19 was observed to block several signaling pathway components, including INSR, IRS-2, and IDE [137]. HnRNPA1 was identified as an interacting partner of H19, highlighting its involvement in various biological processes, such as glucose and lipid metabolism, mitochondrial function, and tumorigenesis. Indeed, HnRNPA1 serves as a positive regulator of lipid metabolism in muscle cells, controlling numerous genes involved in fatty acid oxidation, including CPT1b and PGC1a [137, 138]. Research by Gui and colleagues corroborated these findings that H19 recruits HnRNPA1 to CPT1P and PGC1a transcripts, thereby regulating their translation during lipid metabolism [138]. Therefore, insulin resistance may be associated with H19 downregulation and subsequent suppression of the H19/HnRNPA1 axis [137, 138]. **Table 2** provides an overview of lncRNAs and related metabolic disorders.

7.1.2 lncRNAs and cancer cachexia

Cancer Cachexia, a multifaceted metabolic syndrome, is characterized by the loss of skeletal muscle and adipose tissue mass. Its primary clinical manifestation is weight loss accompanied by anorexia, inflammation, cancer, and insulin resistance. Furthermore, cachexia has been associated with increased muscle protein degradation [141, 149, 150]. LncRNAs exert diverse effects on mRNA turnover, including acting as miRNA sponges and subsequently diminishing their regulatory functions. Notably, lncRNAs have been implicated in various diseases, including cancer and cachexia. However, there are a limited number of studies that explore the involvement of lncRNAs in cancer cachexia and elucidate their underlying mechanisms of action [141, 142].

Recent studies have elucidated the role of the lncRNA muscle anabolic regulator 1 (MAR1) in muscle differentiation and regeneration. Notably, MAR1exhibits high expression in skeletal muscle of mouse and has been positively associated with muscular growth and differentiation both *in vitro* and *in vivo*. This lncRNA impedes

Disease/ biological process	lncRNA	Description	Affected pathways	Regulatory mechanism	Ref
Diabetes	*H19*	LncRNA *H19*	PI3K/AKT pathway	Obesity and type 2 diabetes patients have less muscle mass, which impairs insulin signaling and reduces glucose absorption via the PI3K/AKT pathway because of *Let-7* miRNA.	[137, 138]
	LncRNA UCA1	Urothelial carcinoma associated 1 *(lncRNA UCA1)*	*UCA1* alleviates mitochondrial dysfunction by the miR-143-3p/ FGF21 axis.	*UCA1* upregulates the expression of FGF21 through competitively binding to miR-143-3p, thereby preventing the mitochondrial dysfunction of skeletal muscle in T2DM.	[139]
	LncASIR	Adipose-specific insulin responsive lncRNA *(lncASIR)*	Several metabolic pathways downstream of insulin signaling.	Silencing of *lncASIR* inhibits metabolic pathways downstream of insulin signaling, including PPAR and adipocytokine signaling.	[140]
Cancer cachexia	*MAR1*	lncRNA muscle anabolic regulator 1 *(MAR1)*	Wnt5a (Wnt pathway)	Serves as a sponge for mir-487b in C2C12 cells, preventing a decrease in Wnt5a expression and encouraging myogenesis. In mice, expression falls with age and muscle unloading.	[141–143]
	lncIRS1	LncRNA insulin receptor substrate 1 (*lncIRS1*)	IGF1-PI3K/Akt signaling pathway	Skeletal muscle cell proliferation and differentiation are facilitated by *LncIRS1's* regulation of IRS1 expression, which activates the IGF1-PI3K/Akt signaling pathway and promotes IGF-1 production and phosphorylates AKT	[141, 142]
	HOTAIR	HOX transcript antisense intergenic RNA (*HOTAIR*)	NF-κB signaling pathway.	The fact that HOTAIR may be an important player in cancer cachexia is sustained by its involvement in cancer progression and NF-κB signaling. However, there have not investigated on the role of *HOTAIR* and the signaling pathway involve d in cancer-induced or chemotherapy-induced cachexia.	[144]
	Malat1	Metastasis associated lung adenocarcinomatranscript1 (*Malat1*)	Wnt5a (Wnt family member 5A)	*Malat1* is linked to adipose loss in cancer-associated cachexia through regulating adipogenesis by PPAR-γ.	[145]

Disease/ biological process	lncRNA	Description	Affected pathways	Regulatory mechanism	Ref
Obesity	*Blnc1*	Brown fat lncRNA 1 (*Blnc1*)	BAT thermogenesis and mitochondrial function	Transgenic expression increases lipid metabolism and inhibits proinflammatory cytokines, while suppressing increased tissue inflammation and BAT fibrosis. WAT overexpression increases insulin sensitivity and protects against obesity brought on by an HFD.	[146, 147]
	MIST	Macrophage inflammation-suppressing transcript (*Mist*)	Interaction with histone modifiers is reliable epigenetic mechanism, which PARP1 is identified as a lead Mist-interacting partner.	*Mist* inhibits PARP1 (poly ADP-ribose polymerase-1) recruitment to inflammatory gene promoters by interacting with PARP1 in the nucleus.	[148]

Table 2.
Overview of long noncoding RNAs and related disorders.

miR-487b function by sponging it and consequently regulating Wnt family member 5A (Wnt5a). This promotes muscle differentiation and regeneration [141–143]. Furthermore, overexpression of MAR1 has been demonstrated to reduce muscular atrophy while maintaining muscle mass and strength, indicating its potential as a therapeutic target for cancer cachexia-related muscle atrophy [141].

The lncRNA insulin receptor substrate 1 (lncIRS1) is another lncRNA that demonstrated its activity as a miRNA sponge in mouse skeletal muscles. It has been demonstrated that LncIRS1 controls muscle mass and fiber sectional area *in vivo*, as well as myoblast proliferation and differentiation *in vitro* and *in vivo*. LncIRS1 can modulate IRS1 expression by acting as a sponge of the miR-15 family. Overexpression of LncIRS1 leads to the activation of the IGF1-PI3K/Akt signaling pathway, important to counteract muscle atrophy. Subsequently, this lncRNA promotes skeletal muscle cell proliferation and differentiation, leading to an increase in muscle mass and countering muscle atrophy [141, 142].

7.1.3 LncRNAs and obesity

Obesity, which is a risk factor for metabolic diseases, develops during the adipogenesis process. Obesity is inversely associated with muscle strength, activity, and hypertrophy. Obesity process involves the differentiation of precursor mesenchymal stem cells into adipocytes, their growth in size and accumulation, and their eventually fusion to enhance the storage of fat mass in white adipose tissues. Transcriptomic analyses have identified several lncRNAs that exhibit differential expression in body fat or adipose tissue (AT) in obese people. These lncRNAs play a crucial role in controlling adipogenesis and differentiation [137, 146]. The human body comprises two main types of adipose tissues: insulating energy-storing white AT (WAT) and thermogenic brown AT (BAT). Energy-burning and heat-generating BAT is situated around the shoulders and ribs, while visceral WAT, which surrounds the intra-abdominal organs, and subcutaneous WAT, located under the skin, are essential for energy homeostasis. In obesity, BAT undergoes molecular and morphological changes that lead to whitening appearance, resembling WAT. This transformation subsequently impairs metabolic function and contributes to inflammation [146].

Recent studies have identified several lncRNAs associated with the regulation of BAT, including H19 and brown fat lncRNA1 (Blnc1) [151, 152]. Studies have revealed that Blnc1 contributes to the regulation of thermogenic genes, leading to the increased expression of mitochondrial and uncoupling protein 1 (UCP1) genes [153]. Another lncRNA positively associated with the browning of adipose tissue was H19. It was inversely correlated with body mass index (BMI) [154]. In BAT, H19 has a role in regulating mitochondrial respiration, oxidative metabolism, and adipogenesis (**Table 2**). Therefore, altering lncRNA expression appears to be a promising treatment strategy for metabolic disorders. This approach may be based on the induction of browning of WAT or the enhancing of BAT activity [147]. Although lncRNAs show promise as therapeutic targets for obesity treatment, several challenges must be addressed before clinical applications can be realized. A significant obstacle lies in the lack of knowledge of the precise molecular mechanisms underlying lncRNA expression control in such different but communicating tissues as adipose tissue and skeletal muscle. Additionally, due to their size and possible off-target effects, lncRNA-based therapies remain challenging to deliver to specific tissues [137, 147].

8. Long non-coding RNAs as potential therapeutic targets in muscle atrophy

8.1 Long non-coding RNAs as therapeutic targets

New drug discovery begins with the elucidation of mechanisms underpinning physiological and pathological processes. Several studies (see long non-coding RNAs in muscle physiology and long non-coding RNAs in muscle pathology) have elucidated the role of long non-coding RNAs (lncRNAs) in skeletal muscle. Targeting lncRNAs has emerged as a promising therapeutic strategy for the treatment of a diverse range of diseases [155–157], including muscle disorders [93, 95, 104, 105, 158, 159].

Therapeutic potential of lncRNAs relies on their ability to bind complementary nucleic acid sequences. Consequently, also lncRNAs can be targeted by small interfering RNAs (siRNAs) and antisense oligonucleotides (ASOs) [155]. SiRNAs are short double-stranded RNA (dsRNA) molecules, while ASOs are short single-stranded oligonucleotides (RNA or DNA), both of which range in length from 15 to 25 nucleotides. The binding of siRNAs and ASOs to lncRNAs triggers target (lncRNA) degradation, respectively, through the action of the RISC complex and RNase H cleavage [155, 160, 161]. Recently, lncRNAs have emerged as targets of small molecules (small lncRNA-targeting molecules). In fact, the structural properties of lncRNAs enable them to interact with proteins or small molecules, leading to direct alterations in their conformation and, consequently, effects on their functions by obscuring or disrupting lncRNA interaction sites [155, 156]. Currently, CRISPR/Cas technology is used in research to modulate lncRNAs expression, but off-target effects remain a significant challenge in the therapeutic application [155, 156].

Although lncRNAs present promising opportunities in medicine, there are concerns regarding the potential adverse effects of lncRNA modulation. In fact, off-targeting may lead to undesirable outcomes. Other than this, like in other RNA-based therapies, RNA delivery systems require further development to achieve tissue specificity and mitigate immune response and toxicity issues [155, 162].

Investigation into the therapeutic potential of lncRNAs has gained prominence only in the past decade. Consequently, no lncRNA-targeting therapeutics have yet progressed to clinical trials and entered clinical translation [155].

8.2 The potential of long non-coding RNAs to treat muscle atrophy

Long non-coding RNAs have been recently studied for their role in modulating muscle atrophy, dystrophy, and regeneration (see "long non-coding RNAs in muscle physiology" and "long non-coding RNAs in muscle pathology"). In the past 5 years, numerous studies have elucidated the positive regulatory effects of lncRNAs on muscle regeneration [95, 104, 105, 163], while others have demonstrated detrimental effects [93, 107, 144, 158, 164]. Certain lncRNAs, such as Irm, Mir22hg, and GD2H, are essential for the regeneration process, and their downregulation impairs *in vivo* regeneration [95, 104, 105]. In other articles, authors have demonstrated that the downregulation of lncRNAs like Pvt1, Hotair, Mayoparr, lnc-ORA, and lncRNA-3 attenuates or rescues muscle atrophy [93, 107, 144, 158, 159, 164]. Consequently, these molecules represent promising putative targets for skeletal muscle atrophy.

Different diseases or pathological conditions resulted in muscle wasting and atrophy. Notably, the expression of non-coding RNAs could be specific to different atrophic stimuli [103, 165, 166]. Consequently, the upregulation of lncRNAs could be beneficial or detrimental depending on the context. For instance, H19 is upregulated in amyotrophic lateral sclerosis and after denervation [158, 165], while it is down-regulated in hindlimb suspension-induced [167] and fasting-induced [165] muscle atrophy. Furthermore, H19 administration alleviates dystrophy in mice models for muscular dystrophy caused by disrupting mutations of the dystrophin gene [168]. These differences linked to genetic context should be considered when a putative target is selected to develop a new drug.

Despite the substantial potential of lncRNAs as therapeutic targets, there are currently no approved drugs or clinical trials specifically focused on treating muscle atrophy and dystrophies through lncRNA modulation [155–157, 169]. However, there are a few recent examples demonstrating the potential of *in vivo* lncRNA downregulation to attenuate or even rescue muscle atrophy. For example, *in vivo* knockdown of the lncRNA Myoparr in denervated tibialis anterior (TA) muscles effectively attenuates muscle atrophy [93, 107]. Similarly, the downregulation of Pvt1 in TA muscles protects against atrophy induced by denervation [158].

Other researchers have shown that *in vivo* lncRNA downregulation impairs muscle regeneration, suggesting that these ncRNAs may promote muscle regeneration. During cardiotoxin injury-induced regeneration, downregulation of the lncRNA Irm delays TA muscle regeneration [95]. Similarly, lnc-GD2H silencing significantly impairs muscle regeneration after cardiotoxin treatment [105]. Furthermore, knockdown of Mir22hg in TA muscles inhibits and delays the regeneration of BaCl2-injured skeletal muscle [104]. This year, Zhang and colleagues investigated the role of lncRNA-3 in muscle regeneration during aging. They demonstrated that lncRNA-3 overexpression delays the skeletal muscle regeneration process, while lncRNA-3 silencing by shRNA restores muscle regenerative capacity in the gastrocnemius muscle of aged mice injured by a cardiotoxin injection [159]. The effects of long non-coding RNAs on muscle regeneration and atrophy are summarized in **Table 3**.

8.3 Circular RNAs therapeutic potential in muscle atrophy

Circular RNAs (circRNAs) are covalently closed, non-coding RNAs devoid 5′ and 3′ ends [174]. Although discovered decades ago, circRNAs gained prominence as the widespread class of RNA with the advent of high-throughput RNA sequencing (RNA-seq) and novel bioinformatic tools (see “Circular RNAs”) [175, 176].

Recent studies have revealed the multifaceted biological functions of circRNAs, which can serve as miRNA sponges by binding to miRNAs, thereby preventing mRNA target downregulation. Alternatively, they can function as protein sponges and scaffolds by binding to proteins [174, 176]. The ability of circRNAs to interact with both miRNAs and proteins renders them potential therapeutic agents in a diverse range of diseases. CircRNAs play a crucial role in regulating muscle regeneration and atrophy *in vivo*. For example, the downregulation of circTmeff1 [64] and cicrDdb1 [63] prior to denervation or immobilization has been demonstrated to attenuate muscle atrophy. In addition, the overexpression of Circ2388 [170], CircARID1A [171], and CircFgfr2 [172] promotes muscle regeneration in cardiotoxin-injured muscle, while CircCPE [173] exhibits a detrimental effect on regeneration. These findings collectively demonstrate the role of circRNAs in skeletal muscle and highlight their potential as therapeutic targets for muscle atrophy. The effects of circular RNAs on muscle regeneration and atrophy are summarized in **Table 3**.

ncRNA	Model	Modulation	Effect	Ref
Long non-coding RNA				
Myoparr	Den-induced atrophy	shRNA knockdown	Attenuates atrophy	[93, 107]
Pvt1	Den-induced atrophy	Gampers knockdown	Attenuates atrophy	[158]
Mir22hg	$BaCl_2$ injury-induced regeneration	shRNA knockdown	Delay muscle regeneration	[104]
Lnc-GD2H	CTX injury-induced regeneration	siRNA silencing	Delay muscle regeneration	[105]
LncRNA-3	CTX injury-induced regeneration	Sk-lncRNA-3 knock-in	Delay muscle regeneration	[159]
		shRNA knockdown	Promotes muscle regeneration	
Circular RNA				
CircTmeff1	Den-induced atrophy	shRNA knockdown	Attenuates atrophy	[64]
	Imo-induced atrophy			
CircDdb1	Den-induced atrophy	shRNA knockdown	Attenuates atrophy	[63]
	Imo-induced atrophy			
	Angiotensin II induced atrophy			
Circ2388	CTX injury-induced regeneration	Overexpression	Promotes muscle regeneration	[170]
CircARID1A	CTX injury-induced regeneration	Overexpression	Promotes muscle regeneration	[171]
CircFgfr2	CTX injury-induced regeneration	Overexpression	Promotes muscle regeneration	[172]
CircCPE	CTX injury-induced regeneration	Overexpression	Delay muscle regeneration	[173]

Den, denervation; CTX, cardiotoxin; Imo, immobilization; shRNA, short hairpin RNA.

Table 3.
Long non-coding RNAs and circular RNA effects on muscle regeneration or atrophy in vivo.

9. Modulation of long non-coding RNAs expression in skeletal muscle cells

9.1 Skeletal muscle models

Skeletal muscle tissue is not only composed of myofibers but also satellite cells, smooth muscle cells, vascular cells, motor neurones, fibroblasts, and resident immune cells. All these cell types are organized in a highly specific architecture to form the skeletal muscle structure [177–179]. Common *in vitro* models to study skeletal muscle functions are mouse myoblasts (C2C12), rat myoblasts (L6), and primary myoblasts derived from rodent or human biopsies. C2C12 is an immortalized cell line derived from mouse skeletal muscle myoblasts. It is a commonly used model in medical research to study skeletal muscle atrophy and related diseases [180].

Primary cultures from rodents or human biopsies offer greater representation of their tissue of origin compared to immortalized cell lines such as C2C12 and L6. Primary cells retain the phenotypic and genotypic characteristics of their source tissue, making them widely used in metabolic studies [178, 180]. However, primary cultures have limited proliferative capacity, posing challenges in extensive cell usage. Consequently, efforts are being made to immortalize isolated primary myoblasts. Moreover, in the past decade, several methods have been established to differentiate myogenic cells from induced pluripotent stem cells (iPSCs) [179, 181, 182]. IPSCs derived from healthy donors or patients affected by muscular disorders overcome limitations associated with tissue availability and ethical considerations.

Bidimensional cell cultures lack a physiological extracellular environment and native skeletal muscle organization [179, 183]. Moreover, rodent myoblast cell lines (C2C12 and L6) and skeletal muscles exhibit transcriptomic and metabolic differences compared to human cells and tissue due to the species specificity [178]. To overcome limitations associated with cell lines and animal models, bioengineering human models that provide greater fidelity to native skeletal muscle tissues have been developed in recent years. Three-dimensional models are designed to replicate the complexity of the native muscle tissue and its niche in order to match the morphological and functional features of skeletal muscle [179–181, 183]. These innovative models for skeletal muscle have the potential to advance our understanding of myogenesis and muscle regeneration, but their widespread adoption in research remains limited.

9.2 Transfection of skeletal muscle cells

In the previous paragraphs (see long non-coding RNAs in muscle physiology, long non-coding RNAs in muscle pathology, and long non-coding RNAs as potential therapeutic targets in muscle atrophy), we highlighted the importance of cell and animal models for skeletal muscle to study the impact of long non-coding RNAs (lncRNAs) in this tissue. The majority of the studies presented here consider the modulation of lncRNAs expression by transfecting cells [92, 95, 115] or tissues [93, 104, 105, 107, 158, 159].

As previously mentioned, C2C12 and primary mouse myoblasts are commonly used in laboratories investigating muscle pathophysiology. However, transfecting these cells presents significant challenges [184, 185]. Over the last 10 years, C2C12 cells have predominantly been transfected using cationic-lipid transfection reagents [185], although transfection efficiency is extremely low [186]. Each transfection protocol should be optimized based on cell type being transfected (e.g., immortalized cell lines and primary muscle cells) and the nucleic acid being transfected (e.g., plasmid and siRNA) [185].

The majority of chemical transfection reagents are composed of cationic components, such as cationic polymers and lipids, which interact with negatively charged nucleic acids to form complexes. In addition to cationic reagents, alternative transfection methods can be employed for C2C12 cells (e.g., calcium phosphate [184] and gold nanoparticle-mediated transfection [187]). Regarding cationic components, in 2022, Cocchiararo and colleagues optimized and compared the transfection of C2C12 cells with small (4.15 kb) and large (8.55 kb) plasmids using five commercial transfection reagents. They assessed cytotoxic effects by evaluating cell growth, viability, and differentiation [185]. Other researchers are actively investigating the development of novel polymers to enhance muscle cell transfection efficiency while minimizing cytotoxicity [188, 189].

Studies on lncRNAs typically include cell or tissue transfections. *In vitro* knock-down of specific lncRNAs is commonly performed by transfecting C2C12 cells with small interference RNAs targeted to the lncRNA [25, 49, 90, 92, 100, 107, 127, 164, 190, 191]. Conversely, overexpression of lncRNAs is achieved by introducing in the cells plasmids carrying the lncRNA sequence [95, 110, 164]. *In vivo* modulation of lncRNAs expression in mouse skeletal muscles can be performed through electroporation of plasmids transcribing for siRNAs or the lncRNA. Alternatively, the direct electroporation of short hairpin RNAs (shRNAs) [93, 107, 110, 192] or the injection of shRNA lentiviral particles can be used [92, 104, 127]. Moreover, knock-in [159] and knock-out [106, 163] mice for specific lncRNAs were developed.

10. Conclusions

Currently, the United States Food and Drug Administration has approved six RNAi-based therapeutics: patisiran (Onpattro), lumasiran (Oxlumo), inclisiran (Leqvio), vutrisiran (Amvuttra), nedosiran (Rivfloza), and givosiran (Givlaari) [193]. The clinical application of RNA interference (RNAi) molecular mechanisms exemplifies a burgeoning class of RNA therapeutics capable of treating diseases by modulating the expression of target genes (**Table 4**). The field of RNA-based therapy emerged in the 1990s, with Martinon's pioneering work demonstrating the in vivo induction of anti-influenza cytotoxic T lymphocytes by immunizing mice with liposomes encapsulating mRNA encoding the influenza virus nucleoprotein [194].

Technique	Mechanistic summary	Common side effects	Current approved examples (agency)
Antisense Oligonucleotides (ASOs)	Bind to mRNA and induce degradation, mask splice sites, avoid translation.	Flu-like symptoms, injection site reactions	Mipomersen (Kynamro) FDA 2013 EMA No
			Tofersen FDA 2023 EMA 2024
			Fomivirsen (Vitravene®) FDA 1998 EMA 1999
			Eplontersen (Wainua) FDA 2023 EMA No
			Mipomersen (Kynamro®) FDA 2013 EMA No,
			Olezarsen (Tryngolza) FDA 2024 EMA No
			Inotersen (Tegsedi®) FDA 2018 EMA 2018,
			Volanesorsen (Waylivra®) FDA 2019 EMA 2019

Technique	Mechanistic summary	Common side effects	Current approved examples (agency)
Short inhibitory RNAs (siRNAs)	Bind to mRNA and induce degradation	Injection site reactions, hypersensitivity reactions	Patisiran (Onpattro) FDA 2018 EMA 2018,
			Nedosiran (Rivfloza) FDA 2023 EMA No
			Givosiran (Givlaari) FDA 2019 EMA 2020,
			Lumasiran (Oxlumo) FDA 2020 EMA 2020,
			Inclisiran (Leqvio®) FDA 2021 EMA 2020,
			Vuttrisiran (Amvuttra®) FDA 2022 EMA 2022
mRNA	Mimic endogenous mRNA and produces entire or part of protein	Varies depending on target and delivery method	Not yet approved
miRNA Mimics and Antagomirs	Mimic or inhibit endogenous miRNAs	Varies depending on target and delivery method	Miravirsen (Clinical trials, not yet approved)
Splice-Switching Oligonucleotide (SSO)	Bind to pre-mRNA and modulate splicing	Injection site reactions, flu-like symptoms	Nusinersen (Spinraza®) FDA 2016 EMA 2017,
			Eteplirsen (Exondys®) FDA 2016 EMA No,
			Golodirsen (Vyondys®) FDA 2019
			Viltolarsen (Viltepso®) FDA 2020,
			Casimersen (Amondys) FDA 2021
RNA Aptamers	Bind to target proteins and modulate their activity	Varies depending on target and delivery method	Pegaptanib (Macugen®) FDA 2004 EMA 2006
RNA Vaccines	Encodes antigenic proteins that stimulate an immune response	Pain at injection site, fever, fatigue	Tozinameran (Comirnaty®) FDA 2021,
			BNT162b2 Comirnaty (Pfizer-BioNTech) FDA 2021 EMA 2020
			Elasomeran (Spikevax®) FDA 2022
			mRNA-1273 Spikevax (Moderna) FDA 2022 EMA 2021

Technique	Mechanistic summary	Common side effects	Current approved examples (agency)
CRISPR-Cas9-mediated gene editing	Induces double-strand breaks in DNA, guided by RNA sequence leading to gene editing	Varies depending on target and delivery method	Not yet approved
TANGO (Targeted Augmentation of Nuclear Gene Output)	Oligo binding to exon in pre-mRNA to produce more functional mRNA that leads in increase in functional protein expression	Not yet known	In clinical trials for Dravet.

Table 4.
Classes of RNA therapeutics with the indication of approved one from Food and Drug Administration (FDA) and European Medicines Agency (EMA).

RNA-based therapy is currently focused on modulating coding genes; however, this manuscript demonstrates that also non-coding RNAs participate in skeletal pathology development. In addition to hereditary skeletal muscle pathologies, lncRNAs participate in several other hereditary disease [195], and their modulation may ameliorate therapeutic effects. An example is spinal muscle atrophy (SMA). It is caused by mutations in a gene called survival motor neuron 1 (SMN1). SMN1 deletions are typical of most SMA patients that maintain SMN2 gene as the sole SMN source. Nonetheless, a C → T substitution transforms an exonic-splicing enhancer to a silencer, triggering frequent exon7 skipping in SMN2 pre-mRNA and resulting in a trimmed protein. Patients treatment with antisense oligonucleotides correcting SMN2 splicing improves SMN expression and motor function [196]. After the discovery of a new antisense transcript from the SMN1 locus, it was demonstrated that SMN-AS1 negatively regulates SMN mRNA level recruiting the PRC2 complex. The application of antisense oligonucleotides directed at SMN-AS1, in conjunction with a splicing corrector to increase exon 7 inclusion in SMN2 mRNA, has demonstrated a two-fold enhancement of SMN2 level within neurons derived from mice that possess only the SMN2 gene [195]. This demonstrates that also lncRNA modulation may be used in therapeutical approaches. Open problems in RNA-based therapy remain. For example, delivery methods should be ameliorated to permit to target desired tissue and avoid side effects or off targeting.

Acknowledgements

S.S. writing, preparation of figures and tables and editing; S.K. writing, preparation of figures and tables; S.C. conceived and organized the work, writing preparation of figures, and revised the final version. All authors approved the submitted version.

Conflict of interest

The authors declare that the research was conducted in the absence of any commercial or financial relationships that could be construed as a potential conflict of interest.

Author details

Silvia Scalabrin[1], Sanaz Kavoosi[1] and Stefano Cagnin[2]*

1 Department of Biology, University of Padova, Padova, Italy

2 Department of Biology, Interdepartmental Research Center of Myology (cirMYO), University of Padova, Padova, Italy

*Address all correspondence to: stefano.cagnin@unipd.it

References

[1] Severinsen MCK, Pedersen BK. Muscle–organ crosstalk: The emerging roles of myokines. Endocrine Reviews. 2020;**41**:594-609

[2] Lieber R. Skeletal Muscle Structure, Function, and Plasticity. United States: Lippincott Williams & Wilkins; 2022

[3] Talbot J, Maves L. Skeletal muscle fiber type: Using insights from muscle developmental biology to dissect targets for susceptibility and resistance to muscle disease. WIREs Developmental Biology. 2016;**5**:518-534

[4] Deries M, Gonçalves A, Thorsteinsdóttir S. Skeletal Muscle Development: From Stem Cells to Body Movement2020. pp. 159-185

[5] Williams K, Yokomori K, Mortazavi A. Heterogeneous skeletal muscle cell and nucleus populations identified by single-cell and single-nucleus resolution transcriptome assays. Frontiers in Genetics;**13**. Epub ahead of print 13 May 2022. DOI: 10.3389/fgene.2022.835099

[6] ENCODE Project Consortium. An integrated encyclopedia of DNA elements in the human genome. Nature. 2012;**489**:57-74

[7] Vandevenne M, Delmarcelle M, Galleni M. RNA regulatory networks as a control of stochasticity in biological systems. Frontiers in Genetics. 2019;**10**:403

[8] Vaquerizas JM, Kummerfeld SK, Teichmann SA, et al. A census of human transcription factors: Function, expression and evolution. Nature Reviews. Genetics. 2009;**10**:252-263

[9] Pertea M. The human transcriptome: An unfinished story. Genes (Basel). 2012;**3**:344-360

[10] Li S, Wan C, Zheng R, et al. Cistrome-GO: A web server for functional enrichment analysis of transcription factor ChIP-seq peaks. Nucleic Acids Research. 2019;**47**:W206-W211

[11] Taft RJ, Pheasant M, Mattick JS. The relationship between non-protein-coding DNA and eukaryotic complexity. BioEssays. 2007;**29**:288-299

[12] Ahnert SE, Fink TMA, Zinovyev A. How much non-coding DNA do eukaryotes require? Journal of Theoretical Biology. 2008;**252**:587-592

[13] Alessio E, Bonadio RS, Buson L, et al. A single cell but many different transcripts: A journey into the world of long non-coding RNAs. International Journal of Molecular Sciences. 2020;**21**:302

[14] Carninci P, Kasukawa T, Katayama S, et al. The transcriptional landscape of the mammalian genome. Science. 2005;**309**:1559-1563

[15] Wang M, Zhao W, Gao L, et al. Genome-wide profiling of long non-coding RNAs from tomato and a comparison with mRNAs associated with the regulation of fruit ripening. BMC Plant Biology. 2018;**18**:75

[16] Arnoldi M, Zarantonello G, Espinoza S, et al. Design and delivery of SINEUP: A new modular tool to increase protein translation. Methods in Molecular Biology. 2022;**2434**:63-87

[17] Valentini P, Pierattini B, Zacco E, et al. Towards SINEUP-based therapeutics: Design of an in vitro synthesized SINEUP RNA.

Molecular Therapy - Nucleic Acids. 2022;**27**:1092-1102

[18] Oliviero S. SINEUPs to boost translation. Molecular Therapy - Nucleic Acids. 2023;**33**:47-48

[19] Ding X, Wang X, Sontag S, et al. The polycomb protein Ezh2 impacts on induced pluripotent stem cell generation. Stem Cells and Development. 2014;**23**:931-940

[20] Khalil AM, Guttman M, Huarte M, et al. Many human large intergenic noncoding RNAs associate with chromatin-modifying complexes and affect gene expression. Proceedings of the National Academy of Sciences of the United States of America. 2009;**106**:11667-11672

[21] Schutt C, Hallmann A, Hachim S, et al. Linc-MYH configures INO80 to regulate muscle stem cell numbers and skeletal muscle hypertrophy. The EMBO Journal. 2020;**39**:e105098

[22] Dieci G, Fiorino G, Castelnuovo M, et al. The expanding RNA polymerase III transcriptome. Trends in Genetics. 2007;**23**:614-622

[23] Guo C-J, Xu G, Chen L-L. Mechanisms of long noncoding RNA nuclear retention. Trends in Biochemical Sciences. 2020;**45**:947-960

[24] Hitachi K, Honda M, Tsuchida K. The functional role of long non-coding RNA in myogenesis and skeletal muscle atrophy. Cells. 2022;**11**:2291

[25] Ruan L, Mendhe B, Parker E, et al. Long non-coding RNA MALAT1 is depleted with age in skeletal muscle in vivo and MALAT1 silencing increases expression of TGF-β1 in vitro. Frontiers in Physiology. 2021;**12**:742004

[26] Palazzo AF, Truong M. Single particle imaging of mRNAs crossing the nuclear pore: Surfing on the edge. BioEssays. 2016;**38**:744-750

[27] Espinoza S, Bon C, Valentini P, et al. SINEUPs: A novel toolbox for RNA therapeutics. Essays in Biochemistry. 2021;**65**:775-789

[28] Rackham O, Shearwood A-MJ, Mercer TR, et al. Long noncoding RNAs are generated from the mitochondrial genome and regulated by nuclear-encoded proteins. RNA. 2011;**17**:2085-2093

[29] Villegas J, Zárraga AM, Muller I, et al. A novel chimeric mitochondrial RNA localized in the nucleus of mouse sperm. DNA and Cell Biology. 2000;**19**:579-588

[30] Ren B, Guan M-X, Zhou T, et al. Emerging functions of mitochondria-encoded noncoding RNAs. Trends in Genetics. 2023;**39**:125-139

[31] Rurek M. Participation of non-coding RNAs in plant organelle biogenesis. Acta Biochimica Polonica. 2016;**63**:653-663

[32] Holmes DS, Mayfield JE, Sander G, et al. Chromosomal RNA: Its properties. Science. 1972;**177**:72-74

[33] Shao Z, Raible F, Mollaaghababa R, et al. Stabilization of chromatin structure by PRC1, a Polycomb complex. Cell. 1999;**98**:37-46

[34] Levine SS, Weiss A, Erdjument-Bromage H, et al. The core of the polycomb repressive complex is compositionally and functionally conserved in flies and humans. Molecular and Cellular Biology. 2002;**22**:6070-6078

[35] Farcas AM, Blackledge NP, Sudbery I, et al. KDM2B links the polycomb repressive complex 1 (PRC1) to recognition of CpG islands. eLife. 2012;**1**:e00205

[36] Lee TI, Jenner RG, Boyer LA, et al. Control of developmental regulators by polycomb in human embryonic stem cells. Cell. 2006;**125**:301-313

[37] Boyer LA, Plath K, Zeitlinger J, et al. Polycomb complexes repress developmental regulators in murine embryonic stem cells. Nature. 2006;**441**:349-353

[38] Bernstein BE, Mikkelsen TS, Xie X, et al. A bivalent chromatin structure marks key developmental genes in embryonic stem cells. Cell. 2006;**125**:315-326

[39] Rosti V, Gorini F, Santarelli P, et al. Polycomb bodies detection in murine fibromuscular stroma from skin, skeletal muscles, and aortic tissues. Methods in Molecular Biology. 2023;**2655**:125-146

[40] Ambros V. The functions of animal microRNAs. Nature. 2004;**431**:350-355

[41] Dexheimer PJ, Cochella L. MicroRNAs: From mechanism to organism. Frontiers in Cell and Developmental Biology. 3 June 2020;**8**:409-427. DOI: 10.3389/fcell.2020.00409 [Epub ahead of print]

[42] Salmena L, Poliseno L, Tay Y, et al. A ceRNA hypothesis: The Rosetta stone of a hidden RNA language? Cell. 2011;**146**:353-358

[43] Cesana M, Cacchiarelli D, Legnini I, et al. A long noncoding RNA controls muscle differentiation by functioning as a competing endogenous RNA. Cell. 2011;**147**:358-369

[44] Mcleod JC, Lim C, Stokes T, et al. Network-based modelling reveals cell-type enriched patterns of non-coding RNA regulation during human skeletal muscle remodelling. NAR. Molecular Medicine. 2024;**1**:ugae016

[45] Blackwood EM, Kadonaga JT. Going the distance: A current view of enhancer action. Science. 1998;**281**:60-63

[46] Kim T-K, Hemberg M, Gray JM, et al. Widespread transcription at neuronal activity-regulated enhancers. Nature. 2010;**465**:182-187

[47] Banerji J, Rusconi S, Schaffner W. Expression of a beta-globin gene is enhanced by remote SV40 DNA sequences. Cell. 1981;**27**:299-308

[48] Zhang Y, Lu D, Liu Y, et al. The untold story between enhancers and skeletal muscle development. Journal of Integrative Agriculture. 2020;**19**:2137-2149

[49] So KKH, Huang Y, Zhang S, et al. seRNA PAM controls skeletal muscle satellite cell proliferation and aging through trans regulation of Timp2 expression synergistically with Ddx5. Aging Cell. 2022;**21**:e13673

[50] Zhang W, Liu Y, Zhang H. Extracellular matrix: An important regulator of cell functions and skeletal muscle development. Cell & Bioscience. 2021;**11**:1-13

[51] Schaukowitch K, Joo J-Y, Liu X, et al. Enhancer RNA facilitates NELF release from immediate early genes. Molecular Cell. 2014;**56**:29-42

[52] Koch F, Fenouil R, Gut M, et al. Transcription initiation platforms and GTF recruitment at tissue-specific enhancers and promoters. Nature

Structural & Molecular Biology. 2011;**18**:956-963

[53] Natoli G, Andrau J-C. Noncoding transcription at enhancers: General principles and functional models. Annual Review of Genetics. 2012;**46**:1-19

[54] Sanger HL, Klotz G, Riesner D, et al. Viroids are single-stranded covalently closed circular RNA molecules existing as highly base-paired rod-like structures. Proceedings of the National Academy of Sciences of the United States of America. 1976;**73**:3852-3856

[55] Hsu MT, Coca-Prados M. Electron microscopic evidence for the circular form of RNA in the cytoplasm of eukaryotic cells. Nature. 1979;**280**:339-340

[56] Salzman J, Gawad C, Wang PL, et al. Circular RNAs are the predominant transcript isoform from hundreds of human genes in diverse cell types. PLoS One. 2012;**7**:e30733

[57] Hansen TB, Jensen TI, Clausen BH, et al. Natural RNA circles function as efficient microRNA sponges. Nature. 2013;**495**:384-388

[58] Memczak S, Jens M, Elefsinioti A, et al. Circular RNAs are a large class of animal RNAs with regulatory potency. Nature. 2013;**495**:333-338

[59] Kristensen LS, Andersen MS, Stagsted LVW, et al. The biogenesis, biology and characterization of circular RNAs. Nature Reviews. Genetics. 2019;**20**:675-691

[60] Li Z, Huang C, Bao C, et al. Exon-intron circular RNAs regulate transcription in the nucleus. Nature Structural & Molecular Biology. 2015;**22**:256-264

[61] Su H, Tao T, Yang Z, et al. Circular RNA cTFRC acts as the sponge of MicroRNA-107 to promote bladder carcinoma progression. Molecular Cancer. 2019;**18**:27

[62] Wesselhoeft RA, Kowalski PS, Parker-Hale FC, et al. RNA circularization diminishes immunogenicity and can extend translation duration in vivo. Molecular Cell. 2019;**74**:508-520.e4

[63] Zhu X, Yang T, Zheng Y, et al. EIF4A3-induced circular RNA CircDdb1 promotes muscle atrophy through encoding a novel protein CircDdb1-867aa. Advanced Science. 2024;**11**:e2406986

[64] Chen R, Yang T, Jin B, et al. CircTmeff1 promotes muscle atrophy by interacting with TDP-43 and encoding a novel TMEFF1-339aa protein. Advanced Science. 2023;**10**:e2206732

[65] Chen M, Wei X, Song M, et al. Circular RNA circMYBPC1 promotes skeletal muscle differentiation by targeting MyHC. Molecular Therapy - Nucleic Acids. 2021;**24**:352-368

[66] Shin YJ, Kwon K-S, Suh Y, et al. The role of non-coding RNAs in muscle aging: Regulatory mechanisms and therapeutic potential. Frontiers in Molecular Biosciences. 2023;**10**:1308274

[67] Francis JC, Gardiner JR, Renaud Y, et al. HOX genes promote cell proliferation and are potential therapeutic targets in adrenocortical tumours. British Journal of Cancer. 2021;**124**:805-816

[68] Yoshino S, Yokoyama T, Sunami Y, et al. Trib1 promotes acute myeloid leukemia progression by modulating the transcriptional programs of Hoxa9. Blood. 2021;**137**:75-88

[69] Lu X, Liang B, Li S, et al. Modulation of HOXA9 after skeletal muscle denervation and reinnervation. American Journal of Physiology. Cell Physiology. 2020;**318**:C1154-C1165

[70] Enuka Y, Lauriola M, Feldman ME, et al. Circular RNAs are long-lived and display only minimal early alterations in response to a growth factor. Nucleic Acids Research. 2016;**44**:1370-1383

[71] Pandey PR, Yang J-H, Tsitsipatis D, et al. circSamd4 represses myogenic transcriptional activity of PUR proteins. Nucleic Acids Research. 2020;**48**:3789-3805

[72] Varma E, Burghaus J, Schwarzl T, et al. Translational control of Ybx1 expression regulates cardiac function in response to pressure overload in vivo. Basic Research in Cardiology. 2023;**118**:25

[73] Ingolia NT, Ghaemmaghami S, Newman JRS, et al. Genome-wide analysis in vivo of translation with nucleotide resolution using ribosome profiling. Science. 2009;**324**:218-223

[74] Wang H, Wang Y, Xie S, et al. Global and cell-type specific properties of lincRNAs with ribosome occupancy. Nucleic Acids Research. 2017;**45**:2786-2796

[75] Matsumoto A, Pasut A, Matsumoto M, et al. mTORC1 and muscle regeneration are regulated by the LINC00961-encoded SPAR polypeptide. Nature. 2017;**541**:228-232

[76] Zhang Q, Vashisht AA, O'Rourke J, et al. The microprotein minion controls cell fusion and muscle formation. Nature Communications. 2017;**8**:15664

[77] Wang L, Fan J, Han L, et al. The micropeptide LEMP plays an evolutionarily conserved role in myogenesis. Cell Death & Disease. 2020;**11**:357

[78] Anderson DM, Anderson KM, Chang C-L, et al. A micropeptide encoded by a putative long noncoding RNA regulates muscle performance. Cell. 2015;**160**:595-606

[79] Nelson BR, Makarewich CA, Anderson DM, et al. A peptide encoded by a transcript annotated as long noncoding RNA enhances SERCA activity in muscle. Science. 2016;**351**:271-275

[80] Stein CS, Jadiya P, Zhang X, et al. Mitoregulin: A lncRNA-encoded microprotein that supports mitochondrial supercomplexes and respiratory efficiency. Cell Reports. 2018;**23**:3710-3720.e8

[81] Makarewich CA, Baskin KK, Munir AZ, et al. MOXI is a mitochondrial micropeptide that enhances fatty acid β-oxidation. Cell Reports. 2018;**23**:3701-3709

[82] Spiroski A-M, Sanders R, Meloni M, et al. The influence of the LINC00961/SPAAR locus loss on murine development, myocardial dynamics, and cardiac response to myocardial infarction. International Journal of Molecular Sciences. 2021;**22**:969

[83] Yan Y, Tang R, Li B, et al. The cardiac translational landscape reveals that micropeptides are new players involved in cardiomyocyte hypertrophy. Molecular Therapy. 2021;**29**:2253-2267

[84] Xiao Y, Ren Y, Hu W, et al. Long non-coding RNA-encoded micropeptides: Functions, mechanisms and implications. Cell Death Discovery. 2024;**10**:1-15

[85] Estrada AL, Valenti ZJ, Hehn G, et al. Extracellular vesicle secretion is tissue-dependent ex vivo and skeletal muscle myofiber extracellular vesicles reach the circulation in vivo. American Journal of Physiology-Cell Physiology. 2022;**322**:C246-C259

[86] Ni P, Yang L, Li F. Exercise-derived skeletal myogenic exosomes as mediators of intercellular crosstalk: A major player in health, disease, and exercise. Journal of Physiology and Biochemistry. 2023;**79**:501-510

[87] Luo J, Pu Q, Wu X. Recent advances of exosomes derived from skeletal muscle and crosstalk with other tissues. International Journal of Molecular Sciences. 2024;**25**:10877

[88] Li L, Zuo X, Liu D, et al. Plasma exosomal RNAs have potential as both clinical biomarkers and therapeutic targets of dermatomyositis. Rheumatology (Oxford). 2022;**61**:2672-2681

[89] Dalakas MC, Hohlfeld R. Polymyositis and dermatomyositis. Lancet. 2003;**362**:971-982

[90] Chen W, Chen W, Liu P, et al. Role of lncRNA Has2os in skeletal muscle differentiation and regeneration. Cells. 2022;**11**:3497

[91] Powrózek T, Pigoń-Zając D, Mazurek M, et al. TNF-α induced myotube atrophy in C2C12 cell line uncovers putative inflammatory-related lncRNAs mediating muscle wasting. International Journal of Molecular Sciences. 2022;**23**:3878

[92] Wang S, Tan B, Xiao L, et al. Long non-coding RNA Gm10561 promotes myogenesis by sponging miR-432. Epigenetics. 2022;**17**:2039-2055

[93] Hitachi K, Nakatani M, Tsuchida K. Long non-coding RNA myoparr regulates GDF5 expression in denervated mouse skeletal muscle. Noncoding RNA. 2019;**5**:33

[94] Qi X, Hu M, Xiang Y, et al. LncRNAs are regulated by chromatin states and affect the skeletal muscle cell differentiation. Cell Proliferation. 2020;**53**:e12879

[95] Sui Y, Han Y, Zhao X, et al. Long non-coding RNA Irm enhances myogenic differentiation by interacting with MEF2D. Cell Death & Disease. 2019;**10**:181

[96] Gu T, He H, Han Z, et al. Expression of macro non-coding RNAs Meg8 and Irm in mouse embryonic development. Acta Histochemica. 2012;**114**:392-399

[97] Hunt LC, Gorman C, Kintakas C, et al. Hyaluronan synthesis and myogenesis: A requirement for hyaluronan synthesis during myogenic differentiation independent of pericellular matrix formation. The Journal of Biological Chemistry. 2013;**288**:13006-13021

[98] Simpson RML, Hong X, Wong MM, et al. Hyaluronan is crucial for stem cell differentiation into smooth muscle lineage. Stem Cells. 2016;**34**:1225-1238

[99] Gesteira TF, Sun M, Coulson-Thomas YM, et al. Hyaluronan rich microenvironment in the Limbal stem cell niche regulates limbal stem cell differentiation. Investigative Ophthalmology & Visual Science. 2017;**58**:4407-4421

[100] Chen W-C, Chen W-X, Tan Y-Y, et al. LncRNA 4930581F22Rik promotes myogenic differentiation by regulating the ERK/MAPK signaling pathway. Heliyon. 2024;**10**:e30640

[101] Xie Y, Lin Z, Liu L, et al. Genotype and phenotype distribution of 435 patients with Charcot-Marie-Tooth disease from central south China. European Journal of Neurology. 2021;**28**:3774-3783

[102] Zhu M, Liu J, Xiao J, et al. Lnc-mg is a long non-coding RNA that promotes myogenesis. Nature Communications. 2017;**8**:14718

[103] Li Y, Meng X, Li G, et al. Noncoding RNAs in muscle atrophy. Advances in Experimental Medicine and Biology. 2018;**1088**:249-266

[104] Li R, Li B, Cao Y, et al. Long non-coding RNA Mir22hg-derived miR-22-3p promotes skeletal muscle differentiation and regeneration by inhibiting HDAC4. Molecular Therapy - Nucleic Acids. 2021;**24**:200-211

[105] Chen R, Lei S, She Y, et al. Lnc-GD2H promotes proliferation by forming a feedback loop with c-Myc and enhances differentiation through interacting with NACA to upregulate Myog in C2C12 myoblasts. Frontiers in Cell and Development Biology. 2021;**9**:671857

[106] Lv W, Jin J, Xu Z, et al. lncMGPF is a novel positive regulator of muscle growth and regeneration. Journal of Cachexia, Sarcopenia and Muscle. 2020;**11**:1723-1746

[107] Hitachi K, Nakatani M, Takasaki A, et al. Myogenin promoter-associated lncRNA Myoparr is essential for myogenic differentiation. EMBO Reports. 2019;**20**:e47468

[108] Yao Y, Wang Z, Chen Y, et al. Single-cell analysis reveals the lncRNA-MEG3/miRNA-133a-3p/PRRT2 axis regulates skeletal muscle regeneration and myogenesis. Genes & Diseases. 2023;**10**:359-362

[109] Fu X, Li S, Jia M, et al. Myogenesis controlled by a long non-coding RNA 1700113A16RIK and post-transcriptional regulation. Cell Regeneration. 2022;**11**:13

[110] Li R, Li B, Shen M, et al. LncRNA 2310043L19Rik inhibits differentiation and promotes proliferation of myoblast by sponging miR-125a-5p. Aging (Albany NY). 2020;**12**:5625-5639

[111] Zhai L, Wan X, Wu R, et al. Linc-RAM promotes muscle cell differentiation via regulating glycogen phosphorylase activity. Cell Regeneration. 2022;**11**:8

[112] Scalzitti S, Mariani D, Setti A, et al. Lnc-SMaRT translational regulation of Spire1, a new player in muscle differentiation. Journal of Molecular Biology. 2022;**434**:167384

[113] Lv W, Jiang W, Luo H, et al. Long noncoding RNA lncMREF promotes myogenic differentiation and muscle regeneration by interacting with the Smarca5/p300 complex. Nucleic Acids Research. 2022;**50**:10733-10755

[114] Trewin AJ, Silver J, Dillon HT, et al. Long non-coding RNA Tug1 modulates mitochondrial and myogenic responses to exercise in skeletal muscle. BMC Biology. 2022;**20**:164

[115] Yue Y, Wang Y, Wen C, et al. Lnc-Malat1 promotes slow myofiber-type transformation through sponging miR-129-5p in C2C12 myotubes. Experimental Cell Research. 2023;**431**:113761

[116] McMullen JR, Drew BG. Long non-coding RNAs (lncRNAs) in skeletal and cardiac muscle: Potential therapeutic and diagnostic targets? Clinical Science (London, England). 2016;**130**:2245-2256

[117] Simionescu-Bankston A, Kumar A. Noncoding RNAs in the

regulation of skeletal muscle biology in health and disease. Journal of Molecular Medicine (Berlin, Germany). 2016;**94**:853-866

[118] Zhou L, Sun K, Zhao Y, et al. Linc-YY1 promotes myogenic differentiation and muscle regeneration through an interaction with the transcription factor YY1. Nature Communications. 2015;**6**:10026

[119] Legnini I, Morlando M, Mangiavacchi A, et al. A feedforward regulatory loop between HuR and the long noncoding RNA linc-MD1 controls early phases of myogenesis. Molecular Cell. 2014;**53**:506-514

[120] Yu X, Zhang Y, Li T, et al. Long non-coding RNA Linc-RAM enhances myogenic differentiation by interacting with MyoD. Nature Communications. 2017;**8**:14016

[121] Li Y, Chen X, Sun H, et al. Long non-coding RNAs in the regulation of skeletal myogenesis and muscle diseases. Cancer Letters. 2018;**417**:58-64

[122] Kallen AN, Zhou X-B, Xu J, et al. The imprinted H19 lncRNA antagonizes let-7 microRNAs. Molecular Cell. 2013;**52**:101-112

[123] Jin JJ, Lv W, Xia P, et al. Long noncoding RNA SYISL regulates myogenesis by interacting with polycomb repressive complex 2. Proceedings of the National Academy of Sciences of the United States of America. 2018;**115**:E9802-E9811

[124] Bovolenta M, Erriquez D, Valli E, et al. The DMD locus harbours multiple long non-coding RNAs which orchestrate and control transcription of muscle dystrophin mRNA isoforms. PLoS One. 2012;7:e45328

[125] Neguembor MV, Jothi M, Gabellini D. Long noncoding RNAs, emerging players in muscle differentiation and disease. Skeletal Muscle. 2014;**4**:8

[126] Cabianca DS, Casa V, Bodega B, et al. A long ncRNA links copy number variation to a polycomb/trithorax epigenetic switch in FSHD muscular dystrophy. Cell. 2012;**149**:819-831

[127] Li J, Yang T, Tang H, et al. Inhibition of lncRNA MAAT controls multiple types of muscle atrophy by cis- and trans-regulatory actions. Molecular Therapy. 2021;**29**:1102-1119

[128] Mueller AC, Cichewicz MA, Dey BK, et al. MUNC, a long noncoding RNA that facilitates the function of MyoD in skeletal myogenesis. Molecular and Cellular Biology. 2015;**35**:498-513

[129] Wang L, Zhao Y, Bao X, et al. LncRNA Dum interacts with Dnmts to regulate Dppa2 expression during myogenic differentiation and muscle regeneration. Cell Research. 2015;**25**:335-350

[130] Gong C, Li Z, Ramanujan K, et al. A long non-coding RNA, LncMyoD, regulates skeletal muscle differentiation by blocking IMP2-mediated mRNA translation. Developmental Cell. 2015;**34**:181-191

[131] Wang S, Zuo H, Jin J, et al. Long noncoding RNA Neat1 modulates myogenesis by recruiting Ezh2. Cell Death & Disease. 2019;**10**:505

[132] Han X, Yang F, Cao H, et al. Malat1 regulates serum response factor through miR-133 as a competing endogenous RNA in myogenesis. The FASEB Journal. 2015;**29**:3054-3064

[133] Caretti G, Schiltz RL, Dilworth FJ, et al. The RNA helicases p68/p72 and the

noncoding RNA SRA are coregulators of MyoD and skeletal muscle differentiation. Developmental Cell. 2006;**11**:547-560

[134] Hitachi K, Nakatani M, Kiyofuji Y, et al. An analysis of differentially expressed coding and long non-coding RNAs in multiple models of skeletal muscle atrophy. International Journal of Molecular Sciences. 2021;**22**:2558

[135] Cai B, Ma M, Zhang J, et al. Long noncoding RNA ZFP36L2-AS functions as a metabolic modulator to regulate muscle development. Cell Death & Disease. 2022;**13**:389

[136] Cai B, Ma M, Zhang J, et al. LncEDCH1 improves mitochondrial function to reduce muscle atrophy by interacting with SERCA2. Molecular Therapy - Nucleic Acids. 2022;**27**:319-334

[137] Alipoor B, Nikouei S, Rezaeinejad F, et al. Long non-coding RNAs in metabolic disorders: Pathogenetic relevance and potential biomarkers and therapeutic targets. Journal of Endocrinological Investigation. 2021;**44**:2015-2041

[138] Gui W, Zhu WF, Zhu Y, et al. LncRNAH19 improves insulin resistance in skeletal muscle by regulating heterogeneous nuclear ribonucleoprotein A1. Cell Communication and Signaling: CCS. 2020;**18**:173

[139] Huang S, Wu K, Li B, et al. lncRNA UCA1 inhibits mitochondrial dysfunction of skeletal muscle in type 2 diabetes mellitus by sequestering miR-143-3p to release FGF21. Cell and Tissue Research. 2023;**391**:561-575

[140] Degirmenci U, Li J, Lim YC, et al. Silencing an insulin-induced lncRNA, LncASIR, impairs the transcriptional response to insulin signalling in adipocytes. Scientific Reports. 2019;**9**:5608

[141] Kottorou A, Dimitrakopoulos F-I, Tsezou A. Non-coding RNAs in cancer-associated cachexia: Clinical implications and future perspectives. Translational Oncology. 2021;**14**:101101

[142] Santos JMO, Peixoto da Silva S, Gil da Costa RM, et al. The emerging role of MicroRNAs and other non-coding RNAs in cancer cachexia. Cancers (Basel). 2020;**12**:1004

[143] Zhang Z-K, Li J, Guan D, et al. A newly identified lncRNA MAR1 acts as a miR-487b sponge to promote skeletal muscle differentiation and regeneration. Journal of Cachexia, Sarcopenia and Muscle. 2018;**9**:613-626

[144] Hu C-Y, Su B-H, Lee Y-C, et al. Interruption of the long non-coding RNA HOTAIR signaling axis ameliorates chemotherapy-induced cachexia in bladder cancer. Journal of Biomedical Science. 2022;**29**:104

[145] Han J, Shen L, Zhan Z, et al. The long noncoding RNA MALAT1 modulates adipose loss in cancer-associated cachexia by suppressing adipogenesis through PPAR-γ. Nutrition & Metabolism (London). 2021;**18**:27

[146] Wijesinghe SN, Nicholson T, Tsintzas K, et al. Involvements of long noncoding RNAs in obesity-associated inflammatory diseases. Obesity Reviews. 2021;**22**:e13156

[147] Guo Y-C, Cao H-D, Lian X-F, et al. Molecular mechanisms of noncoding RNA and epigenetic regulation in obesity with consequent diabetes mellitus development. World Journal of Diabetes. 2023;**14**:1621-1631

[148] Stapleton K, Das S, Reddy MA, et al. Novel long noncoding RNA, macrophage

inflammation-suppressing transcript (MIST), regulates macrophage activation during obesity. Arteriosclerosis, Thrombosis, and Vascular Biology. 2020;**40**:914-928

[149] Fox KM, Brooks JM, Gandra SR, et al. Estimation of cachexia among cancer patients based on four definitions. Journal of Oncology. 2009;**2009**:693458

[150] Liu H, Zhou T, Wang B, et al. Identification and functional analysis of a potential key lncRNA involved in fat loss of cancer cachexia. Journal of Cellular Biochemistry. 2018;**119**:1679-1688

[151] Schmidt E, Dhaouadi I, Gaziano I, et al. LincRNA H19 protects from dietary obesity by constraining expression of monoallelic genes in brown fat. Nature Communications. 2018;**9**:3622

[152] Squillaro T, Peluso G, Galderisi U, et al. Long non-coding RNAs in regulation of adipogenesis and adipose tissue function. eLife. 2020;**9**:e59053

[153] Mi L, Zhao X-Y, Li S, et al. Conserved function of the long noncoding RNA Blnc1 in brown adipocyte differentiation. Molecular Metabolism. 2017;**6**:101-110

[154] Knoll M, Lodish HF, Sun L. Long non-coding RNAs as regulators of the endocrine system. Nature Reviews. Endocrinology. 2015;**11**:151-160

[155] Chen Y, Li Z, Chen X, et al. Long non-coding RNAs: From disease code to drug role. Acta Pharmaceutica Sinica B. 2021;**11**:340-354

[156] Winkle M, El-Daly SM, Fabbri M, et al. Noncoding RNA therapeutics - challenges and potential solutions. Nature Reviews. Drug Discovery. 2021;**20**:629-651

[157] Nappi F. Non-coding RNA-targeted therapy: A state-of-the-art review. International Journal of Molecular Sciences. 2024;**25**:3630

[158] Alessio E, Buson L, Chemello F, et al. Single cell analysis reveals the involvement of the long non-coding RNA Pvt1 in the modulation of muscle atrophy and mitochondrial network. Nucleic Acids Research. 2019;**47**:1653-1670

[159] Zhang Z-K, Guan D, Xu J, et al. Long noncoding RNA lncRNA-3 recruits PRC2 for MyoD1 silencing to suppress muscle regeneration during aging. International Journal of Molecular Sciences. 2024;**25**:12478

[160] Chatterjee S, Bhattacharya M, Lee S-S, et al. An insight of different classes of RNA-based therapeutic, nanodelivery and clinical status: Current landscape. Current Research in Biotechnology. 2023;**6**:100150

[161] Zhang C, Zhang B. RNA therapeutics: Updates and future potential. Science China. Life Sciences. 2023;**66**:12-30

[162] Cullis PR, Felgner PL. The 60-year evolution of lipid nanoparticles for nucleic acid delivery. Nature Reviews. Drug Discovery. 2024;**23**:709-722

[163] Li Y, Yuan J, Chen F, et al. Long noncoding RNA SAM promotes myoblast proliferation through stabilizing Sugt1 and facilitating kinetochore assembly. Nature Communications. 2020;**11**:2725

[164] Cai R, Zhang Q, Wang Y, et al. Lnc-ORA interacts with microRNA-532-3p and IGF2BP2 to inhibit skeletal muscle myogenesis. The Journal of Biological Chemistry. 2021;**296**:100376

[165] Hitachi K, Nakatani M, Funasaki S, et al. Expression levels of long

non-coding RNAs change in models of altered muscle activity and muscle mass. International Journal of Molecular Sciences. 2020;**21**:1628

[166] Sawano S, Fukushima M, Akasaka T, et al. Up- and downregulated genes after long-term muscle atrophy induced by denervation in mice detected using RNA-Seq. Life (Basel). 2023;**13**:1111

[167] Zhang H, Wang F, Pang X, et al. Decreased expression of H19/miR-675 ameliorates muscle atrophy by regulating the IGF1R/Akt/FoxO signaling pathway. Molecular Medicine. 2023;**29**:78

[168] Zhang Y, Li Y, Hu Q, et al. The lncRNA H19 alleviates muscular dystrophy by stabilizing dystrophin. Nature Cell Biology. 2020;**22**:1332-1345

[169] Nemeth K, Bayraktar R, Ferracin M, et al. Non-coding RNAs in disease: From mechanisms to therapeutics. Nature Reviews. Genetics. 2024;**25**:211-232

[170] Zhong D, Huang K, Zhang L, et al. Circ2388 regulates myogenesis and muscle regeneration. Cell and Tissue Research. 2023;**393**:149-161

[171] Liu J, Li M, Kong L, et al. CircARID1A regulates mouse skeletal muscle regeneration by functioning as a sponge of miR-6368. The FASEB Journal. 2021;**35**:e21324

[172] Yan J, Yang Y, Fan X, et al. circRNAome profiling reveals circFgfr2 regulates myogenesis and muscle regeneration via a feedback loop. Journal of Cachexia, Sarcopenia and Muscle. 2022;**13**:696-712

[173] Ru W, Qi A, Shen X, et al. The circular RNA circCPE regulates myoblast development by sponging miR-138. Journal of Animal Science and Biotechnology. 2021;**12**:102

[174] Liu X, Zhang Y, Zhou S, et al. Circular RNA: An emerging frontier in RNA therapeutic targets, RNA therapeutics, and mRNA vaccines. Journal of Controlled Release. 2022;**348**:84-94

[175] Das A, Das A, Das D, et al. Circular RNAs in myogenesis. Biochimica et Biophysica Acta-Gene Regulatory Mechanisms. 2020;**1863**:194372

[176] Liu C-X, Chen L-L. Circular RNAs: Characterization, cellular roles, and applications. Cell. 2022;**185**:2016-2034

[177] Frontera WR, Ochala J. Skeletal muscle: A brief review of structure and function. Calcified Tissue International. 2015;**96**:183-195

[178] Abdelmoez AM, Sardón Puig L, Smith JAB, et al. Comparative profiling of skeletal muscle models reveals heterogeneity of transcriptome and metabolism. American Journal of Physiology. Cell Physiology. 2020;**318**:C615-C626

[179] Jalal S, Dastidar S, Tedesco FS. Advanced models of human skeletal muscle differentiation, development and disease: Three-dimensional cultures, organoids and beyond. Current Opinion in Cell Biology. 2021;**73**:92-104

[180] Li L, Huang C, Pang J, et al. Advances in research on cell models for skeletal muscle atrophy. Biomedicine & Pharmacotherapy. 2023;**167**:115517

[181] Iberite F, Gruppioni E, Ricotti L. Skeletal muscle differentiation of human iPSCs meets bioengineering strategies: Perspectives and challenges. npj Regenerative Medicine. 2022;**7**:23

[182] Pinton L, Khedr M, Lionello VM, et al. 3D human induced pluripotent stem cell-derived bioengineered skeletal muscles for tissue, disease and therapy modeling. Nature Protocols. 2023;**18**:1337-1376

[183] Khodabukus A, Prabhu N, Wang J, et al. In vitro tissue-engineered skeletal muscle models for studying muscle physiology and disease. Advanced Healthcare Materials. 2018;**7**:e1701498

[184] Guo L, Wang L, Yang R, et al. Optimizing conditions for calcium phosphate mediated transient transfection. Saudi Journal of Biological Sciences. 2017;**24**:622-629

[185] Cocchiararo I, Cornut M, Soldati H, et al. Back to basics: Optimization of DNA and RNA transfer in muscle cells using recent transfection reagents. Experimental Cell Research. 2022;**421**:113392

[186] Yamano S, Dai J, Moursi AM. Comparison of transfection efficiency of nonviral gene transfer reagents. Molecular Biotechnology. 2010;**46**:287-300

[187] Acharya S, Hill RA. High efficacy gold-KDEL peptide-siRNA nanoconstruct-mediated transfection in C2C12 myoblasts and myotubes. Nanomedicine. 2014;**10**:329-337

[188] Dréan M, Debuigne A, Jérôme C, et al. Poly(N-methylvinylamine)-based copolymers for improved gene transfection. Macromolecular Bioscience. 2018;**18**:e1700353

[189] Gomez J-P, Tresset G, Pichon C, et al. Improved histidinylated lPEI polyplexes for skeletal muscle cells transfection. International Journal of Pharmaceutics. 2019;**559**:58-67

[190] Lin J, Luo Z, Liu S, et al. Long non-coding RNA H19 promotes myoblast fibrogenesis via regulating the miR-20a-5p-Tgfbr2 axis. Clinical and Experimental Pharmacology & Physiology. 2021;**48**:921-931

[191] Xie S-J, Tao S, Diao L-T, et al. Characterization of long non-coding RNAs modified by m6A RNA methylation in skeletal myogenesis. Frontiers in Cell and Development Biology. 2021;**9**:762669

[192] Miller MJ, Gries KJ, Marcotte GR, et al. Human myofiber-enriched aging-induced lncRNA FRAIL1 promotes loss of skeletal muscle function. Aging Cell. 2024;**23**:e14097

[193] Traber GM, Yu A-M. The growing class of novel RNAi therapeutics. Molecular Pharmacology. 2024;**106**:13-20

[194] Martinon F, Krishnan S, Lenzen G, et al. Induction of virus-specific cytotoxic T lymphocytes in vivo by liposome-entrapped mRNA. European Journal of Immunology. 1993;**23**:1719-1722

[195] Sparber P, Filatova A, Khantemirova M, et al. The role of long non-coding RNAs in the pathogenesis of hereditary diseases. BMC Medical Genomics. 2019;**12**:42

[196] Wan L, Dreyfuss G. Splicing-correcting therapy for SMA. Cell. 2017;**170**:5

Section 2

Long Non-Coding RNA's in Diseases

Chapter 5

Exploring Long Non-Coding RNAs as Therapeutic Targets for Foetal Haemoglobin Induction in β-Thalassaemia

Sumaiya Kausar S. Kalaigar, Prashant Vishwanath and Akila Prashant

Abstract

Foetal haemoglobin (HbF) induction is a promising therapeutic strategy for mitigating the clinical severity of β-thalassaemia. However, a limited understanding of globin gene regulation has impeded its progress. Recently, Long non-coding RNAs (lncRNAs) have gained attention as key regulators in biological processes. While lncRNAs have been linked to a variety of diseases and developmental processes, our understanding of the mechanisms underlying their actions is still limited. There are ongoing studies on the hypothesis that lncRNAs are involved in the induction of HbF in β-thalassaemia. A limited number of well-researched lncRNAs have given us important clues, and several noteworthy functional and molecular themes have begun to emerge. Future studies aiming at determining lncRNA-mediated therapeutic targets for HbF reactivation are made possible by new understandings of the lncRNA-mediated regulatory mechanisms. This chapter explores the mechanisms underlying lncRNA function in gene regulation and their potential therapeutic applications in β-thalassaemia.

Keywords: foetal haemoglobin (HbF), β-thalassaemia, long non-coding RNAs (lncRNAs), gene regulation, therapeutic targets

1. Introduction

"β-Thalassaemia" refers to a collection of the most prevalent single-gene inherited haematopoietic disorders characterised by defects in producing β-globin chains of haemoglobin. Due to population migration, approximately 240 million people are reported to be β-thalassaemia carriers, while 15 million people suffer from a major form of β-thalassaemia [1]. The percentage of β-thalassaemia carriers in India alone ranges from 1 to 17% (average of 3.2%), meaning that 1 in every 25 Indians is a thalassaemia carrier. The primary defect in β-thalassaemia is an imbalance between the

IntechOpen

production of α-globin and β-globin chains, caused by diminished or absent synthesis of β-globin [2]. The surplus unpaired α-globin chains aggregate during erythropoiesis, leading to increased mechanical stress and oxidative damage. The clinical severity of β-thalassaemia corresponds to the extent of this α-globin chain imbalance, resulting in variable presentations, including anaemia, chronic haemolysis, ineffective erythropoiesis, and clinical outcomes due to iron overload following repeated blood transfusions [3]. β-Thalassaemia major typically presents between 6 and 24 months of age and is characterised by growth failure and progressively enlarging liver and spleen (hepatosplenomegaly). There are numerous ways to manage β-thalassaemia major, but there are severe obstacles, especially in developing nations, such as high costs, low adherence to iron chelation medication, and restricted access to safe and sufficient red blood cell transfusions. Due to serious consequences such as iron overload, hypercoagulability, pericarditis, osteoporosis, psychological issues, and hepatocellular carcinoma, managing β-thalassaemia major is challenging [4]. Supportive care with red blood cell transfusions, iron chelation to reduce iron overload, and adjunct medicines like hydroxyurea and erythropoietin are traditional therapeutic techniques. Complications are still common and provide major obstacles worldwide despite these therapies [5].

Recent studies have extensively explored the mechanisms regulating haemoglobin switching, given the established role of inducing foetal haemoglobin (HbF) production in alleviating the clinical severity of β-thalassaemia [6]. The patients with β-haemoglobinopathies who maintain higher HbF levels often experience milder disease manifestations compared to those with lower HbF levels. HbF is the primary haemoglobin expressed during foetal development, playing a vital role in facilitating transplacental oxygen transfer. Foetal haemoglobin (HbF), which consists of two α and two γ globin chains, binds oxygen more strongly than adult haemoglobin (HbA). Following birth, definitive erythroid progenitors undergo a transcriptional switch that changes the production of haemoglobin from foetal haemoglobin (HbF, $\alpha 2\gamma 2$) to adult haemoglobin (HbA, $\alpha 2\beta 2$). This transition is governed by genetic regulation involving major transcription factors such as MYB, BCL11A, ZBTB7A, and KLF-1, leading to low HbF levels in healthy adults [7]. HbF levels drastically drop by the time a child is a year old, leaving only trace amounts to be detected. However, during the switch from HbF to HbA, persons with β-globin mutation gene may experience impaired globin chain synthesis or produce abnormal haemoglobin, as seen in β-thalassaemia [8, 9].

Non-coding RNAs (ncRNAs) have a variety of functions in different biological processes. In the human genome, ncRNAs are more prevalent than messenger RNA (mRNA) transcripts that code for proteins. Because of their regulatory influence on gene expression, long non-coding RNAs (lncRNAs) and microRNAs (miRNAs) are gaining increasing attention in the scientific community [10]. LncRNAs can function as signals, scaffolds, guides, decoys, and transcriptional enhancers. Additionally, they can exert indirect post-transcriptional regulation by acting as competitive endogenous RNAs (ceRNAs) and attaching themselves to microRNAs. They can recruit regulatory molecules to modify gene activity or directly affect gene expression [11]. Additionally, lncRNAs can control a variety of cellular functions, potential therapeutic targets, and biomarkers for the prognosis of disease. Erythropoiesis and haematopoiesis are instances in which a large number of tissue-specific lncRNAs are expressed. Because of these crucial functions in biological processes, lncRNAs are believed to have a part in the onset and progression of numerous diseases, such as haematopoietic stem cells, malignancies, neurological disorders and other blood disorders. However, several studies are in progress to shed light on the regulatory roles of lncRNA, which can alter the globin gene expression in β-thalassaemia and other haemoglobinopathies [12]. The targeted induction of HbF through lncRNAs that regulate γ-globin levels has become a promising area of research. However, further exploration is required to advance pharmacological approaches targeting lncRNAs for HbF induction and to identify novel strategies for γ-globin regulation. This chapter focuses on providing an updated overview of lncRNA types implicated in inducing HbF levels in β-thalassaemia.

2. What are lncRNAs?

Recently, studies using high-throughput technologies, such as next-generation sequencing (NGS) and high-resolution microarrays, have made it possible to enumerate the repertoire of non-coding genomes produced by humans over the past 10 years [13]. Approximately 21,000 genes, which is <2% of the human genome, encode proteins, indicating that most of the transcriptome comprises ncRNAs. Based on their transcript length, non-coding transcripts can be broadly classified into three primary groups: long non-coding RNAs, approximately 10,000–32,000; small non-coding RNAs, approximately 9000; and pseudogenes, 11,000 [14, 15]. There is a heterogeneous class of ncRNAs, including lncRNAs, miRNAs, small nucleolar RNAs (snoRNAs), PIWI-associated RNAs, circular RNAs (cRNAs) and small interfering RNAs (siRNAs) [16, 17]. The lncRNAs are considered the most common and functionally varied type of ncRNA. In recent years, lncRNAs have drawn much attention as a novel and important aspect of biological oversight. Although lncRNAs of all types have been linked to a range of developmental and disease processes, little is known about the mechanisms by which they work, and there is ongoing research exploring lncRNAs' function in controlling HbF levels in β-thalassaemia patients [18, 19].

2.1 Biogenesis and biological features of lncRNAs

lncRNAs have transcript lengths >200 nucleotides, and with a few exceptions, lncRNA production is comparable to mRNA synthesis. The tyrosine-phosphorylated enzyme RNA Polymerase II (Pol II) transcribes the canonical pathway of lncRNA in the opposite direction of mRNA transcription, having a polyA tail at the 3′ ends and a methyl guanosine cap at the 5′ end. The non-canonical pathway for the synthesis of lncRNA includes capping at their termini by small nucleolar RNA-protein (snoRNP) complexes, cleavage by ribonuclease P (RNase P) to generate mature 3′ ends and the creation of RNA structures that are circular. Following synthesis, lncRNA is either exported to the cytoplasm for mRNA and protein modulations or stays inside the nucleus for chromatin-gene control [20–22]. LncRNA, such as NEAT1 (Nuclear-Enriched Abundant Transcript 1) and MALAT1 (Metastasis-Associated Lung Adenocarcinoma Transcript 1), undergo post-transcriptional modification by the enzyme Pol II to produce long RNA sequences at the 3′ end. CircRNA may be formed when intergenic lncRNA facilitates the circularisation of mRNA. A lariat loop is created when non-coding RNA breaks during mRNA methylation, but other lncRNAs, like circRNA, are immune to degradation [23, 24]. LncRNA primarily interacts with DNA, RNA, and protein to regulate gene expression at various stages, including transcriptional, translational, post-transcriptional, post-translational, and epigenetic levels. The other biological processes of lncRNA include control of cellular functions such as nuclear transport, transcriptional activation and interference, chromosomal modification, imprinting genomic loci, allosteric regulation of enzyme activity, stem cell pluripotency, and stem cell reprogramming. The biogenesis of lncRNAs is regulated by stimuli that are specific to cell type and developmental stage [20]. lncRNA is broadly classified into six groups based on genomic origins: (1) Antisense lncRNAs (alncRNAs) transcribed from the antisense strand, (2) enhancer lncRNAs (elncRNAs), which are expressed from active enhancers, (3) sRNA-host lncRNAs (shlncRNAs), where lncRNA hosts small RNA (sRNA), such as microRNAs, (4) Intronic lncRNAs (ilncRNAs) situated within intronic regions, (5) Intergenic lncRNAs (lincRNAs) are found intergenically and do not coincide with genes that code for proteins, and (6) Pseudogene lncRNAs (plncRNAs) (**Figure 1**) [25–27].

To govern the expression of lncRNAs, epigenetic changes are essential. For instance, the Chromatin Assembly Factor 1 (CAF-1) complex regulates antisense lncRNA transcription, while H3K56 acetylation and the chromatin remodelling complex SWI/SNF increase it. Furthermore, within imprinted gene clusters, methylation of the paternal allele can dramatically suppress the transcription of lncRNA Air, activating nearby Slc22a2, Slc22a3, and Igf2r, which are protein-coding genes. The nonsense-mediated decay protein UPF1 inhibits the degradation of lncRNAs, which is carried out by exosomes in combination with the TRAMP and the nucleus, Nrd1-Nab3-Sen1 complexes or Xrn1 in the cytoplasm. These findings highlight how important epigenetic changes are to lncRNA biogenesis. However, the precise mechanisms remain unclear, especially when it comes to β-thalassaemia. To clarify how certain epigenetic changes affect lncRNA biogenesis in this situation, more investigation is necessary.

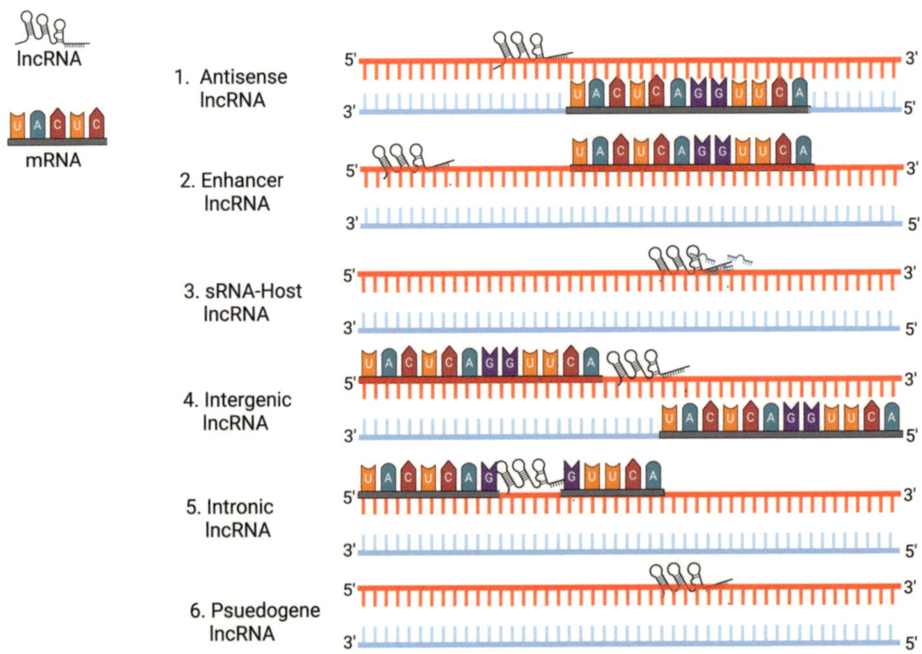

Figure 1.
Classification of lncRNAs based on their genomic location.

2.2 LncRNAs during erythropoiesis

Haematopoiesis can be defined as the continued production of red blood cells (RBC), which begins at the initial embryonic stage and extends through life. In early human embryogenesis, haematopoiesis can be categorised into the primitive phase and the definitive phase. The primitive phase occurs within the blood islands of the yolk sac. The hemangioblast, a progenitor cell precursor of endothelial cells and common myeloid progenitor cells, is the source of primitive haematopoiesis [28]. During the initial transient phase of primitive haematopoiesis, erythroid progenitors with nuclei are generated, followed by the generation of erythro-lymphoid lineages and myeloid lineages during the second transient phase of primitive haematopoiesis. Definitive haematopoiesis arises in the aorta-gonad-mesonephros (AGM) region, which also serves as the site for the generation and renewal of haematopoietic stem cells (HSCs) and their progenitor cells before the entire haematopoietic system transitions to the bone marrow and liver organs. After the embryonic stage of RBC production, in adults, erythropoiesis takes place within the bone marrow; this process includes three different phases: the initiation phase, the differentiation phase, and the maturation phase. During the initiation phase, pluripotent HSCs differentiate into erythroid lineage-committed burst-forming unit erythroid cells (BFU-E), which then progress to colony-forming unit erythroid cells (CFU-E) [29, 30]. Further CFU-E differentiate to form reticulocytes and experience significant alterations, such as a reduction in cell size, haemoglobin synthesis, condensation of chromatin and immature red blood cells without a nucleus. Erythrocytes mature from reticulocytes in the last stage of erythropoiesis, during which the cell adopts its biconcave shape and enters circulation. Mature erythrocytes travel through the bloodstream, but over time, they progressively lose haemoglobin and decrease in size; eventually, they are engulfed by macrophages in the liver or spleen [28]. In cultured mouse foetal erythroblasts, shinc-EC6 lncRNA caused Rac1 to be upregulated, activating the downstream protein PIP5K and inhibiting enucleation. This implies that the Rac1/PIP5K signalling pathway is used by shinc-EC6 lncRNA as a new modulator to control mouse erythropoiesis. LncRNAs play a key role in the regulatory network underlying erythropoiesis [31]. One such lncRNA, alncRNA-EC7, is transcribed from an enhancer and is specifically required to activate the neighbouring gene encoding BAND 3 [32].

2.3 Role of lncRNA in haematopoiesis

In recent years, multiple research teams have conducted in-depth investigations into lncRNAs associated with haematopoiesis. Broad profiling of lncRNA expression during RBC development has been performed using advanced technologies, and comprehensive data sets from RNA-sequence experiments have been obtained and validated using wet lab experiments. Over 1109 lncRNAs, comprising transcript genomes from "megakaryocytic-erythroid progenitors to lineage-committed megakaryocytes and erythroblasts", were found using mouse embryonic day 14.5 foetal liver and bone marrow cells. Among 1109 putative lncRNAs, half of them, that is, 683 lncRNAs, are transcribed in erythroblasts; among these, most of them were not annotated [33]. Another study done by Alvarez-Dominguez *et al.* showed that the transcriptomes of differentiated CFU-E and BFU-E progenitors obtained from the liver of a mouse foetus were compared with TER119-positive erythroblasts. This suggested that lncRNA expression was dynamically regulated during erythroid development, in which 96 lncRNAs were shown to be differently expressed during erythropoiesis [32]. Additionally, RNA-sequence analyses comparing polyadenylated and non-polyadenylated transcripts were performed using TER119-positive cells. A total of 655 lncRNA genes and 9512 coding mRNA genes were found. Many functional lncRNAs have been identified, including those that are essential for the regulation of haematopoiesis. For example, by promoting the survival of erythroid progenitors, lncRNA EPS (erythroid prosurvival) controls the terminal differentiation of erythroid cells and anti-apoptotic ability mediated by erythropoietin (EPO); erythroblast enucleation is controlled by lncRNA-EC1 [33] and lncRNA-EC6 [34]. Myeloid and T cell differentiation are regulated by lncHSC-1 and lncHSC-2, respectively [35], whereas monocyte-derived dendritic cell differentiation is regulated by lnc-DC via STAT3 binding [32]. By controlling Igf2-Igf1r pathway activity [36], the oncofoetal lncRNA gene H19 plays a crucial role in regulating the equilibrium between the silencing and activation of HSCs. The endothelial-to-haematopoietic transition has been reduced when H19 fails to appear throughout the early stages of development.

Additionally, promoter hypermethylation and the concurrent down-regulation of multiple significant haematopoietic transcription factors, including Runx1 and Spi1, have been observed in pre-HSCs with H19 deficiency [36]. Chromosome 10 houses the Fas receptor gene (APO-1/CD95), within whose first intron the lncRNA Fas-antisense 1 (Saf) is transcribed from the antisense strand [37]. The maintenance of RBC production is aided by lncRNA Saf. NF-κB and the essential erythropoiesis-related transcription factors GATA-1 and KLF1 have binding sites in their promoter region. During the initial phases of erythroblast proliferation, NF-κB signalling contributes to the suppression of lncRNA Saf expression. Subsequently, at the advanced stages of RBC maturation, lncRNA Saf expression rises in parallel with increased levels of GATA-1 and KLF1, indicating that lncRNA Saf might be modulated by these transcription factors [38].

The expression of transcription factors that control haematopoiesis can be regulated by lncRNAs. By modifying the primary haematopoietic transcription factor, E2A, the silencing of two LncHSCs has had a distinct effect on lineage commitment and HSC self-renewal [11]. An antisense lncRNA has been found to regulate the PU.1 transcription factor expression. It has been demonstrated that this lncRNA inhibits PU.1 expression by controlling its translation [39]. Polyadenylated lncRNAs are expressed by mouse erythroblast cells, megakaryocyte-erythroid progenitor cells and megakaryocytes. GATA1 and TAL1 are two important transcription factors that regulate lncRNAs [33]. The CpG island that surrounds the HOXA1 gene's start site is the source of HOTAIRM1, which is the most prominent intergenic lncRNA that is overexpressed during induced granulocytic differentiation of haematopoietic cells [40]. Chromatin-related lncRNA that is expressed specifically in helper-T1 cells is called linc-MAF-4. There is an inverse relationship between its expression and that of the transcription factor MAF, which is linked to helper-T 2 [41]. 5-aminolevulinic acid synthase 2 (ALAS2) mediates the first rate-limiting step in heme biosynthesis. GATA1 stimulates transcription by binding to canonical GATA sequences found in the promoter and intronic regions of ALAS2. To create a long-range enhancer loop that promotes increased ALAS2 transcription, the GATA site in intron 1 links the intron 8 to the proximal control region of the ALAS2 promoter. The BLZ-21129 bladder carcinoma cell line is where the lncRNA known as Urothelial carcinoma-associated 1 (UCA1) was first discovered and cloned. ALAS2 mRNA stability is modulated by lncRNA UCA1. According to the study by Liu et al., UCA1 functions as a crucial RNA scaffold, facilitating the recruitment of the RNA-binding protein PTBP1 as a cofactor where GATA1 attaches to the UCA1 promoter and controls its expression. ALAS2 mRNA destabilisation and consequent impairment of heme biosynthesis resulting from the lack of UCA1 results in disruption of erythroid development at the proerythroblast stage [42].

The transcription factor Lymphoma/Leukemia-Related Factor (LRF) facilitates γ-globin gene silencing during erythropoiesis by exhibiting a high binding affinity for CpG-rich DNA sequences, thereby influencing the epigenetic transition from foetal haemoglobin (HbF) to adult haemoglobin (HbA). LncRNAs, which exhibit distinct methylation and expression profiles during terminal erythropoiesis and LRF overexpression, which correspond with the haemoglobin changeover from γ-globin to β-globin, further regulate this process. It is yet unclear how exactly this network of non-coding RNAs contributes to the last phase of haematopoiesis and the transition from the foetal to the adult stage of haemoglobin synthesis. Mice with lncRNA deletion in their blood have developed a very severe myeloproliferative disease that is defined as Leukemia and myelofibrosis. Lack of this lncRNA has caused age-dependent aberrations and aberrant maturation in HSCs [43].

2.4 Role of lncRNA in regulating HbF levels

The phrase "β-haemoglobinopathies" describes anomalies in the β-globin gene's synthesis and functioning, which are primarily inherited as an autosomal recessive pattern. β-Thalassaemia is among the most prevalent medical conditions in the human population that are linked to the Haemoglobin subunit beta (HBB). Patients with β-thalassaemia major rely on frequent blood transfusions to stay alive [44]. Therefore, it is crucial to look for possible β-thalassaemia treatments in addition to supporting and encouraging carriers to perform genetic testing and prenatal screening. Foetal haemoglobin (HbF, α2γ2) is currently being reactivated in an attempt to alleviate the symptoms of β-thalassaemia by binding to the extra α-globin chains and acting as HbA (α2β2) [45]. For β-thalassaemia major, the most effective therapeutic approach involves bone marrow transplantation (BMT) from a well-matched donor. The use of this therapeutic strategy has been restricted due to the scarcity of donors who match the human leukocyte antigen (HLA), the prolonged usage of immunosuppressive regimens, and the increased risk of immunological problems. Additionally, despite advancements in chelation therapy and transfusion procedures, numerous unresolved issues have prompted researchers to create novel therapeutic approaches such as foetal haemoglobin induction. For these attempts to be effective, it is speculated that the process of the haemoglobin switching from foetal (HbF) to adult (HbA) must be understood. Despite much research on haemoglobin flipping, some primarily epigenetic, regulatory principles remain to be clarified [46]. lncRNAs, which can act in cis or trans, have recently come to light as important regulators of gene expression. These regulators have already been demonstrated to have regulatory functions in both healthy and diseased erythropoiesis, including heme metabolism, globin switching and regulation, and erythroid cell survival [47].

The β-globin gene cluster is located on chromosome 11 at 11p15.5 in humans and consists of five genes arranged sequentially in the order of their expression during development: embryonic stage HBE (ε-globin gene), foetal stage HBG1 (γ-globin gene, Aγ) and HBG2 (γ-globin gene, Gγ), adult haemoglobin HBD (δ-globin gene), and HBB (β-globin gene). Regulatory elements like the locus control region (LCR), which is essential for controlling gene expression throughout this cluster, flank these genes. The switch from γ-globin in foetal haemoglobin to β-globin in adult haemoglobin during development is clinically significant. High HbF levels reduce the clinical severity of β-thalassaemia. Naturally occurring mutations in the intergenic gap between the γ-globin and δ-globin genes boost the synthesis of γ-globin, indicating the existence of adverse regulators [48]. Various single-nucleotide polymorphisms (SNPs) have been discovered through genome-wide association studies (GWAS), which include the HBB gene cluster at chr11p15, BCL11A at chr2p16, and polymorphisms in the HBS1L-MYB intergenic region (HMIP) at chromosome 6q23 are quantitative trait loci (QTLs) linked to variations in HbF levels. Together, these QTLs explain around 45% of the variation in HbF among different populations. Additionally, transcription factors regulate HBG expression, including ZBTB7A, MYB, BCL11A, KLF1, NR2C1/NR2C2, KDM1α, and CHD4, as well as other cis-acting components like the intergenic region of HBD-HBBP1 [49–51].

Within the HBB gene cluster, gene expression is regulated by the HBS1L-MYB (126 kb) intergenic region, and controls erythroid cell maturation and proliferation, surrounded by erythroid-specific transcription factor binding sites RUNX1, GATA, KLF, TAL1 and LDB1, a regulatory enhancer situated approximately 84 kb upstream of MYB contains a 3-bp deletion polymorphism (rs66650371) [52]. This enhancer is most likely the functional motif responsible for the majority of the effect this QTL has on the HbF level. MYB expression was downregulated, and HBG expression was upregulated as a result of this enhancer's interaction with the MYB promoter. Additionally, using "ENCODE datasets", RNA Pol II binding and a 50-base pair RNA fragment near rs66650371 were found. This led to the identification of a new 1283 bp long non-coding RNA, HMI-lncRNA, also known as the HBS1L-MYB intergenic long non-coding RNA. γ-globin expression was dramatically increased by 200 times when HMI-lncRNA was reduced to half of its normal level, accompanied by a notable increase in γ-globin production. This impact is probably the result of less interaction between the MYB promoter and the distal enhancer, leading to reduced MYB transcription and subsequent upregulation of γ-globin expression [53].

The transcription factor Lymphoma/Leukemia-Related Factor (LRF) facilitates γ-globin gene silencing during erythropoiesis through mechanisms distinct from those employed by BCL11A [43]. LRF exerts its repressive influence on γ-globin genes by recruiting the Nucleosome Remodelling and Deacetylase (NuRD) complex. This recruitment modulates chromatin structure, leading to a reduction in activating histone markers and subsequent inhibition of γ-globin gene transcription. As an epigenetic regulator, LRF has a repressive effect on erythrocyte formation in its latter stages. The haemoglobin transition from HbF to HbA depends on this activity; this process is further modulated by lncRNAs, which exhibit different patterns of methylation and expression during terminal erythropoiesis and overexpression of LRF, which corresponds to the transition of haemoglobin from γ-globin to β-globin. The precise function of the non-coding RNA network in the late stage of haematopoiesis and the concurrent change from the foetal to the adult stage of haemoglobin production is still unknown. In HUDEP-2 cells, a human immortalised erythroid line that requires doxycycline withdrawal for terminal differentiation, the double knockout of LRF/ZBTB7A shows γ-globin expression is significantly elevated. Conversely, erythroid differentiation is delayed in human CD34+ haematopoietic stem and progenitor cells (HSPCs) that lack LRF [54].

Two regulatory regions, 1.5 kb downstream of HBG1 and 3.5 kb upstream of HBG2, are bound by ERF (Ets2 Repressor Factor), an HbF repressor. RP11-196G18.23 lncRNA interacts with the ERF promoter and facilitates the recruitment of DNMT3A (DNA (Cytosine-5)-Methyltransferase 3 Alpha), leading to DNA hypermethylation at the promoter site. This epigenetic modification suppresses ERF expression, enhancing γ-globin production in individuals with β-thalassaemia [47]. The lncRNA Haemoglobin Subunit Beta Pseudogene 1 (HBBP1), located on chromosome 11 near lncRNA BGLT3 and adult δ-globin, plays a pivotal role in γ-globin regulation. In patients with β-thalassaemia, the single-nucleotide polymorphisms rs10128556 and rs2071348 in the second intron of HBBP1 are strongly linked to increased HbF levels [55, 56]. The transcription factor ELK1, belonging to the ETS family, a crucial modulator of γ-globin expression and HBBP1 expression, interacts with HBBP1. ELK1 attaches itself to the proximal promoter of HBBP1 to directly increase γ-globin along with its downstream effectors C-FOS and SRF [57]. The lncRNA BGLT3 facilitates the positive regulation of γ-globin genes by looping through their genomic region in a transcription-dependent manner and interacting with the MED12 subunit. Both the looping mechanism and interaction with the Mediator complex may contribute to the recruitment or stabilisation of RNA Pol II at the γ-globin genes. The increase in RNA Pol II density and transcription, facilitated by BGLT3, highlights its specific role in activating γ-globin gene expression [19, 58].

Under aberrant epigenetic environments, enzymes such as histone deacetylase 1 (HDAC1) and DNA methyltransferase 3 (DNMT3) play a critical role in chromatin landscape configuration, which in turn affects LRF's function in β-globin gene expression. Chromatin immunoprecipitation sequencing (ChIP-seq) investigations have shown that LRF interacts with BCL11A to suppress the production of γ-globin genes by binding to their promoters (HBG2/1). Furthermore, LRF targets the BGLT3 gene during erythropoiesis, encouraging the γ-δ intergenic region to produce BGLT3 lncRNA. The transcriptional transition from γ-globin to β-globin expression is facilitated by this mechanism. LRF binds to the upstream region of the BGLT3 gene and the HBG2/1 promoter regions in K562 cells, especially when haemin/EPO stimulation is present. BCL11A binding to HBG2/1 is required for HbF repression during the haemoglobin transfer from HbF to HbA, but LRF occupancy is required for complete silence. Additionally, by increasing the expression of "BGLT3-lncRNA", LRF's dual interaction with HBG2/1 and BGLT3 amplifies BCL11A-mediated silencing and facilitates subsequent access of chromatin up to the HBB gene. As an indirect chromatin regulator, LRF/ZBTB7A contributes increasingly to the HbF to HbA transition [58]. Additionally, in K562 cells, a Bcr-Abl-positive erythroleukemia model, which is marked by a chromosomal translocation between chromosomes 9 and 22 that results in the hybrid bcr-abl gene and ultimately induces apoptosis, imatinib treatment suppresses the production of BGLT3-lncRNA. According to research, Bcr-Abl is a regulator of BGLT3-lncRNA that most likely inhibits transcription by hypermethylating the DNA [18].

lncRNA, mRNA, and miRNA microarray expression profiling in reticulocytes from β-thalassaemia patients identified several dysregulated transcripts potentially linked to HbF induction. lncRNA NR_120526 and lncRNA T315543 were identified as regulators of Colony Stimulating Factors (CSF2 and CSF3), whereas lncRNA NR_001589, a lncRNA, was proposed as a regulator of HBE1. ε-globin is encoded by HBE1 and is typically active in the yolk sac of the embryo and was found to retain its expression in β-thalassaemia carriers with elevated HbF levels. The haematopoietic cell lineage pathway, critical for blood cell development, highlighted transcription factors such as CSF2 and CSF3 as significant players. These altered transcripts appear closely associated with HbF production, strongly supporting the role of lncRNA NR_001589, lncRNA NR_120526, and lncRNA T315543 as genetic modulators inducing the γ-globin gene and potential therapeutic targets to elevate HbF levels. However, additional confirmation of these findings is required [59]. MALAT1 lncRNA is a lncRNA that is encoded by the MALAT1 gene, which is found on chromosome 11q13.1. It is extensively expressed in many normal cell types and tissues [60]. Elevated MALAT1 expression in β-thalassaemia is closely associated with an earlier onset of disease, a younger age at the initiation of blood transfusions, and an increased incidence of splenomegaly. These associations suggest a heightened rate of haemolysis, potentially contributing to reduced baseline haemoglobin levels. The lncRNA known as MIAT (Myocardial Infarction Associated Transcript) is encoded by the MIAT gene, which is found on chromosome 22q12. It is believed to interact with splicing factor 1 (SF1) via the SF1-binding motif to regulate gene expression and be involved in RNA splicing [61]. Increased severity of β-thalassaemia is linked to the early initiation of the first blood transfusion, which is indicated by higher expression of MIAT levels. The ANRIL gene, which is part of the cluster of genes CDKN2A-CDKN2B on chromosome 9p21, encodes the lncRNA known as ANRIL (Antisense Noncoding RNA in the INK Locus), which is transcribed in the antisense direction concerning the INK4B-ARF-INK4A locus. After DNA damage, the transcription factor E2F1 upregulates its expression. Additionally, ANRIL contributes to the epigenetic repression of nearby genes within the cluster through its interactions with Polycomb Repressive Complex 1 (PRC1) and PRC2 [18, 62]. In the later phases of the DNA damage response, the lncRNA ANRIL suppresses the expression of the INK4A-ARF-INK4B locus by translating to INK4b as an antisense RNA. Notably, the INK4b-ARF-INK4A locus is essential for controlling the stress response, cell division, and haematopoietic differentiation. After DNA repair is finished, ANRIL helps to restore cellular homeostasis. The long non-coding RNAs ANRIL, MIAT, and MALAT1 were notably upregulated in patients with β-thalassaemia, suggesting their potential involvement in the disease's pathophysiology and progression. This upregulation highlights their promise as novel biomarkers for assessing disease severity [63].

The lncRNA H19 is a maternally expressed and paternally imprinted carcino-embryonic gene. In studies involving erythroid differentiation, H19 overexpression resulted in a reduction of γ-globin mRNA levels in K562 cells. However, in human umbilical cord blood-derived erythroid progenitor (HUDEP-2) cells and CD34+ haematopoietic stem cells, γ-globin gene expression was drastically increased. Interestingly, H19 exhibited distinct subcellular localisation patterns: it was primarily nuclear in K562 cells, whereas it was primarily found in the cytoplasm of HUDEP-2 cells. This cytoplasmic localisation in HUDEP-2 cells raises the possibility that H19 functions as a competitive endogenous RNA (ceRNA), affecting microRNA or mRNA interactions to modify γ-globin expression [64]. Microarray profiling conducted on seven β-thalassaemia patients with elevated HbF levels revealed significantly greater levels of average expression of the lncRNA uc002fcj.1 and lncRNA NR_001589. A positive correlation was identified between these lncRNAs and HbF levels. Additional functional research of lncRNA uc002fcj.1 and lncRNA NR_001589 could provide deeper insights into the molecular mechanisms underlying β-thalassaemia. This study suggests that these lncRNAs may function as competing endogenous RNAs (ceRNAs), enhancing the expression of HBG1/2 by sponging specific miRNAs, thereby contributing to HbF induction in β-thalassaemia (**Figure 2**) [65].

The most recent studies have investigated the use of antisense oligonucleotides for miRNA suppression, "nano vector-encapsulated miRNA mimics" for miRNA restoration, and small interfering RNAs (siRNAs). Although the FDA has approved siRNA-based therapeutics, preclinical and early-phase clinical research is still being conducted to examine lncRNA-based treatments, miRNA inhibitors and mimics. Additionally, two siRNAs that target non-coding RNAs—circRNAs and

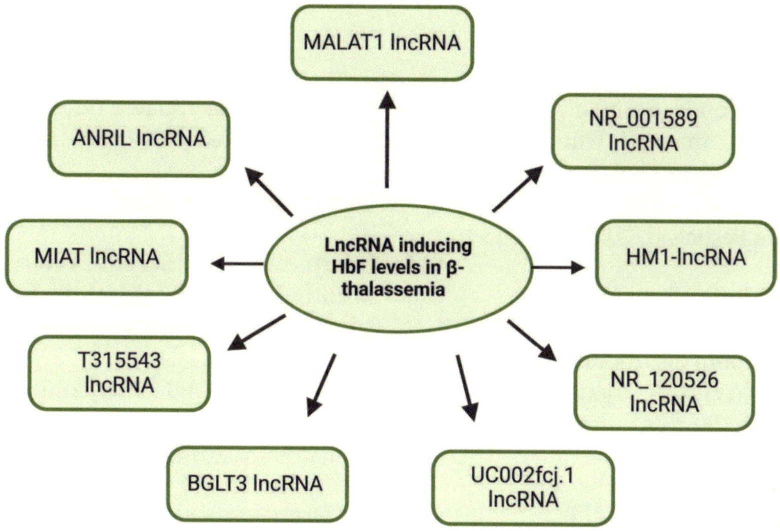

Figure 2.
List of lncRNAs involved in inducing HbF levels in β-thalassaemia.

lncRNAs—have advanced to clinical trials from preclinical research. For instance, LNP-based siRNA targeting the LINC01257 lncRNA has been suggested as a novel and secure treatment approach for acute myeloid Leukemia in the paediatric group [66]. To successfully continue into more advanced stages of clinical testing, challenges such as effective delivery, off-target effects, and safety concerns need to be addressed with further attention. The investigation of lncRNAs as biomarkers for the diagnosis, prognosis, and therapy of β-thalassaemia may advance with the help of this knowledge in the design of future studies [67].

Author details

Sumaiya Kausar S. Kalaigar[1], Prashant Vishwanath[1] and Akila Prashant[1,2]*

1 Department of Biochemistry, JSS Academy of Higher Education and Research, Mysuru, India

2 Department of Medical Genetics, JSS Academy of Higher Education and Research, Mysuru, India

*Address all correspondence to: prashantv@jssuni.edu.in

References

[1] Safiullah Jan N, Safiullah Jan Z, Taj Niazi H, Zameer H, Humayun S. Frequency of Beta Thalassemia Major among Children with Iron Deficiency Anemia in Peshawar. 2022;**15**(1):35

[2] Modell B, Darlison M. Public health reviews global epidemiology of haemoglobin disorders. Bulletin of the World Health Organization. 2008;**86**(6):480-484

[3] Biswas B, Naskar N, Basu K, Dasgupta A, Basu R, Paul B. Transfusion-transmitted infections, its risk factors and impact on quality of life: An epidemiological study among β-thalassemia major children. Asian Journal of Transfusion Science. 2022;**16**(1):99-105

[4] Galanello R, Origa R. Open access review BioMed central beta-thalassemia. Orphanet Journal of Rare Diseases. 2010;**5**:1-5

[5] Pinto VM, Poggi M, Russo R, Giusti A, Forni GL. Management of the aging beta-thalassemia transfusion-dependent population – The Italian experience. Blood Reviews. 2019;**38**:100595

[6] Sankaran VG, Xu J, Orkin SH. Advances in the understanding of haemoglobin switching: Review. British Journal of Haematology. 2010;**149**(2):181-194

[7] Frati G, Miccio A. Genome editing for β-hemoglobinopathies: Advances and challenges. Journal of Clinical Medicine. 2021;**10**:1-17

[8] Bou-Fakhredin R, De Franceschi L, Motta I, Cappellini MD, Taher AT. Pharmacological induction of fetal hemoglobin in β-Thalassemia and sickle cell disease: An updated perspective. Pharmaceuticals. 2022;**15**(6):753

[9] Amato A, Cappabianca MP, Perri M, Zaghis I, Grisanti P, Ponzini D, et al. Interpreting elevated fetal hemoglobin in pathology and health at the basic laboratory level: New and known γ- gene mutations associated with hereditary persistence of fetal hemoglobin. International Journal of Laboratory Hematology. 2014;**36**:13-19

[10] Diamantopoulos MA, Tsiakanikas P, Scorilas A. Non-coding RNAs: The riddle of the transcriptome and their perspectives in cancer. Annals of Translational Medicine. 2018;**6**(12):241-241

[11] Ghafouri-Fard S, Niazi V, Taheri M. Role of miRNAs and lncRNAs in hematopoietic stem cell differentiation. Non-Coding RNA Research. 2021;**6**:8-14

[12] Sharma S, Houfani AA, Foster LJ. Pivotal functions and impact of long con-coding RNAs on cellular processes and genome integrity. Journal of Biomedical Science. 2024;**31**(1):52

[13] Birney E, Stamatoyannopoulos JA, Dutta A, Guigó R, Gingeras TR, Margulies EH, et al. Identification and analysis of functional elements in 1% of the human genome by the ENCODE pilot project. Nature. 2007;**447**(7146):799-816

[14] Mattick JS, Amaral PP, Carninci P, Carpenter S, Chang HY, Chen LL, et al. Long non-coding RNAs: Definitions, functions, challenges and recommendations. Nature Reviews. Molecular Cell Biology. 2023;**24**(6):430-447

[15] JTY K, Colognori D, Lee JT. Long noncoding RNAs: Past, present, and future. Genetics. 2013;**193**:651-669

[16] Kulczyńska K, Siatecka M. A regulatory function of long non-coding RNAs in red blood cell development. Acta Biochimica Polonica. Polskie Towarzystwo Biochemiczne. 2016;**63**:675-680

[17] Amaral PP, Clark MB, Gascoigne DK, Dinger ME, Mattick JS. LncRNAdb: A reference database for long noncoding RNAs. Nucleic Acids Research. 2011;**39**(Suppl. 1):D146-51

[18] Guo G, Kang Q, Zhu X, Chen Q, Wang X, Chen Y, et al. A long noncoding RNA critically regulates Bcr-Abl-mediated cellular transformation by acting as a competitive endogenous RNA. Oncogene. 2014;**34**(14):1768-1779

[19] Ivaldi MS, Diaz LF, Chakalova L, Lee J, Krivega I, Dean A. Fetal g-globin genes are regulated by the BGLT3 long noncoding RNA locus. Blood. 2018;**132**(18):1963-1973

[20] Bridges MC, Daulagala AC, Kourtidis A. LNCcation: lncRNA localization and function. Journal of Cell Biology. 2021;**220**(2):e202009045

[21] Statello L, Guo CJ, Chen LL, Huarte M. Gene regulation by long non-coding RNAs and its biological functions. Nature Reviews Molecular Cell Biology. Nature Research. 2021;**22**:96-118

[22] Guttman M, Amit I, Garber M, French C, Lin MF, Feldser D, et al. Chromatin signature reveals over a thousand highly conserved large non-coding RNAs in mammals. Nature. 2009;**458**(7235):223-227

[23] Quinn JJ, Chang HY. Unique features of long non-coding RNA biogenesis and function. Nature Reviews Genetics. 2016;**17**:47-62

[24] Achkar NP, Cambiagno DA, Manavella PA. miRNA biogenesis: A dynamic pathway. Trends in Plant Science. 2016;**21**:1034-1044

[25] Derrien T, Johnson R, Bussotti G, Tanzer A, Djebali S, Tilgner H, et al. The GENCODE v7 catalog of human long noncoding RNAs: Analysis of their gene structure, evolution, and expression. Genome Research. 2012;**22**(9):1775-1789

[26] Jeong M, Goodell MA. Noncoding regulatory RNAs in hematopoiesis. In: Current Topics in Developmental Biology. San Diego, California: Academic Press Inc.; 2016. pp. 245-270

[27] Liu Y, Ding W, Yu W, Zhang Y, Ao X, Wang J. Long non-coding RNAs: Biogenesis, functions, and clinical significance in gastric cancer. Molecular Therapy Oncolytics. 2021;**23**:458-476

[28] Zivot A, Lipton JM, Narla A, Blanc L. Erythropoiesis: Insights into pathophysiology and treatments in 2017. Molecular Medicine. 2018;**24**:1-5

[29] Lee SH, Crocker PR, Westaby S, Key N, Mason DY, Gordon IS, et al. Isolation and Immunocytochemical Characterization of Human Bone Marrow Stromal Macrophages in Hemopoietic Clusters. 1988;**168**(3):1193-1198

[30] Gregory CJ, Eaves AC. Human marrow cells capable of erythropoietic differentiation In vitro: Definition of three erythroid colony responses. Blood. 1977;**49**:855-864

[31] Wang C, Wu X, Shen F, Li Y, Zhang Y, Yu D. Shlnc-EC6 regulates murine erythroid enucleation by Rac1-PIP5K pathway. Development, Growth and Differentiation. 2015;**57**(6):466-473

[32] Alvarez-Dominguez J, R, Hu W, Yuan B, Shi J, Park SS, Gromatzky AA,

et al. Global discovery of erythroid long noncoding RNAs reveals novel regulators of red cell maturation. Red Cells, Iron, And Erythropoiesis. 2014;**123**(4):570-581

[33] Paralkar VR, Mishra T, Luan J, Yao Y, Kossenkov AV, Anderson SM, et al. Lineage and species-specific long noncoding RNAs during erythro-megakaryocytic development. Blood. 2014;**123**(12):1927-1937

[34] Hu J. lncRNA CRNDE204 Regulates Erythropoiesis and Myelopoiesis by Binding to PUS1. Harvard medical School; 2021.

[35] Luo M, Jeong M, Sun D, Park HJ, Rodriguez BAT, Xia Z, et al. Long non-coding RNAs control hematopoietic stem cell function. Cell Stem Cell. 2015;**16**(4):426-438

[36] Venkatraman A, He XC, Thorvaldsen JL, Sugimura R, Perry JM, Tao F, et al. Maternal imprinting at the H19-Igf2 locus maintains adult haematopoietic stem cell quiescence. Nature. 2013;**500**(7462):345-349

[37] De YM, Hong CC, Lai GM, Cheng AL, Lin YW, Chuang SE. Identification and characterization of a novel gene Saf transcribed from the opposite strand of Fas. Human Molecular Genetics. 2005;**14**(11):1465-1474

[38] Villamizar O, Chambers CB, Mo YY, Torry DS, Hofstrand R, Riberdy JM, et al. Fas-antisense long noncoding RNA is differentially expressed during maturation of human erythrocytes and confers resistance to Fas-mediated cell death. Blood Cells, Molecules and Diseases. 2016;**58**:57-66

[39] Godet AC, Roussel E, Laugero N, Morfoisse F, Lacazette E, Garmy-Susini B, et al. Translational control by long non-coding RNAs. Biochimie. 2024;**217**:42-53

[40] Zhang X, Lian Z, Padden C, Gerstein MB, Rozowsky J, Snyder M, et al. A myelopoiesis-associated regulatory intergenic noncoding RNA transcript within the human HOXA cluster. Blood. 2009;**113**(11):2526-2534

[41] Zhou J, Xu J, Zhang L, Liu S, Ma Y, Wen X, et al. Combined single-cell profiling of lncRNAs and functional screening reveals that H19 is pivotal for embryonic hematopoietic stem cell development. Cell Stem Cell. 2019;**24**(2):285-298.e5

[42] Liu J, Li Y, Tong J, Gao J, Guo Q, Zhang L, et al. Long non-coding RNA-dependent mechanism to regulate heme biosynthesis and erythrocyte development. Nature Communications. 2018;**9**(1):4386

[43] Athanasopoulou K, Chondrou V, Xiropotamos P, Psarias G, Vasilopoulos Y, Georgakilas GK, et al. Transcriptional repression of lncRNA and miRNA subsets mediated by LRF during erythropoiesis. Journal of Molecular Medicine. 2023;**101**(9):1097-1112

[44] Lattanzi F, Bellotti P, Picano E, Chiarella F, Mazzarisi A, Melevendi C, et al. Quantitative Ultrasonic Analysis of Myocardium in Patients with Thalassemia Major and Iron Overload. 1993;**87**(3):748-754

[45] Wienert B, Martyn GE, Kurita R, Nakamura Y, Quinlan KGR, Crossley M. KLF1 drives the expression of fetal hemoglobin in British HPFH. Blood. 2017;**130**(6):803-807

[46] Rahimmanesh I, Boshtam M, Kouhpayeh S, Khanahmad H, Dabiri A, Ahangarzadeh S, et al. Gene

editing-based technologies for beta-hemoglobinopathies treatment. Biology. 2022;**11**(6):862

[47] Bao X, Gao Y, Wang Z, Ye Y, Chen D, Zuo Y, et al. Activation of γ-globin expression by LncRNA-mediated ERF promoter hypermethylation in β-thalassemia. Clinical Epigenetics. 2024;**16**(1):12

[48] Moleirinho A, Seixas S, Lopes AM, Bento C, Prata MJ, Amorim A. Evolutionary constraints in the β-globin cluster: The signature of purifying selection at the δ-globin (HBD) locus and its role in developmental gene regulation. Genome Biology and Evolution. 2013;**5**(3):559-571

[49] Masuda T, Wang X, Maeda M, Canver MC, Sher F, Funnell APW, et al. Gene regulation: Transcription factors LRF and BCL11A independently repress expression of fetal hemoglobin. Science (1979). 2016;**351**(6270):285-289

[50] Kang Y, Kim YW, Yun J, Shin J, Kim AR. KLF1 stabilizes GATA-1 and TAL1 occupancy in the human β-globin locus. Biochimica et Biophysica Acta-Gene Regulatory Mechanisms. 2015;**1849**(3):282-289

[51] Wongborisuth C, Chumchuen S, Sripichai O, Anurathaphan U, Sathirapongsasuti N, Songdej D, et al. Down-regulation of the transcriptional repressor ZNF802 (JAZF1) reactivates fetal hemoglobin in β0-thalassemia/HbE. Scientific Reports. 2022;**12**(1):4952

[52] Stadhouders R, Aktuna S, Thongjuea S, Aghajanirefah A, Pourfarzad F, Van IJcken W, et al. HBS1L-MYB intergenic variants modulate fetal hemoglobin via long-range MYB enhancers. Journal of Clinical Investigation. 2014;**124**(4):1699-1710

[53] Morrison TA, Wilcox I, Luo HY, Farrell JJ, Kurita R, Nakamura Y, et al. A long noncoding RNA from the HBS1L-MYB intergenic region on chr6q23 regulates human fetal hemoglobin expression. Blood Cells, Molecules and Diseases. 2018;**69**:1-9

[54] Chondrou V, Stavrou EF, Markopoulos G, Kouraklis-Symeonidis A, Fotopoulos V, Symeonidis A, et al. Impact of ZBTB7A hypomethylation and expression patterns on treatment response to hydroxyurea. Human Genomics. 2018;**12**(1):1-4

[55] Bhanushali AA, Himani K, Patra PK, Das BR. Hb F levels in Indian sickle cell patients and association with the HBB locus variant rs10128556 (C>T), and the HBG XmnI (Arab-Indian) variant. Hemoglobin. 2017;**41**(4-6):317-320

[56] Ma Y, Liu S, Gao J, Chen C, Zhang X, Yuan H, et al. Genome-wide analysis of pseudogenes reveals HBBP1's human-specific essentiality in erythropoiesis and implication in β-thalassemia. Developmental Cell. 2021;**56**(4):478-493. e11

[57] Ma SP, Xi HR, Gao XX, Yang JM, Kurita R, Nakamura Y, et al. Long noncoding RNA HBBP1 enhances γ-globin expression through the ETS transcription factor ELK1. Biochemical and Biophysical Research Communications. 2021;**552**:157-163

[58] Chondrou V, Shaukat AN, Psarias G, Athanasopoulou K, Iliopoulou E, Damanaki A, et al. LRF promotes indirectly advantageous chromatin conformation via BGLT3-lncRNA expression and switch from fetal to adult hemoglobin. International Journal of Molecular Sciences. 2022;**23**(13):7025

[59] Lai K, Jia S, Yu S, Luo J, He Y. Genome-Wide Analysis of Aberrantly

Expressed lncRNAs and miRNAs with Associated co-Expression and ceRNA Networks in β-Thalassemia and Hereditary Persistence of Fetal Hemoglobin. 2017. Available from: www.impactjournals.com/oncotarget

[60] Zhang B, Arun G, Mao YS, Lazar Z, Hung G, Bhattacharjee G, et al. The lncRNA malat1 is dispensable for mouse development but its transcription plays a cis-regulatory role in the adult. Cell Reports. 2012;**2**(1):111-123

[61] Tsuiji H, Yoshimoto R, Hasegawa Y, Furuno M, Yoshida M, Nakagawa S. Competition between a noncoding exon and introns: Gomafu contains tandem UACUAAC repeats and associates with splicing factor-1. Genes to Cells. 2011;**16**(5):479-490

[62] Wan G, Mathur R, Hu X, Liu Y, Zhang X, Peng G, et al. Long non-coding RNA ANRIL (CDKN2B-AS) is induced by the ATM-E2F1 signaling pathway. Cellular Signalling. 2013;**25**(5):1086-1095

[63] Fakhr-Eldeen A, Toraih EA, Fawzy MS. Long non-coding RNAs MALAT1, MIAT and ANRIL gene expression profiles in beta-thalassemia patients: A cross-sectional analysis. Hematology (United Kingdom). 2019;**24**(1):308-317

[64] Xie D, Han Y, Zhang W, Wu J, An B, Huang S, et al. Long non-coding RNA H19 leads to Upregulation of γ-globin gene expression during Erythroid differentiation. Hemoglobin. 2024;**48**(1):4-14

[65] Jia W, Jia S, Chen P, He Y. Construction and analysis of a long non-coding RNA (lncRNA)-associated ceRNA network in β-thalassemia and hereditary persistence of fetal hemoglobin. Medical Science Monitor. 2019;**25**:7079-7086

[66] Connerty P, Moles E, de Bock CE, Jayatilleke N, Smith JL, Meshinchi S, et al. Development of siRNA-loaded lipid nanoparticles targeting long non-coding RNA LINC01257 as a novel and safe therapeutic approach for t(8,21) pediatric acute myeloid leukemia. Pharmaceutics. 2021;**13**(10):1681

[67] Grillone K, Caridà G, Luciano F, Cordua A, Di Martino MT, Tagliaferri P, et al. A systematic review of non-coding RNA therapeutics in early clinical trials: A new perspective against cancer. Journal of Translational Medicine. 2024;**22**:731

Chapter 6

Role of Circulatory miRNAs in Cardiovascular Disorder: Anti-Sense RNA as Therapeutic Targets

Wahid Ali, Priyanka Sharma, Nida Shabbir and Swati Kumari

Abstract

Consequently, there is a significant need for biomarkers capable of assessing the early stages of atherosclerosis and the likelihood of developing coronary artery disease (CAD). MicroRNAs (miRNAs) are brief, non-coding RNA molecules with a single strand that controls cellular processes by degrading and repressing the translation of messenger RNAs (mRNAs). miRNAs circulating in the bloodstream regulate gene expression post-transcriptionally, playing a crucial role in cell growth, differentiation, and metabolism, as well as in the pathophysiology of numerous disorders. Studies conducted both *in vitro* and *in vivo*, involving the manipulation of miRNA expression, have demonstrated its significant impact on heart function regulation, particularly in areas such as cardiac growth, hypertrophy, and failure. Earlier research has shown that individuals with heart failure (HF) exhibit lower plasma concentrations of miR-126 compared to healthy subjects, and an inverse relationship with brain natriuretic peptide (BNP) was observed, establishing miR-126 as a reliable indicator of heart failure.

Keywords: CVD, miRNA, anti-sense, RNA atherosclerosis, angiography

1. Introduction

CVDs, which is short for cardiovascular diseases, are one of the most prevalent and fatal health complications in the globe. Coronary artery disease begins with atherosclerosis, which in turn precipitates a series of pathological events that lead to significant cardiovascular events. Hence, biomarkers that can capture the first step of atherosclerosis and risk for CAD occurrence are quite valuable. Post-transcriptionally acting, short non-coding RNAs—microRNAs or miRNAs—are regarded to be suitable candidate biomarkers from this point of view. This chapter focuses on circulating miRNAs in cardiovascular diseases and considers their prospection for treatment through anti-sense RNA approaches. MicroRNAs have emerged as promising biomarkers for early detection and monitoring of cardiovascular diseases, particularly in the context of atherosclerosis and coronary artery disease. Their presence in circulation and stability in various biological fluids make them attractive targets for non-invasive diagnostic and prognostic tools. Furthermore, the potential of anti-sense RNA approaches targeting specific miRNAs offers a novel therapeutic strategy for managing CVDs, potentially interrupting disease progression at its earliest stages. While microRNAs show promise as biomarkers and therapeutic targets for cardiovascular diseases, their clinical application faces challenges such as standardization of detection methods, identification of disease-specific miRNA signatures, and development of effective delivery systems for anti-sense RNA therapies.

2. Overview of microRNAs

2.1 Definition and function

Short, microRNAs, which are non-coding RNA molecules consisting of about 22 nucleotides in a single strand, play a role in regulating various biological functions. Their primary mechanism of action involves attaching to the 3′ untranslated regions (UTRs) of target mRNAs, subsequently causing their degradation or inhibiting their translation. Thus, miRNAs can modulate cellular functions such as proliferation, differentiation, apoptosis, and metabolism without compromising the fidelity of their input mRNAs [1, 2]. Research has shown that microRNAs play significant roles in numerous biological and pathological processes, encompassing embryonic growth, immune system responses, and the advancement of cancer. Their ability to regulate gene expression at the post-transcriptional level makes them powerful modulators of cellular behavior and potential therapeutic targets. Research shows that miRNAs exist inside extracellular vesicles enabling their use in cellular communication and their potential use as diagnostic and prognostic biomarkers.

2.2 Biogenesis of miRNAs

DNA transcription by RNA polymerase II starts the formation of pri-miRNA, which becomes the primary miRNA. The microprocessor complex, which includes Drosha and DGCR8, functions within the nucleus to transform pri-miRNAs into pre-miRNAs. Full development of miRNAs leads them to the RNA-induced silencing complex (RISC), which includes Argonaute proteins [1, 3]. Between these stages, Dicer modifies pre-miRNAs and produces mature miRNAs. In RISC assembly, the miRNA receives direction toward its specific mRNA target, so it properly attaches in the 3′ untranslated region. Through the binding of miRNA to its target sequence, the level of compatibility determines mRNA destabilization and translation suppression [3]. The specific section of the seed functions by detecting similar sequences on target mRNA molecules for proper attachment. Several aspects determine how well miRNA controls gene expression including the number and placement of binding sites found in target mRNA as well as which RNA-binding proteins enhance or disrupt miRNA binding.

3. Role of circulating miRNAs in cardiovascular disorders

Scientists have recently identified circulating microRNAs (miRNAs) as crucial elements in cardiovascular disease study. Research indicates that these small non-coding ribonucleic acid molecules are vital in controlling cellular and genetic operations inside the cardiovascular system. The stability of miRNAs circulating in body fluids and blood makes them attractive as biomarkers for cardiovascular disease diagnosis and management purposes. Scientific research demonstrates how unique miRNA profiles can determine the development of multiple heart conditions, starting from myocardial infarction to heart failure and atherosclerosis. Circulating miRNA detection requires no invasive procedures, which makes them suitable for medical applications so healthcare providers may do without more complex diagnostic measures. Toward better treatment of cardiovascular diseases, scientists are currently studying how to manage miRNA expression through interventions to produce beneficial outcomes [4].

3.1 Circulating miRNAs as biomarkers in cardiovascular disorders

Many diseases including cardiovascular disorders can be diagnosed using body fluid miRNAs due to their stable nature in serum and plasma. Certain levels show that a cardiovascular system suffers from pathology. HF patients show decreased plasma concentrations of miR-126 when compared to people who do not have heart failure. Research indicates that miR-126 shows an inverse relationship with brain natriuretic peptide BNP, which serves as a primary marker of heart failure therefore making it a promising diagnostic and prognostic biomarker for HF [4]. Scientific findings indicate that miR-126 has value as a biomarker substitute for BNP in heart failure clinical evaluation. The potential role of miRNAs extends to cardiovascular biomarker detection of myocardial infarction and atherosclerosis, apart from heart failure. Further research is needed to verify miRNA utility as diagnostic indicators and develop standard evaluation techniques for implementing their use in cardiovascular disease management. The use of miRNAs as biomarkers shows promise in diagnosing and predicting multiple cardiovascular diseases as well as heart failure cases. Scientists conduct present-day research to determine the diagnostic accuracy of miRNA tests for tracking cardiovascular disease progression at multiple stages. Research shows that integrating microRNA biomarkers into healthcare practice would advance patient-specific treatment methods for people dealing with cardiovascular diseases.

4. Mechanisms of action in cardiovascular pathophysiology

Cardiac development together with hypertrophy and heart failure functions under the regulation of these miRNAs. These miRNAs place them in a successful position to manage pathways significant for cardiovascular disease progression including inflammation and apoptosis with fibrosis development. The microRNA miR-21 has been associated with cardiac hypertrophy, according to studies that show its capacity to inactivate phosphatase and tensin homolog (PTEN) while promoting survival signaling pathways [5]. Medical tests indicate miR-21 expression control shows promising results for treating cardiac hypertrophy together with heart failure patients. The cellular activities of cardiomyocytes are influenced by additional microRNAs, which have been proven to control both their proliferation and differentiation processes. New therapies for heart conditions show promise because researchers have observed that microRNAs form complex relationships with cardiovascular disease mechanisms.

Three miRNAs have been identified to affect cardiac remodeling alongside miR-21: miR-1, miR-133, and miR-208. Scientists determined that MiR-1 maintains control of cardiomyocyte proliferation and apoptosis and miR-133 blocks cardiac hypertrophy by targeting muscle growth signaling pathways [6]. Scientists have discovered that miR-208 serves as a regulator of cardiac stress responses by controlling intronic part within α-myosin heavy chain gene found at Ref. [7]. Research data demonstrates how miRNAs sustain complex relationships with multiple cellular processes found in cardiac tissue. Multiple scientific laboratories across the world study these miRNAs to develop them as potential therapeutic targets against cardiovascular diseases. Cancer pathologies affect miR-499 expression levels in cardiac cells, where this microRNA maintains an essential role in cardiac differentiation and remodeling [6]. Single miRNAs do not function alone in heart regulation because they perform their complex regulation through collective mechanisms that precisely control both gene expression and cellular responses throughout cardiac tissue. Understanding the complex networks of interactions and cardiac function effects will support the future development of novel therapeutic solutions for heart disease treatment. The research shows that MiR-499 has essential functions during cardiac differentiation and remodeling while displaying altered expression levels in numerous cardiovascular diseases [6]. The discovery of miR-177-3p increases the complexity of miRNA regulatory mechanisms within heart tissue. The collective function of multiple miRNAs that control gene expression patterns requires thorough research into molecular interactions.

5. *In vitro* and *in vivo* studies

Researchers have proven the regulatory effect of specific miRNAs on cellular stress responses by testing cardiomyocyte cell lines *in vitro*. The research showed that the overexpression of miR-21 in cardiomyocytes made them become resistant to apoptosis under oxidative stress which indicates therapeutic potential [5]. At the same time, research showed miR-1 and miR-133 controlled how cardiomyocytes responded to hypertrophy and cardiac remodeling under stress conditions [6]. The stress response mechanism of the heart and its fibrotic development are both influenced by miR-208a [7]. The identified potential of using miRNA-based treatments to manage cardiac stress reactions and boost heart performance becomes more promising through both lab-based research and animal testing. Knockout mice without specific miRNAs display modified heart functions and stress reactions, which helps researchers better understand cardiovascular molecule functions during health. Clinical tests have initiated research to evaluate miRNA-based treatments for patients with heart failure and other cardiac conditions and the assessments yield promising indicators.

Researchers have further proved the roles of miRNAs in disease progression using *in vivo* heart failure models in animal studies. The cardiac fibrosis regulatory function of miR-29a makes it a potential therapeutic target for treating heart failure, according to Van Rooij et al. [7]. Research about miRNAs in heart failure development has sparked new interest in creating miRNA-based therapeutic strategies for heart failure treatment. Multiple research studies involving animals have demonstrated successful results when targeting specific microRNA molecules in order to enhance cardiac performance and lower fibrotic issues. Effective clinical treatments based on these findings face significant barriers because delivery strategies and unintended consequences must receive proper attention. New research investigates different strategies to develop miRNA-based heart failure therapeutic options despite existing implementation hurdles. Nanoparticle delivery systems show great potential to enhance the accuracy of miRNA modulator transport to cardiac tissue. Research investigates combination therapy approaches to target multiple miRNAs that impact heart failure development for possible synergistic treatment effects.

6. Anti-sense RNA as therapeutic targets

6.1 Therapeutic potential of anti-sense RNA

Anti-sense RNA possesses technology that uses designed oligonucleotides to match specific miRNAs, which results in their functional inhibition. The method shows potential to reverse established pathological processes of cardiovascular disorders. Through its inhibition mechanism, anti-sense RNA can stop disease-related miRNAs from expressing while restoring normal gene expression patterns, which leads to improved cardiac function.

Research data indicates promising results from this new approach because it successfully controls specific miRNA targets that contribute to cardiac reshaping and scarring. Scientific trials at present examine anti-sense RNA treatments as a therapeutic approach for patients who have heart disorders such as congestive heart failure and artery plaque buildup. The successful implementation of these therapies should transform cardiovascular disease management by enabling doctors to treat specific molecular pathways for each patient. Custom miRNA inhibition capabilities present clinicians with a strong tool to create treatments that match individual patients' genetic compositions. Anti-sense RNA technology has potential benefits compared to conventional pharmaceutical treatments by minimizing side effects as well as improving long-term care results for patients who have cardiovascular conditions.

Research into anti-sense RNA therapies for cardiovascular conditions has established a major breakthrough for precision medical care. Anti-sense RNA therapies show promise to develop better treatments that produce fewer systemic side effects than traditional treatments. The ongoing research will likely result in higher numbers of anti-sense RNA-based treatments entering clinical trials which will eventually become accessible in the market for cardiovascular diseases. The emergence of anti-sense RNA therapeutic approaches for cardiovascular diseases stands as a major achievement in precision medicine discipline. The treatment method can achieve better results offering patients reduced systemic side effects above traditional therapeutic methods. The field of anti-sense RNA research continues to advance so it is expected more therapies in this category will undergo clinical trials before reaching the market for cardiovascular treatments [8].

7. Application in cardiovascular disease

Researchers have studied different cardiovascular environments to apply therapeutic anti-sense RNA that targets specific miRNAs. The study of anti-sense oligonucleotides directed against miR-21 leads to substantial reductions in cardiac hypertrophy along with fibrosis in animal subjects [8]. The therapeutic application of miRNA expression regulation through this method resulted in reduced cardiac fibrosis and better heart performance [7]. Animal cardiovascular health at a systemic level may improve through anti-sense oligonucleotide treatment of blood miRNA, which reduces circulating levels of these molecules. Medical experts interpret these results as evidence that anti-sense RNA therapy, which targets particular miRNAs, presents promising possibilities for cardiovascular disease treatment. More investigations are necessary to discover complete mechanisms of action behind these therapies and enhance their clinical delivery efficiency. Improved cardiovascular patient care will result from using more specific anti-sense oligonucleotides that deliver better therapeutic effects and produce minimal side effects. Doctors are performing clinical trials to establish safety measures and effectiveness rates of anti-sense RNA treatments for human cardiovascular patients. Scientists use these studies to convert successful animal model results into applications in human patients. The research seeks to combine anti-sense RNA therapy with standard cardiovascular treatments in order to deliver better therapeutic results for patients.

8. Challenges and considerations in therapeutic development

Although the potential of anti-sense RNA as a therapeutic strategy for cardiovascular disease appears promising, challenges must be met. The first problem is a delivery issue as anti-sense oligonucleotides still need to reach the tissues where they are going to work. The stability and uptake of these molecules in the cardiovascular system into the desired target site are necessary for the therapeutic efficacy. They must be delivered through effective delivery systems. Secondly, off-target effects and by products need to be carefully watched for. Most miRNAs target many mRNAs and so inhibition of a particular miRNA could have unintended consequences with respect to other pathways. Extended preclinical investigations must occur before using anti-sense oligonucleotide therapies in cardiovascular medicine since they need evaluation in terms of safety and treatment performance abilities. The potential immunological reactions to these foreign therapeutic molecules require thorough consideration. Scientists must find ways to match appropriate timing and proper dosage levels of delivery to obtain therapeutic benefits and avoid unwanted side effects. A comprehensive study program should establish both effect duration and delayed responses following extended anti-sense RNA therapy applications in cardiovascular medicine. Research on anti-sense RNA therapies for cardiovascular diseases requires ongoing development due to their remaining challenges. Research translation of promising anti-sense RNA treatments requires combined work from academic establishments together with pharmaceutical businesses and regulatory organizations to address current obstacles. Forthcoming innovations in drug delivery methods combined with better oligonucleotide chemistry research plus advanced preclinical model development will enable the complete realization of this therapeutic potential.

9. Future directions

9.1 Biomarker development

Research into circulating miRNAs represents a key academic field for developing biomarkers that detect cardiovascular conditions early and assess cardiovascular risks. Two modern technology platforms, namely Next Generation Sequencing and high-throughput quantitative PCR, enable scientists to discover cardiovascular-related novel miRNAs. Standardization of biomarkers must occur before they become practical for daily clinic use, while researchers must determine strong associations between biomarkers and clinical situations. Scientific research must identify and confirm the clinical worth of particular miRNA biomarkers which would serve in cardiovascular disease detection and risk evaluation. The confirmation of these biomarkers' performance characteristics as testing tools requires extensive studies conducted on various patient demographics. Standardization in miRNA measurement methodologies alongside reference value creation forms the basic requirement for applying these biomarkers in clinical environments. The clinical value of using particular miRNA biomarkers for cardiovascular disease diagnosis requires more research to confirm its applications. Mass-scale prospective studies of these biomarkers are essential to measure their performance rates when dealing with different patient demographic groups. Society will benefit from normalized miRNA measurement approaches alongside baseline reference definitions to achieve broad clinical acceptance of these biomarkers.

9.2 Therapeutic innovations

Future work should look to optimize anti-sense RNA techniques to make them more specific and delivery methods. Alternatively, nanotechnology and lipid-based carriers may provide effective delivery of oligonucleotides to target tissues, improving therapeutic outcomes in cardiovascular diseases. These advancements in delivery methods could potentially enhance the efficacy of anti-sense RNA therapies by increasing their bioavailability and reducing off-target effects. Additionally, combining anti-sense RNA techniques with other emerging technologies, such as CRISPR-Cas9 gene editing, may offer synergistic benefits in treating complex cardiovascular disorders. Further research is also needed to elucidate the long-term safety and potential side effects of these novel therapeutic approaches in clinical settings. Longitudinal studies should be conducted to assess the long-term efficacy and safety profiles of anti-sense RNA therapies in cardiovascular patients. Moreover, personalized medicine approaches could be explored to tailor these therapies based on individual genetic profiles and disease characteristics. Finally, investigating the potential of anti-sense RNA techniques in preventing cardiovascular diseases, rather than just treating existing conditions, could open up new avenues for preventive cardiology.

9.3 Personalized medicine

Research into circulating miRNAs represents a key academic field for developing biomarkers which detect cardiovascular conditions early and assess cardiovascular risks. Two modern technology platforms namely Next Generation Sequencing and high-throughput quantitative PCR enable scientists to discover cardiovascular-related novel miRNAs. Standardization of biomarkers must occur before they become practical for daily clinic use, while researchers must determine strong associations between biomarkers and clinical situations. Scientific research must identify and confirm the clinical worth of particular miRNA biomarkers which would serve in cardiovascular disease detection and risk evaluation. The confirmation of these biomarkers' performance characteristics as testing tools requires extensive studies conducted on various patient demographics. Standardization in miRNA measurement methodologies alongside reference value creation forms the basic requirement for applying these biomarkers in clinical environments. The clinical value of using particular miRNA biomarkers for cardiovascular disease diagnosis requires more research to confirm its applications. Mass-scale prospective studies of these biomarkers are essential to measure their performance rates when dealing with different patient demographic groups. Society will benefit from normalized miRNA measurement approaches alongside baseline reference definitions to achieve broad clinical acceptance of these biomarkers.

10. Conclusion

Circulating miRNAs represent a novel class of biomarkers with great potential to contribute to cardiovascular disorder assessment and management. Little is known about their roles in regulating key biological processes, but they are of special interest as they play central roles in the pathophysiology of diseases like heart failure and atherosclerosis. Through the development of anti-sense RNA therapy to modulate specific miRNAs, this strategy promises to modulate disease processes and benefit cardiovascular patients. Our findings should spawn continued basic research in this area in order to translate them to clinical practice, and thereby enhance our ability to diagnose, treat, and prevent cardiovascular disease. The circulating miRNAs have demonstrated potential for early identification and tracking of heart-related conditions, possibly allowing for more prompt interventions and tailored treatment approaches. Additional research must be conducted to clearly understand the intricate regulatory circuits and signaling networks which result from these miRNAs in cardiovascular disease pathology. Future advancement in miRNA study will enable researchers for creating new therapeutic options that could employ miRNA regulation to control disease progression while improving patient results. Specific miRNA biomarkers require validation through major clinical experiments involving various patient demographics for diagnostic and prognostic applications. The clinical application of miRNA profiling requires standardized methods for specimen handling before testing along with analytical protocols to reach high precision reliability. The translation of miRNA research into clinical settings could transform cardiovascular medicine by providing exact diagnostic methods and well-targeted therapeutic strategies.

Acknowledgements

The authors show their appreciation to the Pathology Department at King George's Medical University. The author acknowledges the use of Paperpal for language polishing of the manuscript.

Conflict of interest

The authors declared no conflict of interest.

Appendix and nomenclature

miRNAs	microRNAs
BNP	brain natriuretic peptide
HF	heart failure
CVD	cardiovascular disease
RISC	RNA-induced silencing complex
pre-miRNAs	precursor miRNAs
PTEN	phosphatase and tensin homolog

Author details

Wahid Ali*, Priyanka Sharma, Nida Shabbir and Swati Kumari
Department of Pathology, King George's Medical University, Lucknow, India

*Address all correspondence to: aliwahid78@gmail.com

References

[1] Bartel DP. MicroRNAs: Genomics, biogenesis, mechanism, and function. Cell. 2004;**116**(2):281-297. DOI: 10.1016/s0092-8674(04)00045-5

[2] Ambros V. The functions of animal microRNAs. Nature. 2004;**431**:350-355. DOI: 10.1038/nature02871

[3] Lee Y, Ahn C, Han J, Choi H, Kim J, Yim J, et al. The nuclear RNase III Drosha initiates microRNA processing. Nature. 2003;**425**(6956):415-419. DOI: 10.1038/nature01957

[4] Yan H, Ma F, Zhang Y, Wang C, Qiu D, Zhou K, et al. miRNAs as biomarkers for diagnosis of heart failure: A systematic review and meta-analysis. Medicine. 2017;**96**(22):e6825. DOI: 10.1097/MD.0000000000006825

[5] Thum T, Catalucci D, Bauersachs J. MicroRNAs: Novel regulators in cardiac development and disease. Cardiovascular Research. 2008;**79**(4):562-570

[6] Song Z, Gao R, Yan B. Potential roles of microRNA-1 and microRNA-133 in cardiovascular disease. Reviews in Cardiovascular Medicine. 2020;**21**(1):57-64

[7] Van Rooij E, Sutherland LB, Thatcher JE, DiMaio JM, Naseem RH, Marshall WS, et al. Dysregulation of microRNAs after myocardial infarction reveals a role of miR-29 in cardiac fibrosis. National Academy of Sciences of the United States of America. 2008;**105**(35):13027-13032

[8] Ben-Nun D, Buja LM, Fuentes F. Prevention of heart failure with preserved ejection fraction (HFpEF): Reexamining microRNA-21 inhibition in the era of oligonucleotide-based therapeutics. Cardiovascular Pathology. 2020;**49**:107243

Section 3

Long Non-Coding RNA's in Biotechnology

Chapter 7

CRISPR Technology: Mechanisms and Applications in Genome Editing and Long Non-Coding RNA Functional Analysis

Poonam Ranga, Varsha Ranga and Anita Mann

Abstract

CRISPR (Clustered Regularly Interspaced Short Palindromic Repeats) has become a groundbreaking tool in genome editing, providing unmatched precision for DNA modification. The system is composed of two main elements: Cas (CRISPR-associated) protein, typically Cas9, and a guide RNA (gRNA). Together, they enable the targeted alteration of specific genomic regions, allowing for gene knockouts, insertions, deletions, and corrections with high accuracy. While methodologies to study non-coding regions, including long non-coding RNAs (lncRNAs), have historically lagged behind those for coding genes, CRISPR has introduced novel techniques for investigating these largely unexplored regions. CRISPR-mediated loss-of-function approaches allow for the disruption or repression of lncRNA transcription, while gain-of-function strategies enable the insertion of custom sequences and activation of transcription. LncRNAs, which make up a significant portion of the non-coding genome, are known to play crucial roles in a variety of biological processes, but their specific functions remain poorly understood due to limited genetic tools. The introduction of CRISPR/Cas9 technology has greatly expanded the ability to screen for novel lncRNAs and study their functional roles. This chapter aims to offer a comprehensive overview of CRISPR-Cas9 technology in genome editing and explore how it can be utilized to gain new insights into the biology of long non-coding RNAs (lncRNAs).

Keywords: genome editing technology, CRISPR/Cas9, workflow of CRISPR/Cas, applications, long non-coding RNA function

1. Introduction

The Clustered Regularly Interspaced Short Palindromic Repeats (CRISPR) and CRISPR-associated protein (Cas) systems have revolutionized the field of genome editing by providing unprecedented precision, versatility, and efficiency. Originally discovered as part of the adaptive immune system in prokaryotes, CRISPR/Cas systems enable bacteria and archaea to fend off invading genetic elements such as viruses and plasmids. The ability to repurpose this natural mechanism for targeted genome editing has opened new frontiers in biological research, biotechnology, and medicine. CRISPR loci were first identified in *Escherichia coli* in 1987, where repetitive DNA sequences were observed but their function remained unclear [1]. Subsequent studies revealed that these loci are part of a defense system that incorporates fragments of foreign DNA (spacers) into the host genome, which are then transcribed into CRISPR RNAs (crRNAs) [2, 3]. When a similar invader reappears, crRNAs guide the Cas proteins to recognize and cleave the target DNA, providing sequence-specific immunity [4]. The most widely used CRISPR/Cas system in genome editing is CRISPR/Cas9 from *Streptococcus pyogenes*. It relies on two main components: a single-guide RNA (sgRNA), which is engineered to match the target sequence, and the Cas9 nuclease, which introduces double-strand breaks (DSBs) at the specified genomic location [5]. Following DSB formation, cellular repair mechanisms such as non-homologous end joining (NHEJ) or homology-directed repair (HDR) can be harnessed to introduce desired genetic modifications.

The specificity of CRISPR/Cas9 is dictated by the sgRNA and the protospacer adjacent motif (PAM) sequence, ensuring precise targeting of the genome. Innovations such as high-fidelity Cas9 variants and modified sgRNAs have further minimized off-target effects, enhancing the precision of this technology [6]. CRISPR technology is highly versatile, supporting applications beyond simple gene knockout. It has been adapted for base editing, where single-nucleotide changes are introduced without generating DSBs [7]; prime editing, which allows for precise insertions, deletions, and substitutions [8]; and epigenome editing, where transcriptional activity is modulated without altering the DNA sequence [9]. Additionally, CRISPR interference (CRISPRi) and CRISPR activation (CRISPRa) systems enable gene repression or activation, respectively, providing powerful tools for functional genomics [10–14].

2. Non-coding RNAs in genomics

Non-coding RNAs (ncRNAs) are increasingly recognized as pivotal players in the regulation of genomic processes, transcending their initial categorization as "junk" or "passive" transcriptional byproducts. Among the ncRNAs, long non-coding RNAs (lncRNAs) have garnered significant attention due to their versatile roles in gene expression regulation, chromatin organization, and cellular signaling pathways. Their involvement spans a wide range of biological processes, including development, differentiation, and stress responses, underscoring their essential functions in maintaining cellular homeostasis (**Figure 1**). The discovery of ncRNAs began with the identification of transfer RNAs (tRNAs) and ribosomal RNAs (rRNAs) in the mid-twentieth century, which were essential for protein synthesis. However, the advent of high-throughput sequencing technologies unveiled a vast landscape of ncRNAs, revealing their abundance and diversity in eukaryotic genomes [15]. LncRNAs, typically defined as transcripts longer than 200 nucleotides with limited protein-coding potential, constitute a significant subset of these ncRNAs [16].

Long non-coding RNAs (lncRNAs) exert their regulatory effects through a variety of mechanisms, including chromatin remodeling and epigenetic regulation. For example, lncRNAs such as XIST and HOTAIR interact with chromatin-modifying complexes, influencing histone modifications and DNA methylation. These interactions can lead to either transcriptional activation or repression [17]. LncRNAs also

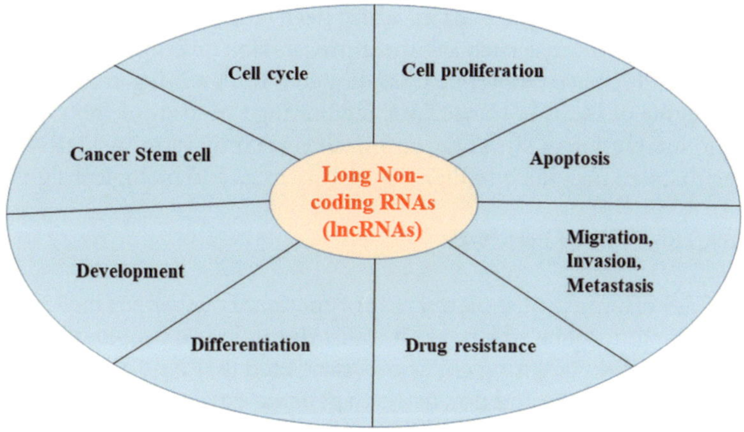

Figure 1.
Biological functions of long non-coding RNAs (lncRNAs).

play a crucial role in transcriptional regulation by functioning as enhancers, scaffolds, or decoys. In these capacities, they modulate the activity of transcription factors and RNA polymerase, thereby influencing the recruitment of these factors to target loci [18]. Beyond transcription, lncRNAs contribute to post-transcriptional regulation. They impact mRNA stability, splicing, and translation by binding to RNA-binding proteins or forming RNA-RNA duplexes, essential for fine-tuning gene expression at the post-transcriptional level [19]. Furthermore, certain lncRNAs act as molecular sponges for microRNAs, regulating their availability and activity within cellular signaling networks [20].

The functional diversity of lncRNAs extends to critical biological processes such as embryogenesis, organogenesis, and immune responses. However, dysregulation of lncRNA expression has been linked to various diseases, including cancer, neuro-degenerative disorders, and cardiovascular diseases [21]. For instance, lncRNAs like MALAT1 are involved in metastasis regulation in cancer, while others, such as MEG3, function as tumor suppressors by activating p53. Thus, lncRNAs are integral to both normal cellular processes and disease pathogenesis.

3. CRISPR-Cas9: A revolutionary tool for studying lncRNAs

CRISPR-Cas9 technology has revolutionized the study of long non-coding RNAs (lncRNAs) by providing precise and efficient tools for manipulating their expression and function. One of its key advantages is targeted gene editing, which allows for precise knockouts of lncRNA genes to assess their biological roles. By introducing small indels at specific loci, researchers can disrupt transcription or alter regulatory sequences to evaluate the functional consequences [22].

CRISPR interference (CRISPRi) uses a catalytically inactive Cas9 (dCas9) fused with repressive domains, enabling the transcriptional silencing of lncRNAs without altering the underlying DNA sequence. This method is particularly useful for studying lncRNAs that overlap with protein-coding genes [23]. In contrast, CRISPR activation (CRISPRa) employs dCas9 fused to transcriptional activators to upregulate lncRNA expression, offering insights into their gain-of-function effects in cellular processes [24].

High-throughput CRISPR screens have also been employed to identify lncRNAs involved in specific pathways, such as cancer progression or drug resistance [23]. Additionally, CRISPR-based tools like Cas9-targeted RNA editing enable precise modifications or tagging of lncRNA transcripts, facilitating the study of their localization, structure, and interactions [25]. These approaches are transforming lncRNA research by providing detailed insights into their regulatory roles and biological significance.

3.1 Functional analysis of lncRNAs

CRISPR-Cas9 offers a robust platform for functional analysis of lncRNAs, allowing researchers to integrate gene editing with transcriptomic and proteomic approaches. For instance, selectively targeting enhancer-associated lncRNAs can reveal their impact on enhancer activity and downstream gene expression [26]. Furthermore, CRISPR-Cas9 facilitates the creation of isogenic cell lines with lncRNA-specific alterations, enabling controlled studies of their contributions to complex phenotypes.

3.2 Emerging tools and applications

Advances in RNA sequencing (RNA-seq) and CRISPR technologies have accelerated the functional annotation of lncRNAs. Techniques such as CRISPRi and CRISPRa enable precise modulation of lncRNA expression, facilitating their study in specific biological contexts [27]. Given their broad functional spectrum and involvement in health and disease, lncRNAs represent a promising frontier in genomic research. The integration of CRISPR-Cas9 technology into this field provides unparalleled opportunities to uncover their roles and translate these findings into clinical applications, such as therapeutic interventions and diagnostic biomarkers.

4. Mechanisms of CRISPR/Cas systems

4.1 Structure and components

The CRISPR/Cas system, derived from a bacterial adaptive immune system, comprises two primary components: the CRISPR-associated protein (Cas) and guide RNA (gRNA). In the CRISPR/Cas9 system, Cas9 functions as an endonuclease that introduces double-strand breaks (DSBs) at specific DNA sites, while the gRNA directs Cas9 to the target sequence with high precision.

4.2 Cas9 protein

Cas9 is a multidomain protein with nuclease activity. It consists of two main nuclease domains: the RuvC and HNH domains. The RuvC domain cleaves the non-target DNA strand, while the HNH domain cleaves the target strand. The protein also contains a PAM-interacting domain, which ensures the correct targeting by recognizing the protospacer adjacent motif (PAM) sequence.

4.3 Guide RNA (gRNA)

The gRNA is a synthetic RNA molecule that combines two components found in the natural bacterial system: the CRISPR RNA (crRNA) and the trans-activating crRNA (tracrRNA). The crRNA contains a spacer sequence complementary to the target DNA, while the tracrRNA stabilizes the complex and facilitates Cas9 binding. In genome editing, a single-guide RNA (sgRNA) is commonly used, which combines these two components into one engineered molecule. The precision of DNA targeting in the CRISPR/Cas9 system is determined by the complementarity between the gRNA spacer sequence and the target DNA, alongside the presence of a PAM sequence immediately downstream of the target site.

4.4 CRISPR/Cas9 workflow

The CRISPR/Cas9 genome editing process involves several steps, each critical for ensuring specificity and efficiency. The workflow for CRISPR/Cas9 is illustrated in **Figure 2**.

a. Design and Synthesis of gRNA: The first step is to identify the target DNA sequence and design a complementary spacer sequence in the gRNA. Bioinformatics tools aid in selecting target sites with minimal off-target effects.

b. Construction of CRISPR/Cas9 Complex: The synthesized gRNA is combined with the Cas9 protein to form a ribonucleoprotein complex. This complex acts as a molecular machine to locate and cleave the target DNA.

c. Target Binding: The gRNA directs the Cas9 complex to the target DNA through base-pairing interactions. The presence of a PAM sequence (e.g., 5′-NGG-3′ for SpCas9) downstream of the target site is essential for binding and activation of the nuclease.

d. DNA cleavage: Cas9 introduces a double-strand break at the target site. The cell's repair machinery then resolves this break through one of two pathways: (i) non-homologous end joining (NHEJ): Error-prone and often results in insertions or deletions (indels) and (ii) Homology-Directed Repair (HDR): Precise repair using a donor DNA template.

The importance of the PAM sequence cannot be overstated, as it serves as a critical checkpoint for Cas9 activity, preventing off-target binding and cleavage in regions lacking the correct PAM motif.

4.5 Alternative Cas systems

CRISPR technology extends beyond Cas9 with the discovery of alternative Cas proteins, broadening the potential for genome and transcriptome engineering. Cas12 proteins, including Cas12a (Cpf1), target DNA using a single RNA molecule and generate staggered double-stranded breaks (DSBs) with sticky ends, in contrast to the blunt ends produced by Cas9. Cas12a recognizes a distinct PAM sequence (e.g., 5′-TTTV-3′) and has a simpler gRNA architecture, making it especially useful for multiplexed genome editing. On the other hand, Cas13 proteins, such as Cas13a (C2c2), target RNA instead of DNA, enabling RNA-specific manipulation. Cas13 proteins cleave single-stranded RNA in a sequence-specific manner, with applications such as RNA knockdown, where specific RNA transcripts are

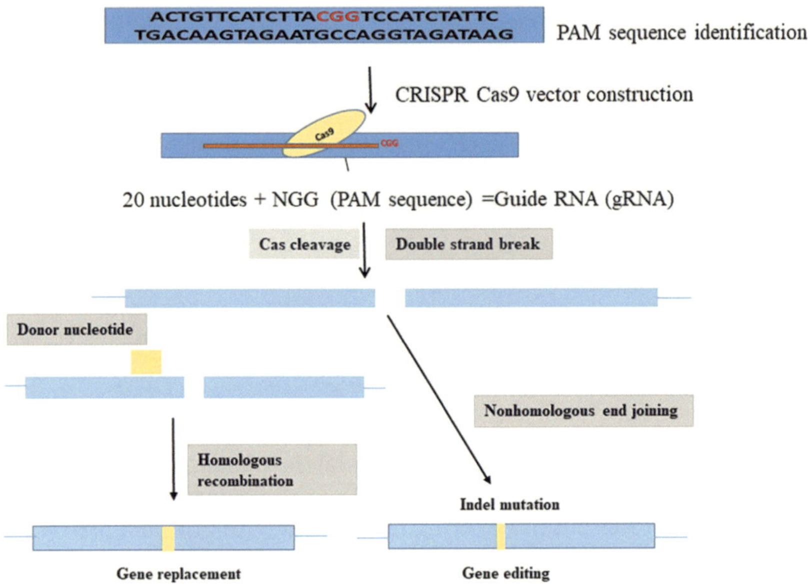

Figure 2.
Workflow of CRISPR/CAS9 editing.

degraded, and RNA editing, which allows for base editing or nucleotide modification in RNA molecules. Additionally, emerging Cas variants like Cas14 and CasPhi offer unique capabilities, such as targeting ultra-small genomes or non-canonical sequences, further enriching the CRISPR toolkit for genetic studies and applications.

5. CRISPR/Cas9 in genome editing-coding regions

The CRISPR/Cas9 system has revolutionized genome editing, particularly in coding regions, and has made significant advancements in functional genomics and biotechnology. One of its key applications is gene knockout, where CRISPR/Cas9 introduces indels through non-homologous end joining (NHEJ), effectively silencing the target gene. This approach is widely used to study gene function and disease mechanisms across various organisms [28]. Additionally, CRISPR/Cas9 facilitates gene insertion through homology-directed repair (HDR), enabling the precise integration of desired sequences at specific loci. This capability is instrumental in developing disease models and therapeutic applications [29]. The system can also introduce point mutations *via* HDR, allowing researchers to mimic disease-causing mutations or correct genetic defects, advancing personalized medicine and functional analysis of protein-coding genes.

CRISPR/Cas9 offers several advantages over traditional genome editing techniques, such as zinc-finger nucleases (ZFNs) and transcription activator-like effector nucleases (TALENs). One major benefit is its high precision, achieved through the complementarity between the guide RNA (gRNA) and target DNA, which minimizes off-target effects when properly designed. The system's simplicity and rapid implementation contribute to its efficiency, making it suitable for generating mutations in a wide range of cell types and organisms. Moreover, CRISPR/Cas9 is highly scalable, enabling high-throughput genome editing and supporting large-scale functional genomic studies and multiplexed editing strategies. Its versatility is another advantage, as it can induce various genetic modifications, ranging from simple knockouts to complex alterations like large insertions or chromosomal rearrangements.

Genome-wide CRISPR screens have emerged as a powerful tool for identifying essential genes and regulatory elements. Pooled libraries of sgRNAs targeting thousands of genes can be used to perform loss-of-function or gain-of-function screens, facilitating the discovery of genes critical for survival, drug resistance, or disease pathogenesis and aiding in the identification of therapeutic targets [30]. Furthermore, advances in single-cell sequencing technologies now allow for detailed analysis of gene perturbations at the single-cell level, offering new insights into cellular heterogeneity and lineage determination. Together, these CRISPR-based approaches are transforming our understanding of gene function and driving innovations in medicine and biotechnology.

5.1 CRISPR/Cas9 applications in lncRNA functional analysis

CRISPR/Cas9 technology has significantly advanced the study of long non-coding RNAs (lncRNAs), which play crucial roles in gene regulation, chromatin modification, and cellular differentiation. Unlike protein-coding genes, lncRNAs exhibit diverse and often tissue-specific functions, making them essential for understanding complex biological processes. For instance, lncRNAs like XIST regulate gene expression through transcriptional, post-transcriptional, and epigenetic mechanisms, as seen in X-chromosome inactivation [31]. Additionally, lncRNAs such as HOTAIR recruit chromatin-modifying complexes like Polycomb Repressive Complex 2 (PRC2) to specific genomic loci, leading to gene silencing through histone modification [32]. LncRNAs like TUG1 and NEAT1 are also involved in cellular differentiation processes such as neuronal development and cancer progression [33, 34]. Despite their importance, much of the functional understanding of lncRNAs remains limited due to their context-dependent roles.

CRISPR/Cas9 provides a powerful platform to investigate lncRNA functions through knockout, repression, activation, and tagging strategies. Using CRISPR-mediated knockout, researchers can introduce targeted loss-of-function mutations or complete deletions of lncRNAs, such as MALAT1, by disrupting key functional domains [35]. Complete deletions of lncRNA loci, such as HOTAIR and XIST, can also be achieved using paired guide RNAs flanking the target region [36]. Another approach, CRISPR interference (CRISPRi), utilizes a catalytically inactive Cas9 (dCas9) fused to transcriptional repressors to inhibit lncRNA transcription without causing DNA breaks. This method has been used to study lncRNAs like NEAT1, revealing their role in paraspeckle formation and stress responses [23]. CRISPRi offers specific, reversible repression of lncRNAs without the risk of off-target mutagenesis associated with DNA cleavage.

Conversely, CRISPR activation (CRISPRa) allows researchers to upregulate lncRNA transcription by using dCas9 fused to transcriptional activators. This technique has been applied to upregulate oncogenic lncRNAs such as LINC00116, providing insights into their roles in cellular processes and diseases [37]. Furthermore, CRISPR/Cas9 facilitates the insertion of functional tags or reporter genes into lncRNA loci to track their expression and localization. For example, inserting fluorescent tags like GFP allows researchers to study the dynamics of lncRNAs in live cells [38], while reporter genes like luciferase enable the measurement of transcriptional activity under different conditions, as demonstrated with GAS5 in glucocorticoid receptor signaling [39].

Overall, CRISPR/Cas9 has revolutionized the study of lncRNAs, providing valuable tools to uncover their roles in gene regulation, chromatin dynamics, and cellular differentiation. As CRISPR technology continues to evolve, it will further enhance our understanding of lncRNA biology and its implications in various health and disease contexts.

5.2 Challenges and limitations in CRISPR editing

While the CRISPR/Cas9 system offers immense potential in genome editing, it is not without challenges. One of the primary limitations is the risk of off-target effects. These occur when the CRISPR-Cas9 system binds to unintended genomic regions, leading to unintended edits. This issue arises because the guide RNA (gRNA) used to direct Cas9 can tolerate mismatches between its sequence and the target DNA, especially in regions distal to the PAM (Protospacer Adjacent Motif) site. This tolerance increases the likelihood of off-target binding to similar, but unintended, sequences elsewhere in the genome [40, 41]. Off-target effects can lead to mutations in unrelated genes, potentially disrupting essential genes or activating oncogenes, posing a significant risk in therapeutic applications [42].

To address these concerns, several strategies have been developed to enhance CRISPR/Cas9 specificity. High-fidelity Cas9 variants, such as eSpCas9, SpCas9-HF1, and HypaCas9, exhibit reduced off-target activity by improving the precision of gRNA-DNA pairing [43]. Improved gRNA design, aided by computational tools, can also minimize sequence similarity with non-target sites, further reducing off-target binding [44]. Another approach involves using paired Cas9 nickases, which introduce single-strand breaks rather than double-strand breaks. This requires two nickases to act near each other, reducing off-target effects [45]. Additionally, chemically modifying gRNAs or using truncated versions (tru-gRNAs) can enhance specificity and reduce unintended interactions [46].

5.3 DNA repair mechanisms

After CRISPR/Cas9 induces double-strand breaks, cellular DNA repair mechanisms dictate the outcome of the editing process. However, controlling these pathways remains a significant challenge. The predominant repair pathway, non-homologous end joining (NHEJ), is error-prone and often introduces insertions or deletions (indels) at the break site. While useful for gene knockout applications, NHEJ limits the precision of genome editing [47]. In contrast, Homology-Directed Repair (HDR) enables precise DNA sequence modifications using a homologous DNA template. However, HDR is less efficient than NHEJ and is most active during the S and G2 phases of the cell cycle, limiting its applicability in non-dividing or slowly dividing cells [48].

Efforts to bias repair toward HDR, such as synchronizing the cell cycle or using small-molecule inhibitors of NHEJ (e.g., SCR7), have shown promise but remain inconsistent across different cell types and conditions [49]. Enhancing HDR efficiency while avoiding cellular toxicity is an ongoing challenge in therapeutic applications [50].

5.4 Delivery challenges

Efficient delivery of the CRISPR/Cas9 system to target cells or tissues is crucial for successful genome editing. Various delivery methods come with their own sets of challenges.

- *Viral delivery methods*: Adeno-associated viruses (AAVs) are commonly used due to their safety and ability to infect a broad range of cell types. However, AAVs have a limited packaging capacity (~4.7 kb), which may be insufficient for delivering both Cas9 and gRNA, particularly when using larger Cas9 variants like SpCas9 [51]. Lentiviruses and retroviruses offer stable integration of CRISPR components, but the risk of insertional mutagenesis and long-term side effects in therapeutic contexts remains a concern [52].

- *Non-viral delivery methods*: Lipid nanoparticles (LNPs) can encapsulate CRISPR components and facilitate their delivery into cells *via* endocytosis. While LNPs are non-integrative and can be tailored for tissue-specific targeting, their efficiency can vary depending on the formulation and cell type [53]. Electroporation, a method that uses electric pulses to introduce CRISPR components into cells, is highly efficient in ex vivo applications but can cause significant cell damage, making it less suitable for *in vivo* use [54]. Physical delivery methods such as microinjection and hydrodynamic injection are labor-intensive and impractical for large-scale or systemic applications [55].

In summary, while CRISPR/Cas9 offers transformative potential for genome editing, overcoming challenges related to off-target effects, DNA repair mechanisms, and efficient delivery remains essential for its widespread and effective application in research and therapy. Continued advancements in protein engineering, delivery systems, and repair pathway modulation will be pivotal in overcoming these limitations and realizing the full potential of CRISPR in research and therapeutic contexts.

5.5 Ethical and regulatory considerations in CRISPR research

CRISPR-Cas9 technology has revolutionized genetic research and molecular biology, offering precise and efficient genome editing capabilities. However, its application raises several ethical concerns, especially in human genome editing and the editing of non-coding regions with unclear functions. Human genome editing, particularly germline editing, has sparked significant debate. Key ethical concerns include the potential risks of off-target mutations, mosaicism, and unforeseen long-term effects, which can be difficult to predict. Additionally, there are challenges regarding consent, as it is impossible to obtain consent from future generations who may be affected by these modifications. Issues of equity also arise, as access to genome editing technologies may be limited by socioeconomic factors, raising the potential for genetic inequality or the creation of "designer babies" that cater only to the wealthy. Furthermore, over-reliance on genome editing could reduce genetic diversity, leading to unforeseen vulnerabilities in human populations. The idea of editing non-disease traits, such as intelligence or physical appearance, could also lead to ethical dilemmas surrounding human enhancement and eugenics. Editing non-coding regions, once considered "junk DNA," presents additional concerns due to the unclear functionality of many of these sequences. The full roles of these regions in gene regulation and chromatin structure are still being uncovered, and modifying them without a comprehensive understanding could lead to unintended consequences, including disruptions in essential regulatory pathways and long-term developmental issues.

In addition to the ethical challenges, the regulatory landscape surrounding CRISPR research varies across countries and focuses on ensuring safety, transparency, and ethical oversight. Regulatory agencies like the US Food and Drug Administration (FDA), the European Medicines Agency (EMA), and the World Health Organization (WHO) oversee the safety and approval of CRISPR-based therapies. Preclinical studies, often using animal models, are required to evaluate potential off-target effects and genomic stability before CRISPR interventions can proceed to clinical trials. Clinical trials, conducted under strict ethical oversight, are necessary to assess the safety and efficacy of these therapies in humans, ensuring that patient consent and long-term effects are carefully monitored. Transparency in CRISPR research is also crucial for maintaining public trust, with regulatory bodies pushing for clear reporting of research findings, including off-target effects and unintended genetic modifications. Open data sharing and ethical review boards are essential components of this transparency, helping to ensure the reproducibility and reliability of research. Finally, CRISPR-based therapeutic applications, such as gene editing for inherited diseases, must undergo rigorous regulatory approval processes, including FDA evaluation and international collaboration to harmonize standards across the globe. These measures aim to ensure that CRISPR technologies are used safely and responsibly, minimizing risks while maximizing their potential benefits.

CRISPR technology holds immense promise in medicine, agriculture, and beyond, but it also raises significant ethical and regulatory concerns. The ethical challenges, especially regarding human genome editing and the modification of non-coding regions, require careful consideration of safety, consent, and the long-term implications of genetic modifications. Simultaneously, the regulatory landscape must evolve to ensure that CRISPR-based interventions are safe, transparent, and ethically sound. As research progresses, it is crucial that regulatory agencies, scientists, and ethicists work together to guide the responsible use of this powerful technology.

5.6 Future directions

The transformative potential of CRISPR-based approaches in lncRNA research is immense, with evolving technologies opening up exciting possibilities for the future. One key area is personalized medicine, where CRISPR tools could be used to target disease-associated lncRNAs. Many diseases, especially cancers, are linked to dysregulated lncRNA expression, and CRISPR could precisely edit or silence these non-coding RNAs to create tailored therapeutic strategies. This approach could go beyond protein-coding gene correction, allowing for the reprogramming of entire non-coding RNA networks to address specific genetic conditions. Additionally, CRISPR-based therapies hold promise for treating diseases caused by lncRNA dysfunction, offering a direct solution where traditional gene-editing methods may struggle.

Another promising avenue is the role of lncRNAs as modulators of epigenetic regulation. LncRNAs play pivotal roles in chromatin modifications and gene silencing, influencing cellular processes such as differentiation, pluripotency, and cancer metastasis. With CRISPR, researchers can manipulate lncRNA expression or function to investigate how these non-coding RNAs interact with epigenetic regulators, which could lead to novel epigenetic therapies targeting diseases like cancer or developmental disorders.

In vivo functional studies of lncRNAs are also becoming more feasible with CRISPR technology. The ability to manipulate lncRNAs directly in living organisms using CRISPR-based tools like CRISPRa and CRISPRi allows researchers to gain a deeper understanding of how these molecules influence gene expression across various tissues and disease models. This will provide valuable insights into their roles at both the transcriptional and post-transcriptional levels.

Overall, CRISPR technology has significantly impacted functional genomics by enabling precise modifications not just in coding genes but also in the realm of non-coding RNAs, such as lncRNAs. Once considered merely transcriptional "noise," lncRNAs are now recognized as critical regulators of gene expression, chromatin remodeling, and key cellular processes like differentiation, development, and disease progression. By allowing targeted knockouts or activation/inhibition of specific lncRNAs, CRISPR has revolutionized our ability to explore these roles in greater detail. This has far-reaching implications for understanding cellular processes and advancing clinical applications, especially in areas like immune response and adaptation, as seen in ongoing research on Drosophila's immune pathways. As CRISPR continues to evolve, its integration into lncRNA research will likely uncover new therapeutic strategies and enrich our understanding of the complex regulatory landscape of the genome. This growing body of research will continue to shape the future of functional genomics, leading to better models of gene regulation and improved disease treatment strategies.

Acknowledgements

The author acknowledges the use of Chat GPT for language polishing of the manuscript.

Author details

Poonam Ranga[1]*, Varsha Ranga[2] and Anita Mann[1]

1 ICAR-Central Soil Salinity Research Institute, Karnal, Haryana, India

2 Rishikul Government College, Hemwati Nandan Bahuguna Garhwal University, Haridwar, Uttarakhand, India

*Address all correspondence to: poonamranga27@yahoo.com

References

[1] Ishino Y, Shinagawa H, Makino K, Amemura M, Nakata A. Nucleotide sequence of the iap gene, responsible for alkaline phosphatase isozyme conversion in *Escherichia coli*, and identification of the gene product. Journal of Bacteriology. 1987;**169**(12):5429-5433

[2] Mojica FJ, Díez-Villaseñor C, García-Martínez J, Soria E. Intervening sequences of regularly spaced prokaryotic repeats derive from foreign genetic elements. Journal of Molecular Evolution. 2005;**60**(2):174-182

[3] Bolotin A, Quinquis B, Sorokin A, Ehrlich SD. Clustered regularly interspaced short palindrome repeats (CRISPRs) have spacers of extrachromosomal origin. Microbiology. 2005;**151**(8):2551-2561

[4] Barrangou R et al. CRISPR provides acquired resistance against viruses in prokaryotes. Science. 2007;**315**(5819):1709-1712

[5] Jinek M et al. A programmable dual-RNA–guided DNA endonuclease in adaptive bacterial immunity. Science. 2012;**337**(6096):816-821

[6] Kleinstiver BP et al. High-fidelity CRISPR–Cas9 nucleases with no detectable genome-wide off-target effects. Nature. 2016;**529**(7587):490-495

[7] Komor AC et al. Programmable editing of a target base in genomic DNA without double-stranded DNA cleavage. Nature. 2016;**533**(7603):420-424

[8] Anzalone AV et al. Search-and-replace genome editing without double-strand breaks or donor DNA. Nature. 2019;**576**(7785):149-157

[9] Hilton IB et al. Epigenome editing by a CRISPR-Cas9-based acetyltransferase activates genes from promoters and enhancers. Nature Biotechnology. 2015;**33**(5):510-517

[10] Qi LS et al. Repurposing CRISPR as an RNA-guided platform for sequence-specific control of gene expression. Cell. 2013;**152**(5):1173-1183

[11] Jaganathan D et al. CRISPR for crop improvement: An update review. Frontiers in Plant Science. 2018;**9**:985

[12] Frangoul H et al. CRISPR-Cas9 gene editing for sickle cell disease and β-thalassemia. New England Journal of Medicine. 2021;**384**(3):252-260

[13] Chen JS et al. CRISPR-Cas12a target binding unleashes indiscriminate single-stranded DNase activity. Science. 2018;**360**(6387):436-439

[14] Broughton JP et al. CRISPR–Cas12-based detection of SARS-CoV-2. Nature Biotechnology. 2020;**38**(7):870-874

[15] Mattick JS, Makunin IV. Non-coding RNA. Human Molecular Genetics. 2006;**15**(R1):R17-R29

[16] Rinn JL, Chang HY. Genome regulation by long noncoding RNAs. Annual Review of Biochemistry. 2012;**81**:145-166

[17] Ponting CP, Oliver PL, Reik W. Evolution and functions of long noncoding RNAs. Cell. 2009;**136**(4):629-641

[18] Engreitz JM et al. The Xist lncRNA exploits three-dimensional genome architecture to spread

across the X chromosome. Science. 2016;**341**(6147):1237973

[19] Kopp F, Mendell JT. Functional classification and experimental dissection of long noncoding RNAs. Cell. 2018;**172**(3):393-407

[20] Salmena L et al. A ceRNA hypothesis: The Rosetta stone of a hidden RNA language? Cell. 2011;**146**(3):353-358

[21] Schmitt AM, Chang HY. Long noncoding RNAs in cancer pathways. Cancer Cell. 2016;**29**(4):452-463

[22] Huarte M. The emerging role of lncRNAs in cancer. Nature Medicine. 2015;**21**(11):1253-1261

[23] Liu SJ et al. CRISPRi-based genome-scale identification of functional long noncoding RNA loci in human cells. Science. 2017;**355**(6321):eaah7111

[24] Konermann S et al. Genome-scale transcriptional activation by an engineered CRISPR-Cas9 complex. Nature. 2015;**517**(7536):583-588

[25] Abudayyeh OO et al. RNA targeting with CRISPR-Cas13. Nature. 2017;**550**(7675):280-284

[26] Fulco CP et al. Systematic mapping of functional enhancer-promoter connections with CRISPR interference. Science. 2016;**354**(6313):769-773

[27] Cong L et al. Multiplex genome engineering using CRISPR/Cas systems. Science. 2013;**339**(6121):819-823

[28] Mali P et al. RNA-guided human genome engineering via Cas9. Science. 2013;**339**(6121):823-826

[29] Shalem O et al. Genome-scale CRISPR-Cas9 knockout screening in human cells. Science. 2014;**343**(6166):84-87

[30] Cerase A et al. XIST regulation of gene expression through transcriptional, post-transcriptional, and epigenetic mechanisms in X-chromosome inactivation. Nature Reviews Molecular Cell Biology. 2015;**16**(4):255-266. DOI: 10.1038/nrm3956

[31] Rinn JL et al. HOTAIR recruits Polycomb repressive complex 2 (PRC2) to specific genomic loci, leading to gene silencing through histone modification. Cell. 2007;**129**(4):1299-1310. DOI: 10.1016/j.cell.2007.05.022

[32] Lin X et al. TUG1 and its involvement in cellular differentiation processes such as neuronal development. The Journal of Neuroscience. 2011;**31**(24):8647-8657. DOI: 10.1523/JNEUROSCI.1227-11.2011

[33] Zhang Z et al. NEAT1 in cancer progression. Oncotarget. 2013;**4**(7):1091-1102. DOI: 10.18632/oncotarget.1140

[34] Gutschner T et al. CRISPR-mediated knockout of MALAT1 in the study of long non-coding RNA functions. Genes & Development. 2013;**27**(2):123-134. DOI: 10.1101/gad.221111.112

[35] Li Y et al. Paired guide RNAs for complete deletions of lncRNA loci such as HOTAIR and XIST. Nature Biotechnology. 2016;**34**(9):994-999. DOI: 10.1038/nbt.3657

[36] Konermann S et al. CRISPR activation (CRISPRa) for upregulation of oncogenic lncRNAs such as LINC00116. Nature Communications. 2015;**6**:7411. DOI: 10.1038/ncomms8411

[37] Sasaki H et al. Insertion of fluorescent tags into lncRNA loci to track their expression and localization in live cells. Nature Methods.

2009;**6**(6):463-464. DOI: 10.1038/nmeth.1332

[38] Weiss JM et al. Reporter genes like luciferase to measure transcriptional activity in GAS5 during glucocorticoid receptor signaling. Gene. 2016;**582**(1):1-6. DOI: 10.1016/j.gene.2016.01.036

[39] Fu Y et al. Off-target effects in CRISPR-Cas9 genome editing. Nature Biotechnology. 2013;**31**(9):822-826. DOI: 10.1038/nbt.2623

[40] Hsu PD et al. Risk of off-target effects in CRISPR/Cas9 gene editing. Nature Biotechnology. 2013;**31**(9):820-822. DOI: 10.1038/nbt.2627

[41] Tsai SQ, Joung JK. Strategies to mitigate off-target effects in CRISPR-Cas9 genome editing. Trends in Genetics. 2016;**32**(5):280-290. DOI: 10.1016/j.tig.2016.02.005

[42] Kleinstiver BP et al. High-fidelity Cas9 variants to reduce off-target activity. Nature. 2016;**529**(7587):490-495. DOI: 10.1038/nature16526

[43] Doench JG et al. Improved gRNA design tools for minimizing off-target binding. Nature Methods. 2014;**12**(10):739-745. DOI: 10.1038/nmeth.3010

[44] Ran FA et al. Paired Cas9 nickases for reducing off-target effects. Cell. 2013;**154**(6):1380-1389. DOI: 10.1016/j.cell.2013.08.022

[45] Cho SW et al. Chemically modified gRNAs for enhancing specificity in CRISPR-Cas9 genome editing. Nature Biotechnology. 2014;**32**(6):569-576. DOI: 10.1038/nbt.2907

[46] Sharma P et al. NHEJ and its impact on genome editing outcomes in CRISPR/Cas9 systems. Molecular Therapy. 2015;**23**(1):23-34. DOI: 10.1038/mt.2014.218

[47] Ceccaldi R et al. HDR and its efficiency in genome editing. Nature Reviews Molecular Cell Biology. 2016;**17**(2):102-114. DOI: 10.1038/nrm.2015.4

[48] Maruyama T et al. Biasing repair pathways toward HDR using small-molecule inhibitors of NHEJ. Nature Communications. 2015;**6**:6171. DOI: 10.1038/ncomms7171

[49] Song Z et al. Enhancing HDR efficiency in therapeutic genome editing. Nature Biotechnology. 2016;**34**(5):557-563. DOI: 10.1038/nbt.3560

[50] Zhang Y et al. Adeno-associated viruses (AAVs) for CRISPR delivery and their packaging limitations. Nature Biotechnology. 2015;**33**(4):312-319. DOI: 10.1038/nbt.3151

[51] Wang X et al. Lentiviruses and retroviruses for CRISPR delivery: Stability and concerns. Frontiers in Genetics. 2018;**9**:366. DOI: 10.3389/fgene.2018.00366

[52] Kowalski PS et al. Lipid nanoparticles for CRISPR delivery in cells. Molecular Therapy. 2019;**27**(5):904-913. DOI: 10.1016/j.ymthe.2019.02.004

[53] Kim D et al. Electroporation for CRISPR/Cas9 delivery in ex vivo applications. Nature Communications. 2017;**8**:14073. DOI: 10.1038/ncomms14073

[54] Mout R et al. Physical delivery methods for CRISPR components: Microinjection and hydrodynamic injection. Advanced Drug Delivery Reviews. 2017;**108**:82-95. DOI: 10.1016/j.addr.2016.12.010

[55] WHO (World Health Organization). Guidelines for CRISPR-based Research and Therapy. n.d. Available from: https://www.who.int/genomics/guidelines/en

Chapter 8

Long Non-Coding RNAs in Stem Cell Regulation and Regenerative Medicine: Stemness, Differentiation, and Therapeutic Innovation

Melika Emarati, Hossein Azizi, Nima Ghasemi and Thomas Skutella

Abstract

Long non-coding RNAs (lncRNAs) have emerged as critical regulators in stem cell biology, influencing cellular functions such as pluripotency, differentiation, and self-renewal. Their unique ability to modulate gene expression at multiple levels—epigenetic, transcriptional, and post-transcriptional—makes lncRNAs powerful tools for controlling cell fate. In regenerative medicine, understanding the roles of specific lncRNAs can enhance therapeutic approaches, particularly in stem cell-based tissue repair and engineering. By modulating lncRNA activity, researchers can potentially direct stem cell differentiation toward desired lineages, facilitating the development of functional tissues for clinical applications. This chapter explores how lncRNAs influence stem cell states, highlights current research in regenerative therapies, and discusses potential future applications where lncRNA-based interventions could drive advancements in tissue engineering and regenerative medicine.

Keywords: long non-coding RNA (lncRNA), stem cell biology, regenerative medicine, pluripotency, bioinformatics

1. Introduction

The human transcriptome is remarkably complex, with a substantial proportion transcribed into non-coding RNAs (ncRNAs) that lack protein-coding potential. These ncRNAs constitute a heterogeneous group with key roles in regulating transcriptional and post-transcriptional processes, including X-chromosome inactivation, epigenetic modulation, genomic imprinting, and mRNA splicing [1–3]. NcRNAs are broadly categorized into two groups based on their size, small ncRNAs (18–200 nucleotides) and long non-coding RNAs (lncRNAs, >200 nucleotides). While small ncRNAs like microRNAs (miRNAs) and small interfering RNAs (siRNAs) have been widely studied in their gene expression ability, lncRNAs remain less understood but increasingly recognized for their pivotal regulatory roles [2]. Long non-coding RNAs (lncRNAs) can be classified in several ways, with one common method being based on their genomic organization. This classification divides lncRNAs into four main categories according to their positions relative to protein-coding genes: (1) intergenic lncRNAs, also called lincRNAs, which are transcribed from regions of DNA located between two protein-coding genes and often play regulatory roles; (2) intronic lncRNAs, derived from the introns of protein-coding genes and sometimes involved in splicing or transcriptional regulation; (3) overlapping lncRNAs, which are transcripts that partially or entirely overlap with known protein-coding genes, suggesting potential interactions with their overlapping counterparts; and (4) antisense lncRNAs, transcribed in the opposite direction of a protein-coding gene, often influencing gene expression through complementary base-pairing or chromatin modification [2, 4–8].

LncRNAs are transcribed by RNA polymerase II and exhibit features such as 5′ capping, splicing, and polyadenylation, akin to mRNAs [1]. Functionally, lncRNAs participate in chromatin remodeling, transcriptional regulation, and post-transcriptional processing, acting as essential regulators in embryonic stem cell pluripotency, development, differentiation, and tumorigenesis. Long non-coding RNAs (lncRNAs) exert significant regulatory influence across transcriptional, post-transcriptional, and translational processes and alternative splicing. Transcriptionally, lncRNAs can affect transcriptional complexes or DNA elements (e.g., promoter). In this regard, LncRNA "PANDA" regulates transcription by interacting with the transcription factor NF-YA. This interaction sequesters NF-YA away from its target gene-associated chromatin [9]. LncRNAs can also transcriptionally control gene expression by acting as scaffolds for histone-modifying enzymes, affecting the expression of genes [10]. HOTAIR, a well-studied lncRNA, alters histone H3K27 methylation patterns by interacting with the PRC2 complex. This modification silences specific genes, facilitating tumor cell invasion and metastasis [11]. Post-transcriptionally, they regulate mRNA stability, such as Linc-RoR stabilizing c-Myc by interacting with AUF1 and hnRNP I [2, 12]. LncRNAs can also modulate alternative splicing by cooperating with hnRNPs [13]. MALAT1 influences serine-arginine proteins [13]. At the translational level, lncRNAs like Linc-RoR suppresses p53 translation by blocking hnRNP I interactions with p53 5′ UTR [14], while others, such as Uchl1, enhance translation through UTR-mediated interactions [15]. Despite their functional diversity, lncRNAs are generally expressed at low levels, exhibit high cell-type specificity, and are often dysregulated in diseases, including cancer.

Stem cells represent a unique cellular state characterized by self-renewal and the ability to differentiate into specialized cell types, and lncRNAs play a pivotal role in maintaining this delicate balance. LncRNAs such as H19, TUNA, and linc-ROR are central to regulating pluripotency and lineage commitment. For example, Linc-ROR modulates the transcription factors OCT4, SOX2, and NANOG, which are essential for maintaining the pluripotent state [16]. Moreover, lncRNAs like MALAT1 and MEG3 are involved in cell fate determination, influencing differentiation pathways through epigenetic and transcriptional regulation [17]. These molecules act as critical modulators of stem cell identity, responding to developmental cues and environmental signals to drive specific differentiation trajectories. Dysregulation of stem cell-associated lncRNAs has been implicated in developmental disorders and cancer, further underscoring their importance in cellular homeostasis and disease. Understanding the role of lncRNAs in stem cell biology offers exciting prospects for regenerative medicine and the development of targeted therapies.

Advancements in high-throughput sequencing and bioinformatics have facilitated the identification and characterization of thousands of lncRNAs, revealing their roles as master regulators of gene expression [18]. LncRNAs can exert their regulatory functions through various mechanisms, including acting as molecular scaffolds, decoys, or guides, forming complex interactions with proteins, RNA, and DNA. While some lncRNAs, such as XIST and H19, have been studied for decades, the field continues to uncover their intricate involvement in health and disease. As lncRNAs are increasingly implicated in cancer biology and stem cell regulation, understanding their mechanisms and functional diversity holds immense potential for novel therapeutic interventions, including precision medicine and regenerative therapies.

2. LncRNAs in stemness and differentiation

Long non-coding RNAs (lncRNAs) are emerging as pivotal regulators of stem cell biology, playing crucial roles in maintaining stemness and differentiation. Stem cells, characterized by their unique ability to self-renew and differentiate into specialized cell types, are central to tissue development, repair, and homeostasis. The regulation of these processes involves intricate molecular networks, where lncRNAs act as key modulators by influencing gene expression, epigenetic landscapes, and signaling pathways [19]. Through interactions with transcription factors, chromatin remodelers, and microRNAs, lncRNAs fine-tune the balance between maintaining a pluripotent state and committing to lineage-specific differentiation [20]. Understanding the role of lncRNAs in these processes sheds light on fundamental aspects of stem cell biology and opens new avenues for therapeutic applications in regenerative medicine and disease treatment.

2.1 LncRNAs in embryonic stem cells

Long non-coding RNAs (lncRNAs) are pivotal regulators in maintaining pluripotency and the differentiation of embryonic stem cells (ESCs). Genome-wide studies in mouse ESCs (mESCs) identified lncRNAs such as AK028326 and AK141205, which Oct4 and Nanog transcriptionally regulate. AK028326, for example, functions as a coactivator of Oct4, while AK141205 is repressed by Nanog, indicating the direct involvement of these lncRNAs in the pluripotency network [21]. Additionally, lncRNAs (lincRNAs) play roles in pluripotency and proliferation, many of which are directly regulated by key transcription factors such as Sox2, Oct4, and Nanog [19]. In human ESCs (hESCs), lncRNAs like RMST and SOX2OT have been implicated in maintaining pluripotency and regulating differentiation. RMST interacts with SOX2 as a transcriptional co-regulator [22], while SOX2OT modulates SOX2 expression [23]. Furthermore, cytoplasmic lncRNAs, such as linc-RoR, act as microRNA sponges, shielding mRNAs of key transcription factors from degradation and supporting ESC maintenance [24]. These findings underscore the dual nuclear and cytoplasmic roles of lncRNAs in ESC biology.

The functional significance of lncRNAs extends to lineage-specific differentiation and chromatin regulation. The majority of lincRNAs in ESCs influence gene expression, with 26 lincRNAs directly affecting Nanog expression [25]. Moreover, lineage-specific differentiation involves the activity of numerous lincRNAs, such as those associated with endoderm, ectoderm, and mesoderm formation. The involvement of pluripotency-associated transcription factors, including Oct4, Sox2, and Nanog, in regulating these lincRNAs shows a possible tightly interconnected regulatory network [21]. Additionally, the association of 30% of lincRNAs with chromatin-modifying proteins highlights their role in remodeling chromatin to maintain ESC identity [21]. Gas5, for instance, maintains pluripotency by regulating OCT4 and SOX2 through the TGFβ pathway, while Digit influences endoderm differentiation *via* SMAD3 [26, 27]. Similarly, Terra promotes self-renewal through the WNT/β-catenin pathway, and its overexpression prevents ESCs from exiting pluripotency [28].

X-chromosome inactivation (XCI) is a critical process in female ESC differentiation with lncRNA involvement. XCI is regulated by non-coding RNAs like Xist, Tsix, and RepA, which are themselves targets of pluripotency transcription factors [29]. For example, Oct4 and Nanog repress Xist expression in mESCs, maintaining two active X chromosomes, a hallmark of pluripotency [30]. Upon differentiation, Xist expression increases, leading to XCI [30]. Furthermore, NEAT1 plays a role in the transition from pluripotency to differentiation. NEAT1 is a nuclear lncRNA absent in undifferentiated ESCs, facilitating paraspeckle formation in differentiated cells [31, 32]. Other lncRNAs, such as Lncenc1, regulate ESC metabolism by interacting with RNA-binding proteins like PTBP1 and HNRNPK to control glycolytic gene expression [33].

Collectively, lncRNAs have different roles in ESC biology. They act as molecular scaffolds, transcriptional regulators, microRNA sponges, and modulators of chromatin and signaling pathways. While significant progress has been made in understanding their functions, many questions remain about the precise mechanisms and interacting partners of these lncRNAs. Future research is needed to explore the dynamic regulation of lncRNAs and their potential uses in regenerative medicine and cell-based therapies.

2.2 LncRNAs in induced pluripotent stem cells

Research about the roles of lncRNAs in induced pluripotent stem cells (iPSCs) can shed light on the molecular mechanisms underlying reprogramming. Several studies have identified distinct sets of lncRNAs that are differentially expressed in iPSCs compared to ESCs, suggesting their functional relevance in achieving and maintaining pluripotency. For instance, 26 lncRNAs have been identified to be overexpressed in both human ESCs and iPSCs, with some showing elevated levels specifically in iPSCs [25]. Notably, lncRNAs, such as Meg3, Meg9, and Rian, located in the Dlk1-Dio3 gene cluster, are typically expressed in ESCs but often silenced in iPSCs [21]. Similarly, transcriptional changes in the X-inactivation center, including the biallelic expression of Tsix and activation of Xite, are important in reactivating both X chromosomes during the transition to the pluripotency state [34, 35].

Furthermore, lncRNAs such as lincRNA-SFMBT2, lincRNA-VLDLR, and lincRNA-ST8SIA3 are linked to the reprogramming process. They have been found to interact with core pluripotency transcription factors Oct4, Sox2, and Nanog [36]. Knockdown studies have demonstrated that these lncRNAs are essential for efficient colony formation, with lincRNA-ST8SIA3 showing strong involvement in regulating p53-mediated apoptotic pathways during reprogramming [37]. LincRNA-RoR (Regulator of Reprogramming), functions as a competing endogenous RNA [38]. Silencing linc-ROR expression in stem cells prompts their exit from the stem cell state, accompanied by elevated levels of ectodermal markers, including SOX1 and vimentin, further highlighting its role in maintaining the pluripotent state [17]. Global epigenetic remodeling and the dynamic expression of over 400 lncRNAs at defined reprogramming stages show their complex regulatory mechanisms [39]. Collectively, lncRNAs are critical in transcriptional and epigenetic regulation which are necessary for iPSC generation and maintenance, offering new strategies for stem cell-based therapies and personalized medicine.

2.3 LncRNAs in mesenchymal stem cell

Mesenchymal stem cells (MSCs) are multipotent cells that can differentiate into various lineages, including mesodermal, ectodermal, and endodermal fates. MSCs support hematopoietic stem cells and can differentiate into bone, cartilage, and adipose tissues. MSCs are also present in various other tissues such as adipose tissue, umbilical cord blood, and the dental pulp of baby teeth [40]. Due to their capacity for multipotent differentiation, MSCs hold significant promise for regenerative medicine. Research has highlighted the important regulatory roles of lncRNAs in MSCs and their promising potential in disease and therapeutic goals. Notably, lncRNAs like ANCR, MALAT1, and H19 have been shown to regulate MSC differentiation processes, influencing cell fate decisions across different lineages. For example, ANCR has been implicated in inhibiting the trans-differentiation of adipose-derived stem cells (ADSCs) into endodermal lineages [41], while MALAT1 plays a role in the trans-differentiation of bone marrow-derived stem cells (BMSCs) into hepatocytes [41]. Furthermore, lncRNAs are involved in regulating MSC differentiation into ectodermal lineages, including neural cells. Recent studies have shown that specific lncRNAs modulate the neural differentiation of BMSCs, such as the negative regulation of neural-like differentiation by H19 through the miR-675-IGFR axis [41].

2.3.1 LncRNAs in adipogenic differentiation

In adipogenesis, MSCs differentiate into adipocytes. It involves complex regulatory mechanisms, including the pivotal role of long non-coding RNAs (lncRNAs). MSCs differentiate into both brown adipose tissue (BAT) and white adipose tissue (WAT), with specific transcription factors, such as peroxisome proliferator-activated receptor γ (PPARγ) and CCAAT/enhancer-binding protein α (C/EBPα), driving adipogenesis [42]. RNA sequencing has shown that lncRNAs are dynamically expressed during adipogenesis. For example, 175 lncRNAs are associated with the differentiation of BAT and WAT [43]. Among these, the brown fat lncRNA 1 (Blnc1) is essential for brown adipogenesis, forming a ribonucleoprotein complex with EBF2 to regulate thermogenic gene expression and promoting a feed-forward regulatory loop [44]. The lncRNA SRA is known for the promotion of adipogenesis by enhancing PPARγ expression, which results in improving insulin sensitivity while inhibiting inflammatory pathways [45]. Another lncRNA, ADINR, positively regulates PPARγ and C/EBPα by recruiting the PA1 complex to the CEBPA promoter, affecting histone modifications to facilitate adipogenic gene expression [46]. Such findings highlight the central role of lncRNAs in adipogenic differentiation and energy metabolism.

The process of adipogenesis is divided into two phases: lineage commitment, where MSCs commit to the adipocyte lineage, and terminal differentiation, where preadipocytes mature into lipid-storing adipocytes [47]. Numerous intracellular and extracellular signals, such as TGF-β, Wnt, and Hedgehog signaling pathways, are present in this process by regulating transcription factors like PPARγ and C/EBPα [47]. LncRNAs have been shown to influence these pathways, often functioning as competing endogenous RNAs (ceRNAs) or interacting with chromatin to modulate gene expression. For instance, during adipogenesis, the levels of the lncRNA GAS5 gradually decrease and impair adipogenesis. It regulates the MSC's adipogenic differentiation by sponging miR-18a, co-suppressing CTGF protein translation [48], while HOTAIR reduces adipogenesis through DNA methylation *via* a triple helix structure [49]. lncRNAs such as MEG3 and ADNCR act differently for suppressing adipogenesis. MEG3 inhibits adipogenic differentiation by sponging miR-140-5p, a microRNA that upregulates PPARγ and C/EBPα [50], while ADNCR functions as a competing endogenous RNA (ceRNA) for miR-204 to promote SIRT1 expression, suppressing PPARγ activity and adipocyte differentiation [51]. These findings show the potential opportunities for targeting lncRNAs for therapeutic applications in obesity and related metabolic disorders.

2.3.2 LncRNAs in osteogenic differentiation

Osteogenesis is a dynamic process involving the differentiation of mesenchymal stem cells (MSCs) into osteoblasts and subsequent bone tissue formation. It is regulated by long non-coding RNAs (lncRNAs), which modulate key signaling pathways, transcription factors, and epigenetic modifications, playing crucial roles in lineage commitment and maturation during osteogenic differentiation. For instance, the lncRNA H19 enhances osteogenesis by sponging miR-141 and miR-22, resulting in promoting the Wnt/β-catenin signaling pathway, which is essential for osteoblast development [52]. Similarly, Linc-ROR activates the Wnt/β-catenin pathway with upregulating osteogenic genes by interacting with miR-138 and miR-145 [53]. In contrast, inhibitory lncRNAs such as MIAT and DANCR suppress osteogenesis by targeting specific pathways. MIAT inhibits osteogenic differentiation by sponging miR-150-5p and modulating AKT activity [54], while DANCR interferes with p38 MAPK signaling and downregulates osteogenic markers [55]. Additionally, lncRNA ANCR maintains the undifferentiated state of MSCs. It represses the osteogenic transcription factor Runx2 through EZH2-mediated H3K27me3 modifications [56]. Beyond these, lncRNAs such as MEG3 and AK141205 enhance osteogenesis *via* BMP4 activation and CXCL13 expression [57]. Given their critical involvement in osteogenic differentiation, lncRNAs represent potential therapeutic targets for treating bone-related disorders such as osteoporosis and osteogenesis imperfecta.

2.3.3 LncRNAs in cardiomyocyte differentiation

Long non-coding RNAs (lncRNAs) are emerging as critical regulators of cardiomyocyte differentiation and cardiovascular development by orchestrating gene expression, chromatin dynamics, and lineage specification. Fendrr and Braveheart (Bvht) are two identified lncRNAs among the several differentially expressed lncRNAs during cardiac differentiation, which play pivotal roles in heart development. Fendrr, expressed in the lateral mesoderm, interacts with the PRC2 and/or TrxG/MILL complex to regulate chromatin signatures at cardiac transcription factor promoters, ensuring proper mesodermal differentiation and heart formation [58]. Its loss leads to severe myocardial dysfunction and embryonic lethality [58]. On the other hand, Braveheart drives cardiomyocyte lineage specification. Studies showed its ability to interact with SUZ12 and CNBP, with its AGIL motif being indispensable for its function in promoting cardiomyocyte differentiation [59, 60].

Other lncRNAs such as HBL1 and CARMEN add further complexity to this regulatory network. HBL1 acts as a suppressor of cardiomyocyte differentiation by sponging miR-1 and downregulating critical cardiac genes [61], while CARMEN, a super-enhancer-associated lncRNA, promotes cardiac progenitor cell (CPC) differentiation into cardiomyocytes and enhances the expression of cardiac markers like GATA4 and NKX2.5 [20, 62]. Additionally, novel lncRNAs such as yylncT have been implicated in mesodermal commitment during cardiac differentiation, regulating the expression of BRACHYURY by inhibiting DNMT3B activity [63]. Emerging evidence also points to the roles of TTN-AS1, ALIEN, and PUNISHER in cardiovascular differentiation, with functions ranging from modulating extracellular matrix remodeling to enhancing endothelial cell maturation [20, 64].

Collectively, studies show the diverse and stage-specific mechanisms that lncRNAs regulate cardiomyocyte differentiation and cardiovascular development. Their ability to modulate transcriptional programs, chromatin dynamics, and lineage specification highlights them as a promising therapeutic target for treating cardiovascular diseases and advancing regenerative medicine strategies. Further research into these molecules promises to unveil novel cardiac repair and regeneration approaches.

2.3.4 LncRNAs in myocyte differentiation

Myogenesis, the highly regulated process of muscle formation, is orchestrated by a dynamic network of epigenetic regulators, transcription factors, and long non-coding RNAs (lncRNAs). Muscle-specific lncRNAs play crucial roles at various stages of myoblast proliferation, differentiation, and myotube formation. Linc-MD1, one of the first identified muscle-specific lncRNAs, is activated during myoblast differentiation and acts as a competing endogenous RNA (ceRNA). It sponges miR-133 and miR-135 and enhances the expression of MEF2C and MAML1, two key transcription factors driving muscle-specific gene expression [65]. Similarly, lncRNA H19 is expressed during myoblast differentiation and enhances skeletal muscle differentiation through multiple mechanisms, including encoding miR-675 to suppress BMP signaling and Cdc6 expression [65]. It also promotes Igf2 expression, a critical myogenic stimulator [66]. H19 also regulates the SIRT1/FoxO1 axis, further underscoring its involvement in muscle-specific pathways [67].

The lncRNAs LncMyoD and MUNC demonstrate their importance in regulating MyoD expression and early myogenic differentiation. LncMyoD, located upstream of the MyoD gene, inhibits myoblast cell cycling and facilitates terminal differentiation by directly interacting with the IMP2 protein [68]. In contrast, MUNC acts as an enhancer RNA (eRNA) transcribed from the MyoD regulatory regions, driving the expression of key myogenic transcription factors like MYOGENIN and MYH3 [36]. Notably, silencing of MUNC delays muscle regeneration, highlighting its pivotal role in muscle repair [69]. Other lncRNAs such as Dum and Yam-1 also illustrate the diversity in lncRNA functions. MyoD induces Dum and facilitates myogenic differentiation by repressing Dppa2 through DNA methylation at its promoter [70]. Conversely, Yam-1 inhibits myogenesis by modulating miR-715 expression, which targets Wnt7b and impairs Wnt/β-catenin signaling [71].

Interestingly, lncRNA Malat1, initially identified in non-muscle contexts, inhibits skeletal muscle differentiation by regulating the miR-181a-Malat1-MyoD/Suv39h1 axis [72]. Malat1 suppression accelerates myoblast differentiation and enhances muscle regeneration which can be used as a therapeutic target in future research. Additionally, SRA (steroid receptor RNA activator) emerges as a coactivator of MyoD in myogenesis [73]. While its non-coding RNA form enhances myogenic transcriptional activity, alternative splicing generates a protein counterpart, SRAP, which antagonizes the RNA's coactivator function, demonstrating a finely tuned regulatory mechanism during muscle differentiation [74]. Collectively, lncRNAs are central regulators of skeletal muscle development and regeneration, offering their potential therapeutic uses for muscle-related disorders.

2.4 LncRNAs in hematopoietic stem cells and hematopoiesis

The hematopoietic system is a complex biological framework that produces blood cells and immune system components. This system plays a critical role in maintaining homeostasis in the human body. Red blood cells (RBCs) are vital for oxygen transportation, platelets are essential for blood coagulation, and white blood cells (WBCs) act as defenders against pathogens, serving as a part of the immune system [75]. In the early 1960s, Till and McCulloch revealed the existence of hematopoietic stem cells (HSCs), which are responsible for blood regeneration. These multipotent stem cells can self-renew and differentiate into all blood cell types [76]. The clinical significance of HSCs is profound, with applications in treating leukemia, lymphoma, and genetic disorders such as anemia and thalassemia. Bone marrow transplants, often derived from HSCs, are a cornerstone of these therapies, replacing a patient's defective blood cells with healthy ones from a compatible donor.

Recent advancements in understanding the regulatory mechanisms of hematopoiesis have highlighted the pivotal role of long non-coding RNAs (lncRNAs). lincRNA-EPS is among the first lncRNAs identified in erythropoiesis (red blood cell production). Hu et al. [77] conducted RNA-seq analysis on embryonic liver cells, with active erythropoiesis, and identified over 400 lncRNAs involved in the process of erythropoiesis. Of these, 163 were upregulated and 42 were downregulated during erythropoiesis. Their work focused on lincRNA-EPS, which was shown to be crucial for erythroid differentiation. Functional studies revealed that lincRNA-EPS depletion led to increased apoptosis and proliferation reduction of erythroid progenitors, even in the presence of erythropoietin. The anti-apoptotic role of lincRNA-EPS was confirmed by the overexpression studies of lincRNA-EPS in erythroid progenitors. Mechanistically, lincRNA-EPS represses the pro-apoptotic gene Pycard, which activates caspase-mediated apoptosis and ensures the survival of erythroid progenitors during differentiation. Another analysis [78] identified approximately 1100 lncRNAs expressed during the erythro-megakaryopoiesis (development of red blood cells and platelets) process in both murine and human models. Intriguingly, the chromatin immunoprecipitation sequencing (ChIP-seq) showed that about 75% of these lncRNAs were transcribed from promoter regions of genes, while 25% originated from enhancer regions. Key transcription factors, such as GATA1 and TAL1, were found to occupy these lncRNA loci. Knockdown experiments revealed that several lncRNAs, including LINCRED1, ERYTHRA, and SCARLETLTR, were critical for the terminal maturation of erythroblasts into reticulocytes. Eosinophils, another lineage derived from myeloid progenitors, play essential roles in parasitic immunity and allergic responses. The lncRNA EGO (eosinophil granule ontogeny) was discovered within the intronic region of the Itpr1 gene and found to be upregulated upon stimulation of CD34+ hematopoietic stem cells with IL-5, a cytokine specific to eosinophil development [79]. The expression of eosinophil proteins such as major basic protein (MBP) and eosinophil-derived neurotoxin (EDN) was weakened through silencing EGO. This indicates that EGO is essential for normal eosinophil differentiation [80]. Another well-studied lncRNA is HOTAIRM1 (HOX antisense intergenic myeloid 1), specifically expressed during granulocytic differentiation [81]. Induction of HOTAIRM1 is lineage-specific and regulated by the transcription factor PU.1, a key player in myeloid differentiation [82]. Silencing HOTAIRM1 impairs myeloid cell maturation. The absence of its activity suppresses the expression of granulocytic markers, including HoxA1 and HoxA4, as well as integrins such as CD11b and CD18 [81]. The lncRNAs also regulate T-cell differentiation and maturation. LincRNAs like LincR-Ccr2-5′AS regulate chemokine receptor expression, influencing T-cell migration during immune responses [83].

Emerging research continues to show the roles of lncRNAs in regulating hematopoiesis, providing deeper insights into their potential as therapeutic targets for hematological disorders. These non-coding RNAs not only regulate lineage-specific differentiation but also ensure the functional integrity of blood and immune cells, paving the way for innovative approaches in regenerative medicine and disease treatment.

2.5 LncRNAs in neural stem cells and neurogenesis

LncRNAs play crucial roles in the complex network governing neural differentiation and neurogenesis, mediating key processes in embryonic and adult neural stem cells (NSCs). Transcriptomic studies have revealed numerous lncRNAs with differential expression patterns during neural differentiation. For instance, genome-wide shRNA screening identified ~20 essential lncRNAs for NSC pluripotency such as TUNA and lncRNA-1604 [84]. They are highly expressed in the central nervous system (CNS). TUNA promotes neural differentiation by recruiting multiprotein complexes to gene promoters, enhancing histone modifications (H3K4me), and activating pluripotency factors like Nanog and Sox2 [84]. Similarly, lncRNA-1604 interacts with miR-200c to suppress ZEB1/2 expression, thereby facilitating ectodermal differentiation and upregulating neural progenitor markers like Sox1 and Nestin [85].

Another pivotal lncRNA, MEG3, is derived from the DLK1-DIO3 locus, which controls neural marker expression and modulates neural lineage differentiation rates [86, 87]. Evf2, transcribed from the Dlx5/6 intergenic region, is critical for hippocampal development, regulating GABAergic neuron production through topological and transcriptional mechanisms [88, 89]. Moreover, RMST collaborates with SOX2 to drive neurogenesis by enabling SOX2's binding to neurogenic transcription factor promoters [90]. PAUPAR and DALI are other lncRNAs that have different functions in NSC maintenance and differentiation. PAUPAR regulates Pax6 expression and gene networks involved in synaptic activity [91], while DALI modulates neuronal differentiation *via* cis and trans-regulatory mechanisms, influencing chromatin organization and DNA methylation [92].

There are several lncRNAs with inhibitory roles in neuronal differentiation. Expressed in NSCs, Pnky interacts with PTBP1 to maintain stem cell states, repressing premature differentiation into neurons. Pnky is critical in balancing neural stem cell self-renewal and differentiation, and its knockdown leads to increased neurogenesis [20, 93]. Collectively, these findings reveal that lncRNAs act as both positive and negative regulators in the intricate orchestration of neural differentiation, offering insights into their potential as therapeutic targets for neurodevelopmental disorders and regenerative medicine.

2.6 LncRNAs in spermatogonial stem cells and differentiation

Spermatogenesis is a highly regulated physiological process that generates millions of mature spermatozoa daily from spermatogonial stem cells (SSCs). These progenitor cells originate from primordial germ cells (PGCs), which migrate to the gonadal ridges during embryonic development and enter mitotic arrest until birth [94]. Recent studies have highlighted the pivotal role of long non-coding RNAs (lncRNAs) in regulating various stages of spermatogenesis, from SSC maintenance to spermatid differentiation.

The MRHL lncRNA (mouse recombination hotspot locus) plays a critical role in SSC physiology by negatively regulating the Wnt signaling pathway. MRHL physically interacts with the helicase protein p68/DDX5, restricting the nuclear translocation of β-catenin and thereby modulating downstream Wnt signaling [95]. Genome-wide occupancy studies revealed that MRHL binds to over 1400 chromatin loci, including 37 genes involved in spermatogenesis and differentiation [95]. Furthermore, Wnt signaling downregulates MRHL expression *via* CTBP1-mediated repressive histone modifications, highlighting a feedback loop in SSC differentiation [96]. LncRNA 033862 is transcribed in an antisense direction from exon 9 of the Gfra1 locus; it physically associates with Gfra1 chromatin and regulates SSC self-renewal and maintenance. Its knockdown leads to apoptosis, reduced expression of self-renewal genes (Bcl6b, Ccnd2, and Pou5f1), and diminished SSC colony size. These findings underscore the critical role of 033862 in maintaining SSC integrity and its regulation by GDNF signaling [97, 98].

TSX, identified through transcriptomic studies, is a testis-specific X-linked lncRNA that exhibits dynamic expression, peaking in pachytene spermatocytes during meiosis [99]. TSX knockout leads to increased apoptosis of germ cells at this stage, suggesting its role in facilitating meiotic progression [100]. Similarly, high-throughput analyses have highlighted lncRNAs AK011429 and AK00574, which correlate with protein-coding genes critical for spermatogonial renewal and differentiation [101]. While many of these lncRNAs remain functionally uncharacterized, their stage-specific expression patterns and chromatin interactions suggest they play intricate roles in germ cell development and testicular homeostasis.

3. LncRNAs in cancer stem cells and cancer

lncRNAs have become significant regulators in various biological processes, including cancer development and progression. Cancer stem cells (CSCs), which possess self-renewal and differentiation capabilities, are key contributors to tumor initiation, metastasis, and chemoresistance. LncRNAs regulate key pathways associated with CSCs, such as stemness, epithelial-mesenchymal transition (EMT), immune escape, and epigenetic reprogramming, highlighting their potential as therapeutic targets. This review explores the roles of lncRNAs in CSCs, their mechanistic involvement in cancer, and their clinical significance.

3.1 LncRNAs in tumorigenesis and CSC maintenance

Long non-coding RNAs (lncRNAs) are increasingly recognized as crucial regulators in cancer stem cells (CSCs), impacting tumor initiation, progression, and metastasis. CSCs are a small subset of cells within tumors capable of self-renewal, differentiation, and resistance to conventional therapies, contributing to tumor recurrence and chemoresistance. LncRNAs influence these functions through various mechanisms, including regulation of stemness, epithelial-mesenchymal transition (EMT), and interaction with the tumor microenvironment. Several lncRNAs are upregulated in CSCs and are associated with regulating critical stemness transcription factors such as Sox2, NANOG, and Oct4. By modulating these pathways, lncRNAs help maintain the CSC phenotype and enhance tumorigenicity [102].

3.1.1 LncRNAs and stemness regulation in CSCs

LncRNAs maintain the stem-like properties of CSCs by regulating key signaling pathways involved in self-renewal and differentiation. For example, HOTAIR, MALAT1, and H19 are among the most studied lncRNAs in CSCs [102]. These lncRNAs promote the expression of stemness-related genes by having interactions with other factors. LncRNAs function as "miRNA sponges," a well-known mechanism that prevents miRNAs from binding to their target mRNAs [103]. This regulatory process can increase the expression of stemness-related transcription factors such as Sox2 and ZEB1/2 [102]. For instance, MALAT1 promotes the activation of the Wnt/β-catenin signaling pathway, which is essential for the maintenance of stemness in various cancers, including liver cancers [104]. Similarly, the lncRNA H19, which is involved in the regulation of miR-675, modulates the expression of stemness factors in various CSCs [105].

3.1.2 LncRNAs in epithelial-mesenchymal transition (EMT) in CSCs

EMT is a crucial process for CSCs giving them migratory and invasive properties, contributing to metastasis. LncRNAs are key regulators of EMT in CSCs, and their expression levels are often linked to cancer progression. For example, HOTAIR, MALAT1, and MEG3 promote EMT by enhancing the expression of key transcription factors such as ZEB1/2 and Twist. These transcription factors drive the loss of epithelial markers and the gain of mesenchymal markers, which are hallmarks of EMT [106]. LncRNA HOTAIR is upregulated in CSCs and induces EMT. It promotes tumorigenesis and metastasis [107, 108]. Additionally, lncRNAs such as linc00617 have been identified as critical regulators of EMT, acting through the activation of Sox2 transcription and promoting cancer progression and metastasis [109].

3.1.3 LncRNAs and chemoresistance in CSCs

CSCs are key factors of chemoresistance. Chemoresistance remains one of the major challenges in cancer treatment. LncRNAs have been implicated in the regulation of chemoresistance by modulating the expression of genes involved in drug efflux, apoptosis resistance, and DNA repair. For example, lncRNAs such as H19 and MACC1-AS1 are effective in drug efflux, thus promoting resistance to chemotherapeutic agents. They enhance the expression of ATP-binding cassette (ABC) transporters [110]. Additionally, lncRNAs like LET, MEG3, and GAS5 counteract chemoresistance by regulating apoptosis and reducing stemness [102]. In bladder cancer, for instance, the lncRNA LET is downregulated by TGF-β1 signaling, leading to enhanced stemness and resistance to gemcitabine [111]. These findings underscore the potential of targeting specific lncRNAs to overcome chemoresistance in CSCs.

3.1.4 LncRNAs and tumor microenvironment interactions

The tumor microenvironment (TME) plays a critical role in the regulation of CSC behavior, and lncRNAs are critical in the interaction between CSC and TME.

Several lncRNAs, including HOTAIR, MALAT1, and H19, are upregulated in CSCs by secreted factors from the TME, such as TGF-β1 [112]. This cytokine, secreted by various cell types within the TME (e.g., cancer-associated fibroblasts and macrophages), promotes stemness, EMT, and metastasis [102]. In gastric cancer, TGF-β1 secreted by mesenchymal stem cells induces the expression of MACC1-AS1, which enhances CSC properties and chemoresistance [110]. Additionally, lncRNAs such as H19 and Sox2ot, which are packaged into exosomes, facilitate communication between CSCs and other cells in the TME, resulting in the support of stemness and EMT [113]. These findings highlight the intricate interplay between CSCs and the TME and the potential of targeting lncRNA-mediated pathways to disrupt CSC functions.

3.1.5 LncRNAs and immune escape in CSCs

CSCs are known to regulate immune suppression within the TME. LncRNAs contribute to immune escape by regulating immune checkpoint molecules, cytokine production, and antigen presentation. CSCs express immune checkpoint molecules such as PD-L1 and Galectin-3, suppressing T-cell activity and promoting immune tolerance [114]. Additionally, lncRNAs like TUC339 and AFAP1-AS1 have been shown to modulate the immune response by influencing the function of immune cells such as myeloid cells and T cells [115, 116]. The release of lncRNAs in exosomes from CSCs further contributes to immune suppression by dampening the inflammatory response. As a result, lncRNAs represent potential therapeutic targets for overcoming immune escape and improving the efficacy of cancer immunotherapies.

3.2 Oncogenic and tumor suppressor LncRNAs

Depending on their cellular context, lncRNAs can act as either oncogenes or tumor suppressors. HOTAIR, a well-known oncogenic lncRNA, promotes breast cancer metastasis by enhancing tumor cell invasion. Overexpression of HOTAIR is associated with poor prognosis in breast cancer, and its knockdown significantly inhibits invasiveness [11]. On the other hand, lncRNAs such as Xist function as tumor suppressors. Knockdown of Xist in hematopoietic cells led to aggressive myeloproliferative neoplasm and myelodysplastic syndrome, suggesting that Xist plays a vital role in preventing genomic instability and cancer development [117]. lincRNA-p21, regulated by the tumor suppressor p53, also is a tumor suppressor lncRNA. It interacts with hnRNP-K to mediate gene repression, which is essential for maintaining cellular homeostasis in response to DNA damage [118].

3.3 LncRNAs in epigenetic reprogramming in cancer

Epigenetic changes have a key role in tumorigenesis and cancer, and lncRNAs have been identified as critical regulators of chromatin remodeling and gene expression. LncRNAs such as HOTAIR and ANRIL are involved in silencing tumor suppressor genes by recruiting polycomb repressive complexes (PRC2) to chromatin. HOTAIR, for instance, promotes tumorigenicity in breast cancer by altering PRC2 localization, which correlates with increased cell invasiveness [119]. Similarly, ANRIL targets the p15 locus, inhibiting its expression and contributing to cancer cell proliferation [120]. SChLAP1, another lncRNA, promotes metastasis by antagonizing the SWI/SNF chromatin-remodeling complex, the importance of lncRNAs in the epigenetic reprogramming of cancer cells [121].

3.4 LncRNAs in microvascular invasion and angiogenesis

The formation of new blood vessels known as angiogenesis, is essential for tumor growth and metastasis. Recent studies have highlighted the role of lncRNAs in promoting angiogenesis. MVIH, a lncRNA associated with microvascular invasion in hepatocellular carcinoma, is overexpressed in tumors and correlates with poor prognosis [122]. MVIH facilitates angiogenesis by inhibiting phosphoglycerate kinase 1, a key enzyme involved in energy metabolism, thus enhancing tumor growth [120, 123]. Additionally, MEG3 has been shown to regulate angiogenesis through p53 signaling [124]. Loss of MEG3 contributes to tumor progression by enhancing the migration and invasion [125].

3.5 Clinical implications of LncRNAs in cancer therapy

LncRNAs are considered potential therapeutic targets. Several lncRNAs have been linked to poor prognosis, chemoresistance, and metastasis in various cancers, making them promising biomarkers for early detection and monitoring. Targeting lncRNAs with antisense oligonucleotides (ASOs) or CRISPR/Cas9 technology has shown promising results in preclinical models, suggesting that lncRNA-based therapies could inhibit CSC functions and improve the treatment process [126–130]. Additionally, the stability of lncRNAs in body fluids, such as serum and plasma, makes them excellent candidates for non-invasive biomarkers [131]. Identifying and targeting specific lncRNAs involved in CSC biology may lead to the development of novel therapeutic strategies that target the root causes of cancer progression and resistance.

4. Conclusions

Long non-coding RNAs (lncRNAs) have emerged as essential regulators in stem cell biology, modulating gene expression at epigenetic, transcriptional, and post-transcriptional levels. Their ability to interact with chromatin, transcription factors, and RNA-binding proteins underscores their functional diversity and specificity. As critical modulators of pluripotency and differentiation, lncRNAs like H19, MALAT1, and linc-ROR exemplify their roles in maintaining stem cell identity and directing lineage commitment. Despite their immense potential, much remains to be uncovered about their context-specific mechanisms and therapeutic applications. Emerging tools like single-cell, spatial transcriptomics, and CRISPR-based functional studies are promising in deepening our understanding of lncRNA biology. Furthermore, therapeutic approaches of lncRNA-targeting technologies, including antisense oligonucleotides and synthetic lncRNAs, could make a great progression in regenerative medicine and cancer therapies. Addressing challenges in delivery, specificity, and safety will be crucial for clinical translation. By integrating bioinformatics, advanced technologies, and collaborative efforts, lncRNAs can unlock new opportunities for precision medicine and advancements in stem cell research and therapeutic uses.

Acknowledgements

This research was sponsored by the Faculty of Biotechnology at Amol University of Special Modern Technologies in the Islamic Republic of Iran and received no external funding.

Author contributions

M.E.: Carried out and edited the article, assembled the data, and designed the experiment; H.A.: Carried out and edited the article; N.G.: Assembled the data, designed the experiment, wrote, and edited the manuscript; T.S.: Carried out and edited the article.

Conflict of interest

The authors declare no conflict of interest.

AI tools

Artificial intelligence tools are strictly prohibited from generating data, composing article sections, formulating scientific conclusions, or engaging in any activities that could result in scientific misrepresentation. Their sole purpose is for editorial correction and grammar checks within the text, aimed at refining and enhancing clarity and accuracy. These tools are not permitted to manipulate data or misrepresent scientific discourse in any way.

Author details

Melika Emarati[1], Hossein Azizi[1]*, Nima Ghasemi[1] and Thomas Skutella[2]

1 Faculty of Biotechnology, Amol University of Special Modern Technologies, Amol, Iran

2 Institute for Anatomy and Cell Biology, Medical Faculty, University of Heidelberg, Heidelberg, Germany

*Address all correspondence to: h.azizi@ausmt.ac.ir

References

[1] Estrada-Meza C et al. Recent insights into the microRNA and long non-coding RNA-mediated regulation of stem cell populations. 3 Biotech. 2022;**12**(10):270

[2] Peng WX, Koirala P, Mo YY. LncRNA-mediated regulation of cell signaling in cancer. Oncogene. 2017;**36**(41):5661-5667

[3] Mattick JS et al. Long non-coding RNAs: Definitions, functions, challenges and recommendations. Nature Reviews Molecular Cell Biology. 2023;**24**(6):430-447

[4] Katayama S et al. Antisense transcription in the mammalian transcriptome. Science. 2005;**309**(5740):1564-1566

[5] Marques AC, Ponting CP. Intergenic lncRNAs and the evolution of gene expression. Current Opinion in Genetics & Development. 2014;**27**:48-53

[6] Mattick JS, Rinn JL. Discovery and annotation of long noncoding RNAs. Nature Structural & Molecular Biology. 2015;**22**(1):5-7

[7] Ulitsky I, Bartel DP. lincRNAs: Genomics, evolution, and mechanisms. Cell. 2013;**154**(1):26-46

[8] Ma L, Bajic VB, Zhang Z. On the classification of long non-coding RNAs. RNA Biology. 2013;**10**(6):925-933

[9] Balas MM, Johnson AM. Exploring the mechanisms behind long noncoding RNAs and cancer. Non-coding RNA Research. 2018;**3**(3):108-117

[10] Meller VH, Joshi SS, Deshpande N. Modulation of chromatin by noncoding RNA. Annual Review of Genetics. 2015;**49**:673-695

[11] Gupta RA et al. Long non-coding RNA HOTAIR reprograms chromatin state to promote cancer metastasis. Nature. 2010;**464**(7291):1071-1076

[12] DeMaria CT, Brewer G. AUF1 binding affinity to A+U-rich elements correlates with rapid mRNA degradation. The Journal of Biological Chemistry. 1996;**271**(21):12179-12184

[13] Malakar P et al. The nexus of long noncoding RNAs, splicing factors, alternative splicing and their modulations. RNA Biology. 2024;**21**(1):1-20

[14] Song J, Cui Q, Gao J. Roles of lncRNAs related to the p53 network in breast cancer progression. Frontiers in Oncology. 2024;**14**:1453807

[15] Carrieri C et al. Long non-coding antisense RNA controls Uchl1 translation through an embedded SINEB2 repeat. Nature. 2012;**491**(7424):454-457

[16] Alsadat Mahmoudian R et al. Interaction between LINC-ROR and stemness state in gastric cancer cells with *Helicobacter pylori* infection. Iranian Biomedical Journal. 2021;**25**(3):157-168

[17] Chen H-F, Wu K-J. LncRNAs and asymmetric cell division: The epigenetic mechanisms. Biomedical Journal. 2024:100774. DOI: 10.1016/j.bj.2024.100774. ISSN 2319-4170. Available from: https://www.sciencedirect.com/science/article/pii/S2319417024000775

[18] Zhao D et al. Advances in the identification of long non-coding RNA binding proteins. Analytical Biochemistry. 2022;**639**:114520

[19] Hu S, Shan G. LncRNAs in stem cells. Stem Cells International. 2016;**2016**(1):2681925

[20] Chen J et al. LncRNA functions as a new emerging epigenetic factor in determining the fate of stem cells. Frontiers in Genetics. 2020;**11**:277

[21] Guttman M et al. lincRNAs act in the circuitry controlling pluripotency and differentiation. Nature. 2011;**477**(7364):295-300

[22] Lu X et al. The retrovirus HERVH is a long noncoding RNA required for human embryonic stem cell identity. Nature Structural & Molecular Biology. 2014;**21**(4):423-425

[23] Shahryari A et al. Long non-coding RNA SOX2OT: Expression signature, splicing patterns, and emerging roles in pluripotency and tumorigenesis. Frontiers in Genetics. 2015;**6**:196

[24] Song Y et al. Functional bidirectionality of ERV-derived long non-coding RNAs in humans. International Journal of Molecular Sciences. 2024;**25**(19):10481

[25] Ghosal S, Das S, Chakrabarti J. Long noncoding RNAs: New players in the molecular mechanism for maintenance and differentiation of pluripotent stem cells. Stem Cells and Development. 2013;**22**(16):2240-2253

[26] Xu C et al. Long non-coding RNA GAS5 controls human embryonic stem cell self-renewal by maintaining NODAL signalling. Nature Communications. 2016;**7**(1):13287

[27] Yang J, Jiang W. The role of SMAD2/3 in human embryonic stem cells. Frontiers in Cell and Development Biology. 2020;**8**:653

[28] Xu X et al. Telomeric noncoding RNA promotes mouse embryonic stem cell self-renewal through inhibition of TCF3 activity. American Journal of Physiology-Cell Physiology. 2018;**314**(6):C712-C720

[29] Minkovsky A, Patel S, Plath K. Concise review: Pluripotency and the transcriptional inactivation of the female mammalian X chromosome. Stem Cells. 2012;**30**(1):48-54

[30] Luchsinger-Morcelle SJ, Gribnau J, Mira-Bontenbal H. Orchestrating asymmetric expression: Mechanisms behind xist regulation. Epigenomes. 2024;**8**(1):6

[31] Keihani S, Kluever V, Fornasiero EF. Brain long noncoding RNAs: Multitask regulators of neuronal differentiation and function. Molecules. 2021;**26**(13):3951

[32] Lu J et al. lncRNAs regulate cell stemness in physiology and pathology during differentiation and development. American Journal of Stem Cells. 2024;**13**(2):59-74

[33] Fan W, Li X. RNA-mediated regulation of glycolysis in embryonic stem cell pluripotency and differentiation. Current Stem Cell Reports. 2024;**10**(3):58-64

[34] Maherali N et al. Directly reprogrammed fibroblasts show global epigenetic remodeling and widespread tissue contribution. Cell Stem Cell. 2007;**1**(1):55-70

[35] Stadtfeld M et al. Defining molecular cornerstones during fibroblast to iPS cell reprogramming in mouse. Cell Stem Cell. 2008;**2**(3):230-240

[36] Pal D, Rao MRS. Long noncoding RNAs in pluripotency of stem cells and cell fate specification. Advances in Experimental Medicine and

Biology. 2017;**1008**:223-252. DOI: 10.1007/978-981-10-5203-3_8

[37] Chaudhary R, Lal A. Long noncoding RNAs in the p53 network. Wiley Interdisciplinary Reviews RNA. 2017;**8**(3)e1410. DOI: 10.1002/wrna.1410

[38] Xu N et al. MicroRNA-145 regulates OCT4, SOX2, and KLF4 and represses pluripotency in human embryonic stem cells. Cell. 2009;**137**(4):647-658

[39] Kim DH et al. Single-cell transcriptome analysis reveals dynamic changes in lncRNA expression during reprogramming. Cell Stem Cell. 2015;**16**(1):88-101

[40] Ding D-C, Shyu W-C, Lin S-Z. Mesenchymal stem cells. Cell Transplantation. 2011;**20**(1):5-14

[41] Yang Y et al. Long non-coding RNA regulation of mesenchymal stem cell homeostasis and differentiation: Advances, challenges, and perspectives. Frontiers in Cell and Development Biology. 2021;**9**:711005

[42] Hutchings G et al. The proliferation and differentiation of adipose-derived stem cells in neovascularization and angiogenesis. International Journal of Molecular Sciences. 2020;**21**(11)

[43] Zhang P et al. LncRNA-mediated adipogenesis in different adipocytes. International Journal of Molecular Sciences. 2022;**23**(13)

[44] Elattar S, Satyanarayana A. Can brown fat win the battle against white fat? Journal of Cellular Physiology. 2015;**230**(10):2311-2317

[45] Sheng L et al. New insights into the long non-coding RNA SRA: Physiological functions and mechanisms of action. Frontiers in Medicine (Lausanne). 2018;**5**:244

[46] Sufianov A et al. The role of long non-coding RNAs in the development of adipose cells. Non-coding RNA Research. 2023;**8**(2):255-262

[47] Guo Q et al. Regulation of bone marrow mesenchymal stem cell fate by long non-coding RNA. Bone. 2020;**141**:115617

[48] Li M et al. The long noncoding RNA GAS5 negatively regulates the adipogenic differentiation of MSCs by modulating the miR-18a/CTGF axis as a ceRNA. Cell Death & Disease. 2018;**9**(5):554

[49] Kalwa M et al. The lncRNA HOTAIR impacts on mesenchymal stem cells via triple helix formation. Nucleic Acids Research. 2016;**44**(22):10631-10643

[50] Rey F et al. Role of long non-coding RNAs in adipogenesis: State of the art and implications in obesity and obesity-associated diseases. Obesity Reviews. 2021;**22**(7):e13203

[51] Li M et al. Long non-coding RNA ADNCR suppresses adipogenic differentiation by targeting miR-204. Biochimica et Biophysica Acta (BBA) - Gene Regulatory Mechanisms. 2016;**1859**(7):871-882

[52] Wang L, Qi L. The role and mechanism of long non-coding RNA H19 in stem cell osteogenic differentiation. Molecular Medicine. 2021;**27**(1):86

[53] Lanzillotti C et al. Long non-coding RNAs and microRNAs interplay in osteogenic differentiation of mesenchymal stem cells. Frontiers in Cell and Development Biology. 2021;**9**:646032

[54] An F et al. Research progress on the role of lncRNA-miRNA networks in regulating adipogenic and osteogenic differentiation of bone marrow mesenchymal stem cells in osteoporosis.

Frontiers in Endocrinology (Lausanne). 2023;**14**:1210627

[55] Zhang J, Tao Z, Wang Y. Long non-coding RNA DANCR regulates the proliferation and osteogenic differentiation of human bone-derived marrow mesenchymal stem cells via the p38 MAPK pathway. International Journal of Molecular Medicine. 2018;**41**(1):213-219

[56] Mishra A et al. Differential expression of non-coding RNAs in stem cell development and therapeutics of bone disorders. Cells. 2023;**12**. DOI: 10.3390/cells12081159

[57] Wang Z et al. Epigenetics: Novel crucial approach for osteogenesis of mesenchymal stem cells. Journal of Tissue Engineering. 2023;**14**:20417314231175364

[58] Grote P et al. The tissue-specific lncRNA Fendrr is an essential regulator of heart and body wall development in the mouse. Developmental Cell. 2013;**24**(2):206-214

[59] Xue Z et al. A G-rich motif in the lncRNA braveheart interacts with a zinc-finger transcription factor to specify the cardiovascular lineage. Molecular Cell. 2016;**64**(1):37-50

[60] Klattenhoff CA et al. Braveheart, a long noncoding RNA required for cardiovascular lineage commitment. Cell. 2013;**152**(3):570-583

[61] Han L, Yang L. Multidimensional mechanistic spectrum of long non-coding RNAs in heart development and disease. Frontiers in Cardiovascular Medicine. 2021;**8**:728746

[62] Ounzain S et al. CARMEN, a human super enhancer-associated long noncoding RNA controlling cardiac specification, differentiation and homeostasis. Journal of Molecular and Cellular Cardiology. 2015;**89**:98-112

[63] Frank S et al. yylncT defines a class of divergently transcribed lncRNAs and safeguards the T-mediated mesodermal commitment of human PSCs. Cell Stem Cell. 2019;**24**(2):318-327.e8

[64] Li Y et al. Dynamic organization of lncRNA and circular RNA regulators collectively controlled cardiac differentiation in humans. eBioMedicine. 2017;**24**:137-146

[65] Li Y et al. Long non-coding RNAs in the regulation of skeletal myogenesis and muscle diseases. Cancer Letters. 2018;**417**:58-64

[66] Wang Y, Hylemon PB, Zhou H. Long noncoding RNA H19: A key player in liver diseases. Hepatology. 2021;**74**(3):1652-1659

[67] Nandwani A, Rathore S, Datta M. LncRNA H19 inhibition impairs endoplasmic reticulum-mitochondria contact in hepatic cells and augments gluconeogenesis by increasing VDAC1 levels. Redox Biology. 2024;**69**:102989

[68] Gong C et al. A long non-coding RNA, LncMyoD, regulates skeletal muscle differentiation by blocking IMP2-mediated mRNA translation. Developmental Cell. 2015;**34**(2):181-191

[69] Pinheiro A, Naya FJ. The key Lnc (RNA)s in cardiac and skeletal muscle development, regeneration, and disease. Journal of Cardiovascular Development and Disease. 2021;**8**. DOI: 10.3390/jcdd8080084

[70] Wang L et al. LncRNA Dum interacts with Dnmts to regulate Dppa2 expression during myogenic differentiation and

muscle regeneration. Cell Research. 2015;**25**(3):335-350

[71] Lu L et al. Genome-wide survey by ChIP-seq reveals YY1 regulation of lincRNAs in skeletal myogenesis. The EMBO Journal. 2013;**32**(19):2575-2588

[72] Chen X et al. Malat1 regulates myogenic differentiation and muscle regeneration through modulating MyoD transcriptional activity. Cell Discovery. 2017;**3**(1):1-23

[73] Caretti G et al. The RNA helicases p68/p72 and the noncoding RNA SRA are coregulators of MyoD and skeletal muscle differentiation. Developmental Cell. 2006;**11**(4):547-560

[74] Hubé F et al. Steroid receptor RNA activator protein binds to and counteracts SRA RNA-mediated activation of MyoD and muscle differentiation. Nucleic Acids Research. 2011;**39**(2):513-525

[75] Jagannathan-Bogdan M, Zon LI. Hematopoiesis. Development. 2013;**140**(12):2463-2467

[76] Till JE, McCulloch CE. A direct measurement of the radiation sensitivity of normal mouse bone marrow cells. Radiation Research. 1961;**14**:213-222

[77] Hu W et al. Long noncoding RNA-mediated anti-apoptotic activity in murine erythroid terminal differentiation. Genes & Development. 2011;**25**(24):2573-2578

[78] Paralkar VR, Weiss MJ. A new 'Linc' between noncoding RNAs and blood development. Genes & Development. 2011;**25**(24):2555-2558

[79] Wagner LA et al. EGO, a novel, noncoding RNA gene, regulates eosinophil granule protein transcript expression. Blood. 2007;**109**(12):5191-5198

[80] Ghafouri-Fard S, Niazi V, Taheri M. Role of miRNAs and lncRNAs in hematopoietic stem cell differentiation. Non-coding RNA Research. 2021;**6**(1):8-14

[81] Zhang X et al. A myelopoiesis-associated regulatory intergenic noncoding RNA transcript within the human HOXA cluster. Blood. 2009;**113**(11):2526-2534

[82] Wei S et al. PU.1 controls the expression of long noncoding RNA HOTAIRM1 during granulocytic differentiation. Journal of Hematology & Oncology. 2016;**9**(1):44

[83] Liu C et al. Long noncoding RNAs as orchestrators of CD4(+) T-cell fate. Frontiers in Cell and Development Biology. 2022;**10**:831215

[84] Lin N et al. An evolutionarily conserved long noncoding RNA TUNA controls pluripotency and neural lineage commitment. Molecular Cell. 2014;**53**(6):1005-1019

[85] Weng R et al. Long noncoding RNA-1604 orchestrates neural differentiation through the miR-200c/ZEB Axis. Stem Cells. 2018;**36**(3):325-336

[86] Yen YP et al. Dlk1-Dio3 locus-derived lncRNAs perpetuate postmitotic motor neuron cell fate and subtype identity. eLife. 2018;**7**

[87] Mo CF et al. Loss of non-coding RNA expression from the DLK1-DIO3 imprinted locus correlates with reduced neural differentiation potential in human embryonic stem cell lines. Stem Cell Research & Therapy. 2015;**6**(1):1

[88] D'haene E, Vergult S. Interpreting the impact of noncoding structural variation in neurodevelopmental disorders. Genetics in Medicine. 2021;**23**(1):34-46

[89] Feng J et al. The Evf-2 noncoding RNA is transcribed from the Dlx-5/6 ultraconserved region and functions as a Dlx-2 transcriptional coactivator. Genes & Development. 2006;**20**(11):1470-1484

[90] Wuebben EL, Rizzino A. The dark side of SOX2: Cancer—A comprehensive overview. Oncotarget. 2017;**8**(27): 44917-44943

[91] Xu Y et al. PAUPAR and PAX6 sequentially regulate human embryonic stem cell cortical differentiation. Nucleic Acids Research. 2021;**49**(4):1935-1950

[92] Chalei V et al. The long non-coding RNA Dali is an epigenetic regulator of neural differentiation. eLife. 2014;**3**:e04530

[93] Shi J et al. Epigenetic regulation in adult neural stem cells. Frontiers in Cell and Development Biology. 2024;**12**:1331074

[94] Ghasemi N et al. The role of Plzf in spermatogonial stem cell maintenance and differentiation: Mapping the transcriptional dynamics and key interactions. Cells. 2024;**13**. DOI: 10.3390/cells13231930

[95] Akhade VS et al. Genome wide chromatin occupancy of mrhl RNA and its role in gene regulation in mouse spermatogonial cells. RNA Biology. 2014;**11**(10):1262-1279

[96] Akhade VS et al. Mechanism of Wnt signaling induced down regulation of mrhl long non-coding RNA in mouse spermatogonial cells. Nucleic Acids Research. 2016;**44**(1):387-401

[97] Adriaens C et al. p53 induces formation of NEAT1 lncRNA-containing paraspeckles that modulate replication stress response and chemosensitivity. Nature Medicine. 2016;**22**(8):861-868

[98] Li L et al. A long non-coding RNA interacts with Gfra1 and maintains survival of mouse spermatogonial stem cells. Cell Death & Disease. 2016;**7**(3):e2140-e2140

[99] Anguera MC et al. Tsx produces a long noncoding RNA and has general functions in the germline, stem cells, and brain. PLoS Genetics. 2011;**7**(9):e1002248

[100] Liu W et al. A novel meiosis-related lncRNA, Rbakdn, contributes to spermatogenesis by stabilizing Ptbp2. Frontiers in Genetics. 2021;**12**:752495

[101] Pal D, Rao MRS. Long Noncoding RNAs in Pluripotency of Stem Cells and Cell Fate Specification. Advances in Experimental Medicine and Biology. 2017;**1008**:223-252. DOI: 10.1007/978-981-10-5203-3_8

[102] Schwerdtfeger M et al. Long non-coding RNAs in cancer stem cells. Translational Oncology. 2021;**14**(8):101134

[103] Ratti M et al. MicroRNAs (miRNAs) and long non-coding RNAs (lncRNAs) as new tools for cancer therapy: First steps from bench to bedside. Targeted Oncology. 2020;**15**(3):261-278

[104] Hussein MA et al. MALAT-1 is a key regulator of epithelial–mesenchymal transition in cancer: A potential therapeutic target for metastasis. Cancers. 2024;**16**. DOI: 10.3390/cancers16010234

[105] Lecerf C et al. Propagation and maintenance of cancer stem cells: A major influence of the long non-coding RNA H19. Cells. 2020;**9**. DOI: 10.3390/cells9122613

[106] Cheng JT et al. Insights into biological role of LncRNAs in epithelial-mesenchymal transition. Cells. 2019;**8**(10)

[107] Yang Z et al. Overexpression of long non-coding RNA HOTAIR predicts tumor recurrence in hepatocellular carcinoma patients following liver transplantation. Annals of Surgical Oncology. 2011;**18**(5):1243-1250

[108] Li X et al. Long non-coding RNA HOTAIR, a driver of malignancy, predicts negative prognosis and exhibits oncogenic activity in oesophageal squamous cell carcinoma. British Journal of Cancer. 2013;**109**(8):2266-2278

[109] Li H et al. Long noncoding RNA linc00617 exhibits oncogenic activity in breast cancer. Molecular Carcinogenesis. 2017;**56**(1):3-17

[110] He W et al. MSC-regulated lncRNA MACC1-AS1 promotes stemness and chemoresistance through fatty acid oxidation in gastric cancer. Oncogene. 2019;**38**(23):4637-4654

[111] Zhuang J et al. TGFβ1 promotes gemcitabine resistance through regulating the LncRNA-LET/NF90/miR-145 signaling axis in bladder cancer. Theranostics. 2017;**7**(12):3053-3067

[112] Ahmadi A et al. Transforming growth factor-β signaling: Tumorigenesis and targeting for cancer therapy. Journal of Cellular Physiology. 2019;**234**(8):12173-12187

[113] Li Z et al. Tumor-derived exosomal lnc-Sox2ot promotes EMT and stemness by acting as a ceRNA in pancreatic ductal adenocarcinoma. Oncogene. 2018;**37**(28):3822-3838

[114] Wei J et al. Glioma-associated cancer-initiating cells induce immunosuppression. Clinical Cancer Research. 2010;**16**(2):461-473

[115] Li X et al. Regulation of macrophage activation and polarization by HCC-derived exosomal lncRNA TUC339. International Journal of Molecular Sciences. 2018;**19**. DOI: 10.3390/ijms19102958

[116] Tang Y et al. Co-expression of AFAP1-AS1 and PD-1 predicts poor prognosis in nasopharyngeal carcinoma. Oncotarget. 2017;**8**(24):39001-39011

[117] Yildirim E et al. Xist RNA is a potent suppressor of hematologic cancer in mice. Cell. 2013;**152**(4):727-742

[118] Huarte M et al. A large intergenic noncoding RNA induced by p53 mediates global gene repression in the p53 response. Cell. 2010;**142**(3):409-419

[119] Kogo R et al. Long noncoding RNA HOTAIR regulates polycomb-dependent chromatin modification and is associated with poor prognosis in colorectal cancers. Cancer Research. 2011;**71**(20):6320-6326

[120] Eades G et al. Long non-coding RNAs in stem cells and cancer. World Journal of Clinical Oncology. 2014;**5**(2):134-141

[121] Prensner JR et al. The long noncoding RNA SChLAP1 promotes aggressive prostate cancer and antagonizes the SWI/SNF complex. Nature Genetics. 2013;**45**(11):1392-1398

[122] Zhao X et al. Roles and molecular mechanisms of biomarkers in hepatocellular carcinoma with microvascular invasion: A review. Journal of Clinical and Translational Hepatology. 2023;**11**(5):1170-1183

[123] Zhang K, Sun L, Kang Y. Regulation of phosphoglycerate kinase 1 and its critical role in cancer. Cell Communication and Signaling. 2023;**21**(1):240

[124] Zhang L et al. The biological roles and molecular mechanisms of

long non-coding RNA MEG3 in the hallmarks of cancer. Cancers. 2022;**14**. DOI: 10.3390/cancers14246032

[125] Tu H et al. Loss of MEG3 contributes to the enhanced migration and invasion in arsenic-induced carcinogenesis through NQO1/FSCN1 pathway. American Journal of Cancer Research. 2023;**13**(6):2307-2322

[126] Jiang MC et al. Emerging roles of lncRNA in cancer and therapeutic opportunities. American Journal of Cancer Research. 2019;**9**(7):1354-1366

[127] Li C et al. CRISPR-CasRx targeting LncRNA LINC00341 inhibits tumor cell growth in vitro and in vivo. Frontiers in Molecular Biosciences. 2021;**8**:638995

[128] Mahato RK et al. Targeting long non-coding RNAs in cancer therapy using CRISPR-Cas9 technology: A novel paradigm for precision oncology. Journal of Biotechnology. 2024;**379**:98-119

[129] Tian L et al. Targeting LncRNA LLNLR-299G3. 1 with antisense oligonucleotide inhibits malignancy of esophageal squamous cell carcinoma cells in vitro and in vivo. Oncology Research. 2023;**31**(4):463

[130] Sarkar S et al. Antisense oligonucleotide embedded context responsive nanoparticles derived from synthetic ionizable lipids for lncRNA targeted therapy of breast cancer. ACS Applied Materials & Interfaces. 2024;**16**(35):45871-45887

[131] Tang J et al. Long non-coding RNA (lncRNA) in oral squamous cell carcinoma: Biological function and clinical application. Cancers. 2021;**13**. DOI: 10.3390/cancers13235944

Section 4

Long Non-Coding RNA's in Plants

Chapter 9

Perspective Chapter: Long Non-Coding RNAs in Plants – Unlocking Hidden Regulators for Stress Tolerance and Crop Resilience

Komal G. Lakhani, Rasmieh Hamid and Gopal V. Marviya

Abstract

Long non-coding RNAs (lncRNAs) are emerging as crucial regulators in plant biology, particularly under abiotic and biotic stress conditions. These molecules, which lack protein-coding potential, exhibit diverse roles in regulating gene expression at transcriptional, post-transcriptional, and epigenetic levels. Recent advancements reveal their involvement in critical stress responses, including drought, salinity, extreme temperatures, and heavy metal toxicity, highlighting their potential in enhancing crop resilience amidst changing climatic conditions. Additionally, lncRNAs influence chromatin remodeling, histone modifications, and DNA methylation, underlining their role in epigenetic regulation. High-throughput sequencing technologies and computational tools have significantly advanced the identification and functional annotation of lncRNAs across various plant species, paving the way for innovative strategies in crop improvement programmes. This chapter explores the evolution, classification, functional mechanisms, and stress-specific roles of lncRNAs, providing a comprehensive understanding of their applications in sustainable agriculture.

Keywords: lncRNAs, abiotic stress, biotic stress, gene regulation, climate resilience

1. Introduction

In the twenty-first century, the global challenge of food scarcity persists and is exacerbated by the profound effects of climate change [1]. It is estimated that the world population will grow to 10.6 billion people by 2050, increasing concerns about food security [2]. However, severe unfavorable climatic conditions such as drought, high temperatures, unpredictable rainfall, salinity, and the increase in CO_2 levels in the atmosphere are the main causes of impeded crop growth and reduced food production. These stress factors lead to changes in plant physiology, including altered photosynthetic efficiency, nutrient uptake, and water-use efficiency, which in turn lead to lower yields [3, 4].

The increase in atmospheric CO_2 levels and temperature leads to a decrease in various essential nutrient compositions such as the protein, vitamin, and mineral content of crops, resulting in malnutrition, which affects human health and is the cause of various diseases [5]. In the current scenario, several studies have found that cultivation of staple crops such as wheat, rice, and maize under various abiotic stress conditions leads to deterioration in the composition of various essential nutrients such as protein, zinc, and iron by reducing the photosynthetic process and down-regulating the expression of various genes in the plants that are necessary for increasing the nutrient content in the food crops. The decline in nutrient quality poses a significant risk to global food shortages.

To counteract the disadvantages of climate change, the scientific community has focused on developing new molecular genetic approaches to climate change resilience. There are many unanswered questions about how plants adapt to stress conditions by altering the expression of different genes in a spatio-temporal manner by regulating biological functions. In this context, various genes have been altered by overexpression and silencing using gene transformation and gene modification to improve their ability to regulate stress. In addition, unknown intergenic spaces between protein-coding genes are found by analysing unclear "experimental artefacts" or "junk DNA" using sequencing technologies such as transcriptome analysis [6].

The non-coding RNAs or non-protein-coding RNAs represent a diverse class of transcripts that differ in their mechanism and genetic genealogy and include both regulatory non-coding RNAs and housekeeping RNAs [7]. Among regulatory ncRNAs, two major types are classified based on nucleotide length: (i) long noncoding RNAs (lncRNAs), which are predominantly over 200 nucleotides in length, and (ii) small noncoding RNAs, which are approximately 50–200 nucleotides in length, including small interfering RNAs (siRNAs), Piwi-interacting RNAs (piRNAs), and microRNAs (miRNAs) [8, 9]. The functional significance of lncRNAs is still uncertain because, in contrast to the well-characterized mRNAs and miRNAs, they show only low expression and are poorly conserved in different species [10, 11]. In plants, long non-coding RNAs (lncRNAs), which help to increase plant production and improve nutrient content by regulating the expression of various genes at different levels, including transcription, epigenetics, and post-transcription, play a role in plant survival by combating various stress conditions [12–14]. Nevertheless, extensive research has uncovered the pivotal roles lncRNAs perform in numerous essential biological processes in plants, including the maintenance of cellular structural integrity, heat shock response, cell cycle progression, genomic imprinting, regulation of splicing, translation, and transcription. Under condition of climate change, these lncRNAs plays vital role for crop improvement programmes in plants with several facets of plant biology by examining complex regulatory networks of stress-responsive lncRNAs adaptation to a dynamic and challenging environment.

2. History/discovery for evolution of lncRNA in plant

In RNA biology of eukaryotes, plant biology has a long history of contributing important findings (**Figure 1**), such as the first reports of stress-induced RNA-protein cytoplasmic aggregates and posttranscriptional gene silencing. During the era of 1990s and 2000s, to evaluate the function of lncRNAs researchers have significantly used approaches such as traditional gene targeting. During the early of 1991, the X Inactive Specific Transcript (XIST) plays a role in controlling the inactivation of the

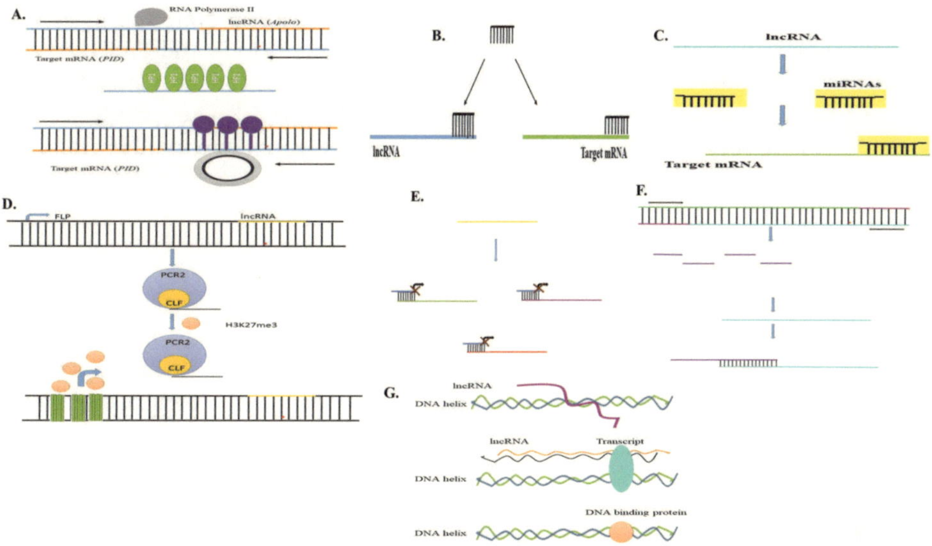

Figure 1.
Biotic and abiotic stress responses and the regulatory functions of pant lncRNA (A) DNA methylation, (B) miRNA mimics, (C) histone modification, (D) miRNA precursor, (E) cis-acting factor, (F) trans-acting factor, and (G) chromatin organization modifications.

X chromosome the first lncRNA, which was discovered in animal cells [15]. In 1999, Kouchi et al. discovered that rice (*Oryza sativa* L.) has the first lncRNAs found in plants [16]. This led to the discovery of numerous lncRNAs in other plant species, including Arabidopsis (*Arabidopsis thaliana*) and tomato (*Solanum lycopersicum*) [17].

Many long non-coding RNAs (lncRNAs) had already been recognized as essential regulators in a wide range of cellular processes before the genomics era and the historic sequencing of the *Arabidopsis thaliana* genome in 2000 (Arabidopsis Genome Initiative 2000). Currently, among 1239 organisms, the NONCODE v3.0 database, which compiles lncRNAs documented in scientific literature, has identified 73,370 lncRNAs. However, only a small fraction fewer than 200 have known functions [18]. The discovery of the initial long non-coding RNAs (lncRNAs) was made possible by advancements in molecular biology and genetic analysis tools. Of ncRNAs, these groundbreaking innovations empowered researchers to uncover previously neglected genomic elements that do not code for proteins but are pivotal in orchestrating gene expression, modulating chromatin dynamics, and regulating fundamental cellular processes. During the year of 2000, researchers identified lncRNA in different five plant spp. such as *Arabidopsis*, tomato, tobacco, cucumber and *Medicago truncatula* related to genes such as *GENE WITH UNSTABLE TRANSCRIPT 15 (GUT15)*, *TOMATO PHOSPHATE STARVATION INDUCED 1 (TPSII)*, *GENE WITH UNSTABLE TRANSCRIPT 15 (GUT15)*, *CYTOKININ REPRESSED 20 (CR20)*, and *EARLY NODULIN 40 (EN40)*, respectively. Among these genes related to lncRNAs, some lncRNAs such as CR20 and GUT15 are conserved among angiosperms but have some limitations. These lncRNAs have been identified as alternatively spliced and regulated by hormones [19, 20]. In addition, scientists have also identified lncRNAs in legumes that are related to ENOD40, which play an important role in root nodulation [21]. Root nodulation in legumes is influenced by ENOD40, the first lncRNA to be independently identified in both soybean (*Glycine max*) and *M. truncatula*. The first lncRNAs identified under environmental stress by transcriptomic approaches analysed their widespread expression. Two lncRNAs, namely MT4 and TPSI1, which respond to phosphate starvation and are conserved in dicotyledonous plants, have been identified by researchers in response to nutrient stress [22]. These studies, particularly in model plants, marked the beginning of research into lncRNAs in plants and their role in regulatory networks.

3. Type of LNCRNAs

Long non-coding RNAs (lncRNAs) function as *riboregulators*, a diverse class of RNA transcripts transcribed by RNA polymerase II or III. Unlike mRNAs, lncRNAs lack protein-coding potential, are typically over 200 nucleotides in length, and are found within the cytoplasm or nucleus. These molecules are integral to various regulatory processes, influencing gene expression, chromatin structure, and cellular signaling pathways [23, 24]. The progenitors of small interfering RNAs (siRNAs) are Pol IV lncRNAs [25]. Both strands of the protein-coding locus can produce Pol V-dependent lncRNAs, which help to modify the local chromatin loop, expressing themselves in a "tissue-specific" way [26, 27]. Long noncoding RNAs (lncRNAs) originate from dedicated promoters as well as from promoters shared with various other transcriptional elements. These origins provide lncRNAs with versatile regulatory potential, allowing them to modulate gene expression networks through diverse transcriptional landscapes [28].

LncRNAs are majorly divided into three categories based on mechanism of action, location of genome, and effect of DNA. Based on their genomic positioning relative to nearby protein-coding genes, lncRNAs can be broadly classified into: (i) long intergenic non-coding RNAs (lincRNAs), (ii) circular long non-coding RNAs (circRNAs), (iii) intronic non-coding RNAs (incRNAs), (iv) natural antisense transcripts (NATs), and (v) each category reflects distinct biogenesis pathways and functional roles, enabling lncRNAs to orchestrate complex layers of gene regulation across the genome [12, 29]. Moreover, according to their mode of action, these lncRNAs can further be categorized based on their specific roles in gene regulation. Some lncRNAs function by interacting with chromatin to modulate gene expression, while others regulate RNA processing or act as scaffolds for protein complexes that influence cellular processes.

Additionally, lncRNAs can influence how pre-mRNAs are spliced into different forms. They can also act as natural decoys, competing with various miRNAs, and can serve as sources for siRNAs and miRNAs [30, 31]. In contrast, the back-splicing of internal exons within pre-mRNA culminates in the generation of circular RNAs (circRNAs), a relatively rare and distinct class of RNA molecules that exhibit exceptional stability and pivotal regulatory functions. Due to their covalently closed-loop structure, circRNAs are impervious to exonuclease degradation, thus conferring them a remarkable resilience. These unique molecules are integral to a variety of biological processes, including the fine-tuning of gene expression, modulation of splicing events, and the formation of complexes with RNA-binding proteins, thereby influencing cellular homeostasis and signalling pathways [32]. These circular lncRNAs serve diverse functional roles within cellular processes also known as scaffolding backbones, sponge molecules, guiding agents, molecular decoys, and signaling molecules. Natural Antisense Transcripts (NATs), a distinct category of lncRNAs, exhibit both cis- and trans-regulatory functions. These transcripts are derived from complementary DNA strands of sense-coding regions, where they exert profound influence on gene expression through mechanisms such as transcriptional interference, epigenetic modulation, and post-transcriptional regulation [33]. Additional functions of lncRNAs encompass modulating the alternative splicing of pre-mRNAs, serving as endogenous target mimics (eTMs) that compete with various miRNAs, and acting as precursors to miRNAs and siRNAs [26]. From the different regions, intronic regions were used for transcribe incRNAs, whereas intergenic regions developed LincRNAs. These generated lncRNAs engage in trans-regulatory functions, modulating distant genes, and display tissue-specific expression patterns. They are polyadenylated but exhibit inefficient splicing [34].

Based on their genomic positioning, these lncRNAs are primarily classified into five subcategories: bidirectional lncRNAs, antisense lncRNAs, sense lncRNAs, intergenic lncRNAs, and intronic lncRNA. In the space between two genes, intergenic lncRNA is developed also known as large intervening ncRNAs (lincRNAs) or large interventional ncRNAs [35]. The transcription initiation site of the protein-coding gene on the antiparallel strand, oriented in the reverse direction, lies in striking proximity to the commencement site of bidirectional lncRNA transcription [36]. The space between two genes produces intergenic lncRNA. Meanwhile, the mRNA precursor sequence or secondary transcript's intron region is the source of intronic lncRNAs. The exon stages of another protein-coding gene on the opposite or same strand coincide with antisense or sense lncRNAs, respectively [37].

Wang and Chekanova [38] delineated lncRNAs into four primary archetypes based on their modes of action: (1) guide molecules: RNA-binding protein (RBP) complexes can be recruited to particular genomic loci by lncRNAs that coordinate their localisation to specific target sites or interact with chromatin remodeling complexes to modify chromatin dynamics [36]; (2) signaling molecules: The expression of numerous genes is orchestrated by specific long non-coding RNAs (lncRNAs) that function as pivotal regulators in signalling pathways that govern these intricate regulatory mechanisms. These lncRNAs, often referred to as signalling lncRNAs, play a crucial role in modulating gene expression by acting as molecular beacons or mediators within cellular communication networks; (3) skeleton molecules: By interacting with protein complexes or other effector molecules, lncRNAs can function as molecular scaffolds, facilitating intricate molecular interactions that fine-tune the precision of signalling processes. These lncRNAs play a critical role in modulating the specificity and efficacy of signalling pathways, acting as essential mediators that influence cellular responses and ensure the accurate transmission of regulatory signals across biological systems; and (4) bait molecules: By binding several proteins such as RNA binding protein to particular targets, these LncRNAs that are involved in the recruitment of RNA-binding proteins act as crucial modulators in cellular processes [39].

4. Features and characteristics of lncRNA

An alternative site for splicing, a 3′ polyadenylated tail, and a 5′ cap m7G caps are structural traits that mRNAs and lncRNAs have in common. These characteristics of lncRNA highlight the diversity between these RNAs in terms of function and allow them to play a variety of functions in gene regulation. The cytoplasm and nucleus are sites for synthesis of lncRNAs, which have a significant impact on cellular homeostasis and signaling pathways by influencing RNA processing, control of transcriptional process, and post-transcriptional regulation [40]. In contrast to mRNA, lncRNA has been identified more abundantly expressed, less conserved, and fewer site of exons. In contrast to mRNA, lncRNA exhibits higher levels of expression, reduced evolutionary conservation, and a more limited number of exonic regions [41]. Their unique features of lncRNAs show their role in protein coding and reflect their adaptability in orchestrating complex gene regulatory networks and modulating cellular processes across diverse biological contexts [42, 43]. Cellular organization and regulation, encompassing processes such as cell growth, RNA transcription, DNA replication, cell differentiation, and protein translation, is significantly influenced by lncRNAs [44]. These molecules often exhibit the capacity to adopt highly stable secondary and tertiary structures, enabling them to act as critical modulators within these intricate biological systems. Several researchers have observed that the folding of the secondary structure of long non-coding RNA (lncRNA) is significantly influenced by environmental conditions and cellular positioning. This intricate process occurs with remarkable efficiency, requiring minimal energy expenditure [45]. The enhanced nuclear retention of lncRNAs is closely associated with an extended distance between the 3′ splice site and the branch point, coupled with attenuated internal splicing signals [46]. Furthermore, the accumulation of lncRNAs within the nucleus is driven by additional factors, including differential expression of key splicing regulators, which intricately modulates the splicing machinery and its efficiency [47]. Distinct patterns of lncRNA distribution have been documented across the genomes of various plant species. In grapevines, lncRNAs exhibit a notably uneven chromosomal distribution, contrasting with their uniform distribution across the maize genome [48]. In maize, however, a marginally reduced density of lncRNAs is observed on chromosome 1, reflecting subtle genomic variation [49]. Several types of RNA molecules such as ceRNA, mRNA, and miRNA can also interact with lncRNAs, another kind of post-transcriptional regulation. According to ceRNA's prior description, several kinds of lncRNA can interact to change the amount of gene expression or transcription isoforms [50].

5. Techniques for identification of lncRNA in plant

The high-throughput sequencing technologies and computational bioinformatics approaches used by researchers to identify role of several genes, and transcription factors plays vital role in plant growth and development [51]. In genome sequence, RNA transcripts are used for identification of lncRNAs after those with large open-reading frames and/or are eliminated, along with those that exhibit traits of mRNAs and other ncRNAs (such as and snRNA) filtered and remaining non-coding RNAs in transcriptomics sequence identified as a lncRNAs [52]. In current scenario, the biotechnological approaches such as the lncRNA microarray technique, RNA-sequencing (RNA-Seq), and degradome sequencing are widely used by scientific community for identification of molecular mechanism behind the plant behavior in stress condition [53]. Among these approaches, the use of next-generation sequencing techniques for transcriptomic or RNA-sequencing is highly used for the finding of lncRNAs [54]. These approaches not completely identify all lncRNAs in plants due to their several unique features such as due to the intricacy of these molecules to code several peptides or lengthy open-reading frames that are not translated [55, 56]. The advancement of biological experimentation hinges on the precise prediction of non-coding peptides (ncPEPs), as this greatly conserves both time and resources. Accurate ncPEP identification streamlines experimental workflows, enabling targeted approaches that optimize efficiency and reduce expenditure, while minimizing the potential for erroneous classification of these molecules as coding genes [55]. For example, earlier

Type of modification	No. of plant species located and identified lncRNAs	Identified monocot plant species	Total identified Eudicotes
AlnC	678	21.7%	69.6%
GreeNC v2.0	93	32.3%	57%
PLncDBv2.0	80	20%	65%
CANTATAdb v2.0	39	—	—
LncPheDB	9	55.6%	44.4%
NONCODE v6.0	23	21.7%	69.6%

Table 1.
Statistics of different database for lncRNA in plants.

Type of modification	Modification site	Transcriptional effect	Group
Acetylation	H2A/H2B, H3, H4	Activation	Methyl
Biotinylation	H4, H3, H2A	Repression	Biotin
Sumoylation	H2A, H4, K17	Repression	SUMO
Phosphorylation	H3	Activation	Phosphate
Ubiquitination	H2A and H2B	Repression and activation	Ubiquitin
Acetylation	H2A/H2B, H3, H4	Activation	Methyl

Table 2.
Effect of transcription through various ways of alteration to histones.

studies anticipated that, within the database of the model plant *Arabidopsis thaliana* (Araport11), certain protein-coding transcripts were misidentified [57, 58]. Notably, the auxin-regulated *APOLO* (Auxin-Regulated Promoter Loop Induced by Phosphate Starvation) [59] and *IPS* (lncRNA IPS), initially classified as protein-coding, were later confirmed to be long non-coding RNAs (lncRNAs) [60, 61]. Despite these challenges, for accurate prediction of lncRNAs in plants from next-generation sequencing data, it is essential to collect diverse transcriptome or RNA sequencing data to capture plant response changes across time, developmental stages, and species-specific regulation, ensuring robust and comprehensive identification [62].

In plant species, two primary strategies are predominantly employed for the annotation and localization of long non-coding RNAs (lncRNAs): (1) prediction of non-coding conserved regions and (2) comparative genomics [10]. A novel and promising approach involves leveraging machine learning to detect lncRNAs through various bioinformatics tools, such as LncADeep [63], CPAT [64], PLEK [65], CPC2 [66], LncFinder [63], and CNCI [67] (see **Table 1**).

These tools leverage sequence and structural features of transcripts, employing advanced computational models such as support vector machines (SVMs) and logistic regression algorithms to robustly assess coding potential and distinguish between coding and non-coding RNAs [68]. For the advancement of lncRNAs prediction in plants need of developing robust ensembles techniques and retraining popular algorithms. A wide array of publicly accessible databases and web-based analytical platforms are available to facilitate advanced research on long non-coding RNAs (lncRNAs), encompassing curated datasets from over 900 plant species, each featuring transcripts exceeding 200 nucleotides in length (see **Table 2**).

6. Modes of action of lncRNAs in plants

By developing complex systems to coordinate exact temporal and spatial patterns of gene expression in response to both internal developmental cues and external environmental stimuli, plants have effectively adapted to their ecological niches [69]. Epigenetic regulation, encompassing mechanisms and the actions of noncoding including histone variants, chromatin remodeling, histone modifications, and DNA methylation, represents a cornerstone of intricate and finely tuned biological processes. In recent decades, plant epigenetics has emerged as a pivotal research frontier, driven by its profound implications for tackling global challenges such as food security and agricultural productivity [70]. In plants, researchers identified involvement of long noncoding RNAs (lncRNAs) in a diverse range of biological processes over the last 10 years [31, 71]. In plants, lncRNAs play a crucial role in combating the challenges posed by climate change such as salinity, temperature fluctuation, and drought stress as well as regulating growth and developmental processes from germination to fruit development [72, 73]. In climate resiliencies, sustainable agriculture, and food scarcity, plant epigenetics study plays a vital role for gained prominence [74]. The interplay between epigenetic modifiers and other epigenetic factors in regulating gene transcription and long non-coding RNAs (lncRNAs) in plants remains insufficiently understood [75]. This gap in knowledge highlights the need for further research to unravel the intricate network of interactions that govern plant gene regulation. lncRNAs involved in various epigenetic regulation process include lncRNA-RNA interactions, chromatin remodeling, splicing regulation, R-loop formation, and DNA methylation [76].

6.1 DNA methylation

In plants, the enzyme DNA methyltransferases (DNMTs) involved in process of DNA methylation an essential epigenetic modification [77, 78]. In tobacco plants, the phenomenon of DNA methylation mediated by RNA molecules was first documented in 1994. This pivotal discovery revealed the intricate interplay between RNA and epigenetic regulation, highlighting RNA molecules as key mediators in guiding sequence-specific DNA methylation. This process also known as RNA-directed DNA methylation (RdDM) plays a vital role in transfer of genetic information from DNA to RNA [79]. RNA-directed DNA methylation (RdDM) plays a pivotal role in regulating a multitude of mechanisms that enhance a plant's adaptability and resilience to fluctuating environmental conditions and pathogen attacks. By directing sequence-specific DNA methylation, RdDM modulates gene expression, fortifies genome integrity, and activates defense pathway [80].

In the process of DdRPs, many enzymes are involved to regulate the process such as Dicer-like protein 3 (DCL3), RNA-dependent RNA polymerase 2 (RDR2), Pol IV, and Pol V [81, 82]. The ncRNA-directed DNA methylation (RdDM) pathway plays a pivotal role in orchestrating *de novo* DNA methylation, encompassing the regulation of asymmetric CHH methylation. This sophisticated epigenetic mechanism employs non-coding RNAs to target specific genomic loci, enabling the establishment and maintenance of precise methylation patterns [83]. Both distinct yet interdependent roles are used in the synthesis of siRNA and ncRNA scaffolds, with Pol II mediating critical steps in this intricate process. Within this framework, long non-coding RNAs (lncRNAs), derived from siRNA precursors, function as precise molecular guides to facilitate targeted DNA methylation in plants [84].

Both Pol V and Pol IV play distinct yet interdependent roles in the biogenesis of lncRNA and siRNA 24 nt, with Pol II mediating critical steps in this intricate process. Within this framework, long non-coding RNAs (lncRNAs), derived from siRNA precursors, function as precise molecular guides to facilitate targeted DNA methylation in plants [85]. Double-stranded RNA (dsRNA) is meticulously synthesized from single-stranded RNA (ssRNA) templates through the synergistic interplay of RDR2 and Pol IV. The strong production of dsRNA molecules is the result of this complex process, which begins with Pol IV starting the transcription of ssRNA and is subsequently continued by the catalytic accuracy of RDR2 to produce complementary RNA strands [86, 87]. The hypothesized ATP-dependent chromatin remodelling factor CLSY1 (CLASSY1) is likely to mediate or catalyse this process. Following DICER-LIKE 3 (DCL3) protein cleavage of this dsRNA, 23 nt and 24 nt siRNAs are produced in which 23 nt siRNA acts as passenger strand and subsequently loaded onto an ARGONAUTE protein, specifically AGO4 (sometimes AGO6) to develop the complex of AGO4-siRNA [88].

IN DE NOVO 2 (IDN2) protein is hypothesized to stabilize and reinforce the duplex formation between siRNA and lncRNA, with its engagement transpiring subsequent to the assembly of the AGO4-siRNA complex. The *de novo* methyltransferase DOMAINS REARRANGED METHYLTRANSFERASE 2 (DRM2) recruitment efficiency is enhanced by this stabilization process. In all cytosine settings (CG, CHG, CHH; where H stands for A, T, or C), DRM2 is in charge of putting methylation marks at particular genomic locations [89].

6.2 Chromatin remodeling by lncRNA

The course of the cell cycle in plants, which is intricately regulated by developmental and environmental cues, relies on chromatin dynamics. In the cell nucleus, the genetic material is derived from chromatin, a complex of DNA and histone proteins that intricately organize into chromosomes during cell division [90]. For proper molecular mechanism, processes such as recombination, plant transcriptional silencing, genome replication, and DNA repair rely on the chromatin structure [91]. The previous studies highlighted the critical role of lncRNAs as key regulators of protein-coding genes, expression of genes, and functioning as influence of both trans- and cis-acting activities [92].

It has been determined that about 38% of the cell's lncRNAs play a crucial role in chromatin remodeling. They accomplish this by coordinating transcriptional suppression and epigenetic regulation by building complexes with the polycomb repressive complex 2 (PRC2) or chromatin-modifying proteins such as SMCX and CoREST [93]. Priming, intricately linked to chromatin modifications and capable of persisting independently of active gene transcription, equips plants with an extraordinary ability to "memorize" prior stress experiences. This adaptation enables them to withstand extreme environmental fluctuations by orchestrating more robust and efficient responses when re-exposed to similar stressors. The three dynamic features of nucleosomes reconstruction, covalent alteration mediated by enzymes, and repositioning are essential for the fundamental mechanism of chromatin remodeling [94]. The remodeling of chromatin mainly relies on nucleosome properties repositioning, reconstruction, and enzyme-induced covalent modification. During chromatin reconstruction, the SWR1 histone-exchange complex facilitates the incorporation of histone variants by replacing canonical histones. The newly formed variant then recruits certain unique regulators (Reg) to regulate a number of biological processes, such as centromere construction and gene activation.

The process of chromatin remodeling occurs mainly *via* two ways (1) ATP-dependent chromatin remodeling and (2) covalent histone modification. The process of chromatin remodeling occurs mainly *via* three ways: (1) ATP-independent chromatin remodeling, (2) ATP-dependent chromatin remodeling, and (3) covalent histone modification. Chromatin remodelling can also occur through ATP-independent mechanisms, which do not rely on energy from ATP hydrolysis. In such cases, histone dissociation from chromatin is facilitated by histone chaperones or DNA-binding transcription factors that aid in nucleosome reorganization.

In ATP-dependent chromatin remodeling, structure of chromatin altered by the use of hydrolysed ATP'S energy and ATPase includes *BRG1*, *BRM*, *CHD1*, *CHD3*, *CHD4*, *Mi-2*, *SWR1*, *SNF2L*, and *SNF2* [95]. The process begins when the complex identifies a specific target region on chromatin, often guided by transcription factors or DNA-binding proteins [96]. Through the activity of ATPase subunits such as Snf2, the complex disrupts histone-DNA interactions, enabling three main outcomes: nucleosome sliding (to modulate accessibility by repositioning nucleosomes along DNA), nucleosome ejection (removing nucleosomes entirely), or histone exchange (replacing histones with variants) [97]. These modifications are pivotal for regulating fundamental nuclear processes such as DNA accessibility, facilitating or repressing transcription, replication, DNA repair, or recombination. In this process, nucleosomes repositioning regulated by complex of remodelers use hydrolysed ATP's energy (12–14 kcal mol^{-1}). In addition, this process also involves the replacement or removal of H2A–H2B dimers, as well as the potential ejection of entire histone octamers from nucleosome core [98]. Such modifications significantly influence the chromatin landscape, dynamically balancing the interplay between chromatin compaction and accessibility, thereby ensuring precise and context-specific regulation of genomic functions.

6.3 Histone modification mediated by lncRNAs

The fundamental structure of chromatin is established through the intricate coiling of nucleic acids (DNA) around nucleosomes that are protein complexes consisting of histone octamers. These octamers comprise two copies each of the histones H2A, H2B, H3, and H4. Numerous stages of plant development, including seed germination, hypocotyl elongation, and the onset of flowering, are intricately regulated by dynamic alterations in histone modifications. These chromatin-based changes play a pivotal role in controlling gene expression, thus driving the precise timing and coordination of key developmental processes [90]. The modifications of histone are essential in the regulation of gene expression because they influence the structure of chromatin, the material that makes up chromosomes [99]. Chromatin structure is a key factor in determining whether a gene is accessible for transcription or not. Histones are proteins around which DNA is wrapped, and chemical modifications to histones can either loosen or tighten this wrapping, affecting the ability of transcription factors and other regulatory proteins to access the DNA [100]. By altering chromatin structure and controlling the recruitment of transcriptional complexes, histone alterations coordinate the transcription of genes. The particular histone variation, the exact amino acid residue targeted, and the type of chemical modification used all influence the transcriptional result, whether it be activation or repression [101]. Under the catalytic influence of specialized enzymes, histones are subject to a broad spectrum of post-translational modifications, including acetylation, phosphorylation, methylation, and ubiquitination (see **Table 2**) [102]. These intricate and dynamic alterations are indispensable for orchestrating gene expression, either by facilitating transcriptional activation or enforcing repression, while simultaneously governing chromatin architecture with precision.

7. Functional implications of lncRNAs in climate resilience for plants

LncRNAs are increasingly recognized for their role in enhancing plant resilience to climate change-induced stresses. By regulating key genes involved in drought, metal stress, salinity, and heat tolerance, lncRNAs allow plants to adjust their physiology by regulating heat shock proteins, controlling ion transporter genes enhancing efficiency of water uptake, and maintaining osmotic balance protecting cellular machinery of plant to improve various resilience.

7.1 lncRNAs associated with abiotic stress in plants

Climate change is a critical global challenge that significantly impacts crop nutritional quality, stunts growth, and hampers development, thereby exacerbating issues of food scarcity and malnutrition. Adverse climatic conditions, such as extreme temperatures, heavy metal toxicity, drought, salinity, cold stress, and nutrient deficiencies, drastically reduce crop yields. By surviving abiotic stresses such as drought, temperature swings, salt, and heavy metals, the algal lineage started terraforming the terrestrial ecosystem about half a billion years ago, allowing green life to thrive in a wide variety of habitats [103]. In response to these environmental stressors, plants activate intricate defense mechanisms, including the production of specialized proteins and enzymes, to mitigate damage and sustain growth. Furthermore, ncRNAs are essential for influencing how plants react to biotic and abiotic stresses, allowing for resilience and adaptation in harsh environmental circumstances. Their diverse roles highlight their importance in systems of stress tolerance and development [104]. In plants, role of lncRNAs as biological regulators helps to adapt biotic and abiotic stressors identified by previous studies and noted involvement of several transcription factors [105].

7.1.1 Response of LncRNAs in condition of temperature stress

Global warming has precipitated profound and unprecedented temperature anomalies across the globe. Recent summers have witnessed record-shattering heat, marking the highest temperatures observed in decades. Conversely, certain regions have endured exceptionally frigid winters, with temperatures plunging to historic lows. These temperature changes affect growth and development of plant and reduce year the round production.

7.1.1.1 Cold stress

An essential abiotic stressor that hinders plant growth and productivity is low temperature stress, which causes freezing and chilling injury. In plants, stress memory has evolved as a sophisticated adaptive strategy, enabling them to withstand recurring stressors with heightened resistance or tolerance. Following an initial exposure to environmental challenges such as drought, salinity, or pathogen attacks, plants undergo molecular, physiological, and epigenetic changes that “imprint” the stress experience. To identify the molecular mechanisms behind plant tolerance to low temperatures, researchers are focusing on the discovery of quantitative trait loci (QTLs) and the expression profiling of genes under cold stress, leveraging advances in high-throughput sequencing technologies [14].

Cold stress is major reason to cause oxidative stress in plants by affecting antioxidant enzyme activity due to damage of cell membrane by the extravasation of osmotic substances, which affects the function of membrane localization proteins and downstream. By reducing the activity of antioxidant enzymes, cold stress significantly contributes to oxidative stress in plants. This happens because osmotic material leakage damages the cell membrane, impairing the integrity and functionality of proteins that are located in the membrane [106]. Plant health and resilience are further jeopardized as a result of this cascade's detrimental effects on downstream physiological and biochemical processes.

To ensuring the survival of plant in adverse environments, a diverse array of regulatory microRNAs (miRNAs) and transcription factors such as *C-repeat Binding Factors* (CBFs) regulate the expression of *Cold-Regulated* (COR) genes, essential for bolstering cold tolerance and synergistically enhancing the plant's resilience to chilling and freezing conditions [107]. Plant gene expression under cold stress is intricately regulated by transcription factors, notably those within the C-repeat Binding Factor (CBF) pathway. Key players, such as the inducer CBF expression 1 (ICE1), bind to specific promoter regions, initiating the transcription of cold-responsive genes [108]. To provide tolerance under cold stress conditions, signaling pathway of CBF plays vital role activated by activity of calmodulin-binding transcriptional activators (CAMTA3 and CAMTA5) and further regulated by CdWRKY2, which directly targets the CdCBF1 promoter that exerts regulatory control by directly binding to the CdCBF1 promoter [109]. To help plants endure freezing conditions, Plant U-Box 25/26 targets the R2R3-MYB transcription factor, a MYB domain protein 15, for degradation. This protein, which plays a role in the ICE-CBF pathway, acts as a negative regulator of transcription [110].

Furthermore, cold resilience in plants regulated by hormones and the reports noted that to enhance plant ability to survive in condition of cold stress. The genes COR47 and RD29A, associated with hormonal pathways, are instrumental in conferring cold tolerance [111]. Their expression is regulated by the binding of specific cis-elements, CACTTG and CACGTG, which are variants of the E-box motif. These motifs serve as binding sites for transcription factors such as Brassinazole-Resistant1 (BZR1) that plays a pivotal role in modulating the activity of cold-responsive promoters like CBF1 and CBF2. This intricate regulatory network underscores the interplay between hormonal signaling and cold stress responses. These coordinated mechanisms synergistically enhance the expression of cold-responsive genes, fortifying plant resilience against low-temperature stress.

7.1.1.2 lncRNAs in high temperature

Global warming becomes worse by industrialization and the overuse of fertilizers and pesticides, which causes temperatures to rise sharply worldwide. The rising temperate and heat wave retard crop productivity by negatively impact on its growth and development. To provide tolerance to stress by changes expression of genes and biochemical pathway, plants adapt several mechanism [112]. Heat stress causes extensive damage to plant cells by compromising membrane integrity and inducing protein denaturation by altering the synthesis of metabolites, phytohormones, activity of enzymes, heat shock proteins (HSPs), and of heat shock transcription factor (HSFs) [112].

To provide tolerance towards varying temperature conditions, HSPs play a major role to and its expression regulated by the HSFs. HSFs are classified into three categories (1) HSFs class A; (2) HSFs class B, and (3). HSFs class C [113]. Notably, HSFs in classes C and B lack transcriptional activators functions. In contrast, transcriptional activation is primarily orchestrated by HSFA1, attributed to the presence of conserved amino acid motif residues. This activation profoundly influences the expression of other transcription factors, including MULTIPROTEIN-BRIDGING FACTOR 1C (MBF1C), DEHYDRATION-RESPONSIVE ELEMENT BINDING PROTEIN 2A (DREB2A), and various heat shock transcription factors such as HSFBs and HSFAs (e.g. HSFA7 and HSFA2) [114]. These factors are integral to the heat stress response, initiating a cascade of gene expressions to mitigate cellular damage and ensure survival under thermal stress.

There are many studies available identified role of lncRNAs to provide tolerance against high temperature by regulating several molecular pathways. In response to high temperatures in this system, the lncRNA TCONS_00016454 displayed antagonistic expression patterns with its target, a BES1/BZR1 homolog, suggesting that it may be involved in providing heat tolerance *via* BR signaling in Chinese cabbage. Wang et al. identified the role of hormones specifically BR-related lncRNA TCONS_00016454 in Chinese cabbage providing tolerance against heat stress with targeting a homolog BES1/BZR1 [115].

In wheat heat stress is a major issue in pollen development due to affecting cell division at meiosis stage and recuing percentage of maturation of pollen and reducing wheat productivity. In wheat heat stress-responsive lncRNAs identified transcriptomic data by Babaei et al. and study noted stress-responsive miRs such as miR156, miR159, miR160, miR167, miR399, miR408, and miR444 and miR1122 [116]. Future advancements to develop engineering heat-tolerant crop plants could be significantly facilitated by strategically manipulating the overexpression or silencing by RNAi and CRISPR strategy of lncRNAs that regulate genes critical to the heat stress response. This targeted approach holds promise for enhancing plant resilience under elevated temperatures.

7.1.2 lncRNAs in metal toxicity

Plants require different nutrients and metal elements for growth and development. In proper amounts, these improve enhance biological and molecular functioning in plants resulting enhancing plant productivity. The metal enhances crop yield and biochemical attributes play a key role by carrying out as cofactor/activation agent in various process such as DNA replication, processing of protein, electron transport, and biosynthesis of chlorophyll [117].

Globally, the indiscriminate discharge of industrial waste and the excessive use of fertilizers and pesticides severely compromise soil fertility and diminish soil properties, ultimately reducing the yield of agricultural and horticultural crops [118]. Unfortunately, this activity done by humans leads to retard crop productivity. Additionally, these contaminants including toxic heavy metals such as lead, copper, arsenic, nickel, and mercury accumulate in the soil and water, infiltrate the food chain, and pose significant risks to human health [119]. The crop production near industrial areas or metal accumulated soil and water sites causes negative impact on human body due to metal accumulation in different body organs or tissue.

To adapt or provide tolerance in plants by molecular mechanism plays a key role to cope with stressors by regulating molecular mechanism following heavy metal movement from soil. In order to lessen the harmful consequences of heavy metals, lncRNAs have also become important molecular actors, intricately modifying a wide range of cellular pathways. Plants have evolved sophisticated regulatory mechanisms to ensure precise homeostasis of essential metal ions, encompassing their absorption, efflux, translocation, and storage [120]. This intricate machinery reflects the critical need to maintain optimal metal ion concentrations for proper physiological and biochemical functions. The excess or overuse of metal ions causes toxicity disrupt the cell and cell membrane and change structure of protein resulting in negative impacts on biochemical and metabolic activity by the production of reactive oxygen species (ROS) [121].

In recent years, the scientific community has increasingly focused on understanding plant responses to heavy metals such as copper, aluminum, and mercury. However, the precise molecular mechanisms underlying these responses remain largely unidentified. Chen et al. provided ground-breaking insights into the role of long non-coding RNAs (lncRNAs) in heavy metal uptake, specifically examining lead accumulation in poplar (*Populus tomentosa*) [122]. Using RNA sequencing (RNA-seq) analysis, they identified 226 lncRNAs associated with lead uptake, exhibiting differential expression linked to crucial pathways, including signal transduction, primary metabolism, energy biosynthesis, and secondary metabolism. Among these, PtoMYB46 was found to positively influence plant growth by downregulating genes such as Auxin Response Factor 2 (ARF2), thereby enhancing transketolase activity and photosynthetic efficiency. Additionally, PtoMYB46 suppressed the expression of PtoMATE, a gene involved in citric acid (CA) secretion, effectively increasing lead uptake efficiency.

Similarly, research on *Medicago truncatula* has revealed molecular responses to aluminum stress mediated by lncRNAs. A study identified 515 differentially expressed lncRNAs from a total of 3284, highlighting their involvement in pathways such as signal transduction, the tricarboxylic acid (TCA) cycle, and hormonal regulation. Functional annotation predicted that these lncRNAs influenced 1254 genes, emphasizing their critical role in modulating molecular and biological pathways to counteract aluminum stress [123].

7.1.3 lncRNAs in drought

For functioning of development process in plant properly water plays a vital role for growth and crop production. In current scenario, water scarcity and quality is a major issue affecting crop yield especially in regions where agriculture heavily relies on irrigation [124]. Stress caused by the decrease in water leads to alterations in the molecular, physiological, ecological, morphological, and biochemical alterations in plant [125].

In plant, water shortage is caused due to extremely high or temperatures, low rainfall, salinity, and high light intensity. Under condition of drought stress, plants undergo profound physiological alterations, including reduced respiration rates, osmotic disequilibrium, disrupted metabolic homeostasis, and retardation of the photosynthetic process by decrease in the activity of enzymes such as Rubisco and phosphoenolpyruvate [126]. These impairments compromise metabolism, and productivity across all developmental stages, ultimately diminishing crop quality and yield [127].

Under stressors such as drought stress, plants invoke a sophisticated cascade of protein phosphorylation pathways, driven by the synthesis of reactive oxygen species, phosphatidylinositol, and secondary messengers analogous to Ca^{2+}, orchestrating precise adaptive defense. In these pathways, transcription factors involved to avoid damage of cell by modulating the expression of genes that provide protection for resilience against stress, thereby bolstering the plant's resilience against stress-induced damage [128]. To confer defense against drought stress, plants have evolved multifaceted mechanisms, including intricate molecular signalling cascades, regulation of extensive gene networks, and activation of biochemical pathways involving protein dynamics, hormone synthesis, and the accumulation of proline and carbohydrates. Additionally, the identification of quantitative trait loci (QTLs) enhances understanding of stress resilience [129]. Recent reports elucidate that among the various strategies conferring tolerance to drought stress, the upregulation of genes involved in the biosynthesis of stress hormones such as jasmonic acid (JA) and abscisic acid (ABA) plays a crucial role in enhancing resilience [130, 131].

The drought stress response is intricately governed by long non-coding RNAs (lncRNAs), which orchestrates the regulation of drought-responsive gene transcription, mediates chromatin remodeling, modulates antisense transcription, and fine-tunes eTM activity, exemplifying their critical role in sophisticated regulatory networks [14, 132]. Plants exhibit a complex regulatory balance under drought stress and produce several proteins involved in inhibition or regulation of a wide range of non-coding RNAs (ncRNAs), improving cellular resilience or affecting the expression of genes to aid in adaptation [133]. To provide tolerance against drought stress lncRNAs cause several changes occur in plants such as secondary metabolites biosynthesis, metabolism of sucrose, and phyto-hormone signal transduction [134, 135]. Many studied evolved the role of drought tolerance lncRNAs in different plant species such as rapeseed [136], cassava [129], *Arabidopsis* [137, 138], rice [138], and tomato [139].

A recent study by Yang et al. aimed to identify drought-responsive RNAs, including miRNAs, lncRNAs, and mRNAs in rice. The analysis revealed a total of 32 miRNAs and 191 lncRNAs, highlighting their significant roles in the plant's adaptive response to drought stress. Notably, Os05g0586700, a drought-responsive gene implicated in plant membrane repair under drought conditions, was identified as a predicted target for as many as 16 drought-specific lncRNAs [140].

The role of lncRNAs to provide tolerance against drought stress in sugar beet by high-throughout RNA-sequencing. Under condition of drought stress, study identified 386 lncRNAs differentially expressed in sugar beet. Among the differential expressed lncRNAs, two lncRNAs *viz.*, TCONS_00038334 and TCONS_00055787, identified major changes in expression approximately 1800-fold downregulation and 6000-fold upregulation, respectively [126].

7.1.4 lncRNAs in salinity

Among the various abiotic and biotic stressors, stress caused by excessive salinity significantly hampers plant productivity. Salt stress not only impairs crop growth but also degrades soil fertility, posing a significant challenge to sustainable agriculture. It is estimated that approximately 6% of the world's agricultural land contains a high concentration of salt elements [141]. The primary cause of salt accumulation in soils is the prolonged use of irrigation water, which deposits dissolved salts as it evaporates [142]. Reports indicate that salinity affects 6% of drylands and 20% of irrigated soils globally, further exacerbating the issue and threatening agricultural productivity. Recent reports indicated that soil salinization decomposing income of people depends on agricultural for survival and recorded approximately US$31 million economic loss [143].

Plants grown in saline soils experience a significant decline in productivity due to heightened oxidative and osmotic stress, nutrient deficiencies, and disruptions in intracellular ion homeostasis [144]. In plants under salt stress condition, there is chance to change metabolic activity of plant and in cell solute ratio of Na^+/K^+ enhanced, which leads to develop several metabolic disorders [145, 146].

To withstand such challenging conditions, plants have developed intricate adaptive mechanisms, including epigenetic modifications and the regulation of critical processes such as transcription and translation. Enhancing salt stress resistance in plant species necessitates a deeper understanding of the underlying molecular mechanisms, a task that calls for focused efforts by agricultural scientists. Recent studies have highlighted the pivotal role of long non-coding RNAs (lncRNAs) in conferring tolerance to salt stresses, offering promising avenues for advancing crop resilience in many plant species such as cotton [147], sorghum [148], rice [149], tomato [146], and chickpea [150]. lncRNAs act as activate change response of several genes such as hormones *viz.*, ABA and GA as well as upregulate the activity of Na^+ transporters to provide tolerance against salt stress.

The expression of various lncRNAs under saline stress in grapevine identified novel lncRNAs (3952) in grapevine. In study after supplementation of salt 12 h under condition of salt stress reports identified differentially expressed lncRNAs (1661) modulated either indirectly or directly several mRNAs with the use of miRNA-mediated, *trans-*, and *cis patterns* [151].

7.2 lncRNAs in biotic stress

In agriculture, apart from abiotic stress crop production is decreased due to microorganism's attack on plants such as fungi, bacteria, insects, and nematodes. Worldwide, approximately $20 billion of agricultural economic loss due to the loss of food production 10–30% by pathogen attack in plants is estimated (https://www.ars.usda.gov, accessed on 26 December 2024). To defense against pathogen attack, several biochemical and cellular changes occur in plant system such as metabolic and protein structural and functional changes in favor to response by activation of innate immune system of plant [152].

PAMP-triggered immunity a first layer of defense, plants' multilayered immune systems are based on which pathogen identified by cell-surface receptors identifies microbe-associated molecular patterns (MAMPs) or pathogen-associated molecular patterns (PAMPs) and triggers PAMP of immune responses (PTI) [153]. In this intricate defense mechanism, plants synthesize phytohormones such as salicylic acid and jasmonic acid, and enhance the activity of protein kinases and reactive oxygen species (ROS) production, and calcium ion (Ca^{2+}) signalling cascades, effectively mitigating the proliferation of pathogenic bacteria [154]. Additionally, the effector-triggered immunity (ETI) pathway orchestrates a formidable immune response through highly specialized resistance (R) proteins [155]. These R-proteins possess exceptional specificity in recognizing pathogen-derived effectors at the infection site, thereby mobilizing precise and robust defensive countermeasures. This R protein developed by the PTI and ETI cascade typically controls downstream signaling networks.

To provide immune response to plant towards biotic stress lncRNAs plays a vital role identified in many plant species by regulating their downstream immune signaling pathways. Late blight is a serious disease in tomato caused by *Phytophthora infestans* and the study by Zhang et al. identified interactions of lncRNA-miRNA providing immune response. The result identified that as a competitive endogenous RNA (ceRNA) for miR394, lncRNA40787 dramatically lowers miR394 levels and prevents it from cleaving the JA biosynthesis gene LCR. This regulatory mechanism fortifies tomato plants against *P. infestans*, thereby augmenting their stress tolerance through elevated JA-mediated defense pathways [156].

8. Conclusions and future prospectives

Long non-coding RNAs (lncRNAs) have emerged as indispensable regulators in plant biology, offering significant insights into how plants adapt to environmental stresses and developmental challenges. Their roles in modulating transcriptional, post-transcriptional, and epigenetic networks highlight their potential as powerful tools for improving plant resilience to abiotic and biotic stressors. Mechanisms such as chromatin remodeling, histone modification, and DNA methylation further underscore their critical involvement in plant stress responses and adaptation. Looking ahead, the integration of advanced sequencing technologies, bioinformatics tools, and gene-editing systems like CRISPR/Cas provides a promising avenue to unravel the precise regulatory mechanisms of lncRNAs. Future research should focus on harnessing lncRNAs for breeding stress-tolerant crops, improving nutrient use efficiency, and addressing the global challenges of food security and climate change. Translating these findings into practical applications can revolutionize sustainable agriculture by enabling the development of climate-resilient crops, ensuring productivity under diverse environmental conditions.

Acknowledgements

We sincerely acknowledged the SERB, DST, and CII to provide prime minister's fellowship for Doctoral Research to Ms. Komal G. Lakhani.

Author contributions

KGL contributes to conceptualization and writing—original draft; RH contributes to conceptualization and writing—review and editing; GVM corrected and edited the manuscript. All authors read and approved the final manuscript.

Conflict of interest

The authors declare no conflict of interest.

Author details

Komal G. Lakhani[1], Rasmieh Hamid[2*] and Gopal V. Marviya[3]

1 Department of Plant Molecular Biotechnology, Junagadh Agricultural University, Junagadh, Gujarat, India

2 Department of Plant Breeding, Cotton Research Institute of Iran (CRII), Agricultural Research, Education and Extension Organization (AREEO), Gorgan, Iran

3 Krishi Vigyan Kendra, Junagadh Agricultural University, Targhadia (Rajkot), India

*Address all correspondence to: r.hamid@areeo.ac.ir; rasihamid@gmail.com

References

[1] Myers S, Fanzo J, Wiebe K, Huybers P, Smith M. Current guidance underestimates risk of global environmental change to food security. BMJ. 2022;**378**:e071533

[2] Mottaleb KA, Kruseman G, Frija A, Sonder K, Lopez-Ridaura S. Projecting wheat demand in China and India for 2030 and 2050: Implications for food security. Frontiers in Nutrition. 2023;**9**:1077443

[3] Chauhan J, Singh P, Choyal P, Mishra UN, Saha D, Kumar R, et al. Plant photosynthesis under abiotic stresses: Damages, adaptive, and signaling mechanisms. Plant Stress. 2023;**10**:100296

[4] Costa MG, de Mello PR, Palaretti LF, de Souza Júnior JP. The effect of abiotic stresses on plant C: N: P homeostasis and their mitigation by silicon. The Crop Journal. 2024;**12**:340-353

[5] Ziska LH. Rising carbon dioxide and global nutrition: Evidence and action needed. Plants. 2022;**11**(7):1000

[6] Doolittle WF. Is junk DNA bunk? A critique of ENCODE. National Academy of Sciences of the United States of America. 2013;**110**(14):5294-5300

[7] Bhartiya D, Scaria V. Genomic variations in non-coding RNAs: Structure, function and regulation. Genomics. 2016;**107**(2-3):59-68

[8] Mattick JS, Amaral PP, Carninci P, Carpenter S, Chang HY, Chen L-L, et al. Long non-coding RNAs: Definitions, functions, challenges and recommendations. Nature Reviews Molecular Cell Biology. 2023;**24**(6):430-447

[9] Wajahat M, Bracken CP, Orang A. Emerging functions for snoRNAs and snoRNA-derived fragments. International Journal of Molecular Sciences. 2021;**22**(19):10193

[10] Derrien T, Johnson R, Bussotti G, Tanzer A, Djebali S, Tilgner H, et al. The GENCODE v7 catalog of human long noncoding RNAs: Analysis of their gene structure, evolution, and expression. Genome Research. 2012;**22**(9):1775-1789

[11] Mercer TR, Gerhardt DJ, Dinger ME, Crawford J, Trapnell C, Jeddeloh JA, et al. Targeted RNA sequencing reveals the deep complexity of the human transcriptome. Nature Biotechnology. 2012;**30**(1):99-104

[12] Patra GK, Gupta D, Rout GR, Panda SK. Role of long non coding RNA in plants under abiotic and biotic stresses. Plant Physiology and Biochemistry. 2023;**194**:96-110

[13] Waseem M, Liu Y, Xia R. Long non-coding RNAs, the dark matter: An emerging regulatory component in plants. International Journal of Molecular Sciences. 2020;**22**(1):86

[14] Jha UC, Nayyar H, Jha R, Khurshid M, Zhou M, Mantri N, et al. Long non-coding RNAs: Emerging players regulating plant abiotic stress response and adaptation. BMC Plant Biology. 2020;**20**:1-20

[15] Li J, Ming Z, Yang L, Wang T, Liu G, Ma Q. Long noncoding RNA XIST: Mechanisms for X chromosome inactivation, roles in sex-biased diseases, and therapeutic opportunities. Genes & Diseases. 2022;**9**(6):1478-1492

[16] Kouchi H, Ki T, So RB, Ladha JK, Reddy PM. Rice ENOD40: Isolation and expression analysis in rice and transgenic soybean root nodules. The Plant Journal. 1999;**18**(2):121-129

[17] Li R, Fu D, Zhu B, Luo Y, Zhu H. CRISPR/Cas9-mediated mutagenesis of lncRNA1459 alters tomato fruit ripening. The Plant Journal. 2018;**94**(3):513-524

[18] Ma L, Bajic VB, Zhang Z. On the classification of long non-coding RNAs. RNA Biology. 2013;**10**(6):924-933

[19] Plewka P, Thompson A, Szymanski M, Nuc P, Knop K, Rasinska A, et al. A stable tRNA-like molecule is generated from the long noncoding RNA GUT15 in Arabidopsis. RNA Biology. 2018;**15**(6):726-738

[20] MacIntosh GC, Wilkerson C, Green PJ. Identification and analysis of Arabidopsis expressed sequence tags characteristic of non-coding RNAs. Plant Physiology. 2001;**127**(3):765-776

[21] Crespi M, Jurkevitch E, Poiret M, d'Aubenton-Carafa Y, Petrovics G, Kondorosi E, et al. enod40, a gene expressed during nodule organogenesis, codes for a non-translatable RNA involved in plant growth. The EMBO Journal. 1994;**13**(21):5099-5112

[22] Franco-Zorrilla JM, Valli A, Todesco M, Mateos I, Puga MI, Rubio-Somoza I, et al. Target mimicry provides a new mechanism for regulation of microRNA activity. Nature Genetics. 2007;**39**(8):1033-1037

[23] Lakhotia SC, Mallick B, Roy J. Non-coding RNAs: Ever-expanding diversity of types and functions. RNA-based Regulation in Human Health and Disease. 2020:5-57

[24] Seal RL, Chen LL, Griffiths-Jones S, Lowe TM, Mathews MB, O'Reilly D, et al. A guide to naming human non-coding RNA genes. The EMBO Journal. 2020;**39**(6):e103777

[25] Altayli E. Regulator non-coding RNAs: miRNA, siRNA, piRNA, lncRNA, circRNA. Journal of Clinical Medicine of Kazakhstan. 2020;**6**(60):29-39

[26] Zhang H, Zhu J-K. Epigenetic gene regulation in plants and its potential applications in crop improvement. Nature Reviews Molecular Cell Biology. 2024;**26**:51-67

[27] Quinn JJ, Chang HY. Unique features of long non-coding RNA biogenesis and function. Nature Reviews Genetics. 2016;**17**(1):47-62

[28] Dahariya S, Paddibhatla I, Kumar S, Raghuwanshi S, Pallepati A, Gutti RK. Long non-coding RNA: Classification, biogenesis and functions in blood cells. Molecular Immunology. 2019;**112**:82-92

[29] Melissari M-T, Grote P. Roles for long non-coding RNAs in physiology and disease. Pflügers Archiv-European Journal of Physiology. 2016;**468**:945-958

[30] Ali A, Khatoon A, Shao C, Murtaza B, Tanveer Q, Su Z. Therapeutic potential of natural antisense transcripts and various mechanisms involved for clinical applications and disease prevention. RNA Biology. 2024;**21**(1):1-18

[31] Meng X, Li A, Yu B, Li S. Interplay between miRNAs and lncRNAs: Mode of action and biological roles in plant development and stress adaptation. Computational and Structural Biotechnology Journal. 2021;**19**:2567-2574

[32] Xu B, Meng Y, Jin Y. RNA structures in alternative splicing and back-splicing.

Wiley Interdisciplinary Reviews: RNA. 2021;**12**(1):e1626

[33] Nadhan R, Isidoro C, Song YS, Dhanasekaran DN. Signaling by LncRNAs: Structure, cellular homeostasis, and disease pathology. Cells. 2022;**11**(16):2517

[34] Goldsworthy S. Functional Roles of Non-coding RNA During Branchial Arch Development: A PhD Thesis Submitted to The University of Manchester (United Kingdom). 2023

[35] Wu L, Liu S, Qi H, Cai H, Xu M. Research progress on plant long non-coding RNA. Plants. 2020;**9**(4):408

[36] Wang X, Fan H, Wang B, Yuan F. Research progress on the roles of lncRNAs in plant development and stress responses. Frontiers in Plant Science. 2023;**14**:1138901

[37] Pan X, Wang Q, Yu Y, Wu W, Chen L, Wang W, et al. Antisense lncRNA NNT-AS1 promoted esophageal squamous cell carcinoma progression by regulating its sense gene NNT expression. Cell Death Discovery. 2022;**8**(1):424

[38] Wang H-LV, Chekanova JA. Long noncoding RNAs in plants. Advances in Experimental Medicine and Biology. 2017;**1008**:133-154

[39] Koziol MJ, Rinn JL. RNA traffic control of chromatin complexes. Current Opinion in Genetics & Development. 2010;**20**(2):142-148

[40] Shah IM, Dar MA, Bhat KA, Dar TA, Ahmad F, Ahmad SM. Long non-coding RNAs: Biogenesis, mechanism of action and role in different biological and pathological processes. In: Recent Advances in Noncoding RNAs. London, UK: IntechOpen; 2022

[41] Guo C-J, Ma X-K, Xing Y-H, Zheng C-C, Xu Y-F, Shan L, et al. Distinct processing of lncRNAs contributes to non-conserved functions in stem cells. Cell. 2020;**181**(3):621-636 e22

[42] Leypold NA, Speicher MR. Evolutionary conservation in noncoding genomic regions. Trends in Genetics. 2021;**37**(10):903-918

[43] Chodroff RA, Goodstadt L, Sirey TM, Oliver PL, Davies KE, Green ED, et al. Long noncoding RNA genes: Conservation of sequence and brain expression among diverse amniotes. Genome Biology. 2010;**11**:1-16

[44] Statello L, Guo C-J, Chen L-L, Huarte M. Gene regulation by long non-coding RNAs and its biological functions. Nature Reviews Molecular Cell Biology. 2021;**22**(2):96-118

[45] Borkiewicz L, Kalafut J, Dudziak K, Przybyszewska-Podstawka A, Telejko I. Decoding LncRNAs. Cancers. 2021;**13**(11):2643

[46] Khan M, Hou S, Chen M, Lei H. Mechanisms of RNA export and nuclear retention. Wiley Interdisciplinary Reviews: RNA. 2023;**14**(3):e1755

[47] Malakar P, Shukla S, Mondal M, Kar RK, Siddiqui JA. The nexus of long noncoding RNAs, splicing factors, alternative splicing and their modulations. RNA Biology. 2024;**21**(1):1-20

[48] Bhatia G, Sharma S, Upadhyay SK, Singh K. Long non-coding RNAs coordinate developmental transitions and other key biological processes in grapevine. Scientific Reports. 2019;**9**(1):3552

[49] Kim E-D, Xiong Y, Pyo Y, Kim D-H, Kang B-H, Sung S. Spatio-temporal

analysis of coding and long noncoding transcripts during maize endosperm development. Scientific Reports. 2017;7(1):3838

[50] Wang C, Wang L, Ding Y, Lu X, Zhang G, Yang J, et al. LncRNA structural characteristics in epigenetic regulation. International Journal of Molecular Sciences. 2017;**18**(12):2659

[51] Magar ND, Shah P, Barbadikar KM, Bosamia TC, Madhav MS, Mangrauthia SK, et al. Long non-coding RNA-mediated epigenetic response for abiotic stress tolerance in plants. Plant Physiology and Biochemistry. 2024;**206**:108165

[52] Chekanova JA. Long non-coding RNAs and their functions in plants. Current Opinion in Plant Biology. 2015;**27**:207-216

[53] Katiyar H. Genome Wide Identification and Characterization of Long Non-coding RNAs (lncRNAs) and miRNA during Sterility Mosaic Disease Infection in Pigeon Pea (*Cajanus cajan*). A Master's Thesis Submitted to Chandra SHekhar Azad University of Agriculture and Technology; 2024

[54] Saleembhasha A, Mishra S. Novel molecules lncRNAs, tRFs and circRNAs deciphered from next-generation sequencing/RNA sequencing: Computational databases and tools. Briefings in Functional Genomics. 2018;**17**(1):15-25

[55] Wang Z, Cui Q, Su C, Zhao S, Wang R, Wang Z, et al. Unveiling the secrets of non-coding RNA-encoded peptides in plants: A comprehensive review of mining methods and research progress. International Journal of Biological Macromolecules. 2023;**242**:124952

[56] Budak H, Kaya SB, Cagirici HB. Long non-coding RNA in plants in the era of reference sequences. Frontiers in Plant Science. 2020;**11**:276

[57] Corona-GomezJA,Coss-NavarreteEL, Garcia-Lopez IJ, Klapproth C, Pérez-Patiño JA, Fernandez-Valverde SL. Transcriptome-guided annotation and functional classification of long non-coding RNAs in Arabidopsis thaliana. Scientific Reports. 2022;**12**(1):14063

[58] Cheng CY, Krishnakumar V, Chan AP, Thibaud-Nissen F, Schobel S, Town CD. Araport11: A complete reannotation of the Arabidopsis thaliana reference genome. The Plant Journal. 2017;**89**(4):789-804

[59] Ariel F, Lucero L, Christ A, Mammarella MF, Jegu T, Veluchamy A, et al. R-loop mediated trans action of the APOLO long noncoding RNA. Molecular Cell. 2020;**77**(5):1055-1065. e4

[60] Corona-Gomez JA, Garcia-Lopez IJ, Stadler PF, Fernandez-Valverde SL. Splicing conservation signals in plant long noncoding RNAs. RNA. 2020;**26**(7):784-793

[61] Zhang Z, Zheng Y, Ham BK, Zhang S, Fei Z, Lucas WJ. Plant lncRNAs are enriched in and move systemically through the phloem in response to phosphate deficiency. Journal of Integrative Plant Biology. 2019;**61**(4):492-508

[62] Wierzbicki AT, Blevins T, Swiezewski S. Long noncoding RNAs in plants. Annual Review of Plant Biology. 2021;**72**(1):245-271

[63] Yang C, Yang L, Zhou M, Xie H, Zhang C, Wang MD, et al. LncADeep: An ab initio lncRNA identification and functional annotation tool based on deep learning. Bioinformatics. 2018;**34**(22):3825-3834

[64] Wang L, Park HJ, Dasari S, Wang S, Kocher J-P, Li W. CPAT: Coding-potential assessment tool using an alignment-free logistic regression model. Nucleic Acids Research. 2013;**41**(6):e74

[65] Li A, Zhang J, Zhou Z. PLEK: A tool for predicting long non-coding RNAs and messenger RNAs based on an improved k-mer scheme. BMC Bioinformatics. 2014;**15**:1-10

[66] Kang Y-J, Yang D-C, Kong L, Hou M, Meng Y-Q, Wei L, et al. CPC2: A fast and accurate coding potential calculator based on sequence intrinsic features. Nucleic Acids Research. 2017;**45**(W1):W12-WW6

[67] Sun L, Luo H, Bu D, Zhao G, Yu K, Zhang C, et al. Utilizing sequence intrinsic composition to classify protein-coding and long non-coding transcripts. Nucleic Acids Research. 2013;**41**(17):e166

[68] Tian X-C, Chen Z-Y, Nie S, Shi T-L, Yan X-M, Bao Y-T, et al. Plant-LncPipe: A computational pipeline providing significant improvement in plant lncRNA identification. Horticulture Research. 2024;**11**(4):uhae041

[69] Yang W, Bai Q, Li Y, Chen J, Liu C. Epigenetic modifications: Allusive clues of lncRNA functions in plants. Computational and Structural Biotechnology Journal. 2023;**21**:1989-1994

[70] Yu Q, Liu S, Yu L, Xiao Y, Zhang S, Wang X, et al. RNA demethylation increases the yield and biomass of rice and potato plants in field trials. Nature Biotechnology. 2021;**39**(12):1581-1588

[71] Urquiaga MCO, Thiebaut F, Hemerly AS, Ferreira PCG. From trash to luxury: The potential role of plant lncRNA in DNA methylation during abiotic stress. Frontiers in Plant Science. 2021;**11**:603246

[72] Gill SS, Khan NA, Agarwala N, Singh K, Sunkar R, Tuteja N. ncRNAs in Plant Development and Stress Responses. Elsevier; 2024. 108950

[73] Yu Y, Zhang Y, Chen X, Chen Y. Plant noncoding RNAs: Hidden players in development and stress responses. Annual Review of Cell and Developmental Biology. 2019;**35**(1):407-431

[74] Gallusci P, Dai Z, Génard M, Gauffretau A, Leblanc-Fournier N, Richard-Molard C, et al. Epigenetics for plant improvement: Current knowledge and modeling avenues. Trends in Plant Science. 2017;**22**(7):610-623

[75] Summanwar A, Basu U, Rahman H, Kav NN. Non-coding RNAs as emerging targets for crop improvement. Plant Science. 2020;**297**:110521

[76] Gonzales LR, Blom S, Henriques R, Bachem CW, Immink RG. LncRNAs: The art of being influential without protein. Trends in Plant Science. 2024;**29**:770-785

[77] Akhter Z, Bi Z, Ali K, Sun C, Fiaz S, Haider FU, et al. In response to abiotic stress, DNA methylation confers epigenetic changes in plants. Plants. 2021;**10**(6):1096

[78] Czajka K, Mehes-Smith M, Nkongolo K. DNA methylation and histone modifications induced by abiotic stressors in plants. Genes & Genomics. 2021;**44**:279-297

[79] Xie G, Du X, Hu H, Du J. Molecular mechanisms of the RNA polymerases in plant RNA-directed DNA methylation. Trends in Biochemical Sciences. 2024;**49**:247

[80] Erdmann RM, Picard CL. RNA-directed DNA methylation. PLoS Genetics. 2020;**16**(10):e1009034

[81] Loffer A, Singh J, Fukudome A, Mishra V, Wang F, Pikaard CS. A DCL3 dicing code within Pol IV-RDR2 transcripts diversifies the siRNA pool guiding RNA-directed DNA methylation. eLife. 2022;**11**:e73260

[82] Wang Q, Xue Y, Zhang L, Zhong Z, Feng S, Wang C, et al. Mechanism of siRNA production by a plant Dicer-RNA complex in dicing-competent conformation. Science. 2021;**374**(6571):1152-1157

[83] Rowley MJ, Rothi MH, Böhmdorfer G, Kuciński J, Wierzbicki AT. Long-range control of gene expression via RNA-directed DNA methylation. PLoS Genetics. 2017;**13**(5):e1006749

[84] Zhang H, He X, Zhu J-K. RNA-directed DNA methylation in plants: Where to start? RNA Biology. 2013;**10**(10):1593-1596

[85] Yadav A, Mathan J, Dubey AK, Singh A. The emerging role of non-coding RNAs (ncRNAs) in plant growth, development, and stress response signaling. Non-coding RNA. 2024;**10**(1):13

[86] Singh J. Biochemical Mechanisms of RNA-Dependent DNA Methylation in Plants. A Thesis Submitted to Indiana University; 2019

[87] Haag JR, Pikaard CS. Multisubunit RNA polymerases IV and V: Purveyors of non-coding RNA for plant gene silencing. Nature Reviews Molecular Cell Biology. 2011;**12**(8):483-492

[88] Jing X, Xu L, Huai X, Zhang H, Zhao F, Qiao Y. Genome-wide identification and characterization of Argonaute, Dicer-like and RNA-dependent RNA polymerase gene families and their expression analyses in Fragaria spp. Genes. 2023;**14**(1):121

[89] Sethuraman S. Genome-Wide Identification of Non-coding Transcription by RNA Polymerase V and Its Involvement in Transcriptional Gene Silencing. A Thesis Submitted to University of Michigan; 2021

[90] Numan M, Sun Y, Li G. Exploring the emerging role of long non-coding RNAs (lncRNAs) in plant biology: Functions, mechanisms of action, and future directions. Plant Physiology and Biochemistry. 2024;**212**:108797

[91] Kim J-H. Chromatin remodeling and epigenetic regulation in plant DNA damage repair. International Journal of Molecular Sciences. 2019;**20**(17):4093

[92] Bhat SA, Ahmad SM, Mumtaz PT, Malik AA, Dar MA, Urwat U, et al. Long non-coding RNAs: Mechanism of action and functional utility. Non-coding RNA Research. 2016;**1**(1):43-50

[93] Fejes-Toth K, Sotirova V, Sachidanandam R, Assaf G, Hannon GJ, Kapranov P, et al. Post-transcriptional processing generates a diversity of 5′-modified long and short RNAs: Affymetrix/cold Spring Harbor Laboratory ENCODE Transcriptome Project. Nature. 2009;**457**(7232):1028

[94] Bäurle I, Trindade I. Chromatin regulation of somatic abiotic stress memory. Journal of Experimental Botany. 2020;**71**(17):5269-5279

[95] Sahu RK, Singh S, Tomar RS. The mechanisms of action of chromatin remodelers and implications in development and disease. Biochemical Pharmacology. 2020;**180**:114200

[96] Ojolo SP, Cao S, Priyadarshani S, Li W, Yan M, Aslam M, et al. Regulation of plant growth and development: A review from a chromatin remodeling perspective. Frontiers in Plant Science. 2018;**9**:1232

[97] Eustermann S, Patel AB, Hopfner K-P, He Y, Korber P. Energy-driven genome regulation by ATP-dependent chromatin remodellers. Nature Reviews Molecular Cell Biology. 2024;**25**(4):309-332

[98] Candela-Ferre J, Diego-Martin B, Pérez-Alemany J, Gallego-Bartolomé J. Mind the gap: Epigenetic regulation of chromatin accessibility in plants. Plant Physiology. 2024;**194**(4):1998-2016

[99] Carlberg C. Gene Expression and Chromatin. Gene Regulation and Epigenetics: How Science Works. Cham, Switzerland: Springer; 2024. pp. 13-29

[100] Pirrotta V. Chromatin and Epigenetics: An Introduction to Epigenetic Mechanisms. Singapore: World Scientific; 2023

[101] Bure IV, Nemtsova MV, Kuznetsova EB. Histone modifications and non-coding RNAs: Mutual epigenetic regulation and role in pathogenesis. International Journal of Molecular Sciences. 2022;**23**(10):5801

[102] Ramazi S, Allahverdi A, Zahiri J. Evaluation of post-translational modifications in histone proteins: A review on histone modification defects in developmental and neurological disorders. Journal of Biosciences. 2020;**45**(1):135

[103] Rensing SA. How plants conquered land. Cell. 2020;**181**(5):964-966

[104] Yang H, Cui Y, Feng Y, Hu Y, Liu L, Duan L. Long non-coding RNAs of plants in response to abiotic stresses and their regulating roles in promoting environmental adaption. Cells. 2023;**12**(5):729

[105] Moison M, Pacheco JM, Lucero L, Fonouni-Farde C, Rodríguez-Melo J, Mansilla N, et al. The lncRNA APOLO interacts with the transcription factor WRKY42 to trigger root hair cell expansion in response to cold. Molecular Plant. 2021;**14**(6):937-948

[106] Liu G, Liu F, Wang Y, Liu X. A novel long noncoding RNA CIL1 enhances cold stress tolerance in Arabidopsis. Plant Science. 2022;**323**:111370

[107] Guangchao L, Liu F-X, Wang Y, Liu X. A novel long non-coding RNA CIL1 enhances cold stress tolerance in Arabidopsis

[108] Tang K, Zhao L, Ren Y, Yang S, Zhu JK, Zhao C. The transcription factor ICE1 functions in cold stress response by binding to the promoters of CBF and COR genes. Journal of Integrative Plant Biology. 2020;**62**(3):258-263

[109] Huang X, Cao L, Fan J, Ma G, Chen L. CdWRKY2-mediated sucrose biosynthesis and CBF-signalling pathways coordinately contribute to cold tolerance in bermudagrass. Plant Biotechnology Journal. 2022;**20**(4):660-675

[110] Wang X, Ding Y, Li Z, Shi Y, Wang J, Hua J, et al. PUB25 and PUB26 promote plant freezing tolerance by degrading the cold signaling negative regulator MYB15. Developmental Cell. 2019;**51**(2):222-235 e5

[111] Ritonga FN, Chen S. Physiological and molecular mechanism involved in cold stress tolerance in plants. Plants. 2020;**9**(5):560

[112] Janni M, Gullì M, Maestri E, Marmiroli M, Valliyodan B, Nguyen HT, et al. Molecular and genetic bases of heat stress responses in crop plants and breeding for increased resilience and productivity. Journal of Experimental Botany. 2020;**71**(13):3780-3802

[113] Zhao J, Lu Z, Wang L, Jin B. Plant responses to heat stress: Physiology, transcription, noncoding RNAs, and epigenetics. International Journal of Molecular Sciences. 2020;**22**(1):117

[114] Ohama N, Sato H, Shinozaki K, Yamaguchi-Shinozaki K. Transcriptional regulatory network of plant heat stress response. Trends in Plant Science. 2017;**22**(1):53-65

[115] Wang A, Hu J, Gao C, Chen G, Wang B, Lin C, et al. Genome-wide analysis of long non-coding RNAs unveils the regulatory roles in the heat tolerance of Chinese cabbage (*Brassica rapa* ssp. *chinensis*). Scientific Reports. 2019;**9**(1):5002

[116] Babaei S, Bhalla PL, Singh MB. Identifying long non-coding RNAs involved in heat stress response during wheat pollen development. Frontiers in Plant Science. 2024;**15**:1344928

[117] Tanwar UK, Stolarska E, Rudy E, Paluch-Lubawa E, Grabsztunowicz M, Arasimowicz-Jelonek M, et al. Metal tolerance gene family in barley: An in silico comprehensive analysis. Journal of Applied Genetics. 2023;**64**(2):197-215

[118] Adepoju AO, Femi-Adepoju A, Jalloh A, Faeflen S. Soil pollution and management practices. In: Environmental Pollution and Public Health. Amsterdam, Netherlands: Elsevier; 2024. pp. 187-236

[119] Vanisree C, Sankhla MS, Singh P, Jadhav EB, Verma RK, Awasthi KK, et al. Heavy metal contamination of food crops: Transportation via food chain, human consumption, toxicity and management strategies. In: Environmental Impact and Remediation of Heavy Metals. London, UK: IntechOpen; 2022

[120] Rahman SU, Han J-C, Ahmad M, Ashraf MN, Khaliq MA, Yousaf M, et al. Aluminum phytotoxicity in acidic environments: A comprehensive review of plant tolerance and adaptation strategies. Ecotoxicology and Environmental Safety. 2024;**269**:115791

[121] Fasani E, Giannelli G, Varotto S, Visioli G, Bellin D, Furini A, et al. Epigenetic control of plant response to heavy metals. Plants. 2023;**12**(18):3195

[122] Chen P, Song Y, Liu X, Xiao L, Bu C, Liu P, et al. LncRNA PMAT–PtoMYB46 module represses PtoMATE and PtoARF2 promoting Pb2+ uptake and plant growth in poplar. Journal of Hazardous Materials. 2022;**433**:128769

[123] Gui Q, Yang Z, Chen C, Yang F, Wang S, Dong R. Identification and characterization of long noncoding RNAs involved in the aluminum stress response in *Medicago truncatula* via genome-wide analysis. Frontiers in Plant Science. 2022;**13**:1017869

[124] Naz H, Akram NA, Ashraf M, Hefft DI, Jan BL. Leaf extract of neem (*Azadirachta indica*) alleviates adverse effects of drought in quinoa (*Chenopodium quinoa* Willd.) plants through alterations in biochemical attributes and antioxidants. Saudi Journal of Biological Sciences. 2022;**29**(3):1367-1374

[125] Wang W-n, Min Z, Wu J-r, Liu B-c, Xu X-l, Fang Y-l, et al. Physiological and transcriptomic analysis of Cabernet Sauvginon (*Vitis vinifera*

L.) reveals the alleviating effect of exogenous strigolactones on the response of grapevine to drought stress. Plant Physiology and Biochemistry. 2021;**167**:400-409

[126] Zou C, Guo Z, Zhao S, Chen J, Zhang C, Han H. Genome-wide analysis of long non-coding RNAs in sugar beet (*Beta vulgaris* L.) under drought stress. Frontiers in Plant Science. 2023;**14**:1118011

[127] Liu Y, Zhang M, Meng Z, Wang B, Chen M. Research progress on the roles of cytokinin in plant response to stress. International Journal of Molecular Sciences. 2020;**21**(18):6574

[128] Farooq M, Wahid A, Kobayashi N, Fujita D, Basra SM. Plant drought stress: Effects, mechanisms and management. In: Sustainable Agriculture. Dordrecht, Netherland: Springer; 2009. pp. 153-188

[129] Dong S-M, Liang X, Li Z-B, Jie S, Yan H-B, Li S-X, et al. A novel long non-coding RNA, DIR, increases drought tolerance in cassava by modifying stress-related gene expression. Journal of Integrative Agriculture. 2022;**21**(9):2588-2602

[130] Iqbal S, Wang X, Mubeen I, Kamran M, Kanwal I, Díaz GA, et al. Phytohormones trigger drought tolerance in crop plants: Outlook and future perspectives. Frontiers in Plant Science. 2022;**12**:799318

[131] Muhammad Aslam M, Waseem M, Jakada BH, Okal EJ, Lei Z, Saqib HSA, et al. Mechanisms of abscisic acid-mediated drought stress responses in plants. International Journal of Molecular Sciences. 2022;**23**(3):1084

[132] Weidong Q, Hongping C, Zuozhen Y, Biaolin H, Xiangdong L, Bing A, et al. Systematic characterization of long non-coding RNAs and their responses to drought stress in Dongxiang wild rice. Rice Science. 2020;**27**(1):21-31

[133] Tang X, Li Q, Feng X, Yang B, Zhong X, Zhou Y, et al. Identification and functional analysis of drought-responsive long noncoding RNAs in maize roots. International Journal of Molecular Sciences. 2023;**24**(20):15039

[134] Shaw J, Ghosh P, Bhattacharya S, Kundu A. Combined drought and salinity trigger unique lncRNA regulatory signatures in sweet sorghum (*Sorghum bicolor* L. Moench) compared to individual stress responses. Plant Growth Regulation. 2024;**104**:1399-1416

[135] Gelaw TA, Sanan-Mishra N. Non-coding RNAs in response to drought stress. International Journal of Molecular Sciences. 2021;**22**(22):12519

[136] Tan X, Li S, Hu L, Zhang C. Genome-wide analysis of long non-coding RNAs (lncRNAs) in two contrasting rapeseed (*Brassica napus* L.) genotypes subjected to drought stress and re-watering. BMC Plant Biology. 2020;**20**:1-20

[137] Cai J, Zhang Y, He R, Jiang L, Qu Z, Gu J, et al. LncRNA DANA1 promotes drought tolerance and histone deacetylation of drought responsive genes in Arabidopsis. EMBO Reports. 2024;**25**(2):796-812

[138] Chen K, Huang Y, Liu C, Liang Y, Li M. Transcriptome profile analysis of arabidopsis reveals the drought stress-induced long non-coding RNAs associated with photosynthesis, chlorophyll synthesis, fatty acid synthesis and degradation. Frontiers in Plant Science. 2021;**12**:643182

[139] Lamin-Samu AT, Zhuo S, Ali M, Lu G. Long non-coding RNA

transcriptome landscape of anthers at different developmental stages in response to drought stress in tomato. Genomics. 2022;**114**(4):110383

[140] Yang X, Liu C, Niu X, Wang L, Li L, Yuan Q, et al. Research on lncRNA related to drought resistance of Shanlan upland rice. BMC Genomics. 2022;**23**(1):336

[141] Bouras H, Mamassi A, Devkota KP, Choukr-Allah R, Bouazzama B. Integrated effect of saline water irrigation and phosphorus fertilization practices on wheat (*Triticum aestivum*) growth, productivity, nutrient content and soil proprieties under dryland farming. Plant Stress. 2023;**10**:100295

[142] Devkota KP, Devkota M, Rezaei M, Oosterbaan R. Managing salinity for sustainable agricultural production in salt-affected soils of irrigated drylands. Agricultural Systems. 2022;**198**:103390

[143] Roșca M, Mihalache G, Stoleru V. Tomato responses to salinity stress: From morphological traits to genetic changes. Frontiers in Plant Science. 2023;**14**:1118383

[144] Hao S, Wang Y, Yan Y, Liu Y, Wang J, Chen S. A review on plant responses to salt stress and their mechanisms of salt resistance. Horticulturae. 2021;7(6):132

[145] Mostofa MG, Saegusa D, Fujita M, Tran L-SP. Hydrogen sulfide regulates salt tolerance in rice by maintaining Na+/K+ balance, mineral homeostasis and oxidative metabolism under excessive salt stress. Frontiers in Plant Science. 2015;**6**:1055

[146] Li N, Wang Z, Wang B, Wang J, Xu R, Yang T, et al. Identification and characterization of long non-coding RNA in tomato roots under salt stress. Frontiers in Plant Science. 2022;**13**:834027

[147] Deng F, Zhang X, Wang W, Yuan R, Shen F. Identification of *Gossypium hirsutum* long non-coding RNAs (lncRNAs) under salt stress. BMC Plant Biology. 2018;**18**:1-14

[148] Sun X, Zheng H, Li J, Liu L, Zhang X, Sui N. Comparative transcriptome analysis reveals new lncRNAs responding to salt stress in sweet sorghum. Frontiers in Bioengineering and Biotechnology. 2020;**8**:331

[149] Rehman OU, Uzair M, Farooq MS, Saleem B, Attacha S, Attia KA, et al. Comprehensive insights into the regulatory mechanisms of lncRNA in alkaline-salt stress tolerance in rice. Molecular Biology Reports. 2023;**50**(9):7381-7392

[150] Kumar N, Bharadwaj C, Sahu S, Shiv A, Shrivastava AK, Reddy SPP, et al. Genome-wide identification and functional prediction of salt-stress related long non-coding RNAs (lncRNAs) in chickpea (*Cicer arietinum* L.). Physiology and Molecular Biology of Plants. 2021;**27**:2605-2619

[151] Jin Z, Gao S, Ma W, Lyu X, Cao X, Yao Y. Identification and functional prediction of salt stress-related long noncoding RNAs in grapevine roots. Environmental and Experimental Botany. 2020;**179**:104215

[152] Bhar A, Roy A. Emphasizing the role of long non-coding RNAs (lncRNA), circular RNA (circRNA), and micropeptides (miPs) in plant biotic stress tolerance. Plants. 2023;**12**(23):3951

[153] Nishad R, Ahmed T, Rahman VJ, Kareem A. Modulation of plant defense system in response to microbial

interactions. Frontiers in Microbiology. 2020;**11**:1298

[154] Yuan M, Ngou BPM, Ding P, Xin X-F. PTI-ETI crosstalk: An integrative view of plant immunity. Current Opinion in Plant Biology. 2021;**62**:102030

[155] Li Q, Shen H, Yuan S, Dai X, Yang C. miRNAs and lncRNAs in tomato: Roles in biotic and abiotic stress responses. Frontiers in Plant Science. 2023;**13**:1094459

[156] Zhang Y-Y, Hong Y-H, Liu Y-R, Cui J, Luan Y-S. Function identification of miR394 in tomato resistance to *Phytophthora infestans*. Plant Cell Reports. 2021;**40**:1831-1844

Meet the editors

Dr. Stavros Hatzopoulos has been a faculty member of the Audiology & ENT Clinic, University of Ferrara, Italy, since 1994. His background is in biomedical, engineering, audiological engineering, and hearing science (BSc degree from University of Southern California, MSc from Texas A&M, PhD from Worcester Polytechnic Institute in Massachusetts). He received his habilitation as an associate professor in 2014. He is the author of more than 195 book chapters, peer review papers, and congress presentations. He has participated in numerous European Concerted Action projects in the areas of otoacoustic emissions, genetics, and nanotechnologies. He currently serves as the editor of the Portal on Otoacoustic Emissions (www.otoemissions.org) and as the audiology section editor in the Journal of Hearing Science.

Andrea Ciorba, MD, PhD, has been a member of the Audiology & ENT Clinic, University of Ferrara, Italy, since 2004. After earning his degree in Medicine in 1999 and a specialization in Otolaryngology and Head and Neck Surgery in October 2003, he obtained a PhD in Audiology in March 2008. He received his habilitation as an associate professor in 2014 and he is presently a member of the Audiology & ENT Clinic, University of Ferrara, Faculty Board. He is the author of approximately 130 peer-reviewed papers and book chapters and serves as a member of the editorial board or as a reviewer for a number of scientific journals. His clinical field of interest is audiology and vestibology, and he is also involved in research activities in the field of otoacoustic emissions and inner ear electrophysiology.

Prof. Piotr H. Skarzynski, MD, PhD, MS is occupied with the implementation of scientific, clinical and surgical work at the World Hearing Center, Medical University of Warsaw and Institute of Sensory Organs in Warsaw, Poland. He is a member of 29 scientific societies. He has been a Representative Board Member (until 2019) and Vice-President and Institutional Representative (since 2019) of ISfTeH; Vice-President of the Hearing Group; Member of Hearing Committee of AAO–HNS; an auditor of EFAS; a member of the Consultant Committee of International Experts of CPAM-VBMS (by special invitation); an Honorary Member of ORL Danube Society; and a member of the Roster of Experts on Digital Health of WHO. He also participated in the 3rd Stakeholders Consultation meeting during which the World Hearing Forum of WHO was announced. He has been an author and co-author of over 1767 documents (invited speaker presentations – 115, round table presentations – 118, as an instructor – 118 courses) and 843 publications (IF – 166,554, IC – 44450,31). He is a reviewer of 30 national and international scientific journals. He is currently involved with numerous international projects, connected with hearing screening and telemedicine in European, Asian, and African countries.

Contents

Preface

The objective of this book is to present information on the basic features of the auditory system, as well as hearing loss, which is among the most prevalent chronic disabilities worldwide. Nowadays, it is clear that the identification and the rehabilitation of hearing impairment, when possible, has to be adequately and promptly managed, since hearing loss can seriously interfere with psychosocial development, family dynamics, and social interactions. This book presents a number of basic characteristics of the auditory system and a number of novel approaches to issues pertaining to hearing deficits.

This book was made possible through the substantial contribution of several authors.

The format of this volume targets graduate courses in audiology, speech pathology and hearing science, neurosciences, and basic graduate courses in otolaryngology.

The volume is divided in two main sections as follows:

1. **Part I:** Anatomical and Pathophysiological Features. In these chapters, several aspects of the auditory system are discussed, particularly focusing on the pathophysiology of the auditory pathways.

2. **Part II:** Updates on Diagnosis and Therapy. This part offers an update on several innovative approaches for a modern evaluation and therapy of the ear, such as endoscopic ear surgery.

Acknowledgements

We would like to thank all the contributing authors, and the IntechOpen personnel for the continuous and generous assistance during the preparation of the volume.

Stavros Hatzopoulos PhD and Dr. Andrea Ciorba
Clinic of Audiology and ENT,
University of Ferrara,
Ferrara, Italy

Piotr H. Skarzynski MD PhD
Department of Teleaudiology and Screening,
World Hearing Center,
Institute of Physiology and Pathology of Hearing,
Warsaw, Poland

Section 1

Anatomical and Pathophysiological Features of the Auditory Pathways

Chapter 1

Attention and Working Memory in Human Auditory Cortex

Brian Barton and Alyssa A. Brewer

Abstract

Human sensory systems are organized into processing hierarchies within cortex, such that incoming sensory information is analyzed and compiled into our vivid sensory experiences. Computations that are common to these sensory systems include the abilities to maintain enhanced focus on particular aspects of incoming sensory information (i.e., attention) and to retain sensory information in a short-term memory store after such sensory information is no longer available (i.e., working memory). In at least the auditory and visual systems, the necessary computational steps to create these experiences take place in cloverleaf clusters of cortical field maps (CFMs). The human auditory CFMs represent the spectral (i.e., tones) and temporal (i.e., period) aspects of sound, which are represented along the cortical surface as two orderly gradients that are physically orthogonal to one another: tonotopy and periodotopy, respectively. Knowledge of the properties of such CFMs is the foundation for understanding the specific sensory computations carried out in particular cortical regions. This chapter reviews current research into auditory nonverbal attention, auditory working memory, and auditory CFMs, and introduces the next steps to measure the effects of attention and working memory across the known auditory CFMs in human cortex using functional MRI.

Keywords: human auditory cortex, fMRI, tonotopy, periodotopy, cloverleaf cluster, cortical field maps, attention, working memory

1. Introduction

Mammalian sensory systems are composed in cortex of many functionally specialized areas organized into hierarchical networks [1–6]. The most fundamental sensory information is embodied by the organization of the sensory receptors, which is maintained throughout most of the cortical hierarchy of sensory regions with repeating representations of this topography in cortical field maps (CFMs) [5, 7–13]. Accordingly neurons with receptive fields situated next to one another in sensory feature space are positioned next to one another in cortex within a CFM.

In auditory cortex, auditory field maps (AFMs) are identified by two orthogonal sensory representations: tonotopic gradients from the spectral aspects of sound (i.e., tones), and periodotopic gradients from the temporal aspects of sound (i.e., period or temporal envelope) [5, 10, 14]. On a larger scale across cortex, AFMs are grouped into cloverleaf clusters, another fundamental organizational structure also common to visual cortex [8, 10, 15–20]. CFMs within clusters tend to share properties such as receptive field distribution, cortical magnification, and processing specialization (e.g., [18, 19, 21]).

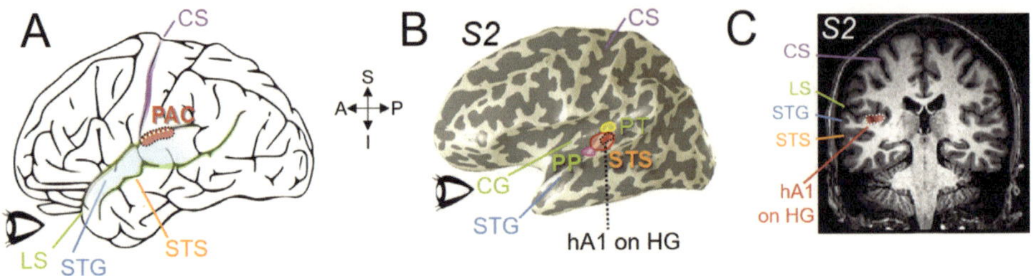

Figure 1.
Primary auditory cortex. (A) The lateral view of the left hemisphere is shown in the schematic. Major sulci are marked by black lines. The approximate position of primary auditory cortex (PAC) is shown with the red overlay inside the black dotted line. The white dotted line within the red region indicates the extension of PAC into the lateral sulcus (LS) along Heschl's gyrus (HG; hidden within the sulcus in this view). Inset refers to anatomical directions as A: anterior; P: posterior; S: superior; I: inferior. PAC: primary auditory cortex (red); LS: lateral sulcus (green; also known as the lateral fissure or Sylvian fissure); CS: central sulcus (purple); STG: superior temporal gyrus (blue); STS: superior temporal sulcus (orange). (B) The cortical surface of the left hemisphere of one subject (S2) is displayed as a typical inflated 3-D rendering created from high-resolution, anatomical MRI measurements. Light gray regions denote gyri; dark gray regions denote sulci. The exact location of this subject's hA1 auditory field map is shown in red within the black dotted lines. Note that HG in S2 is composed of a double peak, seen here as two light gray stripes, rather than the more common single gyrus. The locations of the three cloverleaf clusters composed of the core and belt AFMs are shown along HG by three colored overlays as yellow: hCM/hCL cluster; red: HG cluster including hA1, hR, hRM, hMM, hML, hAL; and magenta: hRTM/hRT/hRTL cluster (cite?). Additional cloverleaf clusters are under investigation along PP, PT, STG, and the STS. Green-labeled anatomical regions are sections within the lateral sulcus—CG: Circular gyrus (green); PP: planum polare (green); PT: planum temporale (green). (C) This single T1 image shows a coronal view of hA1 on HG (red within dotted white line). Adapted from Refs. [5, 12].

Across the cortical hierarchy, there is generally a progressive increase in the complexity of sensory computations from simple sensory stimulus features (e.g., frequency content) to higher levels of cognition (e.g., attention and working memory) [6, 13, 22]. CFM organization likely serves as a framework for integrating bottom-up inputs from sensory receptors with top-down attentional processing [12, 17]. With the recent ability to measure AFMs in the core and belt regions of human auditory cortex along Heschl's gyrus (HG) using high-resolution functional magnetic resonance imaging (fMRI), the stage is now set for investigation into this integration of basic auditory processing with higher-order auditory attention and working memory within human AFMs (**Figure 1**) [5, 12, 15, 23].

This chapter first provides a brief history of research into models of auditory nonverbal attention and working memory, with comparisons to their visual counterparts. Next, we discuss the current state of research into AFMs within human auditory cortex. Finally, we propose directions of future research investigating auditory attention and working memory within these AFMs to illuminate how these higher-order cognitive processes interact with low-level auditory processing.

2. Attention and working memory in human audition

2.1 Models of attention and working memory

Attention, the ability to select and attend to aspects of the sensory environment while simultaneously ignoring or inhibiting others, is a fundamental aspect of human sensory systems (for reviews, see [24–27]). Given the limited resources of the human brain, attention allows for greater resources to be allocated to processing of important incoming sensory stimuli by diverting precious resources from currently unimportant stimuli. Such allocation can be controlled cognitively, in what is generally referred to as 'top-down' attentional control in models of attention, in reference to the higher-order cognitive processes controlling attention from the 'top'

DOI: http://dx.doi.org/10.5772/intechopen.85537

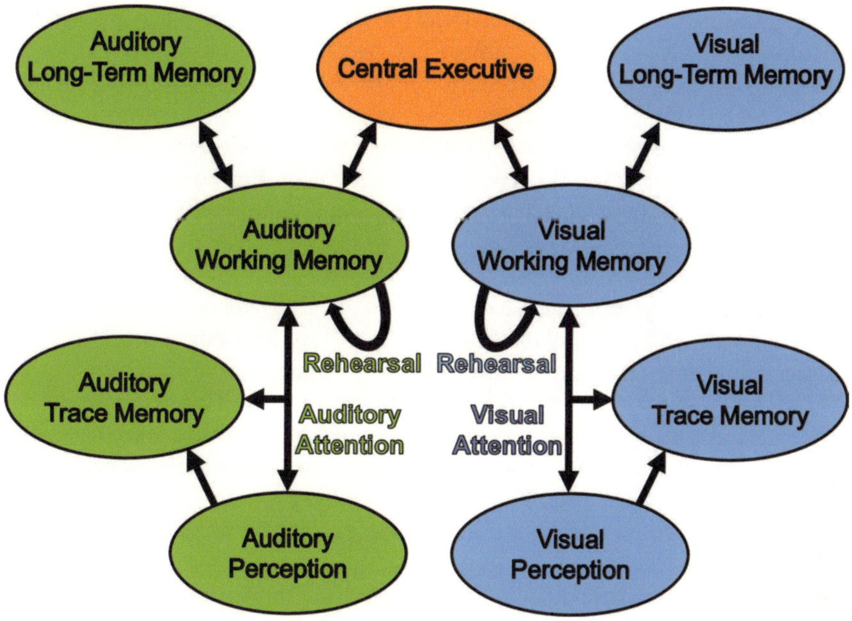

Figure 2.
Attention and working-memory model. A model of the interactions between perception, trace memory, attention, working memory, and long-term memory in the visual and auditory systems, as well as the central executive. Ovals represent neural systems. Arrows represent actions of one system on another. Attention is the term for the action of perception and trace memory on working memory and vice versa. Rehearsal is the term for maintaining information in working memory. This model is not intended to indicate that these systems are discrete or independent; within each sense, they are in fact highly integrated.

of the sensory-processing hierarchy and acting 'down' on the lower levels (**Figure 2**) [24, 28–31]. Despite lower priority being assigned to the currently unimportant stimulus locations, change is constant, so the resource diversion to attended stimuli is not absolute, allowing for the sensory environment to continue to be monitored. If, instead, processing resources were evenly distributed throughout the sensory field, without regard to salience, more resources would be wasted on unimportant aspects of the field. If something in the unattended sensory field should become important, the system requires a mechanism to reorient attention to that aspect of the field. Such stimulus-driven attentional control is referred to as 'bottom-up', referring to the ability of incoming sensory input at the bottom of the hierarchy to orient the higher-order attention system. This broad framework of attentional models is common at least to the senses most commonly studied, vision and audition [25, 27, 31, 32].

In the effort to elucidate the parameters of auditory attention, researchers have taken a myriad of approaches in numerous contexts. Researchers have attempted to decipher at what level of the sensory-processing hierarchy stimulus-driven attention occurs (after which sensory-processing steps does attention act) [24, 30, 31, 33–35], how attention can be deployed (to locations in space or particular sensory features) [36–40], and how can attention be distributed (to how many 'objects' or 'streams' can attention be simultaneously deployed) [41–44]. Many studies have narrowed the range of possibilities without precisely answering these questions, and so remain active areas of research. Modern models of attention generally agree that stimuli are processed to some degree before attention acts, accounting for the stimulus-driven 'bottom-up' attentional shifts, though it is unclear to precisely which degree [24, 30, 33]. Neuroscientific evidence suggests that attention acts throughout sensory-processing hierarchies, so the idea of attention being located at a particular 'height' in the hierarchy may not be a particularly useful insight for identifying the cortical locus of attentional control [45, 46]. Modern attentional models also generally agree that attention can be deployed to locations in or features of sensory space, both of which are fundamental aspects to the sensory-processing hierarchy [24, 35]. Finally, modern models of attention agree that attention is very limited, but not about precisely how it is limited. Some models are still fundamentally 'spotlight' models [25, 44], in which attention is limited to a single location or feature set, while others posit that attention can be divided between a small number of locations or features [41, 47]. Based on related working-memory research, the latter theory is gaining prominence as likely correct.

Working memory (i.e., a more accurate term for 'short-term memory') is the ability to maintain and manipulate information within the focus of attention over a short period of time after the stimulus is no longer perceptible (for reviews, see [48–51]). Without explicit maintenance, this retention period is approximately 1–2 s, but is theoretically indefinite with explicit maintenance. Working memory should not to be confused with 'sensory memory', also known as 'iconic memory' in vision and 'echoic memory' in audition [52]. Sensory memory is a fundamental aspect of sensory systems in which a sensory trace available to attention and working-memory systems persists for less than ~100 ms after stimuli are no longer perceptible. Models of working memory are nearly indistinguishable from models of attention; the key difference is that working memory is a 'memory' of previously perceptible stimuli, whereas attention is thought to act on perceptible stimuli or sensory traces thereof. Working-memory models posit, by definition, that working memory acts after perception processing has occurred (**Figure 2**; for review, see [53]). However, it has been difficult to isolate exactly where working-memory control resides along the cortical hierarchy of sensory processing, likely because low-level perceptual cortex is recruited at least for visual working memory and attention [40, 46, 54, 55].

Like attention, working-memory models also posit that working memory is a highly limited resource, in which a small set of locations or objects (e.g., 3–4 items on average) can be simultaneously maintained [42, 49]. In fact, some modern measures of attention and working memory are nearly identical. The change-detection task is a ubiquitous one in which subjects are asked to view a sensory array, then compare that sensory array to a second one in which some aspect of the array may have changed, and indicate whether a change has occurred (**Figure 3**) [56–60]. A short delay period (i.e., retention interval) is included during each array, which may include a neutral presentation or, if desired, a mask of the sensory stimuli to prevent the use of 'sensory memory'. The length of the delay period can be then be altered to either measure attention or working memory. If the delay period is on the order of ~0–200 ms, it is considered an attentional task; if it is longer, on the order of 1–2 s, it is considered a working-memory task [53]. Therefore, attention and working-memory systems are at a minimum heavily intertwined and very likely the same system studied in slightly different contexts, with attention being a component of a larger working-memory framework.

With the relatively recent invention of fMRI, researchers have been able to begin to localize these models of attention and working memory to their cortical underpinnings (e.g., [6, 37, 40, 50, 55, 61, 62]). FMRI, through its exquisite ability to localize blood oxygenation-level dependent (BOLD) signals (and thus the underlying neural activity) to just a couple of millimeters is the best technology available for such research [63, 64]. Two broad approaches have been employed for studying these high-order cognitive processes: model-based and perception-based. Model-based investigations tend to use tasks based on behavioral investigations into attention and working memory, adapt them to the strict parameters required of fMRI, and compare activity in conditions when attention or working memory are differentially deployed [61, 62]. Perception-based investigations tend to measure low-level perceptual cortex that has already been mapped in detail and measure the effects of attention or working memory within those regions [50, 55, 65]. Both approaches are important and should be fully integrated to garner a more complete and accurate localization of these attentional and working-memory systems.

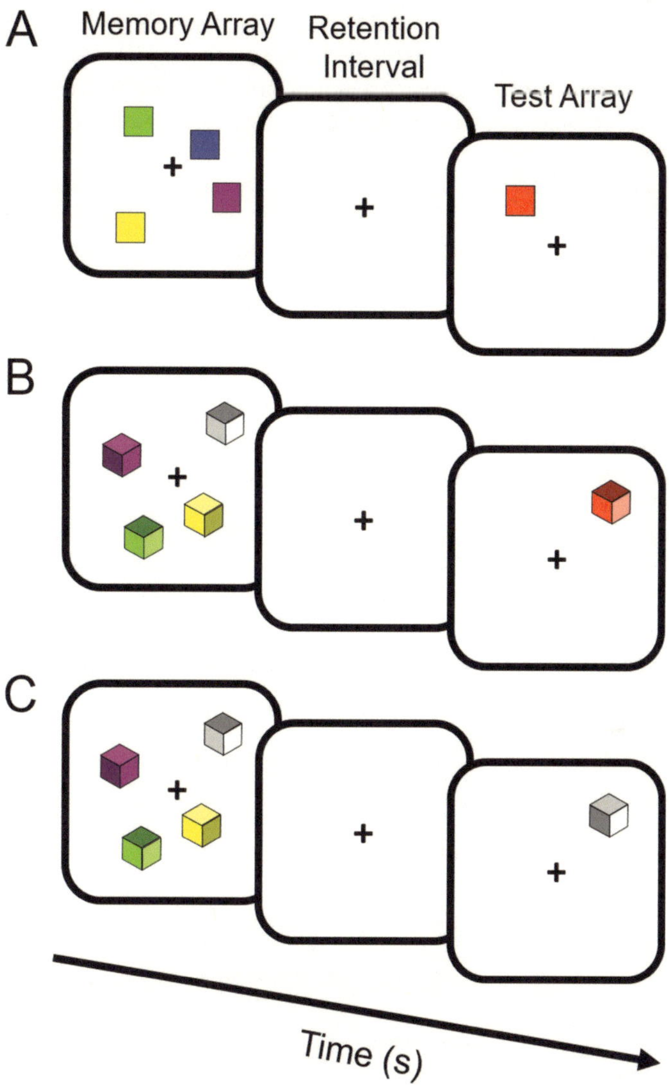

Figure 3.
Visual change-detection task. This task can be used to probe visual attention or working memory and is very similar to its auditory counterpart. Such tasks have three phases: first is encoding, when subjects are given ~100–500 ms to view the sample array; next is maintenance, which is short (~0–200 ms) for measuring attention and longer (~1000 ms) for working memory; last is the probe (lasting until the subject responds or with a time limit, often ~2000 ms). In this example, a set size of four is presented for the sample array and a probe array of one is used, though different set sizes are commonplace and often the probe array will be the same set size as the encoding array with a possibility of one object being changed. Typically there is an equal chance (50%) of the probe array containing a change or not. Generally subjects will be required to fixate centrally, particularly if fMRI, EEG, or PET recordings are being made. (A) Simple colored square stimuli are depicted here, often drawn from a small set of easily distinguished hues (in this case, 6). As a result, changes are always low in similarity, requiring low resolution to make accurate comparisons between encoding and test arrays, which is important at least for visual working-memory measurements. More complex stimuli can also be used as in (B) and (C). These stimuli are shaded cubes with the same hue set as in (A), but also have 6 possible shading patterns with the dark, medium, and light shaded sides on each cube. Changes between hues, as in (B), are equivalently low similarity to (A) and result in similar performance under visual working-memory conditions. Changes in shading patterns, as in (C), result in worse performance than (B) despite having the same number of possible pattern changes as hue changes in (A) or (B), because such changes require higher resolution representations in visual working memory. Adapted from Barton and Brewer [50].

2.2 Overview of auditory and visual attention research

Research into attention began in earnest in the auditory system after World War II with a very practical motivation. It had been noted that fighter pilots sometimes failed to perceive auditory messages presented to them over headphones despite the fact that the messages were completely audible. To solve this problem, Donald Broadbent began studying subjects with an auditory environment similar to the pilots, with multiple speech messages presented over headphones [34]. Based on his findings, he proposed a selective theory of attention, which was popular and persuasive, but ultimately required modification. Environments such as the one Broadbent studied are more commonly encountered at cocktail parties, in which multiple audible conversations are taking place, and people are able to attend to one or a small set of speech streams while attenuating the others. To study the 'cocktail party phenomenon,' the dichotic listening task was developed in the 1950s by Colin Cherry [66, 67]. Subjects were asked to shadow the speech stream presented to one ear of a set of headphones while another stream was presented to the other ear, and they demonstrated little knowledge of the nonshadowed (unattended) stream (**Figure 4**).

A host of studies followed up on the basic finding, revealing several attentional parameters within the context of that type of task (e.g., [30, 35, 40, 68–71]). Importantly, preferential processing of the attended stream relative to the unattended streams is not absolute; for example, particularly salient information, such as the name of the subject, could sometimes be recalled from an unattended stream, presumably by reorienting attention [39, 66, 67, 69]. The streams were typically differentiated spatially (e.g., to each ear through a headset), indicating a spatial aspect to attentional selection and therefore the attentional system. Similarly, the streams

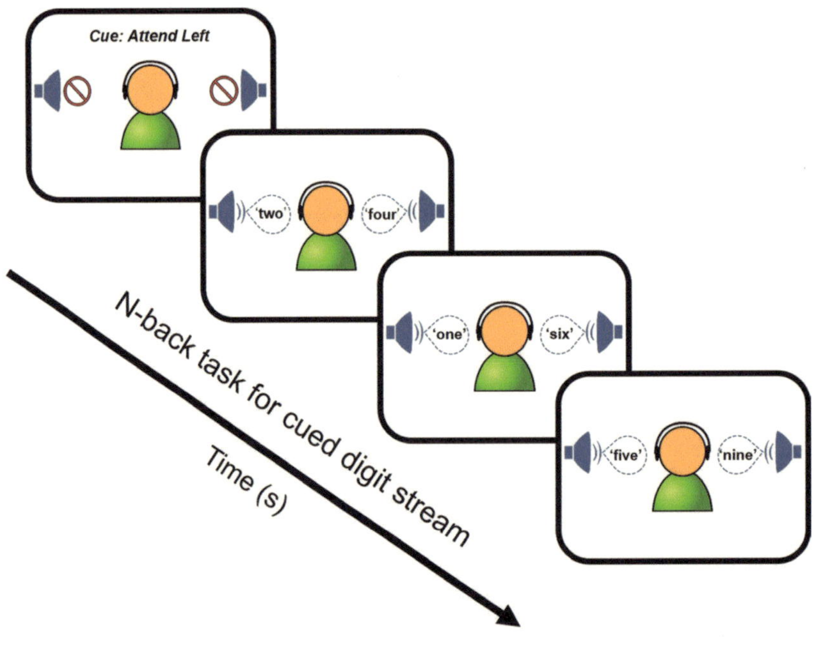

Figure 4.
Auditory spatial attention. Schematic of an example auditory spatial attention task (e.g., see [35, 40, 66, 67]). Each block typically starts with cue (auditory or visual) for the subject to attend left or right on the upcoming trial. Two simultaneous auditory streams of digits are presented as binaural, spatially lateralized signals. Behavioral studies in an anechoic chamber often use speaker physically located to the left and right of the subject; fMRI measurements do not have the option of such a set up, but instead can use differences in the interaural time difference (ITD) to produce a similarly effective lateralization for the two digit streams. The subjects attend to the cued digit stream and perform a 1-back task.

were also typically differentiated by the voice of the person speaking, indicating attentional selection based on the spectrotemporal characteristics of the speaker's voice such as the average and variance of pitch and speech rate (often reflecting additional information about the speaker, such as gender) [66–68, 72].

These findings are very similar to findings in the visual domain, indicating that attentional systems across senses are similarly organized. Visual attention can similarly be deployed to a small set of locations or to visual features with very little recall of nonattended visual stimuli [41]. Roughly analogous to speech shadowing are multiple-object-tracking tasks, which require subjects to visually track a small set of moving objects out of a group [47, 73]. Visual change-detection tasks are also very common, and they demonstrate very similar results as their auditory counterparts [50, 74, 75]. In sum, the evidence suggests that attentional systems are organized very similarly, perhaps identically, between at least vision and audition.

Despite these broad contributions, these types of tasks are of limited utility when tying behavior to cortical activity because the types of stimuli used are rather high-order (e.g., speech) with relatively uncontrolled low-level parameters. For example, the spectrotemporal profile of a stream of speech is complex, likely activating broad swaths of low-level sensory cortex in addition to higher-order regions dedicated to speech comprehension, including working and long-term memory [68, 72, 76, 77]. If one were to compare fMRI activity across auditory cortex in traditional dichotic listening tasks, the differences would have far too many variables for which to account before meaningful conclusions can be made about attentional systems. It may seem intuitive to compare cortical activity between conditions where identical speech stimuli have been presented and the subject either attended to the stimuli or did not. However, areas that have increased activity when the stimuli were attended could simply reflect higher-order processing that only occurs when attention is directed to the stimuli rather than directly revealing areas involved in attentional control. For example, recognition of particular words requires comparison of the speech stimulus to an internal representation, which requires activation of long-term memories of words [77]. Long-term memory retrieval does not happen if the subject never perceived the word due to attention being maintained on a separate speech stream, so such memory-retrieval activity would be confounded with attentional activity in the analysis [70].

Thus, simpler stimuli that are closer in nature to the initial spectrotemporal analyses performed by primary auditory cortex (PAC) are better suited for experiments intended to demonstrate attentional effects in cortex [24]. Reducing the speech comprehension element is a good first step, and research approached this by using a change-detection task and arrays of recognizable animal sounds (cow, owl, frog, etc.; **Figure 5**) [59]. These tests revealed what the researchers termed 'change deafness,' in which subjects often failed to identify changes in the sound arrays. Such inability to detect changes is entirely consistent with very limited attentional resources, and very similar to results of working-memory change-detection tasks [30, 53, 60, 78].

However, even these types of stimuli are not best suited to fMRI investigation at this stage of understanding due to their relative complexity compared to the basic spectrotemporal features of sounds initially processed in auditory cortex [12, 50]. As discussed in detail below, the auditory system represents sounds in spectral and temporal dimensions, and stimuli similar to those used to define those perceptual areas would be best suited now to evaluating the effects of attention in the auditory system (**Figure 6**) [5, 10].

2.3 Overview of auditory and visual working-memory research

Visual and auditory working memory were discovered in quick succession and discussed together in a very popular and influential model by Baddeley and Hitch

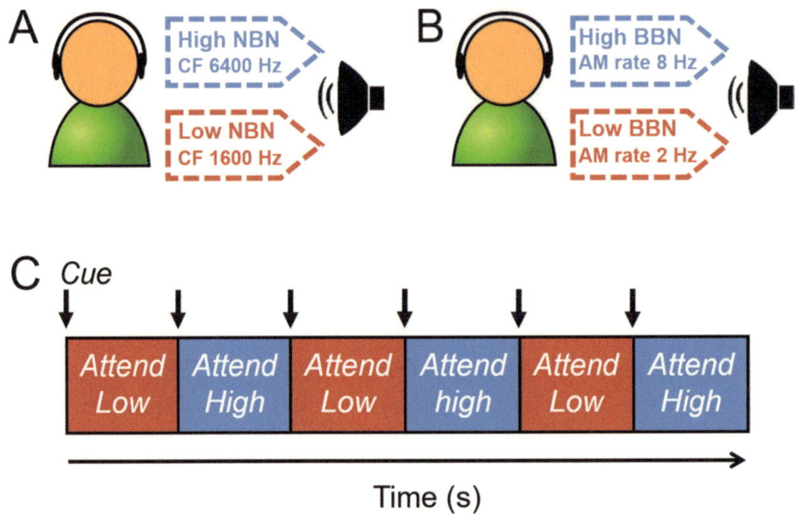

Figure 5.
Auditory feature attention. Schematic outlines a simple proposed attention task utilizing spectral (narrowband noise) and temporal (broadband noise) stimuli taken from the stimuli used by [10] to define auditory field maps. Subjects are asked to attend to one of two simultaneously presented stimuli, which are either (A) narrowband noise, in this case with central frequencies of 6400 and 1600 Hz and the same amplitude modulation (AM) rate of 8 Hz, or (B) broadband noise, in this case with AM rates of 2 and 8 Hz. (C) A proposed task that varies auditory feature attention, in which subjects are instructed to attend to each of the stimuli in an alternating pattern, cued by a short sound at the beginning of each block.

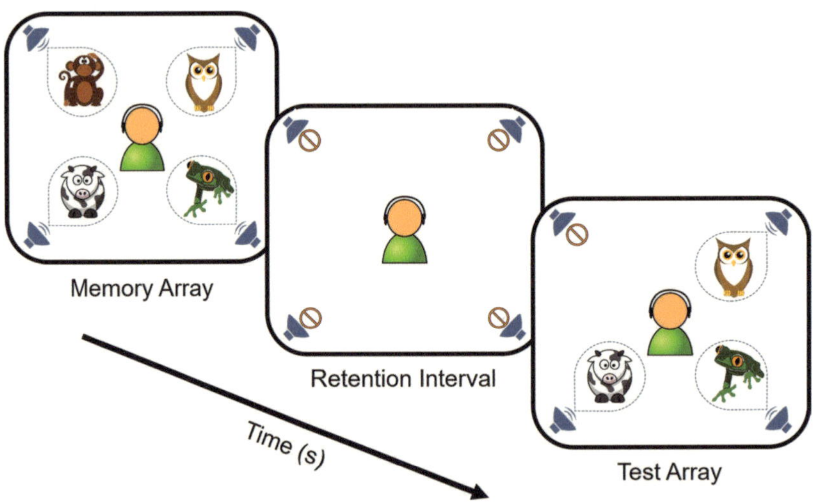

Figure 6.
Auditory object attention and working memory. Schematic of one trial in an auditory change-detection task (e.g., see change-deafness experiments in [59]). Subjects are first presented with an array of four distinct auditory objects (e.g., four different recordings of real animal sounds, randomized each trial from a larger set of iconic animal calls). In the initial memory array, the four animal sounds are initially presented binaurally and are temporally overlapped for a short time (e.g., 2 s). Within an anechoic chamber setup often used in psychoacoustic studies, these speakers may be physically positioned at the corners of a square; fMRI measurements do not have the option of such a set up, but instead can use differences in the interaural time difference (ITD) and interaural level difference (ILD) to produce a similarly effective virtual space. The subject's goal is typically to identify and remember all four animal sounds. The interstimulus interval is commonly filled with silence or white noise and can be varied in length to create shorter or longer retention intervals for attention or working-memory tasks, respectively. During the subsequent test array, subjects attempt to identify which one of the four auditory objects is now missing from the simultaneous animal sound presentations. In such auditory change-deafness paradigms, subjects fail to notice a large proportion of the changes introduced between the initial and test arrays.

linking sensory perception, working memory, and executive control [79–81]. The generally accepted modern model of working memory has changed somewhat from the original depiction, but the vast majority of research has been working within the framework (for reviews, see [30, 51, 53, 79, 81]). Each sense is equipped with its own perceptual system and three memory systems: sensory memory, working memory, and long-term memory. Direct sensory input, gated by attentional selection, is one of the two primary inputs into working memory. Sensory memory is a vivid trace of sensory information that persists after the information has vanished for a short time and is essentially equivalent to direct sensory input into working memory, again gated by attentional selection; one can reorient attention to aspects of the sensory trace as if it were direct sensation. Long-term memory is the second primary input into working memory, which is gated by an attention-like selection, generally referred to as selective memory retrieval. Working memory itself is a short-term memory workspace lasting a couple of seconds without rehearsal, in which sensory information is maintained and manipulated by a central executive [82]. The central executive is a deliberately vague term with nebulous properties; as a colleague often quips, "All we know of the central executive is that it's an oval," after its oval-shaped depiction in the Baddeley and Hitch model. There is ongoing debate as to the level of the hierarchy at which each system is integrated into that of the other senses, with no definitive solutions.

Visual working memory and visual sensory memory (i.e., 'iconic memory') were fundamentally measured by George Sperling in 1960 [52]. He presented arrays of simple visual stimuli for short periods of time and asked subjects to report what they had seen after a number of short delays. He discovered that subjects could only recall a small subset of stimuli in a large array, representing the limited capacity of visual working memory. Furthermore, they could recall a particular subset of the stimuli when cued after the presentation but before the sensory trace had faded ($\leq$100 ms), indicating that visual sensory memory exists and that visual attention can be deployed to stimuli either during sensation or sensory memory. Over the next decade, George Sperling went on to perform similar measurements in the auditory system, delineating very similar properties for auditory perception, sensory memory, and working memory [83].

Without directly measuring brain activity, researchers concluded that sensory systems must be operating independently with dual-task paradigms in which subjects were asked to maintain visual, auditory, or both types of information in working memory. It was shown that subjects could recall ~3–4 'chunks' of information (which may not precisely reflect individual sensory locations or features) of each type, regardless of whether they were asked to maintain visual, auditory, or both types of information [49, 78]. If the systems were integrated, one would be able to allocate multisensory working-memory 'slots' to either sense, with a maximum number (e.g., 6–8) that could be divided between the senses as desired. Instead, subjects can maintain on average ~3–4 visual chunks and ~3–4 auditory chunks, without any ability to reallocate any 'slots' from one sense to the other.

While electroencephalogram (EEG) and positron emission topography (PET) recordings could broadly confirm the contralateral organization of the visual system and coarsely implicate the parietal and frontal lobes in attention and working memory, it was not until the advent of high-resolution fMRI that researchers could begin localizing attention and working memory in human cortex with any detail [6, 17, 37, 50, 84–90]. Model-based fMRI investigations have attempted to localize visual working memory by comparing BOLD activity in conditions where subjects are required to hold different numbers of objects in working memory [50, 62, 91, 92]. The logic goes that, because visual-working-memory models posit that a maximum of ~3–4 objects can be held in visual working memory on average, areas that increase their activity with arrays 1, 2, 3 objects and remaining constant with arrays of 4 or more objects should be areas controlling visual working memory. Such areas were found bilaterally in parietal cortex by multiple laboratories [57, 62, 91, 93], but activity related to visual working memory has also been measured in early visual cortex (e.g., V1 and hV4) [55, 65, 94], prefrontal cortex [95], and possibly in object-processing regions in lateral occipital cortex [62], indicating that working-memory tasks recruit areas throughout the visual-processing hierarchy. (We note that the report of object-processing regions is controversial, as the cortical coordinates reported in that study are more closely consistent with the human motion-processing complex, hMT+, than the lateral occipital complex [15, 17, 96, 97]). However, little has been done to measure visual-working-memory activity in visual field maps, and so these studies should be considered preliminary rather than definitive. Measurements within CFMs would, in fact, help to clear up such controversies.

Auditory-working-memory localization with fMRI has been quite limited compared to its visual counterpart, and largely concentrated on speech stimuli rather than fundamental auditory stimuli [30, 68]. As noted above with attention localization with fMRI, too many variables exist with highly complex stimuli, and as such, a different approach is necessary. Furthermore, even low-level auditory sensory areas have only very recently been properly identified [5, 10].

3. Auditory processing in human cortex

3.1 Inputs to auditory cortex

Auditory processing is essential for a wide range of our sensory experiences, including the identification of and attention to environmental sounds, verbal communication, and the enjoyment of music. The intricate sounds in our daily environments are encoded by our auditory system as the intensity of their individual component frequencies, comparable to a Fourier analysis [98]. This spectral sound information is thus one fundamental aspect of the auditory feature space (**Figure 7A, C**). The basilar membrane of the inner ear responds topographically to incoming sound waves with higher frequencies transduced to neural signals near the entrance to the cochlea and progressively lower frequencies transduced further along the membrane. This organized gradient of frequencies (i.e., tones) is referred to as tonotopy (i.e., a map of tones); this topography may also be termed cochleotopy, referring to a map of the cochlea. Tonotopic organization is maintained as auditory information is processed and passed on from the inner ear through the brainstem, to the thalamus, and into PAC along Heschl's gyrus (HG; **Figure 1**; for additional discussion, see [2, 5, 6, 12, 99, 100]). The preservation of such topographical organization from the basilar membrane of the inner ear to auditory cortex allows for a common reference frame across this hierarchically organized sensory system [6, 7, 12, 13, 22, 23].

A second fundamental aspect of the auditory feature space is temporal sound information, termed periodicity (**Figure 7B, D**) [10, 101, 102]. Human psychoacoustic studies indicate that there are separable filter banks (i.e., neurons with distinct receptive fields) for not only frequency spectra—as expected given tonotopy, but also temporal information [103–105]. The auditory nerve likely encodes such temporal information through activity time-locked to the periodicity of the amplitude modulation (i.e., the length of time from peak-to-peak of the temporal envelope) [101, 106]. Temporally varying aspects of sound are thought to preferentially active neurons selective for the onset and offset of sounds and for sounds of certain durations. Organized representations of periodicity in primates have been

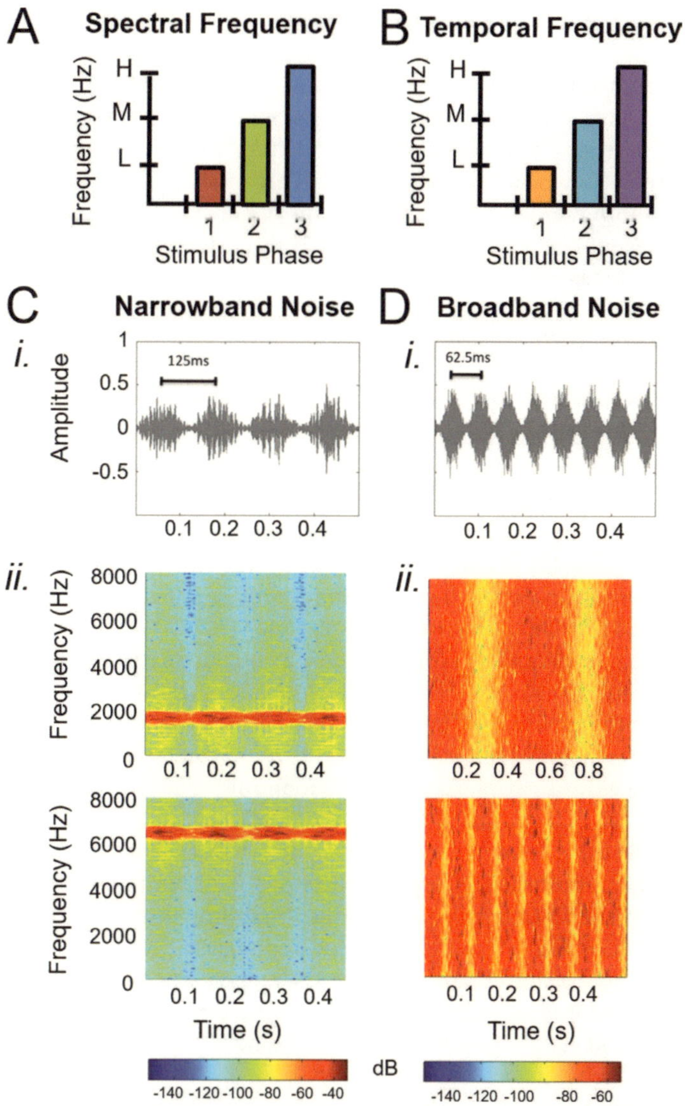

Figure 7.
Example tonotopic and periodotopic stimuli for auditory field mapping. (A) Three stimulus values for one dimension of auditory feature space (e.g., tonotopy) are depicted in the graph: 1—low (L, red); 2—medium (M, green); 3—high (H, blue). (B) Three stimulus values for a second dimension of auditory feature space (e.g., periodotopy) are depicted in the second graph: 1—low (L, orange); 2—medium (M, aqua); 3—high (H, purple). (C) Tonotopic representations can be measured using narrowband noise stimuli, which hold periodicity constant and vary frequency. (i) Sound amplitude (arbitrary units) for this stimulus set as a function of time in seconds. (ii) Sound spectrograms for two example narrowband noise stimuli with center frequencies (CF) of 1600 Hz (top) and 6400 Hz (bottom). Higher amplitudes in decibels (dB) are represented as 'warmer' colors (see dB legend below). (D) Periodotopic representations can be measured using broadband noise stimuli, which maintain constant frequency information and vary periodicity. (i) Sound amplitude (arbitrary units) for this stimulus set as a function of time in seconds. (ii) Sound spectrograms for two example broadband noise stimuli with amplitude modulation (AM) rates of 2 Hz (top) and 8 Hz (bottom). Higher amplitudes are again depicted as 'warmer' colors (see dB legend on bottom).

measured to date in the thalamus and PAC of macaque and human, respectively, and are termed periodotopy, a map of neurons that respond differentially to sounds of different temporal envelope modulation rates [5, 10, 107]. Repeating periodotopic gradients exist in the same cortical locations as, but are orthogonal to, tonotopic gradients, which allows researchers to use measurements of these two acoustic dimensions to identify complete AFMs.

3.2 fMRI measurements of auditory field maps

Measurements of the structure and function of human PAC and lower-level auditory cortex have been relatively few to date, with many studies hampered by methodological issues (for reviews, see [5, 23]. Precise measurements of AFMs across primary and lower-level auditory cortex are vital, however, for studying the neural underpinnings of such prominent auditory behaviors as attention and working memory. Recent research has now successfully applied fMRI methods commonly used to measure visual field maps to the study of AFMs in human auditory cortex.

3.2.1 Phase-encoded fMRI

The phase-encoded fMRI paradigm provides highly detailed *in vivo* measurements of CFMs in individual subjects [9, 10, 15, 108–111]. This technique measures topographical representations using stimuli that periodically repeat a set of values in an orderly sequence (**Figure** 7). The phase-encoded methods are specialized for AFM measurements by combining this periodic stimulus with a sparse-sampling paradigm (**Figure 8**) [10, 112–115]. Sparse-sampling separates the auditory stimulus presentation from the noise of the MR scanner during data acquisition to avoid contamination of the data by nonstimulus sounds [116–118].

The periodic stimulus allows for the use of a Fourier analysis to determine the value of the stimulus (e.g., 800 Hz frequency for tonotopy) that most effectively drives each cortical location [110]. The cortical response at a specific location is said to be 'in phase' throughout the scan with the stimulus value that most effectively activates it, hence the term 'phase-encoded' mapping. The alternate term 'traveling-wave' mapping arises from the consecutive activation of one neighboring cortical location after the other to create a wave-like pattern of activity across the CFM during the stimulus presentation. The phase-encoded paradigm only captures cortical activity that is at the stimulus frequency, thus excluding unrelated cortical activity and other sources of noise. Similarly, cortical regions that are not organized topographically will not be significantly activated by phase-encoded stimuli, as there would be no differential activation across the cortical representation [8, 15, 16]. The statistical threshold for phase-encoded cortical activity is commonly determined by coherence, which is a measure of the amplitude of the BOLD signal modulation at the frequency of the stimulus presentation (e.g., six stimulus cycles per scan), divided by the square root of the power over all other frequencies except the first and second harmonic (e.g., 12 and 18 cycles per scan) [15, 17, 110].

Measurement and analysis of phase-encoded CFM data must be performed within individual subjects rather than across group averages to avoid problematically blurring together discrete CFMs and their associated computations (for extended discussions, see [5, 15, 17]). CFMs may differ radically in size and anatomical position among individual subjects independent of brain size; this variation is reflected in associated shifts in cytoarchitectural and topographic boundaries [119–124]. In the visual system, for example, V1 can differ in size by at least a factor of three despite its location on the relatively stable calcarine sulcus [120]. Accordingly, when such data are group-averaged across subjects, especially through such approaches as aligning data from individual brains to an average brain with atlases such as Talairach space [125] or Montreal Neurological Institute (MNI) coordinates [126], the measurements will be blurred to such a degree that the measured topography of the CFMs is inaccurate or even lost. Blurring from such whole-brain anatomical co-alignment will thus cause different CFMs to be incorrectly averaged together into a single measurement, mixing data together from adjacent CFMs within each subject and preventing the analysis of the distinct computations of each CFM.

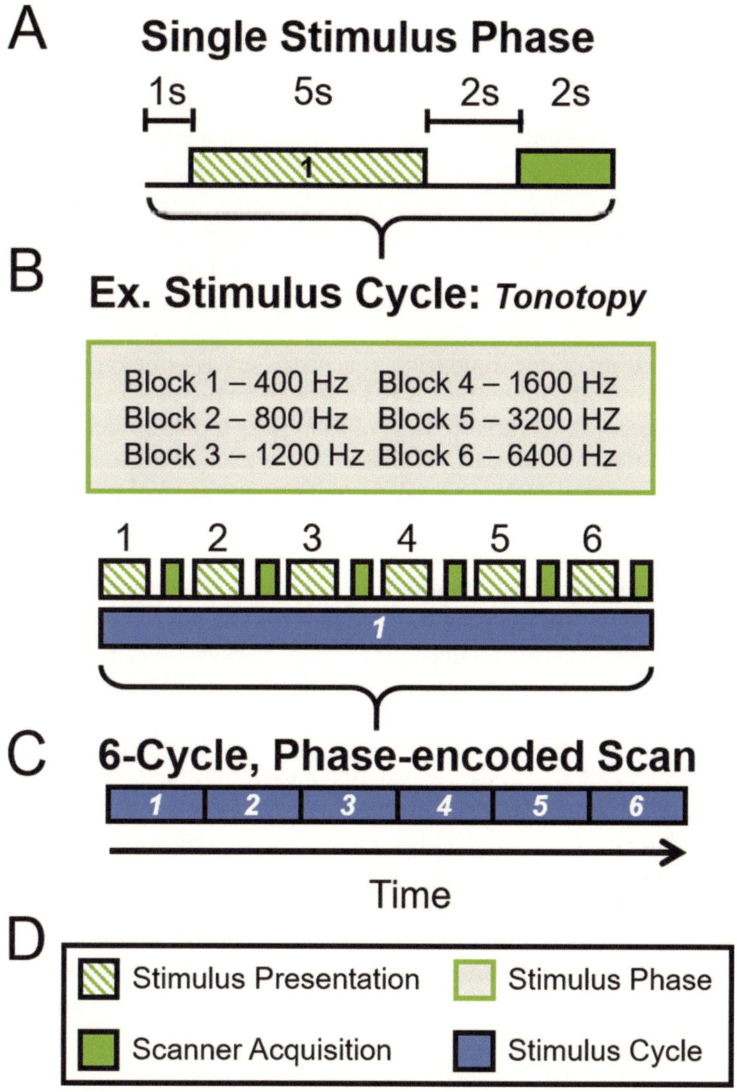

Figure 8.
Schematic of phase-encoded fMRI paradigm for auditory field mapping experiments. (A) Diagram of a single stimulus phase shows the components of a single block of one auditory stimulus presentation (striped green) followed by an fMRI data acquisition period (solid green). This sparse-sampling paradigm separates the auditory stimulus presentation from the noisy environment of the MR scanner acquisition. The timing of the acquisition (2 s delay) is set to collect the approximate peak response of auditory cortex to the stimulus, in accordance with the estimated hemodynamic delay. (B) Each phase (block) of an example tonotopic stimulus is displayed within the gray box above the colored blocks; one block thus represents one stimulus position in the 'phase-encoded' sequence. The diagram of an example stimulus cycle below this depicts six presentation blocks (striped green+ solid green) grouped together into one stimulus cycle (blue). Each block, or stimulus phase, in each cycle represents a specific frequency; e.g., for tonotopic measurements, the stimulus that is presented sequentially changes to each of the Hz listed in the gray box. The term 'traveling-wave' is also used to describe this type of phase-encoded stimulus presentation, as the stimuli produce a sequential activation of representations across a topographically organized cortical region. (C) Diagram shows a full, single scan comprising six cycles. (D) Legend denotes color-coding for diagrams above.

3.2.2 Criteria for auditory field map identification

In order to avoid the imprecise application of the term 'map' to topographical gradients or other similar patterns of cortical organization, the designation of an AFM—and CFMs in general—should be established according to several key criteria (**Figure 9**) (for reviews, see [5, 8, 15]). First, by definition, each AFM must contain at least the two orthogonal, nonrepeating topographical representations of fundamental acoustic feature space described above: tonotopy and periodotopy (**Figure 9A**) [10, 17, 21, 108, 110, 111]. When this criterion is ignored and the measurement of only one topographical representation is acquired (e.g., tonotopy), it is impossible to correctly identify boundaries among cortical regions. Measurements of the organization and function of specific regions of early auditory cortex in human long have mostly relied on tonotopic measurements alone, which has resulted in variable, conflicting, and ultimately unusable interpretations of the organization of human PAC and surrounding regions (for detailed reviews, see [5, 23]).

The representation of one dimension of sensory space—one topographical gradient along cortex like tonotopy—is not adequate to delineate an AFM, or CFMs in any sensory system. The measurement of a singular topographical dimension merely demonstrates that this particular aspect of sensory feature space is represented along that cortical region. The CFMs within that cortical region cannot be identified without measuring an orthogonal second dimension: a region of cortex with a large, confluent gradient for one dimension could denote a single CFM (**Figure 9Ai, ii**) or many CFMs (**Figure 9Ai, iii**), depending upon the organization of the overlapping second topography. Similarly, the two overlapping gradients must be approximately orthogonal, as they will otherwise not represent all the points in sensory space uniquely (**Figure 9B**) [15, 16, 127, 128]. As the complexity of adjacent gradients increases, the determination of the emergent CFM organization grows increasingly complicated.

Due to the relatively recent measurements of periodotopic representations in human auditory cortex and monkey midbrain, AFMs in core and belt regions can now be identified [10, 102]. The identification of periodotopy as the second key dimension of auditory feature space is strengthened by psychoacoustic studies, which show that separable filter banks occur not only for frequency spectra, but also temporal information, indicating the presence of neurons with receptive fields tuned to ranges of frequencies and periods [14, 103–105]. Additionally, representations of temporal acoustic information (i.e., periodicity) have been measured in the auditory system of other model organisms, including PAC in domestic cat and inferior colliculus in chinchilla [129, 130].

A second AFM criterion is that each of its topographical representations must be organized as a generally contiguous and orderly gradient [16, 128]. For such a gradient to develop, the representation must be organized such that it covers a full range of sensory space, in order from one boundary to the other (e.g., from lower to upper frequencies for tonotopy; **Figure 9C**). A topographical gradient is thus one of the most highly structured features of the cortical surface that can be measured using fMRI. The odds of two orderly, orthogonal gradients arising as a spurious pattern from noise in an overlapping section of cortex is extraordinarily low (for a calculation of the probability of spurious gradients arising from noise, see [19]).

Third, each CFM should contain representations of a considerable amount of sensory space. Differences in cortical magnification are likely among CFMs with different computational needs, but a large portion of sensory space is still expected to be represented (e.g., [15, 16, 19, 21, 97, 127, 131]). A high-quality fMRI measurement of the topography is necessary to adequately capture the sensory range and magnification. The quality of the measurement is dependent upon choosing an appropriate set of phase-encoded stimuli. The sampling density and range of values in the stimulus set both affect the accuracy and precision of the measurement. For example, the intensity (i.e., loudness) of the tonotopic stimulus alone can alter the width of the receptive fields of neurons in PAC and consequently increase the

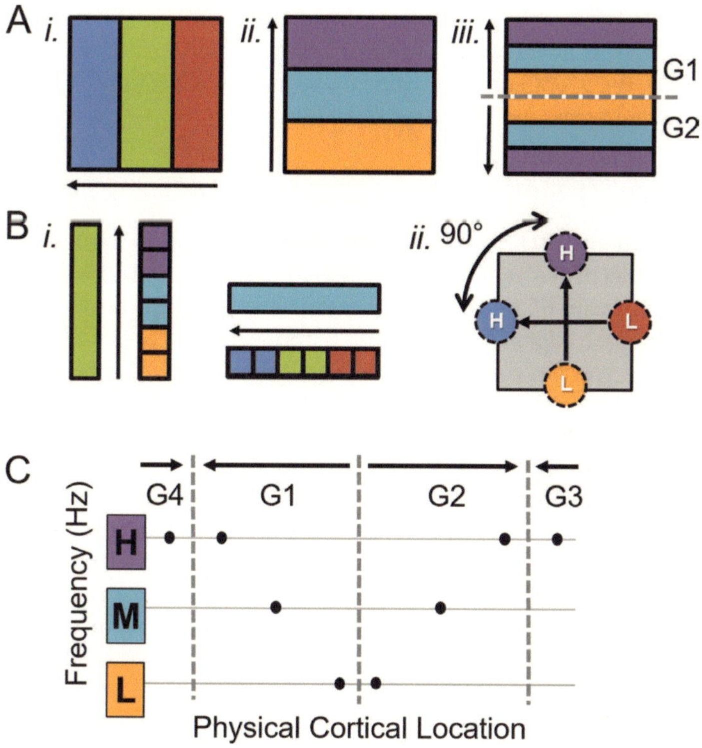

Figure 9.
Definition of auditory field maps (AFMs). (A) (i) Schematic of a single gradient of dimension 1 (e.g., tonotopy). Black arrow shows the low-to-high gradient for this tonotopic gradient. With only measurements of the single dimension of tonotopy, it cannot be determined whether the region within dimension 1 contains one or more cortical field maps without measuring a second, orthogonal gradient. (ii) Schematic of a single gradient of dimension 2 (e.g., periodotopy) overlapping the tonotopic gradient in (i) to form a single AFM like hA1. Black arrow shows the low-to-high gradient for this periodotopic gradient. Note the orthogonal orientation of the two gradients (i vs. ii) composing this AFM. (iii) schematic of an alternative gradient organization for periodotopy overlapping the same tonotopic gradient in (i). Black arrows now show two low-to-high gradients (G1: gradient 1, G2: gradient 2) of this second dimension within the same territory as the orthogonal low-to-high gradient in (i). The gray dotted line marks the boundary dividing this region into two AFMs. (B) (i) In a properly defined AFM, measurements along the cortical representation of a single value of tonotopy (e.g., green) span all values of periodotopy (e.g., orange to cyan to purple), and vice versa. (ii) Schematic of vectors drawn along a single CFM from centers of low-stimulus-value regions of interest (ROIs) to high-stimulus-value ROIs for dimensions 1 (e.g., red to blue) and 2 (e.g., orange to purple). The offset measured between the low-to-high vectors for each dimension should be approximately 90° to be considered orthogonal and thus allow for each voxel/portion of the map to represent a unique combination of dimension 1 and dimension 2 values. (C) The diagram demonstrates how gradient boundaries for one dimension of an AFM are determined. Black dots denote hypothetical measurement points along the cortical surface shown in (A, iii). Black arrows note gradient directions (low, L, to medium, M, to high, H). Dashed gray lines mark gradient reversals. Two gradients that span the full range of dimension 2 measurements can be divided into G1 and G2, with the representations of stimulus values increasing from low to high across the cortical surface in one gradient to the boundary where the representations in the next map then reverse back from high to low along the cortical surface in the next gradient. G3 and G4 (gradients 3 and 4, respectively) denote additional gradients continuing at reversal to regions outside the diagram. (for review, see [23]).

lateral spread of the BOLD signal measured in neuroimaging [132]. In addition, some degree of blurring in the measurements of the topography is expected due to such factors as the overlapping broad receptive fields, the inherent spatial spread of the fMRI signal, and measurement noise [64, 109, 133, 134]. The stimulus parameters and how they may affect the cortical responses should therefore be given careful consideration.

Fourth, the general features of the topographies composing the CFMs and the pattern of CFMs across cortex should both be consistent among individuals. It is essential to remember, nevertheless, that cytoarchitectural and topographic boundaries in PAC vary dramatically in size and anatomical location independent of overall brain size [119, 121–124, 135], as do CFMs across visual cortex [16, 17, 120, 136]. Regardless of these variations, the overall organization among specific CFMs and cloverleaf clusters will be maintained across individuals.

3.2.3 Definition of auditory field map boundaries

The measurement of AFMs is one of the few reliable *in vivo* methods to localize the distinct borders of the auditory core and belt regions in individual subjects [5, 10, 12, 23]. The boundaries of an AFM—and of CFMs in general—are determined by carefully defining the edges of overlapping sections of tonotopic and periodotopic gradients within a specific cortical region in an individual hemisphere (**Figure 9**). If a set of overlapping representations of the two dimensions is present in isolation, the boundary of the AFM can be estimated to be where the gradient responses end, although there will likely be some spatial blurring or spreading of the representation along these edges (**Figure 9Ai, ii**) [16, 17, 110, 137]. For multiple, adjacent representations that each span the full range of one dimension (e.g., low-to-high frequencies of tonotopy) can be divided into two sections at the point at which the gradients reverse (**Figure 9Ai, iii**). At the gradient reversals, the representations of stimulus values increase from low to high (or vice versa) across the cortical surface in one section to the boundary where the representations in the next AFM then reverse back from high to low (or vice versa) along the cortical surface in the next section (**Figure 9C**). Such phase-encoded fMRI measurements of the boundaries of the AFMs in human auditory cortex have been shown to be closely related to those determined by invasive human cytoarchitectural studies and nonhuman primate cytoarchitectural, connectivity, and tonotopic measurements [2, 5, 10, 121, 138–144].

At a scale of several centimeters, groups of adjacent CFMs are organized within both auditory and visual cortex into a macrostructural pattern called the cloverleaf cluster, named for the similarity of the organization of the individual CFMs composing a cluster to the leaves of a clover plant [8, 10, 15–20]. Within a cluster, one dimension of sensory topography is represented in concentric, circular bands from center to periphery of the cluster, and the second, orthogonal dimension separates this confluent representation into multiple CFMs with radial bands spanning the cluster center to periphery. In AFM clusters, a confluent, concentric tonotopic representation is divided into specific AFMs by reversal in the orthogonal periodotopic gradients. Neighboring cloverleaf clusters are then divided along the tonotopic reversals at the cluster boundaries.

While CFM clusters have consistent positions relative to one another across the cortical surface, CFMs within each cluster may be oriented differently among individuals as if rotating about a cluster's central representation. This inter-subject is consistent with the variability in molecular gradient expression that gives rise to the development of cortical topographical gradients [145–149]. This unpredictability of cluster anatomical location and rotation emphasizes the need for careful data analysis to be performed in individual subjects, in which common CFMs can be identified by analyzing the pattern of CFMs and cloverleaf clusters within that sensory system.

3.3 Organization of human auditory field maps

3.3.1 Auditory cortex organization in macaque monkey vs. human

Auditory processing in human cortex and in nonhuman primates occurs bilaterally along the temporal lobes near the lateral sulcus (**Figure 1**; e.g., [5, 10, 115, 121, 139–142, 144, 150–153]). In the macaque monkey model system upon which much of our understanding of human audition is based, converging evidence from cytoarchitectural, connectivity, electrophysiological, and neuroimaging studies have generally identified 13 auditory cortical areas grouped into core, medial and lateral belt, and parabelt regions that are associated with primary, secondary, and tertiary levels of processing, respectively (for extended discussions, see [2, 5, 154]). Auditory processing in macaque cortex begins along the superior temporal gyrus (STG) within three primary auditory areas: A1, R, and RT [140]. In contrast to early visual processing in which primary visual cortex is composed of V1 alone, primary auditory cortex is considered to be a core region composed of these three AFMs; all three areas contain the expanded layer IV arising from dense thalamic inputs and the high expression of cytochrome oxidase, acetylcholinesterase, and parvalbumin distinctive to primary sensory cortices [2, 142, 143, 150, 152, 154–157]. The eight belt regions are divided into four areas along both the lateral (CL, ML, AL, RTL) and medial (CM, RM, MM, RTM) sides of the core [158–160]. Along the lateral belt, two additional areas create the parabelt, which allocates auditory information to neighboring auditory cortex as well as to multimodal cortical regions [2, 161].

Based on cytoarchitectural, connectivity, and neuroimaging measurements, early auditory processing in human cortex has been shown to resemble the organization of lower-level macaque auditory processing [10, 23, 121, 144, 151–153, 162]. Over the ~25 million years of evolutionary separation between the species, the core, belt, and parabelt areas have rotated from the STG to Heschl's gyrus (HG), an anatomical feature unique to humans [11, 163]. The specific structure of HG differs across individuals, variably existing as a single or double gyrus. PAC is then either mostly centered on the single HG or overlapping both gyri in the case of two (**Figure 1B, C**) [122, 135, 136]. Core, belt, and parabelt areas have thus shifted in orientation from a strictly rostral-caudal axis for A1 to R to RT along macaque STG to a medial-lateral axis along human HG for hA1, hR, and hRT. The naming of the AFMs in human is based on the likely homology to macaque, but adds an 'h' to signify human [10].

3.3.2 Eleven human AFMs compose three cloverleaf clusters overlapping Heschl's gyrus

With our new understanding of periodotopic representations overlapping the previously identified tonotopic gradients, *in vivo* fMRI measurements can now identify the 11 AFMs that compose the core and belt regions of human auditory cortex (**Figure 10**) [5, 10, 12, 23]. Running from STG to the circular sulcus (CiS) along HG are three distinct, concentrically organized, tonotopic representations. The primary circular tonotopic gradient is one dimension of the HG cloverleaf cluster, with a confluent low-tone representation located centrally and expanding smoothly to high-tone representations at the outer edge (**Figure 10B, C**) [5]. The HG cluster is divided along the orthogonal periodotopic reversals into two AFMs each of core, medial belt, and lateral belt: hA1, hR, hMM, hRM, hML, and hAL (**Figure 10D, E**). Positioned at the tip of HG, hA1 is the largest of these core and belt AFMs, with the posterior/lateral region representing low tones and the anterior/medial region representing high ones. HA1 is involved in the most basic of cortical auditory computations, which is reflected in its representations of broad ranges of tonotopy and periodotopy [2].

A reversal in the tonotopic gradient along the anteromedial edge of the HG cluster divides it from the CM/CL cluster just past the tip of HG (**Figure 10B, C**). A high-periodicity gradient reversal splits this tonotopic gradient into hCM, and hCL, two regions associated with early language and speech processing as well as audiovisual integration (**Figure 10D, E**) [164]. Finally, the reversal in the tonotopic gradient along the posteriolateral edge of the HG cluster separates it from the RT cluster positioned where HG meets STG (**Figure 10B, C**). Two reversals in

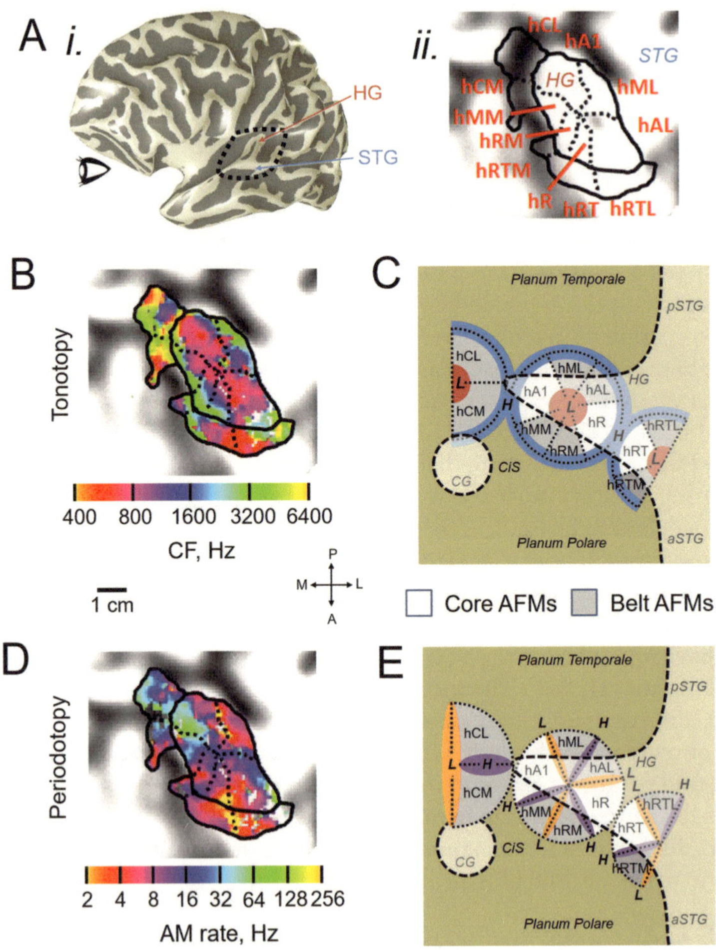

Figure 10.
Auditory field maps and cloverleaf clusters in human cortex. (A) Anatomical views of Heschl's gyrus (HG), superior temporal gyrus (STG) and surrounding auditory cortex in an individual subject's left hemisphere (S2). (i) Inflated 3-D rendering of the cortical surface. Light gray denotes gyri; dark gray denotes sulci. The approximate region presented in the other panels is indicated by the dotted black line. Note that this subject has a double peak along HG. (ii) flattened cortical surface of the region indicated by the dotted black line in (i). AFM boundaries between maps along tonotopic reversals are indicated by solid black lines. These tonotopic reversals constitute the separation of cloverleaf clusters from one another. AFM boundaries along periodotopic reversals are indicated by dotted black lines. These periodotopic reversals compose the separation between maps within a cloverleaf cluster. Red text indicates AFM names. (B) Tonotopic gradients measured using narrowband noise stimuli with a phase-encoded fMRI paradigm (example single-subject data from [10]). Color overlay indicates the preferred frequency range for each voxel. CF: center frequency in Hz. For clarity, only voxels within the core and belt AFMs are shown. Solid and dotted black lines are as in (A). Coherence ≥0.20. Inset scale bar designates 1 cm along the flattened cortical surfaces in (B, D). Inset legend indicates anatomical directions for (B-E). M: medial; L: lateral; A: anterior; P: posterior. (C) Diagram is based on individual-subject data measured by [10] in multiple phase-encoded fMRI experiments. Approximate positions of core AFMs (hA1, hR, hRT) are shown in white, and approximate positions of belt AFMs (hML, hAL, hRTL, hRTM, hRM, hMM, hCM, hCL) are shown in gray. Darker beige background indicates the plane of the lateral sulcus, while lighter beige overlay indicates gyri. Gyri are also marked with dashed black lines. HG: Heschl's gyrus. CG: circular gyrus; CiS: circular sulcus; a/p STG: anterior/posterior superior temporal gyrus. Diagram depicts the locations of tonotopic representations overlaid along the core and belt AFMs, with low (L) and high (H) tonotopic representations are marked in red and blue, respectively. Dotted black lines designate the boundaries between AFMs within three cloverleaf clusters: HG cluster with hA1; hCM/hCL cluster (partial cluster defined to date); hRTM/hRT/hRTL cluster (partial cluster defined to date). (D) Periodotopic representations measured using broadband noise stimuli with a phase-encoded fMRI paradigm. Data are from the same subject as shown for tonotopy in (B), with the color overlay now indicating the preferred period range for each voxel. AM rate: amplitude modulation rate in Hz. Other details are as in (B). (E) Diagram depicts periodotopic representations overlaid on the same example region of cortex as in (C). L and H now designate to the approximate locations of low (orange) or high (purple) periodotopic representations, respectively. Adapted from Barton et a.[10]. For a detailed review, see [5].

the periodotopic representations here divide the RT cluster into hRT, hRTM, and hRTL (**Figure 10D, E**). In macaque, these AFMs along STG are thought to subserve lower-level processing of auditory stimuli like temporally modulated environmental sounds [158, 159]. More research is needed to determine how what other AFMs form the CM/CL and RT clusters. Based on emerging data, it is likely that AFMs will also be a fundamental organization of auditory cortex adjacent to these cloverleaf clusters, such as planum temporale (PT), planum polare (PP) and STG.

3.4 Measuring attention and working memory in human AFMs

The characterization of AFMs and cloverleaf clusters will be crucial for the study of the structure and function of human auditory cortex, as these *in vivo* measurements allow for the systematic exploration of computations across a sensory system (for reviews, see [5, 17]). Such AFM organization provides a basic framework for the complex processing and analysis of input from the sensory receptors of the inner ear [5, 12, 17, 23]. The cloverleaf cluster organization of AFMs may also play a role in coordinating neural computations, with neurons within each cluster sharing computational resources such as common mechanisms to coordinate neural timing or short-term information storage [8, 12]. Similarly, vision studies suggest that functional specializations for perception are organized by cloverleaf clusters, as a particular cloverleaf cluster can be functionally differentiated from its neighbors by its pattern of BOLD responses, surface area, cortical magnification, processing specialization, and receptive field sizes [12, 16, 18, 19, 21, 165]. These distinctions indicate that CFMs within individual cloverleaf clusters are not only anatomically but also functionally related [15, 18, 20, 166].

The cluster organization is not necessarily thought to be driving common sensory functions, but rather reflects how multiple stages in a sensory processing pathway might arise during development across individuals and during evolution across species. It is likely that this cluster organization, like the topographic organization of CFMs, allows for efficient connectivity among neurons that represent neighboring aspects in sensory feature space [166–169]. Since the axons contained within one cubic millimeter of cortex can extend 3-4 km in length, efficient connectivity is vital for sustainable energetics in cortex [170].

The definitions of AFMs and the cloverleaf clusters they compose using phase-encoded fMRI will thus serve as reliable, independent localizers for investigations of attention and working memory in early auditory cortex across individuals. Measurements of individual AFMs along the cortical hierarchy will help reveal the distinct stages of top-down and bottom-up auditory processing. In addition, changes in AFMs can be tracked to study how auditory cortex changes under various attentional and working memory tasks and disorders (e.g., [145, 171–177]).

4. Conclusion

The human brain has sophisticated systems for perception, trace memory, attention, and working memory for audition and vision, and likely the other senses as well. These systems appear to be organized in a very similar manner for each sense, despite the inputs to each system and information content being quite different. Behavioral measures of the last several decades have led to the development of well-defined models of each system. These models form the basis for the investigation of their underlying architecture in the cortical structures of the human brain. EEG and PET have allowed for spatially coarse investigation of cortical activity, but with the advent of fMRI, it has become possible to make exceptionally detailed spatial measurements. The methods of investigation must be carefully crafted to best elicit activity reflecting the desired aspects of each system; not only must the tasks be appropriate for fMRI, the stimuli and task must be closely matched not just to the system being studied, but to the inputs into that system as well.

For both audition and vision, the sensory processing in cortex happens in cloverleaf clusters of CFMs. This organizational pattern has clearly been demonstrated in the lower tiers of the processing hierarchy and very likely is organized as such throughout. Because the CFMs across the entire hierarchy (or at least, most) of one sense can be measured in just one session in the fMRI scanner, they make incredibly efficient localizers. CFMs are be measured in individual subjects, and serve as functional localizers that can be used to average more accurately across subjects than anatomical localizers. As such, due to the pervasive and fundamental role CFMs play in sensory systems, they are also excellent candidates for measuring the effects of attention and working memory in cortex. To best accomplish this feat, it is proposed that stimuli that are similar to those used to measure CFMs are excellent candidates for use in traditional tasks used to define attentional and working-memory models.

Acknowledgements

This material is based upon work supported by the National Science Foundation under Grant Number 1329255 and by startup funds from the Department of Cognitive Sciences at the University of California, Irvine.

Conflict of interest

The authors declare that the research was conducted in the absence of any commercial or financial relationships that could be construed as a potential conflict of interest.

Author details

Brian Barton[1] and Alyssa A. Brewer[1,2*]

1 Department of Cognitive Sciences, University of California, Irvine, CA, USA

2 Department of Language Science, University of California, Irvine, CA, USA

*Address all correspondence to: aabrewer@uci.edu

References

[1] Van Essen DC, Felleman DJ, DeYoe EA, Olavarria J, Knierim J. Modular and hierarchical organization of extrastriate visual cortex in the macaque monkey. Cold Spring Harbor Symposia on Quantitative Biology. 1990;**55**:679-696

[2] Kaas JH, Hackett TA. Subdivisions of auditory cortex and processing streams in primates. Proceedings of the National Academy of Sciences of the United States of America. 2000;**97**(22):11793-11799. DOI: 10.1073/pnas.97.22.11793

[3] Krubitzer L. The magnificent compromise: Cortical field evolution in mammals. Neuron. 2007;**56**(2):201-208. DOI: 10.1016/j.neuron.2007.10.002

[4] Schreiner CE, Winer JA. Auditory cortex mapmaking: Principles, projections, and plasticity. Neuron. 2007;**56**(2):356-365. DOI: 10.1016/j.neuron.2007.10.013

[5] Brewer AA, Barton B. Maps of the auditory cortex. Annual Review of Neuroscience. 2016;**39**:385-407. DOI: 10.1146/annurev-neuro-070815-014045

[6] Wessinger CM, VanMeter J, Tian B, Van Lare J, Pekar J, Rauschecker JP. Hierarchical organization of the human auditory cortex revealed by functional magnetic resonance imaging. Journal of Cognitive Neuroscience. 2001;**13**(1):1-7

[7] Kaas JH. Topographic maps are fundamental to sensory processing. Brain Research Bulletin. 1997;**44**(2):107-112

[8] Wandell BA, Brewer AA, Dougherty RF. Visual field map clusters in human cortex. Philosophical Transactions of the Royal Society of London. Series B, Biological Sciences. 2005;**360**(1456):693-707. DOI: 10.1098/rstb.2005.1628

[9] Sanchez-Panchuelo RM, Francis S, Bowtell R, Schluppeck D. Mapping human somatosensory cortex in individual subjects with 7T functional MRI. Journal of Neurophysiology. 2010;**103**(5):2544-2556. DOI: 10.1152/jn.01017.2009

[10] Barton B, Venezia JH, Saberi K, Hickok G, Brewer AA. Orthogonal acoustic dimensions define auditory field maps in human cortex. Proceedings of the National Academy of Sciences of the United States of America. 2012;**109**(50):20738-20743. DOI: 10.1073/pnas.1213381109

[11] Krubitzer LA, Seelke AM. Cortical evolution in mammals: the bane and beauty of phenotypic variability. Proceedings of the National Academy of Sciences. 2012;**109**(Supplement 1):10647-10654

[12] Brewer AA, Barton B. Cloverleaf clusters: A common macrostructural organization across human visual and auditory cortex. In: Heinbockel T, editor. Sensory Nervous System. London, UK: IntechOpen; 2018. pp. 127-160. DOI: 10.5772/intechopen.77964

[13] Felleman DJ, Van Essen DC. Distributed hierarchical processing in the primate cerebral cortex. Cerebral Cortex. 1991;**1**(1):1-47

[14] Santoro R, Moerel M, De Martino F, Valente G, Ugurbil K, Yacoub E, et al. Reconstructing the spectrotemporal modulations of real-life sounds from fMRI response patterns. Proceedings of the National Academy of Sciences of the United States of America. 2017;**114**(18):4799-4804. DOI: 10.1073/pnas.1617622114

[15] Brewer AA, Barton B. Visual field map organization in human visual cortex. In: Molotchnikoff S, Rouat J, editors. Visual Cortex—Current Status and Perspectives. Rijeka, Croatia: InTech; 2012. pp. 29-60. DOI: 10.5772/51914

[16] Brewer AA, Liu J, Wade AR, Wandell BA. Visual field maps and stimulus selectivity in human ventral occipital cortex. Nature Neuroscience. 2005;**8**(8):1102-1109. DOI: 10.1038/nn1507

[17] Wandell BA, Dumoulin SO, Brewer AA. Visual field maps in human cortex. Neuron. 2007;**56**(2):366-383. DOI: 10.1016/j.neuron.2007.10.012

[18] Kolster H, Peeters R, Orban GA. The retinotopic organization of the human middle temporal area MT/V5 and its cortical neighbors. The Journal of Neuroscience. 2010;**30**(29):9801-9820. DOI: 10.1523/JNEUROSCI.2069-10.2010

[19] Barton B, Brewer AA. Visual field map clusters in high-order visual processing: Organization of V3A/V3B and a new cloverleaf cluster in the posterior superior temporal sulcus. Frontiers in Integrative Neuroscience. 2017;**11**:4. DOI: 10.3389/fnint.2017.00004

[20] Kolster H, Mandeville JB, Arsenault JT, Ekstrom LB, Wald LL, Vanduffel W. Visual field map clusters in macaque extrastriate visual cortex. The Journal of Neuroscience. 2009;**29**(21):7031-7039. DOI: 10.1523/JNEUROSCI.0518-09.2009

[21] Press WA, Brewer AA, Dougherty RF, Wade AR, Wandell BA. Visual areas and spatial summation in human visual cortex. Vision Research. 2001;**41**(10-11):1321-1332

[22] Van Essen DC. Organization of Visual Areas in macaque and human cerebral cortex. In: Chalupa LM, Werner JS, editors. The Visual Neurosciences. Boston: Bradford Books; 2003. pp. 507-521

[23] Brewer AA, Barton B. Human auditory cortex. In: Hickok G, Small SL, editors. Neurobiology of Language. Cambridge: Academic Press, Elsevier; 2016. pp. 49-58. DOI: 10.1016/B978-0-12-407794-2.00005-5

[24] Kaya EM, Elhilali M. Modelling auditory attention. Philosophical Transactions of The Royal Society B: Biological Sciences. 2017;**372**(1714): 1-10. DOI: 10.1098/rstb.2016.0101

[25] Fritz JB, Elhilali M, David SV, Shamma SA. Auditory attention—Focusing the searchlight on sound. Current Opinion in Neurobiology. 2007;**17**(4):437-455

[26] Driver J. A selective review of selective attention research from the past century. British Journal of Psychology. 2001;**92**(1):53-78

[27] Dalton P, Spence C. Selective attention in vision, audition, and touch. In: Byrne J, editor. Learning and Memory: A Comprehensive Reference. Oxford: Elsevier; 2008. pp. 243-257

[28] Corbetta M, Shulman GL. Control of goal-directed and stimulus-driven attention in the brain. Nature Reviews. Neuroscience. 2002;**3**(3):201-215. DOI: 10.1038/nrn755

[29] Corbetta M, Kincade JM, Shulman GL. Neural systems for visual orienting and their relationships to spatial working memory. Journal of Cognitive Neuroscience. 2002;**14**(3):508-523. DOI: 10.1162/089892902317362029

[30] Spence C, Santangelo V. Capturing spatial attention with multisensory cues: A review. Hearing Research. 2009;**258**(1-2):134-142. DOI: 10.1016/j.heares.2009.04.015

[31] Santangelo V, Olivetti Belardinelli M, Spence C, Macaluso E. Interactions between voluntary and stimulus-driven spatial attention mechanisms across sensory modalities. Journal of Cognitive Neuroscience. 2009;**21**(12):2384-2397. DOI: 10.1162/jocn.2008.21178

[32] Hurlstone MJ, Hitch GJ, Baddeley AD. Memory for serial order across domains: An overview of the literature

and directions for future research. Psychological Bulletin. 2014;**140**(2):339-373. DOI: 10.1037/a0034221

[33] Treisman AM. Strategies and models of selective attention. Psychological Review. 1969;**76**(3):282

[34] Broadbent DE. Perception and Communication. New York: Pergamon Press; 1958. p. 338

[35] Bharadwaj HM, Lee AK, Shinn-Cunningham BG. Measuring auditory selective attention using frequency tagging. Frontiers in Integrative Neuroscience. 2014;**8**:6. DOI: 10.3389/fnint.2014.00006

[36] Woods DL, Alain C, Diaz R, Rhodes D, Ogawa KH. Location and frequency cues in auditory selective attention. Journal of Experimental Psychology: Human Perception and Performance. 2001;**27**(1):65

[37] Zatorre RJ, Mondor TA, Evans AC. Auditory attention to space and frequency activates similar cerebral systems. NeuroImage. 1999;**10**(5):544-554

[38] Zatorre RJ, Bouffard M, Ahad P, Belin P. Where is' where'in the human auditory cortex? Nature Neuroscience. 2002;**5**(9):905

[39] Maddox RK, Atilgan H, Bizley JK, Lee AK. Auditory selective attention is enhanced by a task-irrelevant temporally coherent visual stimulus in human listeners. eLife. 2015;**4**:1-11. DOI: 10.7554/eLife.04995

[40] Kong L, Michalka SW, Rosen ML, Sheremata SL, Swisher JD, Shinn-Cunningham BG, et al. Auditory spatial attention representations in the human cerebral cortex. Cerebral Cortex. 2014;**24**(3):773-784. DOI: 10.1093/cercor/bhs359

[41] Awh E, Pashler H. Evidence for split attentional foci. Journal of Experimental Psychology: Human Perception and Performance. 2000;**26**(2):834

[42] Conway AR, Cowan N, Bunting MF. The cocktail party phenomenon revisited: The importance of working memory capacity. Psychonomic Bulletin & Review. 2001;**8**(2):331-335

[43] Alain C, Izenberg A. Effects of attentional load on auditory scene analysis. Journal of Cognitive Neuroscience. 2003;**15**(7):1063-1073

[44] Shinn-Cunningham BG. Object-based auditory and visual attention. Trends in Cognitive Sciences. 2008;**12**(5):182-186

[45] Saygin AP, Sereno MI. Retinotopy and attention in human occipital, temporal, parietal, and frontal cortex. Cerebral Cortex. 2008;**18**(9):2158-2168. DOI: 10.1093/cercor/bhm242

[46] Lauritzen TZ, D'Esposito M, Heeger DJ, Silver MA. Top-down flow of visual spatial attention signals from parietal to occipital cortex. Journal of Vision. 2009;**9**(13):18.1-18.14. DOI: 10.1167/9.13.18

[47] Cavanagh P, Alvarez GA. Tracking multiple targets with multifocal attention. Trends in Cognitive Sciences. 2005;**9**(7):349-354. DOI: 10.1016/j.tics.2005.05.009

[48] Baddeley A. Working memory: Looking back and looking forward. Nature Reviews. Neuroscience. 2003;**4**(10):829-839

[49] Cowan N. The magical number 4 in short-term memory: A reconsideration of mental storage capacity. Behavioral and Brain Sciences. 2001;**24**(01):87-114

[50] Barton B, Brewer AA. Visual working memory in human cortex. Psychology (Irvine). 2013;**4**(8):655-662. DOI: 10.4236/psych.2013.48093

[51] Postle BR. Working memory as an emergent property of the mind and brain. Neuroscience. 2006;**139**(1):23-38. DOI: 10.1016/j.neuroscience.2005.06.005

[52] Sperling G. The information available in brief visual presentations. Psychological Monographs: General and Applied. 1960;**74**(11):1-29. DOI: 10.1037/h0093759

[53] Awh E, Vogel EK, Oh S-H. Interactions between attention and working memory. Neuroscience. 2006;**139**(1):201-208

[54] Hitch GJ, Hu Y, Allen RJ, Baddeley AD. Competition for the focus of attention in visual working memory: Perceptual recency versus executive control. Annals of the New York Academy of Sciences. 2018;1424(1): 64-75. DOI: 10.1111/nyas.13631

[55] Serences JT, Ester EF, Vogel EK, Awh E. Stimulus-specific delay activity in human primary visual cortex. Psychological Science. 2009;**20**(2):207-214

[56] Postle BR, Zarahn E, D'Esposito M. Using event-related fMRI to assess delay-period activity during performance of spatial and nonspatial working memory tasks. Brain Research Protocols. 2000;**5**(1):57-66

[57] Sereno AB, Amador SC. Attention and memory-related responses of neurons in the lateral intraparietal area during spatial and shape-delayed match-to-sample tasks. Journal of Neurophysiology. 2006;**95**(2):1078-1098. DOI: 10.1152/jn.00431.2005

[58] Pashler H. Familiarity and visual change detection. Perception & Psychophysics. 1988;**44**(4):369-378

[59] Pavani F, Turatto M. Change perception in complex auditory scenes. Perception & Psychophysics. 2008;**70**(4):619-629

[60] Simons DJ, Ambinder MS. Change blindness. Current Directions in Psychological Science. 2005;**14**(1):44-48

[61] Todd JJ, Marois R. Capacity limit of visual short-term memory in human posterior parietal cortex. Nature. 2004;**428**(6984):751

[62] Xu Y, Chun MM. Dissociable neural mechanisms supporting visual short-term memory for objects. Nature. 2006;**440**(7080):91-95. DOI: 10.1038/nature04262

[63] Logothetis NK. The neural basis of the blood-oxygen-level-dependent functional magnetic resonance imaging signal. Philosophical Transactions of the Royal Society of London. Series B, Biological Sciences. 2002;**357**(1424):1003-1037

[64] Logothetis NK, Wandell BA. Interpreting the BOLD signal. Annual Review of Physiology. 2004;**66**:735-769. DOI: 10.1146/annurev.physiol.66.082602.092845

[65] Sligte IG, Scholte HS, Lamme VAF. V4 activity predicts the strength of visual short-term memory representations. Journal of Neuroscience. 2009;**29**(23):7432-7438. DOI: 10.1523/Jneurosci.0784-09.2009

[66] Cherry EC. Some experiments upon the recognition of speech with one and two ears. Journal of the Acoustical Society of America. 1953;**25**:975-979

[67] Cherry EC. Some further experiments upon the recognition of speech, with one and with two ears. Journal of the Acoustical Society of America. 1954;**26**:554-559

[68] Kristensen LB, Wang L, Petersson KM, Hagoort P. The interface between language and attention: Prosodic focus marking recruits a general attention network in spoken language comprehension. Cerebral Cortex.

2013;**23**(8):1836-1848. DOI: 10.1093/cercor/bhs164

[69] Wood N, Cowan N. The cocktail party phenomenon revisited: How frequent are attention shifts to one's name in an irrelevant auditory channel? Journal of Experimental Psychology. Learning, Memory, and Cognition. 1995;**21**(1):255-260

[70] Wood NL, Cowan N. The cocktail party phenomenon revisited: Attention and memory in the classic selective listening procedure of Cherry (1953). Journal of Experimental Psychology. General. 1995;**124**(3):243-262

[71] Kaya EM, Elhilali M. Correction to 'Modelling auditory attention' Philosophical Transactions of The Royal Society B: Biological Sciences. 2017;**372**(1727):1. DOI: 10.1098/rstb.2017.0194

[72] Alho J, Green BM, May PJ, Sams M, Tiitinen H, Rauschecker JP, et al. Early-latency categorical speech sound representations in the left inferior frontal gyrus. NeuroImage. 2016;**129**:214-223. DOI: 10.1016/j.neuroimage.2016.01.016

[73] Culham JC, Brandt SA, Cavanagh P, Kanwisher NG, Dale AM, Tootell RB. Cortical fMRI activation produced by attentive tracking of moving targets. Journal of Neurophysiology. 1998;**80**(5):2657-2670

[74] Barton B, Ester EF, Awh E. Discrete resource allocation in visual working memory. Journal of Experimental Psychology. Human Perception and Performance. 2009;**35**(5):1359

[75] Awh E, Barton B, Vogel EK. Visual working memory represents a fixed number of items regardless of complexity. Psychological Science. 2007;**18**(7):622-628. DOI: 10.1111/j.1467-9280.2007.01949.x

[76] Hickok G, Poeppel D. The cortical organization of speech processing. Nature Reviews. Neuroscience. 2007;**8**(5):393-402. DOI: 10.1038/nrn2113

[77] Chomsky N. Knowledge of Language: Its Nature, Origin, and Use. Westport, Connecticut: Praeger Publishers; 1986

[78] Luck SJ, Vogel EK. The capacity of visual working memory for features and conjunctions. Nature. 1997;**390**(6657):279-281. DOI: 10.1038/36846

[79] Baddeley AD, Hitch GJ. Working memory. In: Bower GA, editor. ThePsychology of Learning and Motivation. Advances in Research and Theory. New York: Academic Press; 1974. p. 47-89. DOI: 10.1016/S0079-7421(08)60452-1

[80] Baddeley AD. The episodic buffer: A new component of working memory? Trends in Cognitive Sciences. 2000;**4**(11):417-423

[81] Baddeley AD, Hitch GJ, Allen RJ. From short-term store to multicomponent working memory: The role of the modal model. Memory & Cognition. 2018:1-14. https://doi.org/10.3758/s13421-018-0878-5

[82] Baddeley A. Exploring the central executive. The Quarterly Journal of Experimental Psychology Section A: Human Experimental Psychology. 1996;**49**(1):5-28

[83] Sperling G, Speelman RG. Acoustic similarity and auditory short-term memory: Experiments and a model. In: Norman DA, editor. Models of human memory. New York: Academic Press; 1970. p. 151-202

[84] Niemeier M, Goltz HC, Kuchinad A, Tweed DB, Vilis T. A contralateral preference in the lateral occipital area: Sensory and attentional mechanisms.

Cerebral Cortex. 2005;**15**(3):325-331. DOI: 10.1093/cercor/bhh134

[85] Luck SJ, Woodman GF, Vogel EK. Event-related potential studies of attention. Trends in Cognitive Sciences. 2000;**4**(11):432-440

[86] Postle BR, Berger JS, Taich AM, D'Esposito M. Activity in human frontal cortex associated with spatial working memory and saccadic behavior. Journal of Cognitive Neuroscience. 2000;**12**(supplement 2):2-14. DOI: 10.1162/089892900564028

[87] Swisher JD, Halko MA, Merabet LB, McMains SA, Somers DC. Visual topography of human intraparietal sulcus. The Journal of Neuroscience. 2007;**27**(20):5326-5337. DOI: 10.1523/JNEUROSCI.0991-07.2007

[88] Kastner S, DeSimone K, Konen CS, Szczepanski SM, Weiner KS, Schneider KA. Topographic maps in human frontal cortex revealed in memory-guided saccade and spatial working-memory tasks. Journal of Neurophysiology. 2007;**97**(5):3494-3507. DOI: 10.1152/jn.00010.2007

[89] Szczepanski SM, Konen CS, Kastner S. Mechanisms of spatial attention control in frontal and parietal cortex. The Journal of Neuroscience. 2010;**30**(1):148-160. DOI: 10.1523/JNEUROSCI.3862-09.2010

[90] Zeki S, Bartels A. Toward a theory of visual consciousness. Consciousness and Cognition. 1999;**8**(2):225-259. DOI: 10.1006/ccog.1999.0390

[91] Xu Y, Chun MM. Visual grouping in human parietal cortex. Proceedings of the National Academy of Sciences of the United States of America. 2007;**104**(47):18766-18771. DOI: 10.1073/pnas.0705618104

[92] Xu Y, Chun MM. Selecting and perceiving multiple visual objects. Trends in Cognitive Sciences. 2009;**13**(4):167-174. DOI: 10.1016/j.tics.2009.01.008

[93] Phan ML, Schendel KL, Recanzone GH, Robertson LC. Auditory and visual spatial localization deficits following bilateral parietal lobe lesions in a patient with Balint's syndrome. Journal of Cognitive Neuroscience. 2000;**12**(4):583-600

[94] Harrison SA, Tong F. Decoding reveals the contents of visual working memory in early visual areas. Nature. 2009;**458**(7238):632-635. DOI: 10.1038/nature07832

[95] Rypma B, Berger JS, D'Esposito M. The influence of working-memory demand and subject performance on prefrontal cortical activity. Journal of Cognitive Neuroscience. 2002;**14**(5):721-731. DOI: 10.1162/08989290260138627

[96] Larsson J, Heeger DJ. Two retinotopic visual areas in human lateral occipital cortex. The Journal of Neuroscience. 2006;**26**(51):13128-13142. DOI: 10.1523/JNEUROSCI.1657-06.2006

[97] Huk AC, Dougherty RF, Heeger DJ. Retinotopy and functional subdivision of human areas MT and MST. Journal of Neuroscience. 2002;**22**(16):7195-7205. DOI: 20026661

[98] Spoendlin H. Sensory neural organization of the cochlea. The Journal of Laryngology and Otology. 1979;**93**(9):853-877

[99] Ress D, Chandrasekaran B. Tonotopic organization in the depth of human inferior colliculus. Frontiers in Human Neuroscience. 2013;**7**:586. DOI: 10.3389/fnhum.2013.00586

[100] Chang KH, Thomas JM, Boynton GM, Fine I. Reconstructing tone sequences from functional magnetic resonance imaging blood-oxygen level dependent responses within human primary auditory cortex.

Frontiers in Psychology. 2017;**8**:1983. DOI: 10.3389/fpsyg.2017.01983

[101] Langner G, Sams M, Heil P, Schulze H. Frequency and periodicity are represented in orthogonal maps in the human auditory cortex: Evidence from magnetoencephalography. Journal of Comparative Physiology. A. 1997;**181**(6):665-676

[102] Baumann S, Griffiths TD, Sun L, Petkov CI, Thiele A, Rees A. Orthogonal representation of sound dimensions in the primate midbrain. Nature Neuroscience. 2011;**14**(4):423-425. DOI: 10.1038/nn.2771

[103] Dau T, Kollmeier B, Kohlrausch A. Modeling auditory processing of amplitude modulation. II. Spectral and temporal integration. The Journal of the Acoustical Society of America. 1997;**102**(5 Pt 1):2906-2919

[104] Ewert SD, Dau T. Characterizing frequency selectivity for envelope fluctuations. The Journal of the Acoustical Society of America. 2000;**108**(3 Pt 1):1181-1196

[105] Hsieh IH, Saberi K. Detection of sinusoidal amplitude modulation in logarithmic frequency sweeps across wide regions of the spectrum. Hearing Research. 2010;**262**(1-2):9-18. DOI: 10.1016/j.heares.2010.02.002

[106] Langner G, Schreiner CE. Periodicity coding in the inferior colliculus of the cat. I. Neuronal mechanisms. Journal of Neurophysiology. 1988;**60**(6):1799-1822

[107] Baumann S, Joly O, Rees A, Petkov CI, Sun L, Thiele A, et al. The topography of frequency and time representation in primate auditory cortices. eLife. 2015;**4**:1-15. DOI: 10.7554/eLife.03256

[108] DeYoe EA, Carman GJ, Bandettini P, Glickman S, Wieser J, Cox R, et al. Mapping striate and extrastriate visual areas in human cerebral cortex. Proceedings of the National Academy of Sciences of the United States of America. 1996;**93**(6):2382-2386

[109] Engel SA, Glover GH, Wandell BA. Retinotopic organization in human visual cortex and the spatial precision of functional MRI. Cerebral Cortex. 1997;**7**(2):181-192

[110] Engel SA, Rumelhart DE, Wandell BA, Lee AT, Glover GH, Chichilnisky EJ, et al. fMRI of human visual cortex. Nature. 1994;**369**(6481):525. DOI: 10.1038/369525a0

[111] Sereno MI, Dale AM, Reppas JB, Kwong KK, Belliveau JW, Brady TJ, et al. Borders of multiple visual areas in humans revealed by functional magnetic resonance imaging. Science. 1995;**268**(5212):889-893

[112] Humphries C, Liebenthal E, Binder JR. Tonotopic organization of human auditory cortex. NeuroImage. 2010;**50**(3):1202-1211. DOI: 10.1016/j.neuroimage.2010.01.046

[113] Talavage TM, Sereno MI, Melcher JR, Ledden PJ, Rosen BR, Dale AM. Tonotopic organization in human auditory cortex revealed by progressions of frequency sensitivity. Journal of Neurophysiology. 2004;**91**(3):1282-1296. DOI: 10.1152/jn.01125.2002

[114] Petkov CI, Kayser C, Augath M, Logothetis NK. Optimizing the imaging of the monkey auditory cortex: Sparse vs. continuous fMRI. Magnetic Resonance Imaging. 2009;**27**(8):1065-1073. DOI: 10.1016/j.mri.2009.01.018

[115] Joly O, Baumann S, Balezeau F, Thiele A, Griffiths TD. Merging functional and structural properties of the monkey auditory cortex. Frontiers in Neuroscience. 2014;**8**:198. DOI: 10.3389/fnins.2014.00198

[116] Bandettini PA, Jesmanowicz A, Van Kylen J, Birn RM, Hyde JS. Functional MRI of brain activation induced by scanner acoustic noise. Magnetic Resonance in Medicine. 1998;**39**(3):410-416

[117] Gaab N, Gabrieli JD, Glover GH. Assessing the influence of scanner background noise on auditory processing. II. An fMRI study comparing auditory processing in the absence and presence of recorded scanner noise using a sparse design. Human Brain Mapping. 2007;**28**(8):721-732. DOI: 10.1002/hbm.20299

[118] Scarff CJ, Dort JC, Eggermont JJ, Goodyear BG. The effect of MR scanner noise on auditory cortex activity using fMRI. Human Brain Mapping. 2004;**22**(4):341-349. DOI: 10.1002/hbm.20043

[119] Clarke S, Morosan P. Architecture, connectivity, and transmitter receptors of human auditory cortex. In: Poeppel D, Overath T, Popper A, Richard R, editors. The Human Auditory Cortex. New York: Springer; 2012. pp. 11-38

[120] Dougherty RF, Koch VM, Brewer AA, Fischer B, Modersitzki J, Wandell BA. Visual field representations and locations of visual areas V1/2/3 in human visual cortex. Journal of Vision. 2003;**3**(10):586-598. DOI: 10:1167/3.10.1

[121] Galaburda A, Sanides F. Cytoarchitectonic organization of the human auditory cortex. The Journal of Comparative Neurology. 1980;**190**(3):597-610. DOI: 10.1002/cne.901900312

[122] Morosan P, Rademacher J, Schleicher A, Amunts K, Schormann T, Zilles K. Human primary auditory cortex: Cytoarchitectonic subdivisions and mapping into a spatial reference system. NeuroImage. 2001;**13**(4):684-701. DOI: 10.1006/nimg.2000.0715

[123] Rademacher J, Caviness VS Jr, Steinmetz H, Galaburda AM. Topographical variation of the human primary cortices: Implications for neuroimaging, brain mapping, and neurobiology. Cerebral Cortex. 1993;**3**(4):313-329

[124] Rademacher J, Morosan P, Schormann T, Schleicher A, Werner C, Freund HJ, et al. Probabilistic mapping and volume measurement of human primary auditory cortex. NeuroImage. 2001;**13**(4):669-683. DOI: 10.1006/nimg.2000.0714

[125] Talairach J, Tournoux P. Col-Planar Stereotax Atlas of the Human Brain. New York: Thieme Medical Publishers; 1988

[126] Collins DL, Neelin P, Peters TM, Evans AC. Automatic 3D intersubject registration of MR volumetric data in standardized Talairach space. Journal of Computer Assisted Tomography. 1994;**18**(2):192-205

[127] Wade AR, Brewer AA, Rieger JW, Wandell BA. Functional measurements of human ventral occipital cortex: Retinotopy and colour. Philosophical Transactions of the Royal Society of London. Series B, Biological Sciences. 2002;**357**(1424):963-973. DOI: 10.1098/rstb.2002.1108

[128] Tyler CW, Wade AR. Extended concepts of occipital retinotopy. Current Medical Imaging Review. 2005;**1**:3190329

[129] Langner G, Dinse HR, Godde B. A map of periodicity orthogonal to frequency representation in the cat auditory cortex. Frontiers in Integrative Neuroscience. 2009;**3**:27. DOI: 10.3389/neuro.07.027.2009

[130] Langner G, Albert M, Briede T. Temporal and spatial coding of periodicity information in the inferior colliculus of awake chinchilla

(Chinchilla laniger). Hearing Research. 2002;**168**(1-2):110-130

[131] Arcaro MJ, McMains SA, Singer BD, Kastner S. Retinotopic organization of human ventral visual cortex. The Journal of Neuroscience. 2009;**29**(34):10638-10652. DOI: 10.1523/JNEUROSCI.2807-09.2009

[132] Tanji K, Leopold DA, Ye FQ, Zhu C, Malloy M, Saunders RC, et al. Effect of sound intensity on tonotopic fMRI maps in the unanesthetized monkey. NeuroImage. 2010;**49**(1):150-157. DOI: 10.1016/j.neuroimage.2009.07.029

[133] Hart HC, Hall DA, Palmer AR. The sound-level-dependent growth in the extent of fMRI activation in Heschl's gyrus is different for low- and high-frequency tones. Hearing Research. 2003;**179**(1-2):104-112

[134] Menon RS, Kim SG. Spatial and temporal limits in cognitive neuroimaging with fMRI. Trends in Cognitive Sciences. 1999;**3**(6):207-216

[135] Leonard CM, Puranik C, Kuldau JM, Lombardino LJ. Normal variation in the frequency and location of human auditory cortex landmarks. Heschl's gyrus: Where is it? Cerebral Cortex. 1998;**8**(5):397-406

[136] Amunts K, Schleicher A, Burgel U, Mohlberg H, Uylings HB, Zilles K. Broca's region revisited: Cytoarchitecture and intersubject variability. The Journal of Comparative Neurology. 1999;**412**(2):319-341

[137] Dumoulin SO, Wandell BA. Population receptive field estimates in human visual cortex. NeuroImage. 2008;**39**(2):647-660. DOI: 10.1016/j.neuroimage.2007.09.034

[138] de la Mothe LA, Blumell S, Kajikawa Y, Hackett TA. Cortical connections of the auditory cortex in marmoset monkeys: Core and medial belt regions. The Journal of Comparative Neurology. 2006;**496**(1):27-71. DOI: 10.1002/cne.20923

[139] Petkov CI, Kayser C, Augath M, Logothetis NK. Functional imaging reveals numerous fields in the monkey auditory cortex. PLoS Biology. 2006;**4**(7):e215. DOI: 10.1371/journal.pbio.0040215

[140] Kaas JH, Hackett TA. Subdivisions of auditory cortex and levels of processing in primates. Audiology & Neuro-Otology. 1998;**3**(2-3):73-85

[141] Pandya DN, Sanides F. Architectonic parcellation of the temporal operculum in rhesus monkey and its projection pattern. Zeitschrift für Anatomie und Entwicklungsgeschichte. 1973;**139**(2):127-161

[142] Galaburda AM, Pandya DN. The intrinsic architectonic and connectional organization of the superior temporal region of the rhesus monkey. The Journal of Comparative Neurology. 1983;**221**(2):169-184. DOI: 10.1002/cne.902210206

[143] Morel A, Garraghty PE, Kaas JH. Tonotopic organization, architectonic fields, and connections of auditory cortex in macaque monkeys. The Journal of Comparative Neurology. 1993;**335**(3):437-459. DOI: 10.1002/cne.903350312

[144] Fullerton BC, Pandya DN. Architectonic analysis of the auditory-related areas of the superior temporal region in human brain. The Journal of Comparative Neurology. 2007;**504**(5):470-498. DOI: 10.1002/cne.21432

[145] Muckli L, Naumer MJ, Singer W. Bilateral visual field maps in a patient with only one hemisphere. Proceedings of the National Academy of Sciences of the United States of America. 2009;**106**(31):13034-13039. DOI: 10.1073/pnas.0809688106

[146] Brewer AA, Barton B. Developmental plasticity: FMRI investigations into human visual cortex. In: Papageorgiou TD, Christopoulos G, Smirnakis S, editors. Advanced Brain Neuroimaging Topics in Health and Disease - Methods and Applications. Rijeka, Croatia: InTech; 2014. pp. 305-334. DOI: 10.5772/58256

[147] Brewer AA. Visual maps: To merge or not to merge. Current Biology. 2009;**19**(20):R945-R947. DOI: 10.1016/j.cub.2009.09.016

[148] Katz LC, Shatz CJ. Synaptic activity and the construction of cortical circuits. Science. 1996;**274**(5290):1133-1138

[149] Polleux F, Ince-Dunn G, Ghosh A. Transcriptional regulation of vertebrate axon guidance and synapse formation. Nature Reviews. Neuroscience. 2007;**8**(5):331-340. DOI: 10.1038/nrn2118

[150] Merzenich MM, Brugge JF. Representation of the cochlear partition of the superior temporal plane of the macaque monkey. Brain Research. 1973;**50**(2):275-296

[151] Sweet RA, Dorph-Petersen KA, Lewis DA. Mapping auditory core, lateral belt, and parabelt cortices in the human superior temporal gyrus. The Journal of Comparative Neurology. 2005;**491**(3):270-289. DOI: 10.1002/cne.20702

[152] Dick F, Tierney AT, Lutti A, Josephs O, Sereno MI, Weiskopf N. In vivo functional and myeloarchitectonic mapping of human primary auditory areas. The Journal of Neuroscience. 2012;**32**(46):16095-16105

[153] Rivier F, Clarke S. Cytochrome oxidase, acetylcholinesterase, and NADPH-diaphorase staining in human supratemporal and insular cortex: Evidence for multiple auditory areas. NeuroImage. 1997;**6**(4):288-304. DOI: 10.1006/nimg.1997.0304

[154] Hackett TA. Information flow in the auditory cortical network. Hearing Research. 2011;**271**(1-2):133-146. DOI: 10.1016/j.heares.2010.01.011

[155] Hackett TA, Stepniewska I, Kaas JH. Subdivisions of auditory cortex and ipsilateral cortical connections of the parabelt auditory cortex in macaque monkeys. The Journal of Comparative Neurology. 1998;**394**(4):475-495

[156] Jones EG, Dell'Anna ME, Molinari M, Rausell E, Hashikawa T. Subdivisions of macaque monkey auditory cortex revealed by calcium-binding protein immunoreactivity. The Journal of Comparative Neurology. 1995;**362**(2):153-170. DOI: 10.1002/cne.903620202

[157] Molinari M, Dell'Anna ME, Rausell E, Leggio MG, Hashikawa T, Jones EG. Auditory thalamocortical pathways defined in monkeys by calcium-binding protein immunoreactivity. The Journal of Comparative Neurology. 1995;**362**(2):171-194. DOI: 10.1002/cne.903620203

[158] Kusmierek P, Rauschecker JP. Functional specialization of medial auditory belt cortex in the alert rhesus monkey. Journal of Neurophysiology. 2009;**102**(3):1606-1622. DOI: 10.1152/jn.00167.2009

[159] Rauschecker JP, Tian B, Hauser M. Processing of complex sounds in the macaque nonprimary auditory cortex. Science. 1995;**268**(5207):111-114

[160] Tian B, Rauschecker JP. Processing of frequency-modulated sounds in the lateral auditory belt cortex of the rhesus monkey. Journal of Neurophysiology. 2004;**92**(5):2993-3013. DOI: 10.1152/jn.00472.2003

[161] Kajikawa Y, Frey S, Ross D, Falchier A, Hackett TA, Schroeder CE. Auditory properties in the parabelt regions of the superior temporal gyrus in the awake macaque monkey:

An initial survey. The Journal of Neuroscience. 2015;**35**(10):4140-4150. DOI: 10.1523/JNEUROSCI.3556-14.2015

[162] von Economo C, Koskinas G. Die Cytoarchitectonik der Hirnrinde des erwachsenen menschen. Berlin: Julius-Springer; 1925

[163] Hedges S, Kumar S. Genomic clocks and evolutionary timescales. Trends in Genetics. 2003;**19**:200-206

[164] Kayser C, Petkov CI, Augath M, Logothetis NK. Functional imaging reveals visual modulation of specific fields in auditory cortex. The Journal of Neuroscience. 2007;**27**(8):1824-1835. DOI: 10.1523/JNEUROSCI.4737-06.2007

[165] Bartels A, Zeki S. The architecture of the colour Centre in the human visual brain: New results and a review. The European Journal of Neuroscience. 2000;**12**(1):172-193

[166] Moradi F, Heeger DJ. Inter-ocular contrast normalization in human visual cortex. Journal of Vision. 2009;**9**(3):13.1-13.22. DOI: 10.1167/9.3.13

[167] Mitchison G. Neuronal branching patterns and the economy of cortical wiring. Proceedings of the Biological Sciences. 1991;**245**(1313):151-158. DOI: 10.1098/rspb.1991.0102

[168] Chklovskii DB, Koulakov AA. Maps in the brain: What can we learn from them? Annual Review of Neuroscience. 2004;**27**:369-392. DOI: 10.1146/annurev.neuro.27.070203.144226

[169] Shapley R, Hawken M, Xing D. The dynamics of visual responses in the primary visual cortex. Progress in Brain Research. 2007;**165**:21-32. DOI: 10.1016/S0079-6123(06)65003-6

[170] Braitenberg V, Schüz A. Cortex: Statistics and Geometry of Neuronal Connectivity. 2nd ed. Berlin: Springer; 1998

[171] Baseler HA, Brewer AA, Sharpe LT, Morland AB, Jagle H, Wandell BA. Reorganization of human cortical maps caused by inherited photoreceptor abnormalities. Nature Neuroscience. 2002;**5**(4):364-370. DOI: 10.1038/nn817

[172] Hoffmann MB, Tolhurst DJ, Moore AT, Morland AB. Organization of the visual cortex in human albinism. The Journal of Neuroscience. 2003;**23**(26):8921-8930

[173] Hoffmann MB, Kaule FR, Levin N, Masuda Y, Kumar A, Gottlob I, et al. Plasticity and stability of the visual system in human achiasma. Neuron. 2012;**75**(3):393-401. DOI: 10.1016/j.neuron.2012.05.026

[174] Baseler HA, Gouws A, Haak KV, Racey C, Crossland MD, Tufail A, et al. Large-scale remapping of visual cortex is absent in adult humans with macular degeneration. Nature Neuroscience. 2011;**14**(5):649-655. DOI: 10.1038/nn.2793

[175] Fine I, Wade AR, Brewer AA, May MG, Goodman DF, Boynton GM, et al. Long-term deprivation affects visual perception and cortex. Nature Neuroscience. 2003;**6**(9):915-916. DOI: 10.1038/nn1102

[176] Brewer AA, Barton B. Visual cortex in aging and Alzheimer's disease: Changes in visual field maps and population receptive fields. Frontiers in Psychology. 2014;**5**:74. DOI: 10.3389/fpsyg.2014.00074

[177] Brewer AA, Barton B. Changes in visual cortex in healthy aging and dementia. In: Moretti DV, editor. Update on Dementia. Rijeka, Croatia: InTech; 2016. pp. 273-310. DOI: 10.5772/61983

Chapter 2

The Influence on Cortical Brainwaves in Relation to Word Intelligibility and ASW in Room

Chiung Yao Chen

Abstract

The influence of indoor speech intelligibility and apparent source width (ASW) on the response of cortical brainwaves was studied using two variables, the time gap between direct and the first reflection (Δt_1, ms) and the initial (<80 ms) interaural cross-correlation function ($IACC_{E3}$). Comparisons were performed based on autocorrelation function (ACF) of continuous brainwave (CBW) and slow vertex response (SVR). The results are: (1) the effective delay time of ACF (τ_e) of β-waves (13–30 Hz) in the left hemisphere under changes in Δt_1 was significantly and positively correlated with speech intelligibility ($p < 0.001$). (2) As ASW increased, the relative amplitude of left hemisphere A (P2-N2) tended to decrease ($p < 0.05$) in SVRs, while N2 latency tended to increase ($p < 0.05$); the lateral lemniscus in the auditory nerve was suggested to be the reactive site. (3) With regard to hemispheric specialization in brain, speech intelligibility, the main temporal factor, was found to be controlled by the left hemisphere. A subjective spatial factor, ASW, the relative amplitude of SVR was also found to decrease in the left hemisphere; nevertheless, they are coherent while the N2 latency of SVR significantly prolonged in both left and right hemisphere under changes in $IACC_{E3}$.

Keywords: brainwaves, ACF, apparent sound width, speech intelligibility, subjective diffuseness, hemispheric specialization

1. Introduction

In human speech cognition, speech intelligibility integrates short-term memory and cerebral feedback [1]. However, important factors constituting the spatial impressions of sound also include certain related evaluation indicators, such as the listener's judgment of sound source direction (sense of direction) and distance (sense of proximity), apparent source width (ASW), and lateral envelopment (LEV). As suggested by Ando [2] and Beranek [3], the composition of such spatial impressions mainly depends on fluctuations of the magnitude of the interaural cross-correlation (IACC) and is especially affected by the degree of subjective diffusion of the sound field. However, listeners differ in their needs and perceptions regarding subjective diffusion and ASW.

With regard to neuron-psychology, Sperry [4] discovered the phenomenon of hemispheric disconnect. The cerebral specialization theory distinguishes between "speech functions" and "non-speech functions." Certain symbols in architectural design belong to non-speech functions. For instance, the range of non-speech functions includes aesthetic perception and the feeling of balance. In particular, many non-speech symbols can be observed in environmental design. Earlier research on audio and cerebral correlations found that such common medical problems as aphasia and disturbances in tone judgment originate in the left cerebral hemisphere. Therefore, this study suggested that cerebral responses to speech and non-speech symbol in the physical environment effectively substitute for the semantic differences (SD) caused by age-related and cultural differences. Cerebral responses to communication stimuli are a direct cross-cultural and cross-age reference indicator, which is similar to the principle behind polygraph tests performed by police to examine physiological responses.

This study suggested that cerebral responses can be used to clearly and consistently examine responses to change in "speech functions" of the physical environment, or speech intelligibility, when designing a sound field. Ando [2] considered "speech functions" to be an important temporal factor and the result of autocorrelation function (ACF) evaluations in the brain. Therefore, the environmental effects of temporal factors were examined in this study based on the influence of speech intelligibility on the correlation between "subjective perceptions" and cerebral responses, which served as the basis for the objective design of an acoustic environment. Akita et al. [5] indicated that when the sensory information received by listeners is analyzed by brainwaves, this does not represent their direct experience of changes in the environment, but rather the interaction between physiology and the environment. This phenomenon is common in daily life. The intensity of cerebral evoked responses is the optimal evaluation tool [6]. Soeta et al. [7] studied the effects of sound source features on subjective psychological responses and cerebral responses measured by magnetoencephalography (MEG) and reported that at different delay times of reflection sounds (Δt_1 = 0, 5, 20, 60, and 100 ms) and 50 alternations, the ACF effective delay time of α-waves recorded by MEG indicated subjective preferences regarding sound fields. The methods used in this study can be summarized as follows:

1. The first reflection delay (Δt_1) was changed to change speech intelligibility. The degrees (or process) of subjective recognition of Chinese monosyllables were determined by comparing ACF calculation results related to α-waves and β-waves among cerebral continuous brainwaves (CBW).

2. The $IACC_{E3}$ was changed to change subjective ASW. Changes in the waveforms of auditory evoked potentials (AEPs) during listeners' perceptions of spatial ASW were analyzed.

2. Empirical methods

2.1 Psychological test of intelligibility

This study used monosyllabic speech sound articulation and $IACC_{E3}$ to quantify changes in two subjective experiences, namely, speech intelligibility and ASW. With regard to speech intelligibility, the fifth group of common Chinese monosyllabic speech sounds used in Taiwan [8] (female voice, **Table 1**) was used. Test results related to this group of monosyllabic sounds are characterized by the largest disparity in error rates because most related sounds belong to "fricative sounds" (i.e., apical vowels, such as "zh," "ch," "sh," "r," "z," "ci," and "si" in Bopomofo system). The amounts of fricative and non- fricative rhymes are balance (eight versus ten, respectively).

Chinese monosyllable					
1	shy0	7	yu2	13	ching3
2	iur1	8	leau3	14	tzuen1
3	li0	9	shoou3	15	cha2
4	meei3	10	ian1	16	shuo4
5	tsac3	11	tsong1	17	he4
6	ru2	12	guang1	18	chye2

Note: The pronunciation of each syllable depends on the tone (one of five pitch contours) used, which is indicated by a number attached to the end of the syllable. For example, 0 denotes monosyllables pronounced with a soft puff of air.

Table 1.
List of the term of monosyllables [8].

The sound structure of Mandarin differs from that of other languages. In Mandarin, each character is pronounced as a monosyllable with one of five tones (i.e., types of pitch contour). Each of these tones (0–4), when used with a given monosyllable, causes the monosyllable to convey a meaning distinct from those conveyed when the monosyllable is used with the other four tones. Utterance lengths in the experiment were set to 400–500 ms. Monosyllabic presents were separated by 2.5 s. The experiment was arranged according to the arrangement used in the study by Chen et al. [9].

The experiment was conducted in front of two overlapping loudspeakers in a semi-anechoic room (4 × 3 and 4 m in height) at Chaoyang University of Technology. The loudspeakers (Fostex NF-1A) were located at 1.5 m right front of the center of a listener's head. The first reflected sound was given off by the upper loudspeaker (η = 15) while another gave off the direct sound (η = 0). To vary speech intelligibility, the speech signal was assumed that emitted from the stage with a direct and a reflection sound reflected through the ceiling of the stage. The listening level was adjusted to a usual communicative sound volume of 62 dB(A) at the center of the room. The level of background noise in the semi-anechoic room was 32–42 dB(A), then the S/N ratio are approximate to 30–40 dB. The setup of the instrumental diagram (EEG recordings) could be referred to **Figure 1**, since they were same as that in the spatial ASW experiment stated below. The settings of the physical parameters used in the experiment are shown in **Table 2**. **Figure 2** shows the experimental results that indicate 62 listeners who were significantly able to distinguish sounds using percentage syllable articulation (PSA) tests [10]. To determine PSA, those written syllables are compared with the original syllables to find the percentage of syllables written correctly.

2.2 Psychological quantification test of ASW

The paired-comparison method [11] was used in the psychological quantification test of subjective ASW. The experiment was conducted in the same venue as the first experiment. Three loudspeakers (one for direct sounds and two for reflected sounds) were located at 1.5 m from the center of a listener's head; the incidence combinations (ξ, η) are: (0°, ±15°), (0°, ±55°), (0°, ±90°) and (0°, +15°, −55°) on the horizontal plane. 2 kHz pure-tone (1 ms) sounds were produced. The $IACC_{E3}$ (0.35, 0.57, 0.68, and 0.81) [12, 13] of the sound field was changed by changing the angle of incidence stated above and the sound pressure level. As a result, different subjective ASWs were generated (**Table 3**). The instrumental setup of testing spatial ASW and the process of AEPs recordings are interpreted in **Figure 1**.
The participants (80 students) determined ASWs using paired comparisons. The interval between sound prompts within one group was 2 s and the interval between groups was 10 s; in total, six groups were used. The participants were asked to

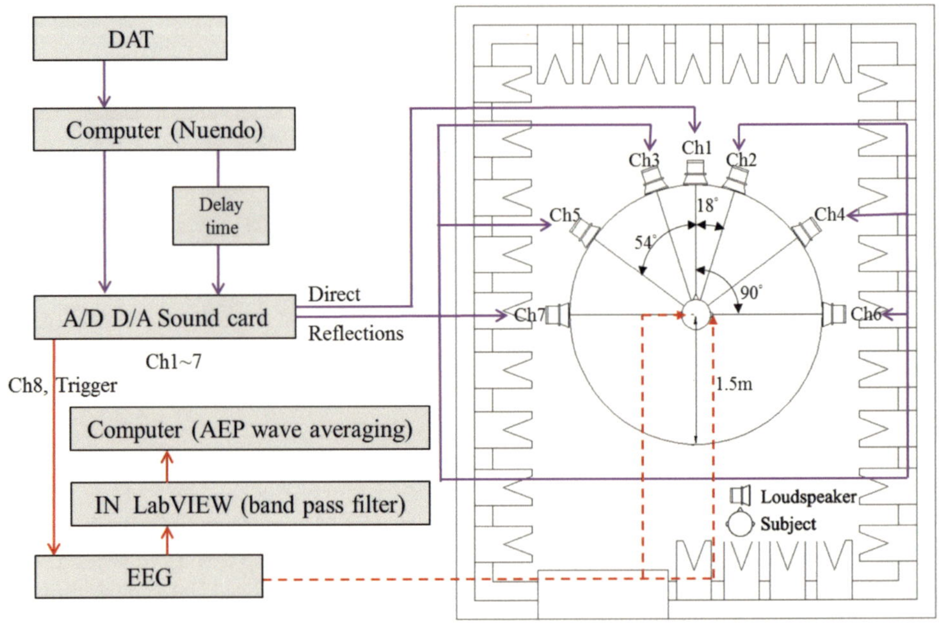

Figure 1.
The setup of the instrumental diagram (audio arrangement and EEG recordings).

Item	Conditions of experiments
Δt_1 (ms)	Delay gap: 0 ms, 35 ms, 100 ms, 150 ms, 200 ms
SPL of individual loudspeakers	Direct sound: 60 dB(A); first reflection, Δt_1: 55 dB(A)
Reverberation times	$RT \doteqdot 0.1$ s

Table 2.
The setting of the physical parameters in subjective articulation test of monosyllables.

immediately determine and record the relative probability of ASWs. Each questionnaire was conducted for 1 min. The psychological scale values of ASWs are shown in **Figure 3** calculated using Thurstone's Case V [11]. Non-linear correlation was observed in the $IACC_{E3}$ result [14].

2.3 Brainwave physiological experiment methods

2.3.1 Brainwave analysis method

After the fast Fourier transform (FFT) was applied to the brainwaves, ACF of CBW calculations were performed for the α-waves (8–13 Hz) and β-waves (13–30 Hz) of the left and right hemispheres. In the earlier study by Chen and Ando [15], 100 Hz α-waves and 500 Hz β-waves were sampled according to the sampling frequency laws and, after A/D conversion (16 bits), input into a computer to calculate the effective duration (τ_e) of CBWs' ACF (**Figure 4**). In ACF calculations of τ_e values in the study by Chen and Chan [16], the 0.3 s integration time (2 *T*) of monosyllabic speech sounds was suggested to be the most effective. Eventually, the monosyllabic signals were played in this study included simulation of the first delay time [17]. Therefore, the integration time (2 *T*) of ACF of continuous brainwaves (CBW) used in calculation was adjusted to 0.5 s. As shown in **Figure 4**, substantial differences were observed in the ACF waveforms of α-waves and β-waves under the same first delay time settings.

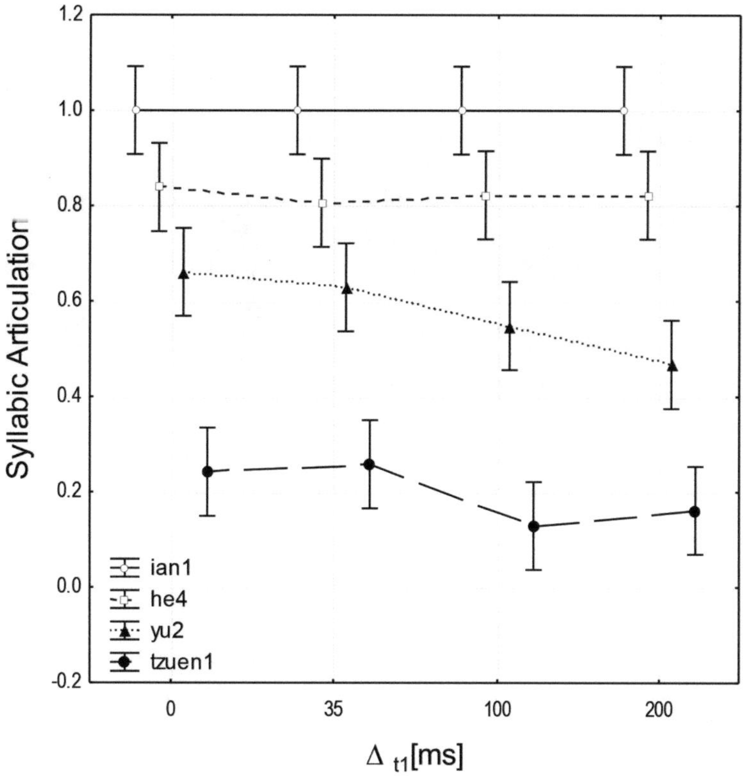

Figure 2.
The percentage syllabic articulation of monosyllable functioning initial time delay of a sound field.

$IACC_{E3}$ (setup values)	Amplitude of direct sound (A_0)	I-1, SPL/dB(A)	Amplitude of first reflection (A_1)
0.35	1	62.6	0.8
0.57	1	62.6	0.8
0.68	1	55.4	0.4
0.81	1	64.0	0.2
I-2, SPL/dB(A)	**Amplitude of second reflection (A2)**	**I-3, SPL/dB(A)**	**Σ*L* Total SPL dB(A)**
59.4	0.8	59.4	65
59.4	0.8	59.4	65
53.4	0.4	55.4	65
53.4	0.2	53.4	65

Note: I-1, I-2 and I-3 denote the sound intensity of direct sound and 1st and 2nd reflections sound measured at the location of the head top of the participants. a denotes the amplitude of the direct sound, 1st reflective and 2nd reflective sound by A_0, A_1 and A_2.

Table 3.
The parameters of subjective source apparent width (ASW) test arranged by 2 kHz pure tone burst.

To explore the changes in subjective perceptions of ASW, AEPs of nine participants were induced, recorded and analyzed as in the psychological intelligibility experiment. However, a spatial impression of a sound signal is a short-term memory phenomenon. Therefore, waveforms induced by the brain AEPs are normally used to observe changes in responses to weak brainwave signals (about 10–100 μV in

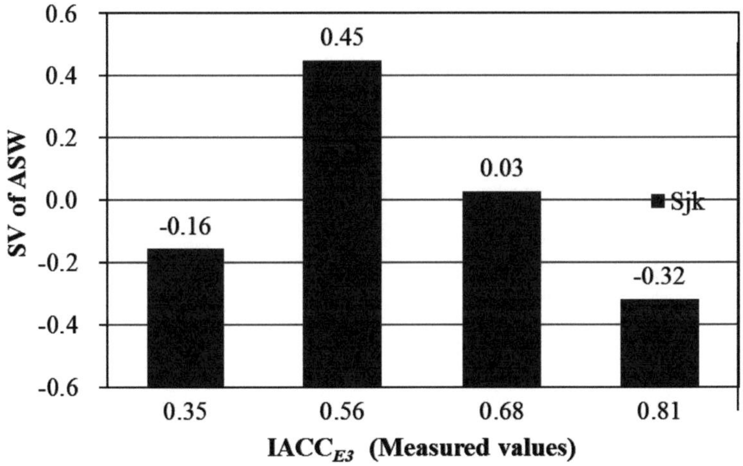

Figure 3.
The scale values of subjective ASW test functioning IACC$_{E3}$.

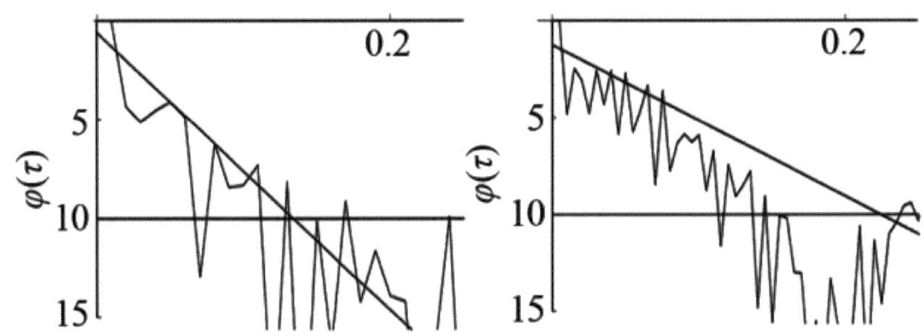

Figure 4.
The ACF curve of α-wave (left) and β-wave (right) recorded in relation to the monosyllable "tzuen1" were announcing.

amplitude when measured from the scalp). Clear consistent brain waveforms are usually obtained by applying the signal averaging method [18] to responses that occur within 500 ms after auditory stimulation (**Figure 5**). In this study, 180 times of averaging process was applied here since the wave form of slow vertex responses (SVR) were clearly obtained. The movements (latency) of waveform peaks and troughs in the wave relative amplitude can reflect the activation of different parts of auditory nerves [19, 20]. As shown in **Figure 5**, this study changed ASW perceptions by changing the sound arrival orientation and energy while fixed reflection delay and echo times ($\Delta t_1 < 15.5$ ms, $RT \doteqdot 0.1$ s) (**Table 3**). This study suggested that brain waveforms could be observed using SVR because the preset reaction time exceeded 10 ms. In such way, AEPs were obtained. Consequently, when potentials P1, P2, N1, and N2 of posterior waveforms (30–200 ms) among different AEPs were generated by different sound stimulus, the relative amplitudes of P1-N1 and P2-N2, and P1, P2, N1, and N2 latency were observed [18].

2.3.2 Brainwave recording method

With regard to brainwave recording, eight participants (and other nine in the AEPs experiments) sat on comfortable office chairs in the semi-anechoic room at Chaoyang University of Technology and their brainwaves were induced and

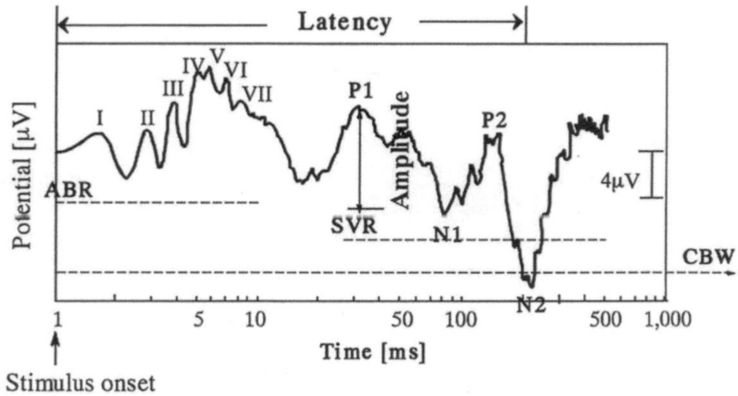

Figure 5.
The diagram illustrates brain waves in different time domains and their index at the peak or trough by Ichikawa [17].

recorded. The room temperature was maintained at 22 ± 2°C. All subjects were prohibited from drinking any alcohol for a period of 3 days before the brainwave recordings were conducted, and they refrained from smoking for 1 h before both experiments. They were instructed to concentrate on listening to the signals during the presentation. The participating subjects were eight male students (plus another nine male students in the AEPs experiments) aged 22–24 years old with normal hearing ability, as confirmed by an audiometry test and right-handed test (self-administered). The audiometry test detects sensorineural hearing loss (damage to the nerve or cochlea) and conductive hearing loss (damage to the eardrum or the tiny ossicle bones). Pure-tone subjective audiometry, in which air conduction hearing thresholds in decibels (dB) for a frequency range of 250–8000 Hz are plotted on an audiogram for each ear independently, was applied. All of the subjects had to be qualified as normal with a pure-tone audiogram (less than 25 dB) for both ears prior to the brainwave experiments and questionnaires.

This procedure has been applied in many studies, such as those by Chen et al. [9], Ando et al. [19], and Ando et al. [20], among others.

Electrodes used to explore brainwaves were positioned at the participants' T3 and T4 head points according to the international 10–20 system [21]. Electric potentials were examined using eardrops on the left and right sides. Unipolar induction of continuous brainwaves in the left and right hemispheres was performed. The G2 electrode was attached between the eyebrows for eye movement reference. The electrode system was grounded each time the brainwaves were recorded in order to avoid external electric interference. The settings of the simulated sound field were similar to that in the aforementioned psychological experiment [16]. The collected brainwave data was analyzed and processed by NI LabVIEW software. The setup of the instrumental diagram is shown in **Figure 1**. During the brainwave experiments, the subjects had to be relaxed while paying close attention to the sound stimuli. Brainwaves are extremely sensitive to any incoming stimuli or stress. For the purpose of this study, a relaxed state but one also focused on environmental variations was considered the best condition for the subjects during the brainwave recording process. For the recordings, periods of blinking had to be disregarded. Thus, a monitor was set up in the anechoic chamber to identify these periods, and these sections were later removed from the recordings.

3. Empirical results and discussion

3.1 Influence of speech intelligibility on brainwaves

Monosyllabic speech sounds had a major effect on both α-waves (F = 12.96 (9, 2488), $p < 0.001$) and β-waves ($F(9, 2488) = 5.21$, $p < 0.001$) at different first reflection delay times. As shown in **Figures 6** and **7**, the ACF of β-waves recorded in the left hemisphere was positively correlated with subjective perceptions of speech intelligibility. With regard to α-waves in the left hemisphere, brainwave

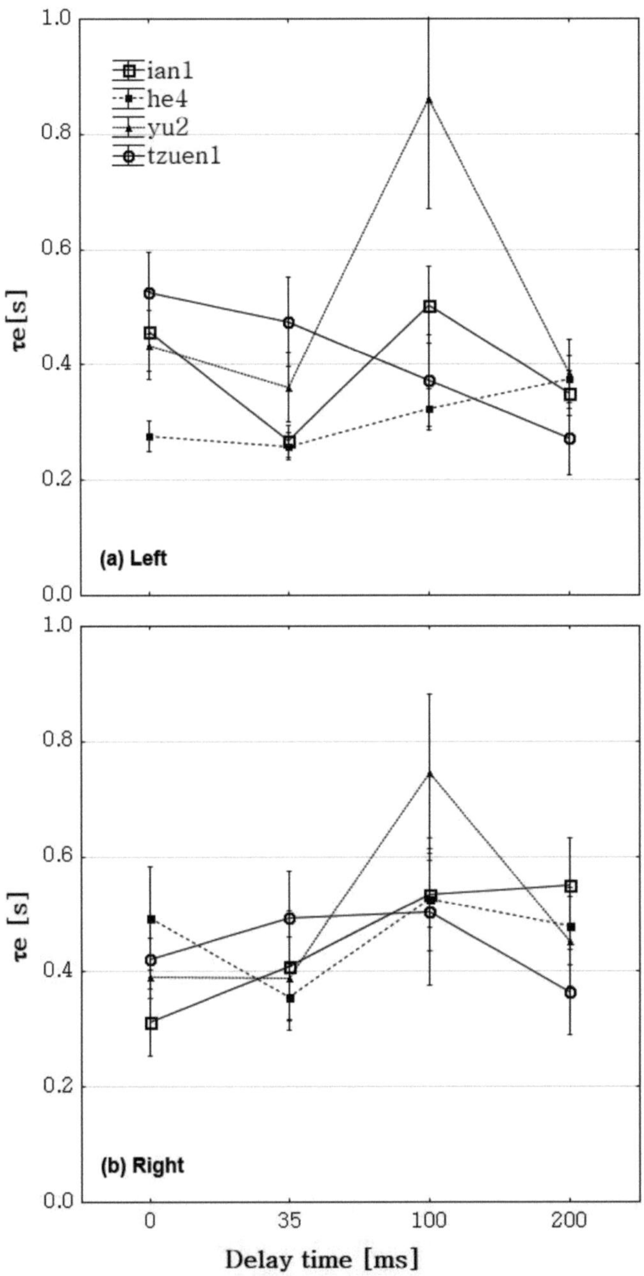

Figure 6.
Relationship between τ_e of ACF, α-wave and Δt_1 of sound field.

responses tended to increase at 100 ms for all sounds apart from "tzuen1." It is not clear whether these results were related to the nasal sounds "uen." The psychological experiment results (**Figure 2**) showed that the lowest articulation rates were observed for four sounds at 100 ms reflection delays. An opposite tendency was

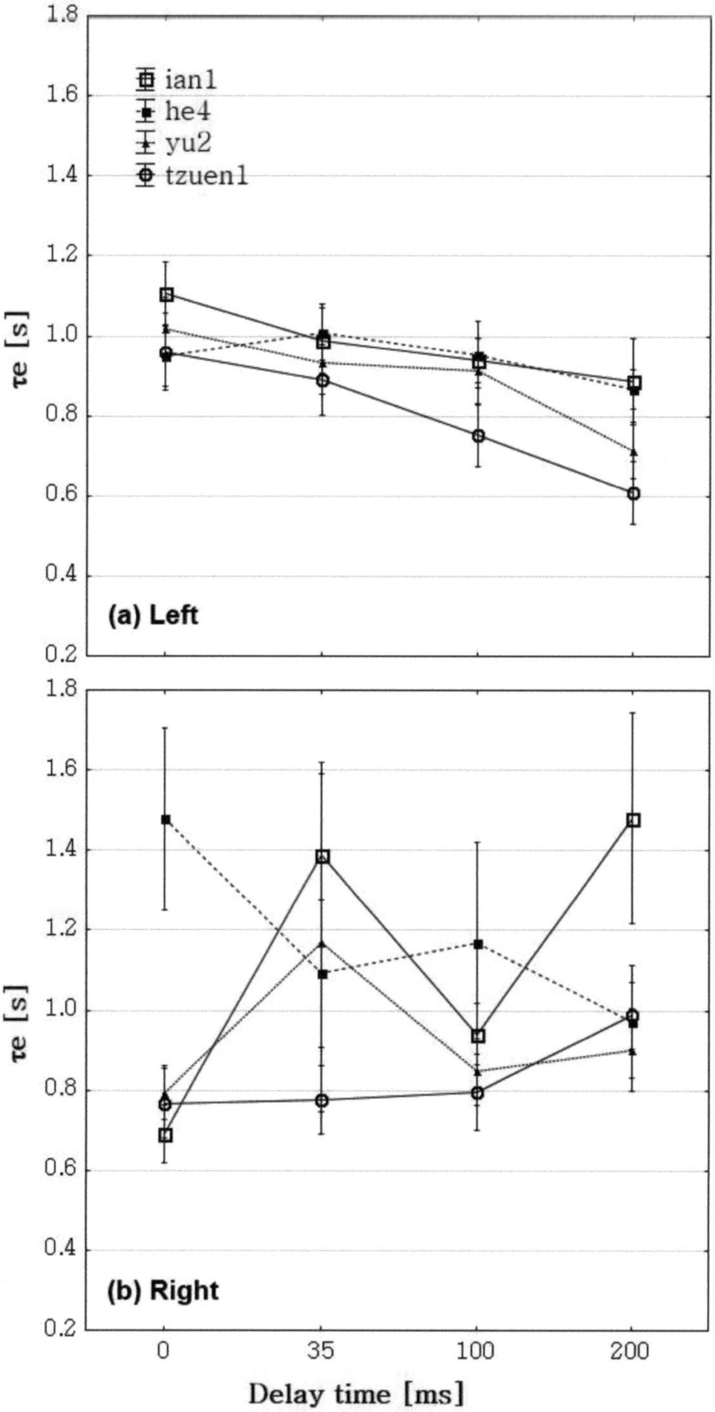

Figure 7.
Relationship between τ_e of ACF, β-wave and Δt_1 of sound field.

detected in the articulation rate results related to "yu2," "he4," and "ian1." The 100 ms delay is close to the 135 ms slow response delay proposed in Ando's [22] study on sound field preferences (echo disturbance). The displeased response of α-waves to the delay time of reflection [14] requires further investigation.

Figure 7 shows changes in β-waves. Consistent results were obtained with regard to the influence of the delay time of reflection on the left hemisphere ($F(9, 2488) = 5.21, p < 0.001$). However, no significant differences were observed in psychological reactions to speech intelligibility. A significant relation between articulation rates and the order of reactions was detected in the mean values related to the right hemisphere. Thus, left hemisphere showed changes in β-waves in relation to the order of delay time on reflection but speech intelligibility reactions.

3.2 Changes in subjective perception of ASW and brainwaves

The findings related to SVR to evoked potentials of nine participants are shown in **Figure 8**. With regard to the left hemisphere, SVR relative amplitude were consistently and inversely related to quantified psychological scale values ($F(1, 16) = 4.90$, $p < 0.05$). However, clear results were difficult to obtain due to the small difference between ASW (−0.16) and ASW (0.03).

Latency changes in the left and right hemispheres indicated the presence of a significant difference between ASW (0.03) and ASW (0.45) only at N2 in the left hemisphere ($F(1, 16) = 11.09, p < 0.05$). The tendency of ASW (0.45) latency being smaller than ASW (0.03) latency in the left hemisphere can be seen from **Figure 9**, whereas in the right hemisphere, ASW (0.45) latency was consistently

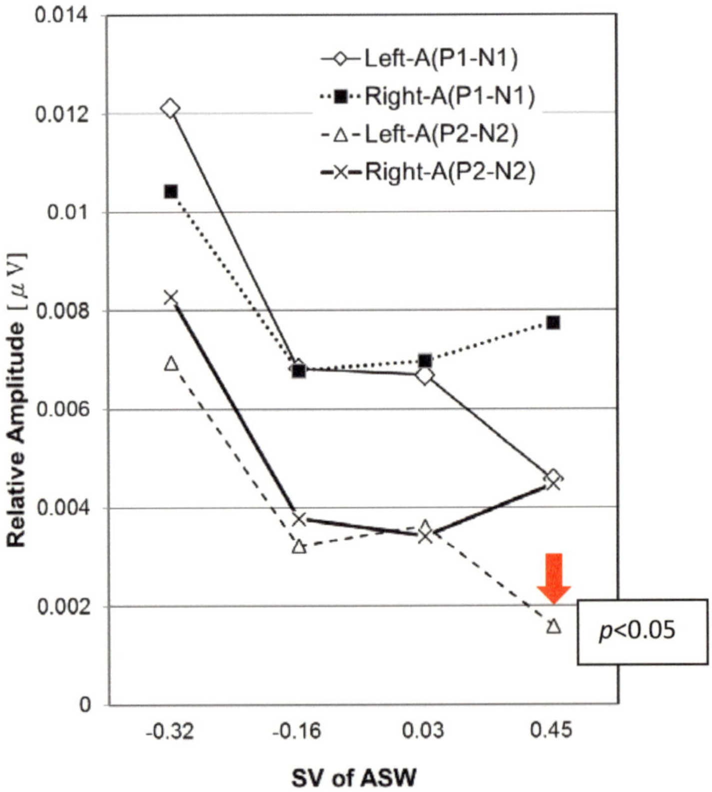

Figure 8.
Relationship between potential, SVR and ASW of sound field.

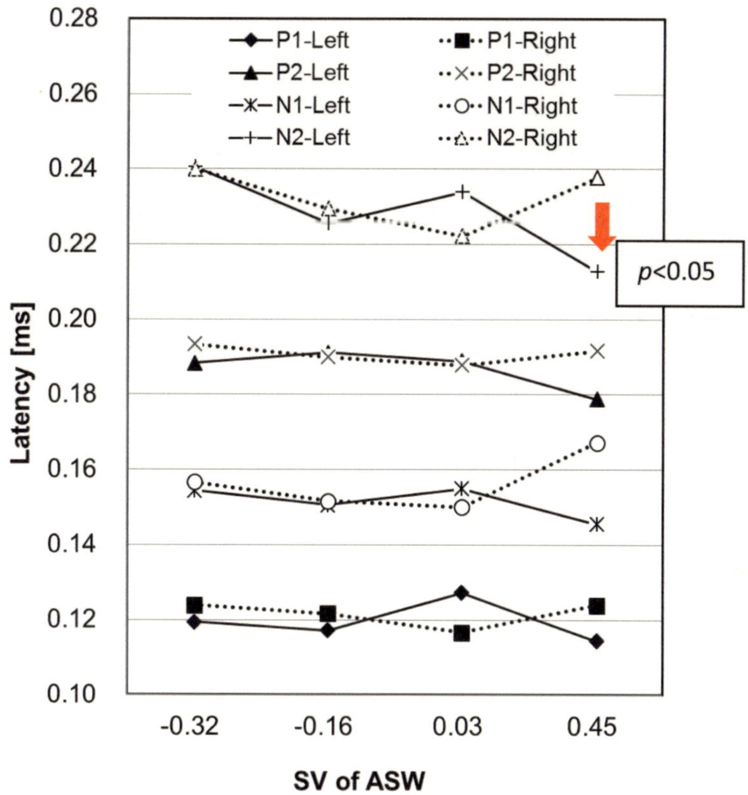

Figure 9.
Relationship between latency, SVR and ASW of sound field.

larger than ASW (0.03) latency. The results showed that relative amplitude in the left hemisphere were caused by subjective perceptions of ASW, which influenced the participants' preference toward a sound field. The consistency of latency at N2 was due to the activation of neural sites, which was clearly observed between ASW (0.03) and ASW (0.45), as well as at $IACC_{E3}$ of 0.56–0.68. Thus, the brain did not have a major effect on the corresponding changes at the extreme $IACC_{E3}$ values of 0.35 and 0.81, which corresponded to the psychological reaction results presented in **Figure 3**.

4. Conclusions

The arrangements and results of the aforementioned brainwave experiments indicated that when simple physical changes in a sound field and complex psychological feedbacks affect cerebral brainwave reactions, the correspondence of the cerebral specialization theory with the results becomes very complicated. In general, in this study, the left hemisphere tended to be activated in both temporal and spatial aspects based on the sound field. When the participants' brainwaves were recorded during the judgment task, the brain activation in the right hemisphere tended to reflect the discriminated object more closely. When CBW were observed during research on speech intelligibility, the left hemisphere showed clear reactions to the first reflection delay time of sound field (**Figure 7**). However, the degree of speech intelligibility is a reflection of the complex thinking process that occurs in the right hemisphere (cerebral feedback). This phenomenon was supported by the subjective ASW experiment. With regard to changes in spatial factors, the left hemisphere received information about sound field changes when the $IACC_{E3}$ value changed. ASW changes between (ASW (0.03) and ASW (0.45)), which were more evident in the right hemisphere, affected both right and left hemispheres. They are coherent while the N2 latency of SVR significantly prolonged in both left and right hemispheres under changes of subjective diffuseness in $IACC_{E3}$ found by Ando et al. [20]. Different sites are activated by brainwaves during focused and ambient use of the brain.

Cerebral specialization has been reported to be determined by focused conscious decisions. For instance, Floel et al. [23] conducted a spatial—visual focus experiment and used a Doppler ultrasound system and magnetic resonance imaging (MRI) equipment to observe the brain reactions of right-handed participants; the researchers found that both spatial recognition and speech functions were activated in the right hemisphere, which corresponded to clinical experiment results.

Nevertheless, for CBW researches, we conclude that α-waves (8–12 Hz) mainly responds to the emotional reactions; β-waves (13–30 Hz) reacts to the auditory matter drift (**Figures 6** and 7). But the left hemisphere leads focus or attention on the varying of situational conditions (**Figures 8** and **9**), and the right one blends with imaginable feeling and experience. Hemispheric specialization has to pay attention to the conditioned response, conscientious and careful detail to setup each brainwaves' experiment.

Acknowledgements

I would like to express my thanks to the graduated students, Yong-Shang Chen and Qi-Wen Lin who have, in brainwaves' experiments, helped me in the course of preparing this study. In particular, I wish to thank Professor Em. Yoichi Ando, who kindly gave me directions on my brain research and data analysis methodology. I am also indebted to the Ministry of Science and Technology Taiwan, for their 2 years period (2007 and 2012) of financial support to complete this research. Meanwhile, the application conformed to academic ethics for the conduct of research. Moreover, during the brainwave experiments, the subjects were assured of their safety and told the procedure was non-invasive. Special thanks are due to my many colleagues for their participation in the experiments involving the subjective judgments and the brainwaves' recordings.

Glossary of symbols

ASW	apparent source width, a sound perception of the subjective diffuseness occurred from beginning to 80 ms of stimulus
Δt_1	delay gap between direct and first reflection in a defuse sound field
$IACC_{E3}$	binaural initial (<80 ms) interaural cross-correlation function
ACF	autocorrelation function
τ_e	effective delay of autocorrelation function (ACF)
CBW	continuous brainwave, a term to distinguish from an evoked potential (EP) or evoked response within EEG
SVR	slow vertex response, an evoked potential is a direct result after a specific sensory stimulus in the period of 10–500 ms
LEV	listener envelopment, a sound perception of the subjective diffuseness occurred after 80 ms of stimulus
SD	semantic differences, a method of questionnaire employed the scale of responses caused by a psychological affection

/yu2/	example of a monosyllable in Taiwanese's life speech
η	vertical angles at a median plane, 0° started from the front of head at ear height
ξ	angles at clockwise horizontal plane, 0° started from the front of head at ear height
SPL	sound pressure level measured by a sound level meter in a fast time-weighting mode
AEP	auditory evoked potential
PSA	percentage syllable articulation

Author details

Chiung Yao Chen
Department of Architecture, Chaoyang University of Technology, Taiwan

*Address all correspondence to: chychen@cyut.edu.tw

References

[1] Schmidt RA, Wrisberg CA. Motor Learning and Performance: A Problem-Based Learning Approach. 2nd ed. Champaign, IL: Human Kinetics, California - Los Angeles; 2000. pp. 255-287. Chapter 10

[2] Ando Y. Concert Hall Acoustics. Berlin Heidelberg, New York: Springer; 1985. pp. 41-44. DOI: 10.1007/978-3-642-69810-1

[3] Beranek LL. Concert hall acoustics. The Journal of the Acoustical Society of America. 1992;**92**(1):1-39. DOI: 10.1121/1.404283

[4] Sperry RW. Lateral preference in the identification of patterned stimuli. The Journal of the Acoustical Society of America. 1970;**47**:574-578. DOI: 10.1121/1.1911932

[5] Akita T, Fujii T, Hirate K, Yasuoka M. A research on auditory information processing in the brain when a person at his task by eans of measurement and analysis of auditory evoked potential. Journal of Architecture, Planning and Environmental Engineering. 1998;**507**:61-69

[6] Jian YH, Chen JS, Huang SH, et al. Measurements on emotional response using EEG. Journal of Advance Engineering Taiwan. 2006;**33**(1):123-131 (in Chinese)

[7] Soeta Y, Nakagawa S, Yano T, et al. Magnetoencephalographic responses in relation to the factors extracted from autocorrelation and inter-aural cross-correlation functions. Journal of South China University of Technology (Natural Science Edition). 2007;**35**(s):80-83

[8] Chen CY, Chen LS, Lin W. A study on evaluation method of Chinese articulation standard of speech intelligibility for sound field in Taiwan. Journal of Architecture, Architectural Institute of the Republic of China. 2002;**43**:27-36 (in Chinese)

[9] Chen CY, Soeta H, Ando Y. Analysis of time-varying continuous brainwaves: Cross correlation function of cortical alpha waves. In: 20th Symposium of Taiwan Architecture Institute, Taipei. 1999 (in Chinese)

[10] Benesty J, Sondhi MM, Huang Y. Springer Handbook of Speech Processing. Berlin, Germany: Springer-Verlag; 2007. ISBN: 978-3-540-49125-5. DOI: 10.1007/978-3-540-49127-9_1

[11] Thurstone LL. A law of comparative judgment. Psychological Review. 1927;**34**:273-289

[12] Chen CY, Chang CH. Sound localization in respect of magnitude of inter-aural cross-correlation function: Two reflections in simulated sound field as example. Chaoyang Journal of Design. 2000;**1**:101-114 (in Chinese)

[13] Chen CY, You CR. A comparative study of subjectively apparent source width between traditional Chinese and Western classic musical instruments recital. Chaoyang Journal of Design. 2011;**12**:96-116 (in Chinese)

[14] Ando Y, Kurihara Y. Nonlinear response in evaluating the subjective diffuseness of sound fields. The Journal of the Acoustical Society of America. 1986;**80**(3):833-836. DOI: 10.1121/1.393906

[15] Chen CY, Ando Y. On the relationship between the autocorrelation function of the α-waves on the left and right hemispheres and subjective preference for the reverberation time of music sound field. Journal

of Architecture, Planning and Environmental Engineering. 1996;**489**:73-80

[16] Chen CY, Chan MH. A study of the Chinese speech intelligibility in halls in relation to the autocorrelation function model: The case of Chinese in Taiwan. Journal of Architecture, Architectural Institute of the Republic of China. 2006;**57**:55-68 (in Chinese)

[17] Chen YS. Analysis of cortical continuous brainwaves in relation to speech intelligibility in room [thesis of Master Degree]. 2010. pp. 33-38

[18] Ichikawa G. The logarithmical presentation for auditory evoked potential. Audiology Japan. 1983;**26**:735-739

[19] Ando Y, Kang SH, Morita K. On the relationship between auditory-evoked potential and subjective preference for sound field. Journal of the Acoustical Society of Japan (E). 1987;**8**:197-204. DOI: 10.1250/ast.8.197

[20] Ando Y, Kang SH, Nagamatsu H. On the auditory-evoked potential in relation to the IACC of sound field. Journal of the Acoustical Society of Japan (E). 1987;**8**:183-190. DOI: 10.1250/ast.8.183

[21] Jasper HH. The 10/20 electrode system of the international federation. Electroencephalography and Clinical Neurophysiology. 1958;**10**:371-375

[22] Ando Y. Architectural Acoustics. New York, Inc.: Springer-Verlag; 1998. pp. 89-95. DOI: 10.1121/1.423953

[23] Floel A, Jansen A, Deppe M, et al. A typical hemispheric dominance for attention: Functional MRI topography. Journal of Cerebral Blood Flow and Metabolism. 2005;**25**:1197-1208

Chapter 3

Neuroplasticity and the Auditory System

Caroline Donadon, Milaine Dominici Sanfins, Piotr Henry Skarzynski, Maria Francisca Colella dos Santos and Stavros Hatzopoulos

Abstract

This chapter will present information on the central auditory nervous system with a special focus in the auditory pathways. The intrinsic and extrinsic aspects of neuroplasticity will be described, and the neuroplasticity of the auditory system will be presented in detail. These topics are the basis of the auditory training (AT) program for central auditory processing disorders.

Keywords: auditory pathways, neuroplasticity, auditory training

1. Introduction

1.1 The central auditory nervous system

The central nervous system is bilaterally symmetrical, and it is composed of seven main regions: the spinal cord, the bulb, the bridge, the cerebellum, the midbrain, the diencephalon, and the cerebral hemispheres. Each of these neural regions performs a number of specific functions. Additionally, each function, whether it is a sensory, a motor, or another integrative task, is performed by more than one neural pathway [1, 2].

The linguistical cerebral functions are mainly located in the auditory cortex, which is divided into four anatomically distinct lobes: the frontal, the parietal, the occipital, and the temporal lobe. The latter is responsible for the function of hearing as well as for various aspects of learning, of memory, and of emotions (**Figure 1**).

The cerebral hemispheres are characterized by two important organizational features:

i. Each hemisphere is primarily related to specific sensory and motor processes on the opposite (contralateral) side of the body. The structure connecting the two hemispheres is the corpus callosum [2].

ii. Although both hemispheres appear to be similar (in humans), they are not structurally fully symmetrical nor they have equivalent functions.

Broca, Wernicke, and Penfield were pioneers in unraveling the functions of the temporal lobe. Penfield found that a stimulation of primary auditory areas produced gross auditory sensations, whereas stimulation in the superior temporal gyrus produced altered perception of auditory sounds, illusions, and hallucinations.

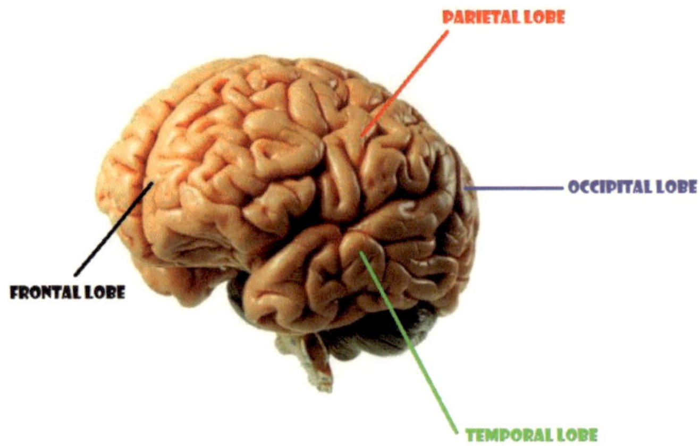

Figure 1.
The four cerebral lobes of the auditory cortex.

Studies on epileptic patients, whose hemispheres were separated by a section of the corpus callosum, have allowed us to understand numerous details related to the concept of hemispheric specialization [3].

An important contribution for the understanding of hemispheric function was achieved through the development of the Wada test [4]. The latter was developed in order to determine the dominant hemisphere for speech, so that inadvertent lesions of the speech centers, during neurosurgical procedures, could be avoided. During the test, the patient is instructed to count aloud, while sodium nitrite, a fast-acting barbiturate, is injected into the left or right carotid artery. The drug preferentially accesses the hemisphere on the same side as the injection, causing a brief speech dysfunction. If the dominant hemisphere's speech center is affected, the patient usually stops counting. With this test, the relationship between hemispheric specialization and laterality can be assessed, especially in left-handed subjects. Data in the literature [4] suggest that 96% of the right-handed people have dominant speech centers in the left hemisphere, while 15% of left-handed people have dominant speech centers in the right hemisphere. In some left-handed people, speech is controlled by both hemispheres. In such cases the administration of sodium nitrite does not suppress speech. Similar results were observed in trials involving hearing. When sounds were presented at the same time in both ears, it was found that in right-handed people the left ear performed better with nonverbal sounds.

The auditory system can be considered a high-performance signal-processing region, presenting a complex built-in hierarchy. Within the system each structure has a specific function, and progressing upwards along the pathways, these functions become more specific and dependent on the functional and physiological integrity of the previous structures. The auditory system plays an essential role for the communication among the members of the same species. Additionally, humans use the sensory inputs of the auditory system to identify different sounds leading to immediate actions, such as the status of alertness caused by the perception of siren sounds from police cars and ambulances.

From a simplified point of view, the auditory system consists of two areas, namely, the auditory periphery and the central auditory system. In each there are several structures which are stimulated when a sound stimulus is presented. Some researchers [5], however, have disputed this simple division of the auditory system suggesting instead a three-stage depiction:

i. The periphery, which captures and converts the acoustic sound stimuli into electrical neural pulses.

ii. The brainstem, which performs the initial processing of the information through modulation and interaction of the signals.

iii. The thalamocortical region, which is responsible for more advanced functions and produces emotional, cognitive, and linguistic responses from the acoustical stimuli

The efficiency of this set of interconnected structures depends primarily on the auditory experience. The simplest auditory task is influenced by high-level functions that include motivation, memory, and decision-making [6].

From a functional point of view, the following functions (among others) are assigned to CANS: the ability to detect and discriminate sound sources, the separation of acoustic stimuli from the background noise, the process of understanding the incoming stimuli, and the process of recognizing the stimuli as something familiar, though memory connections [7].

There are two pathways in the CANS: the ascending (afferent) and the descending (efferent) pathway. The afferent pathway is the path that an impulse, generated in the hair cells, travels along to the auditory cortex), whereas the efferent auditory pathway is a similar path, but in the opposite direction, conducting impulses from the auditory cortex to the hair cells [8].

The afferent and efferent pathways act in an integrated way. The afferent auditory pathway has a bilateral and predominantly contralateral auditory representation. The propagation of auditory information occurs via the cochlear nuclei, superior olivary complex, lateral lemniscus, inferior colliculus, and medial geniculate body up to the auditory area of the temporal lobe in the cerebral cortex. The efferent auditory pathway is composed by the medial and lateral olivocochlear bundles which have anatomical and physiological differences, coordinating the independent function of the two ears [9]. The function of the efferent auditory feedback includes the electrical modulation of the outer hair cells in the cochlea, the reduction of the cochlear nerve action potentials, the protection against noise, the localization of a sound source, the improvement in the detection of sound sources in noisy environments, and the focusing of attention to the incoming acoustic stimuli, which is less effective in patients with tinnitus [9, 10] (**Figure 2**).

It should be emphasized that information from sound stimuli is sent to the brain through both ipsilateral and contralateral pathways, the result of which provides data on signal timing and stimulus intensity. These are passed on to associative areas which process the data in a differentiated way, with the left hemispheric dominance for the processing of language and the right hemisphere for the process of melody. The data on the hemispheric dominance are derived from studies in patients with cortical lesions revealing a loss of recognition of family songs and prosody in patients with right-sided injury and poor recognition of verbal language and symbolic noise in patients with a left lesion [11].

Rees [12] identified auditory perception deficits as one of the causes of language disorders and concluded that auditory abilities seem to play a major role in language and learning. Lubert [13] reported that a deficiency in the ability to detect acoustic characteristics of an auditory signal was overwhelmingly important that affected children never achieved good performance in language tasks.

There is a strong interest in the literature on the impact of auditory processing deficits on language skills and reading [14]. However, the nature of the relationship between auditory and speech processing continues to be debated [15]. The starting point for differentiating auditory processing from that of language consists of a knowledge of acoustic, phonemic, and linguistic characteristics in a behavioral and neurological way. Individuals who have primary deficits in their auditory perceptual

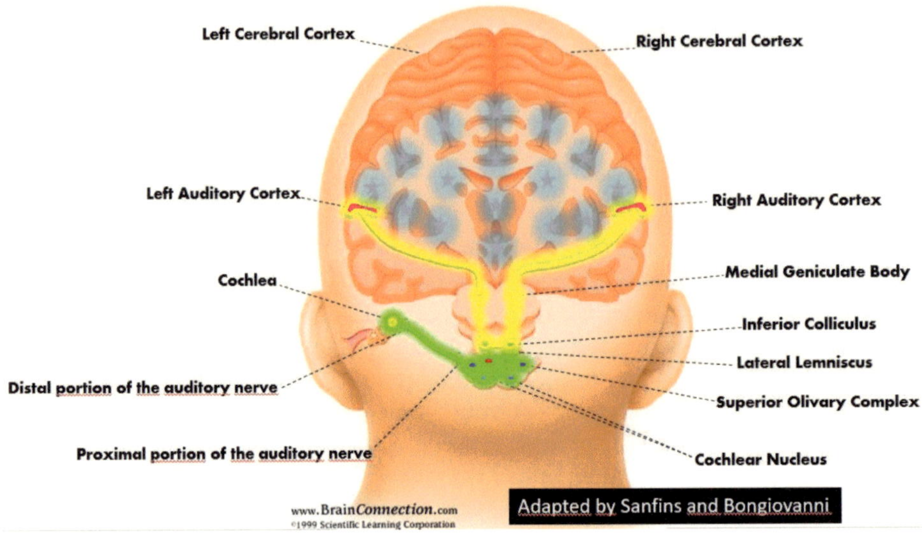

Figure 2.
Schematic representation of the auditory pathways.

abilities therefore have similar symptoms to those who have other pathologies such as dyslexia and attention deficit hyperactivity disorder, and as a consequence they may have attention and executive deficits as well [16].

1.2 Neuroplasticity

The human brain has certain time periods—called critical periods—during which it is conducive to neuroplasticity. In these critical periods, capacities are shaped, perfected, or altered as a result of experience.

In humans, cortical neuroplasticity is most pronounced in the first years of life. During this developmental period, cortical neurons are extensively stimulated, and in this way, synapses mature and developed. In addition, various sensory and cognitive systems interact and adjust their functional properties based on prior experience and learning. Younger brains seem to be more able to change as a result of persisting stimuli [17]. These are usually related to changes (i) in behavior, (ii) in the environment, and (iii) in the neural processes.

Over the course of a lifetime, a lack of experience during critical developmental periods can hinder learning [19]. A critical period can be described by the simultaneous presence of these three conditions:

i. The information must be reliable and extremely precise.

ii. The neural connections need to be intact to be able to process information, either through inhibitory and/or excitatory connections [18].

iii. Mechanisms must be in place to sustain the plasticity process, such as modifications in the morphologies of axons and dendrites and modification to synaptic connections.

It should be emphasized that simple skills require the use of less specialized neural circuits, while more complex abilities depend on the use of more specialized ones. The simplest neural circuits need first to be activated and to be efficient, before new neural circuits can be made reliable [20, 21].

Stimulation of any skill during the critical period of development is an extremely important factor for the success of any intervention process. However, it is important to note that adult brains also have a proven ability to change. Thus, different neuronal systems can be activated regardless of the age of the individual [22, 23].

Activities which depend on the integration of different neuronal systems engage in a multimodal cerebral activation, and thus they can enhance neuroplasticity. A good example of such multisensory stimulation is the process of learning to play music [24–26]. Paraskevopoulos et al. [25] demonstrated that musicians who started their training as young adults had a greater activation of the prefrontal cortex than musicians with only short-term training. Data in the literature suggest that a wide range of beneficial effects can be manifested by elderly musical students, including improvements in attention, memory, motor function, executive function, creativity, anxiety reduction, and visual scanning [27–30].

Intrinsic and extrinsic factors can cause changes in brain cells. Data from the literature suggest that new neurons are present after 6–8 weeks from the time an adult undertakes a new skill [31, 32]. It is therefore suggested that learning and maintaining a new activity should be encouraged in order to activate neural circuits and create new synapses.

Neuroplasticity has been associated with a delayed onset of dementia. Broolmeyer et al. [33] state that brain plasticity should be made a priority in dealing with individuals who have dementia. Concomitantly, age-related cognitive decline can be delayed, interrupted, or even reversed by introducing tasks that involve multimodal neuronal stimulation.

By recognizing the importance of neuroplasticity, professionals involved in rehabilitation are encouraged to turn their efforts toward stimulating, motivating, creating, and developing new strategies for the treatment of their patients.

1.3 Neuroplasticity of the auditory system

Like other systems, the development of the central auditory nervous system depends on a critical period during the first years of life when responses to different stimuli and sound environments are gradually established. In the auditory system the capacity for anatomical and functional modification is called auditory neuroplasticity [34].

The cortical areas that encompass the auditory system develop rapidly in the first years of life, due to an abundance of neuronal connections [35, 36]. At approximately 4 years of age, the neurons responsible for hearing go through a process which is called pruning, where neurons and synapses which are not activated are eliminated from the system [37].

Although the plasticity due to experience is far greater in the first years of life, it is known that the auditory system has some malleability throughout life [37]. Sharma et al. [38] established that there was a difference between what is known as a critical period and a sensitive period. According to Sharma, the critical period ends suddenly, and the neural system is unable to adapt to stimuli; in contrast, the sensitive period is an ideal neuroplastic period during which sound can be introduced into the auditory cortex and promote normal age-appropriate development.

Preterm infants who remain long periods in a neonatal intensive care unit (NICU) are often exposed to high ambient noise levels, generated by the hospital equipment. The high-frequency sounds can cause acoustic trauma and hamper the proper development of the central auditory nervous system [39]. According to Zhang et al., excessive noise at critical periods of development can lead to impaired cortical tonotopic maps, resulting in a reduction in neural synchrony and a decreased sensitivity to particular frequencies [40]. In addition, the extra noise can mask speech sounds, thereby impoverishing the auditory experience. As a result, infants can become more sensitive to noise and focus their attention on this type of sound stimulus instead of ignoring it and focusing on speech [41]. Among preterm infants there is a high rate of impairment of hearing, language, and attention; on the other hand, a home environment rich in post-NICU auditory and linguistic stimuli favors auditory neuroplasticity, meaning that premature infants then have a good chance of developing normal speech, language, and learning [42].

One way to observe plasticity in the auditory system is by monitoring patients undergoing cochlear implantation. Even after a period of auditory deprivation due to hearing loss, it is possible for the brain's auditory system to reorganize and develop better hearing abilities. Research on children implanted at the age of 3, 5, and 7 years has demonstrated that cortical auditory development can be mixed, with some children presenting cortical auditory evoked potential responses (notably P1) within normal limits, while others do not seem to achieve normal central auditory maturity. These findings are consistent with positron emission tomography (PET) imaging tests performed before and after cochlear implantation. It appears that 3.5 years of age is the end of the sensitive period for cochlear implantation in children with congenital deafness; this age is approximately when the observed exponential increase in synaptic density ends and begins to decrease [35]. Beyond 7 years of age, neuroplasticity in the central auditory system is significantly reduced; if new sounds are introduced after this time, the auditory cortex is unable to process auditory information normally [38, 43]. Research on the development of speech and language skills in children has indicated significantly better outcomes in those who received cochlear implants at younger ages [44, 45].

Neuroplasticity can be observed in individuals with central auditory processing disorder (CAPD) who have undergone auditory training. Training is a therapeutic procedure involving auditory stimulation that leads to reorganization (remapping) of the cortex and brainstem, improving synaptic efficiency and increasing neural density. These neurophysiological changes, reflected on behavioral changes, have encouraged the use of this rehabilitation strategy [46–48].

1.4 Auditory training in central auditory processing disorder

It is well established that listeners with CAPD exhibit diverse behaviors such as poor listening skills, difficulty learning through the auditory modality, difficulty following auditory instructions, difficulty in understanding when there is background noise, requesting information to be repeated, poor auditory attention, easily distracted, deficits with phonological awareness and phonic skills, weak auditory memory, delayed response to verbal stimuli, and difficulty with spelling, reading, and learning [49].

A diagnosis of impaired central auditory processing is done by applying a battery of behavioral and electrophysiological procedures. The results provide information about the physiological mechanisms in the auditory system and a profile of abilities that are altered and those that are preserved. Based on this diagnostic information, rehabilitation should start as soon as possible in order to minimize the effects of CAPD on language development. One strategy is the use of auditory training (AT), defined as the set of (acoustic) conditions and/or tasks designed to activate auditory and related systems such that neural connections and the associated auditory behavior is improved [50].

The general aim of AT when applied to individuals with CAPD is to improve auditory skills such as sound localization and lateralization, auditory discrimination, auditory pattern recognition, temporal aspects of audition, and auditory discrimination against competing acoustic signals [51]. Formal and informal AT procedures are conducted by audiologists in the clinical environment. The difference between them is that formal training needs to be acoustically controlled, with a strict control over the stimulus generation and presentation. The combination of formal and informal AT procedures offers a flexible approach which presents positively effective outcomes [50].

The management of CAPD requires a multidisciplinary team, since the pathology commonly appears with other disorders (attention deficit/hyperactivity disorder), learning and language disabilities, or dyslexia. The multidisciplinary team members are often speech-language pathologists, psychologists, neuropsychologists, neuropediatric specialists, teachers and parents, or other specialists involved in the child's overall care [52] (**Figure 3**).

The therapy to enhance auditory skills should be evidence-based, individualized, and segmented into bottom-up and top-down treatments. A bottom-up approach is based on the premise that difficulties in central auditory processing (CAP) lead to impaired auditory perception, language, reading, and communication. The objective of the bottom-up therapy is to improve speech perception. The top-down approach includes auditory cohesion, auditory attention, and metacognitive and metalinguistic activities [52, 53].

The AT program should follow some important principles:

- It should be frequent, challenging, and motivating, using age and language appropriate for the patient.
- It should include diverse tasks to maintain motivation.
- It should be gradual in difficulty over time.
- It should employ a follow-up on acquired responses (achieving response rates >70% is an indication that the task needs to be more demanding).
- It should use monitoring and feedback based on psychophysical, electrophysiological, and questionnaire-based information [50, 52].

The results obtained in the diagnostic battery will guide the therapeutic planning, which should include tasks aimed at discriminating sound intensity, frequency, and duration; phoneme discrimination; time perception discrimination; temporal ordering and sequencing; pattern recognition, location, and lateralization; and recognition of auditory information in the presence of competitive signals. Other aspects may include study of interhemispheric information transfer and binaural listening [51, 54].

In addition, modifications are important depending on the environment. To improve access to auditory information outside the therapy room, teachers and parents also need to help with CAPD treatment strategies. Simple changes may bring many benefits to learning. Options may include:

i. Preferential seating

ii. Addition of visual cues

iii. Clear language

iv. Making frequent checks for understanding

v. Repetition or rephrasing

vi. Multimodality cues and hands-on demonstrations

vii. Pre-teaching of new information and new vocabulary

viii. Provision of a notetaker

ix. Gaining attention prior to speaking

x. Positive reinforcement

xi. Reduce background noise

xii. FM systems

Monitoring progress of the patient is important since it allows the therapist to measure the appropriateness of the AT program and provides a basis for feedback to the patient and parents [50]. Ideally, three types of monitoring should be employed to measure auditory changes: psychophysical, electrophysiological, and questionnaires. These measures should be obtained before and after hearing training. Several questionnaires are available and can be answered by the patient and/or individuals interacting with him or her, such as parents, teachers, and other professionals.

Several questionnaires are described in the literature, such as the Children's Auditory Performance Scale (CHAPS) [55, 56], Screening Instrument for Targeting Educational Risk (SIFTER) [56, 57], Children's Home Inventory of Listening Difficulties (CHILD) [58], and the Scale of Auditory Behaviors (SAB) [58].

A large number of studies provide definitive evidence for the plasticity of the auditory system evidenced by behavioral changes in both animals [59–61] and in humans [62–68]. A recent study by Donadon et al. [69], whose objective was to investigate auditory training in children and adolescents suffering from otitis media

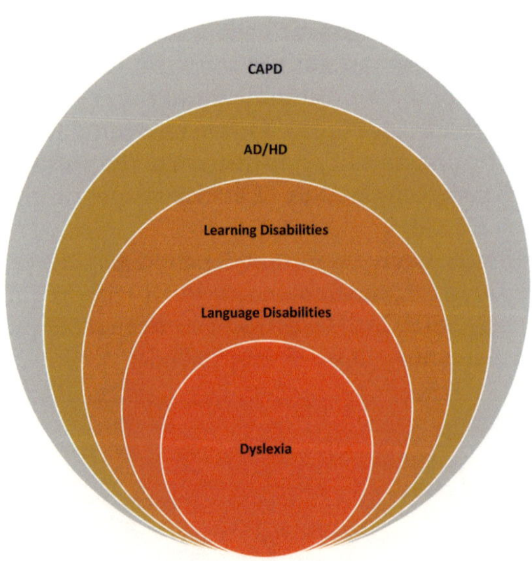

Figure 3.
CAPD and associated pathologies.

with a documented history of bilateral ventilation tube insertion, highlighted some aspects of auditory neuroplasticity. According to the data from the study, the participants were randomly divided into two groups: (i) auditory training and (ii) visual training. In the behavioral tests during the pre-intervention evaluation, no statistical differences were detected. However, after the auditory training program, there was an improvement in the subjects' performance for auditory abilities. In addition, comparing the two types of intervention (visual vs. auditory), the behavioral tests revealed better responses to the post-intervention auditory training. The results, assessed through behavioral tests on subjects with a history of bilateral otitis media, suggest that auditory training provided beneficial gains for all auditory abilities.

2. Conclusion

The central auditory nervous system is responsible for the processing of auditory information. It is highly complex and plastic, being able to reorganize itself in response to auditory stimulation. Auditory training promotes behavioral and electrophysiological changes due to the neurophysiology of the brain's plasticity. The latter enables the positive performance of the auditory training, which is an important rehabilitation strategy for individuals with central auditory processing disorders.

Abbreviations

AD/HD	attention deficit/hyperactivity disorder
AT	auditory training
CANS	central auditory nervous system
CAP	central auditory processing
CAPD	central auditory processing disorder
CNS	central nervous system
critical period	the time during which the neural system is unable to adapt
sensitive period	the ideal period for neuroplasticity to occur
Wada test	a test for determining the dominant hemisphere for speech

Author details

Caroline Donadon[1], Milaine Dominici Sanfins[2*], Piotr Henry Skarzynski[3,4,5], Maria Francisca Colella dos Santos[6] and Stavros Hatzopoulos[7]

1 Program in Child and Adolescent Health, Faculty of Medical Sciences, University of Campinas, Campinas, Brazil

2 Centro de Eletrofisiologia e Neuroaudiologia Avançada (CENA), São Paulo, Brazil

3 Department of Teleaudiology and Hearing Screening, World Hearing Center, Institute of Physiology and Pathology of Hearing, Warsaw/Kajetany, Poland

4 Department of Heart Failure and Cardiac Rehabilitation, Warsaw, Poland

5 Institute of Sensory Organs, Warsaw, Poland

6 Department of Human Development and Rehabilitation, State University of Campinas, Campinas, Brazil

7 Clinic of Audiology and ENT, University of Ferrara, Ferrara, Italy

*Address all correspondence to: msanfins@uol.com.br

References

[1] Donadon C, Dominici Sanfins M, Skarzynski PH, Colella dos Santos MF, Hatzopoulos S. Neuroplasticity and the auditory system. In: Hatzopoulos S, Ciorba A, Skarzynski PH, editors. The Human Auditory System – Basic Features and Updates on Audiological Diagnosis and Therapy. London: IntechOpen; 2019. DOI: 10.5772/intechopen. 90085

[2] Kandel E, Schwartz J, Jessell T. Fundamentos da Neurociência e do Comportamento. Rio de Janeiro: Guanabara Koogan S.A; 2000

[3] Sperry R. Cerebral organization and behavior. The American Association for the Advancement of Science. Science. 1960;**133**:1749-1757

[4] Rasmussen T, Milner B. The role of early left-brain injury in determining lateralization of cerebral speech functions. Annals of the New York Academy of Sciences. 1977;**30**:355-369

[5] Finger S. Origins of Neuroscience: A History of Explorations into Brain Function. New York: Oxford University Press; 1994

[6] Bellis T. Differential diagnosis of (central) auditory processing disorder in older listeners. In: Musiek F, Chermak G, editors. Handbook of (Central) Auditory Processing Disorder: Auditory Neuroscience and Diagnosis. San Diego: Plural Publishing; 2007. pp. 319-346

[7] Pickles J. An Introduction to the Physiology of Hearing. Berkeley: Academic Press; 1988

[8] Berne A, Delrieu G, Creutin J, Obled C. Temporal and spatial resolution of rainfall measurements required for urban hydrology. Journal of Hydrology. 2004;**299**:166-179

[9] Guinann J. Olivocochlear efferents: Anatomy, physiology, function, and the measurement of efferent effects in humans. Ear and Hearing. 2006;**27**(6):589-607

[10] Breuel M, Sanchez T, Bento R. Vias auditivas eferentes e seu papel no sistema auditivo. Arquivos de Otorrinolaringologia. 2001;**5**(2):62-67

[11] Musiek F, Reeves A. Asymmetries of the auditory areas of the cerebrum. Journal of the American Academic of Audiology. 1990:240-245

[12] Rees N. Auditory processing factors in language disorders: The view from procrustes'bed. Journal of Speech and Hearing Disorders. 1973;**38**:304-315

[13] Lubert N. Auditory perceptual impairments in children with specific language disorders: A review of the literature. The Journal of Speech and Hearing Disorders. 1981;**46**(1):1-9

[14] Sharma M, Purdy S, Kelly A. Comorbidity of auditory processing, language, and reading disorders. Journal of Speech, Language, and Hearing Research. 2009:706-722

[15] Richard G. The role of the speech-language pathologist in identifying and treating children with auditory processing disorder. Language, Speech, and Hearing Services in Schools. 2011;**42**(3):241-245

[16] Richard G. Language processing versus auditory processing. In: Geffner D, Ross-Swain D, editors. Auditory Processing Disorders: Assessment, Management and Treatment. 2a ed. San Diego: Plural Publishing; 2013

[17] Kennard M. Reorganization of motor function in the cerebral cortex of monkeys deprived of motor and premotor areas in infancy. Journal of Neurophysiology. 1938;**1**:477-496

[18] Fagiolini M, Hensch T. Inhibitory threshold for critical-period activation in primary visual cortex. Nature. 2000;**404**:183-186

[19] Immelmann K. Sexual imprinting in birds. Advances in the Study of Behavior. 1972;**4**:147-174

[20] Jones E. Cortical and subcortical contributions to activity-dependent plasticity in primate somatosensory cortex. Annual Review of Neuroscience. 2000;**23**:1-37

[21] Daw N. Critical periods and strabismus: What questions remain? Optometry and Vision Science. 1997;**74**:690-694

[22] Irvine D, Fallon J, Kamke M. Plasticity in the adult central auditory system. Acoustics Australia. 2006;**34**:13-17

[23] Mecklenburg D, Babighian G. Cochlear performance as an indicator of auditory plasticity in humans. In: Salvi R, Henderson D, Fiorino F, Colletti V, editors. Auditory System Plasticity and Regeneration. New York: Tieme Medical; 1996. pp. 395-404

[24] Kuchenbuch A, Paraskevopoulos E, Herholz SC, Pantev C. Audio-tactile integration and the influence of musical training. PLoS One. 2014;**9**(1):e85743

[25] Paraskevopoulos E, Kuchenbuch A, Herholz SC, Pantev C. Multisensory integration during short-term music reading training enhances both uni- and multisensory cortical processing. Journal of Cognitive Neuroscience. 2014;**26**(10):2224-2238

[26] Pantev C, Paraskevopoulos E, Kuchenbuch A, Lu Y, Herholz SC. Musical expertise is related to neuroplastic changes of multisensory nature within the auditory cortex. The European Journal of Neuroscience. 2015;**41**(5):709-717

[27] Benz S, Sellaro R, Hommel B, Colzato LS. Music makes the world go round: The impact of musical training on non-musical cognitive functions—A review. Frontiers in Psychology. 2015;**6**:2023

[28] Hars M, Herrmann F, Gold G, Rizzoli R, Trombetti A. Effect of music-based multitask training on cognition and mood in older adults. Age and Ageing. 2014;**43**:196-200

[29] Bottiroli S, Rossi A, Russo R, T V, Cavallini E. The cognitive effects of listening to background music on older adults: Processing speed improves with upbeat music, while memory seems to benefit from both upbeat and downbeat music. Frontiers in Aging Neuroscience. 2014;**6**:284-291

[30] Seinfeld S, Figueroa H, Ortiz-Gil J, Sanchez-Vives MV. Effects of music learning and piano practice on cognitive function, mood and quality of life in older adults. Frontiers in Psychology. 2013;**4**:810

[31] Ge S, Yang C, Hsu K, Ming G, Song H. A critical period for enhanced synaptic plasticity in newly generated neurons of the adult brain. Neuron. 2007;**54**:559-566

[32] Dranovsky A, Picchini A, Moadel T, Sisti A, Yamada A, Kimura S, et al. Experience dictates stem cell fate in the adult hippocampus. Neuron. 2011;**70**:908-923

[33] Brookmeyer R, Johnson E, Ziegler-Graham K, HM A. Forecasting the global burden of Alzheimer's disease. Alzheimer's & Dementia. 2007;**3**:186-191

[34] Pascual-Leone A, Amedi A, Fregni F, Merabet L. The plastic human brain cortex. Annual Review of Neuroscience. 2005;**28**:377-401

[35] Huttenlocher P, Dabholkar A. Regional differences in synaptogenesis in human cerebral cortex. Journal of Comparative Neurology. 1997;**387**:167-178

[36] Huttenlocher P, de Courten C. The development of synapses in striate cortex of man. Human Neurobiology. 1987;**6**:1-9

[37] Sanes D, Woolley S. A behavioral framework to guide research on central auditory development and plasticity. Neuron. 2011;**72**:912-929

[38] Sharma A, Glick H, Campbell J, Torres J, Dorman M, Zeitler DM. Cortical plasticity and reorganization in pediatric single-sided deafness pre- and postcochlear implantation: A case study. Otology & Neurotology. 2016;**37**(2):e26-e34

[39] Lahav A, Skoe E. An acoustic gap between the NICU and womb: A potential risk for compromised neuroplasticity of the auditory system in preterm infants. Frontiers in Neuroscience. 2014;**8**:1-8

[40] Zhang L, Bao S, Merzenich M. Persistent and specific influences of early acoustic environments on primary auditory cortex. Nature Neuroscience. 2001;**4**:1123-1130

[41] Skoe E, Chandrasekaran B. The layering of auditory experiences in driving experience-dependent subcortical plasticity. Hearing Research. 2014;**311**:36-48

[42] Skoe E, Krizman J, Kraus N. The impoverished brain: Disparities in maternal education affect the neural response to sound. The Journal of Neuroscience. 2013;**33**(44):17221-17231

[43] Sharma A, Glick H. Cross-modal re-organization in clinical populations with hearing loss. Brain Sciences. 2016;**6**(1)

[44] Geers AE, Nicholas J, Tobey E, Davidson L. Persistent language delay versus late language emergence in children with early cochlear implantation. Journal of Speech, Language, and Hearing Research. 2016;**59**(1):155-170

[45] Holt R, Svirsky M. An exploratory look at pediatric cochlear implantation: Is earliest always best? Ear and Hearing. 2008;**29**:492-511

[46] Knudsen E. Capacity of plasticity in the adult owl auditory system expanded by juvenile experience. Science. 1998;**279**:1531-1533

[47] de Boer J, Thornton ARD. Neural correlates of perceptual learning in the auditory brainstem: Efferent activity predicts and reflects improvement at a speech-in-noise discrimination task. The Journal of Neuroscience. 2008;**28**:4929-4937

[48] Johnson KL, Nicol T, Zecker SG, Kraus N. Developmental plasticity in the human auditory brainstem. The Journal of Neuroscience. 2008;**28**:4000-4007

[49] Keith RW. Random Gap Detection Test. Missouri: Auditec of Saint Louis; 2000

[50] Musiek FE, Chermak GD, Weihing J. Auditory training. In: Chermak GD, Musiek FE, editors. Handbook of Central Auditory Processing Disorder—Comprehensive Intervention. San Diego: Plural Publishing; 2014

[51] American Speech-Language-Hearing Association. (Central) Auditory Processing Disorders [Technical Report]. Available at: www.asha.org/policy.2005

[52] Bellis TJ. Assessment and Management of Central Auditory Processing Disorders in the Educational Setting from Science to Practice. Plural Publishing; 2011

[53] Chermak GD, Musiek FE. Central Auditory Processing Disorders: New Perspectives. San Diego: Singular Publishing Group; 1997. 374p

[54] AAA (American Academy of Audiology). Guidelines for the

diagnosis, treatment and management of children and adults with central auditory processing disorder. Retrieved from: http://www.audiology.or/resources/documentlibrary/documents/CAPD%20guidelines%208-2010.pdf

[55] Lam E, Sanchez L. Evaluation of screening instruments for auditory processing disorder (APD) in a sample of referred children. The Australian and New Zealand Journal of Audiology. 2007;**29**:26-39

[56] Wilson W, Jackson A, Pener A, Rose C, Wilson J, Heine C, et al. The CHAPS, SIFTER and TAPS-R as predictors of (C)AP skills and (C)APD. Journal of Speech, Language and Hearing Research. 2011;**54**(2):278-291

[57] Emanuel DC. The auditory processing battery: Survey of common practices. Journal of the American Academy of Audiology. 2002;**13**(2):93-117

[58] Schow RL, Seikel JA. Screening for (central) auditory processing disorder. In: Chermak G, Musiek F, editors. Handbook of (Central) Auditory Processing Disorder: Auditory Neuroscience and Diagnosis. San Diego, CA: Plural Pub; 2006. pp. 137-161

[59] Likenhoker B, Knudsen E. Incremental training increases the plasticity of the auditory space map in adult barn owls. Nature. 2002;**419**:293-296

[60] von der Likenhoker E, Ohe C, Knudsen E. Anatomical traces of juvenile learning in the auditory system of adult barn owls. Nature Neuroscience. 2004;**8**:93-98

[61] Recanzone G, Schreiner C, Merzenich M. Plasticity in the frequency representation of primary auditory cortex following discrimination training in adult owl monkeys. Journal of Neuroscience. 1993;**13**(97-103)

[62] Hayes E, Warrier C, Nicol T, Zecker S, Kraus N. Neural plasticity following auditory training in children with learning problems. Clinical Neurophysiology. 2002;**114**:673-684

[63] Bao S, Chang E, Woods J, Merzenich M. Temporal plasticity in the primary cortex induced by operant perceptual learning. Nature Neuroscience. 2004;**7**:974-981

[64] Musiek FE. Temporal (auditory) training for (C)APD. Hearing Journal. 2005;**58**:46

[65] Alonso R, Schochat E. The efficacy of formal auditory training in children with (central) auditory processing disorder: Behavioral and electrophysiological evaluation. Brazilian Journal of Otorhinolaryngology. 2009;**75**:726-732

[66] Kraus N, Chandrasekaran B. Music training for the development of auditory skills. Nature Review Neuroscience. 2010;**11**:599-605

[67] Barcroft J, Sommers M, Tye-Murray N, Mauze E, Schroy C, Spehar B. Tailoring auditory training to patient needs with single and multiple talkers: Transfer appropriate gains on a four-choice discrimination test. International Journal of Audiology. 2011;**50**:802-808

[68] Sharma M, Purdy SC, Kelly AS. A randomized control trial of interventions in school-aged children with auditory processing disorders. International Journal of Audiology. 2012;**51**:506-518

[69] Donadon C, Sanfins MD, Borges LR, Colella-Santos MF. Auditory training: Effects on auditory abilities in children with history of otitis media. International Journal of Pediatric Otorhinolaryngology. 2019;**118**:177-180

Chapter 4

Otitis Media, Behavioral and Electrophysiological Tests, and Auditory Rehabilitation

Milaine Dominici Sanfins, Piotr Henryk Skarzynski and Maria Francisca Colella-Santos

Abstract

For speech and language to develop, an intact and active auditory system is of fundamental importance. The central auditory nervous system (CANS) can be hampered by several occurrences, including otitis media (OM) originating from inflammation in the middle ear and which is often associated with the accumulation of infected (or sometimes noninfected) fluid. OM can have a diffuse effect on cognitive and linguistic abilities, affecting both speech and phoneme perception through a failure to discriminate, store, and reproduce the acoustic contrasts necessary for comprehension. It is especially common in the first years of school. In addition, OM can generate internal noise from the presence of middle ear fluid near the cochlea, which can lead to changes in speech perception, distortion in acoustic images, and a reduction in the speed and accuracy of decoding speech. Evaluating the effectiveness of the CANS is recommended in cases where there have been repeated episodes of OM. Very useful information can be gained from behavioral and electrophysiological tests. The tests allow functional diagnoses to be made and can also reveal clinical and subclinical changes. In this way, they allow information to be collected, which can help in making a prognosis and planning intervention strategies.

Keywords: otitis media, auditory processing, electrophysiology, frequency following response, evoked auditory potential; long latency auditory evoked potential

1. Introduction

Otitis media (OM) is a common childhood disease. Research has shown that recurrent episodes can induce changes or delay the development of the central auditory nervous system, leading to central auditory processing disorder (CAPD). In this chapter, we present results obtained in the behavioral and electrophysiological evaluation of the auditory processing of children and adolescents with OM over the first few years of life. In addition, we discuss aspects of the auditory rehabilitation process itself.

2. Auditory system and otitis media

Language plays an essential role in perceptual organization, including the reception and structuring of information, learning, and social interactions. Language enables us to communicate with each other and acquire and transmit experience and knowledge. The development of speech and language requires a functional auditory system capable of detecting sound, paying attention, remembering, discriminating, and perceiving location. Any interruption to development will lead to significant functional impairments, not only in language but also in cognitive, intellectual, cultural, and social development [1, 2].

Central auditory processing (CAP) is defined as the efficiency and effectiveness with which the central auditory nervous system uses auditory information. It refers to the perceptual processing of auditory information and to the neurobiological activity underlying this processing that gives rise to electrophysiological auditory potentials [3, 4]. The efficient analysis and interpretation of normal auditory information involves several subprocesses and skills, and includes neural mechanisms underlying a range of auditory behaviors such as sound localization and lateralization; auditory discrimination; recognition of auditory patterns; temporal aspects of the hearing (integration, discrimination, resolution, temporal masking); auditory performance in the presence of competing acoustic signals (which includes dichotic listening); and decoding degraded acoustic signals [5, 6].

This whole process involves a complex system of neurons located in several stations of the auditory system. The initial analysis of the stimulus occurs in the peripheral auditory system, constituted by the external and middle ear, responsible for the capture, transduction, and processing of the sound stimulus. The stimulus arrives first at the cochlear nucleus and encephalic trunk, followed by the upper olivary complex, lateral lemniscus, inferior colliculus, and medial geniculate body, and finally reaches the primary area of auditory reception in the temporal lobe of each hemisphere. From the primary auditory cortex of each hemisphere, the signals travel to other regions of the brain-the association areas-both in the same hemisphere and in the opposite hemisphere. As the auditory information travels by ipsi- and contralateral routes, it undergoes increasingly complex levels of processing. This processing occurs both hierarchically and serially, as well as in parallel or overlapping. The result of combining serial and parallel processing makes the system highly efficient and redundant. In addition to ascending pathways, there are also descending pathways that can moderate the response to a received acoustic stimulus [7–8].

Central auditory processing disorder (CAPD) is a dysfunction of the central auditory nervous system that leads to hearing difficulties. It can lead to, or be associated with, changes in language, learning, cognition, or other communicative functions [3–5, 9]. In the pediatric population, there are several possible causes of the disorder, among them otitis media [10, 11].

Otitis media with effusion (OME) is a clinical entity characterized by the presence of effusion in the middle ear, without perforation of the tympanic membrane, but with an acute infection that lasts for a period of at least 3 months. The condition is common enough to be called an "occupational hazard of early childhood" [12] because about 90% of children have OM before school age and they develop, on average, four episodes of OM per year. OM may occur during an upper respiratory infection or occur spontaneously because of poor Eustachian tube function or an inflammatory response following a previous OM, most often between the ages of 6 months and 4 years [13, 14]. In the first year of life, 50% of children will experience OM, increasing to 60% by age 2. When primary school children aged 5–6 years were screened for OM, about 1 in 8 was found to have fluid in one or both ears [15] **Figure 1a–d**.

Most episodes of OM resolve spontaneously within 3 months, but about 30–40% of children have repeated OM episodes and 5–10% of episodes last 1 year [13]. At least 25% of OM episodes persist for 3 months and may be associated with hearing loss which is usually noticed by parents or teachers as inattention, needing to ask several times, disinterest, and poor school achievement.

OM impairs sound transmission to the inner ear by reducing mobility of the tympanic membrane and ossicles, thereby reflecting acoustic energy back into the ear canal instead of allowing it to pass freely to the cochlea.

Diagnosis is performed by otoscopy and confirmed by a basic audiological evaluation. Under otoscopy, a retracted, opaque tympanic membrane with reduced mobility is seen. In the vast majority of cases, a yellowish liquid line, sometimes with air bubbles, is visible through the tympanic membrane. In the audiological evaluation, the result can range from normal hearing to moderate conductive hearing loss (HL of 0–55 dB) [16]. The mean hearing loss associated with OM in children is 28 dB, while a lesser proportion (~20%) exceeds 35 dB, with a type B tympanometric curve characteristic of effusion. Auditory losses are characterized by being fluctuating, temporary, and asymmetrical [17]. The mild degree of loss is sufficient to impair certain auditory functions, and the fluctuating nature (which may change to periods of normal hearing) leads to variable stimulation of the central auditory nervous system. The effect is to make it difficult to perceive sounds, and leads to diffuse cognitive and linguistic abilities affecting both speech and the perception of phonemes; school performance also suffers [18]. In addition, the fluid in the middle ear can cause noise near the cochlea, producing a distorted perception of sounds.

Depending on the clinical history and functional conditions of the child's middle ear, treatment involves either clinical or surgical management. In small children with OME, the most common surgical procedure is tympanotomy with ventilation tube placement, which drains fluid from the middle ear and thus restores hearing. Diagnosis and treatment is essential, since in an acute episode of OM fluids can remain in the middle ear for 3–12 months; in 10–30% of children, the fluid remains for 2–3 months. Thus, a child who has had three to four OME episodes before the

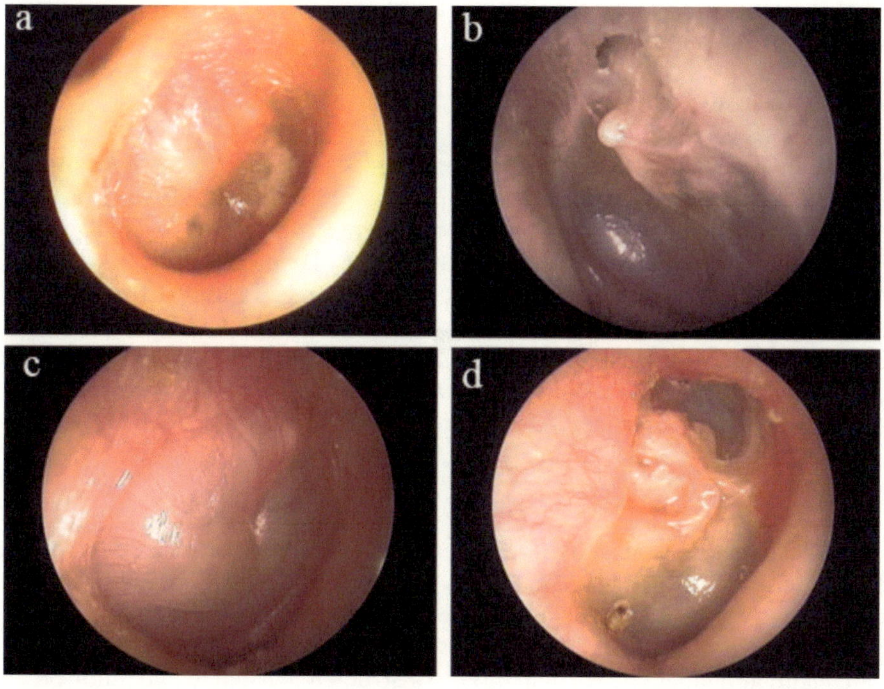

Figure 1
(a–d) Otitis media with effusion (OME). Personal collection.

age of three can have had 12 months of conductive hearing loss, which is a third of the period considered critical for development and learning [19]. The periods of auditory deprivation during the active periods of OME over the first years of life can delay the maturation of the structures in the CANS and consequently impair auditory abilities associated with central auditory processing.

Therefore, evaluation of auditory processing is fundamental in children with a history of otitis media in order to allow diagnosis, intervention, and guidance.

3. Testing the central auditory processing of children with a history of otitis media

To evaluate central auditory processing in children with a history of OM, it is recommended that a battery of test procedures be used by which the mechanisms and auditory abilities involved in the analysis and interpretation of sounds can be investigated. Due to the complexity of CANS, no single test is sufficient to explore its nature [3, 4]. Since the 1950s, numerous tests have been developed to evaluate central hearing function. These tests differ in that each presents different types of stimuli (verbal or nonverbal) and involves presentation to one or both ears (monaural or binaural). Each test is designed to evaluate a particular auditory mechanism or auditory ability and consequently probes different areas and functions of the CANS. Below the tests are divided into categories according to the way in which the stimuli are presented to the ears, the nature of the auditory tasks involved, and the method or approach used. Other currently accepted classifications involve categorizing them as binaural interaction tests, dichotic tests with verbal and non-verbal sounds (binaural integration and separation), monaural tests using low redundancy stimuli, time processing tests, and electroacoustic and electrophysiological procedures [20].

A comprehensive assessment allows for correct quantification and qualification of the various CANS mechanisms and dysfunctions and provides important information for planning and managing treatment.

3.1 Behavioral evaluation

Research by Colella-Santos et al. [11] involved 50 children (28 boys, 22 girls, mean age 11.2 years) with a documented history of bilateral SOM in the first 6 years of life and who had bilateral tympanostomy tube insertion (experimental group, EG); a control group (CG) consisted of 40 children (17 boys, 23 girls, mean age 10.7 years) with no history of otitis media. All children had auditory thresholds within normal limits on the day of evaluation and had a type A tympanometric curve. They were all evaluated with the tests described below [21–23]. The tests were the dichotic digits test, synthetic sentence identification test, gaps-in-noise test, and frequency pattern test. Details are as follows **Figure 2a** and **b**.

3.1.1 Dichotic digits (DD)

The DD test as developed in Brazil consists of four presentations of a list of two-syllable digits in Brazilian Portuguese, in which four different digits are presented simultaneously, two in each ear. The list contains 40 randomly arranged pairs of digits presented at 50 dB HL. The digits used to form the numbers are four, five, seven, eight, and nine. The participants are instructed to listen to two numbers in each ear and repeat all the numbers they hear. The order does not matter. The dichotic digits test verifies binaural integration ability [21].

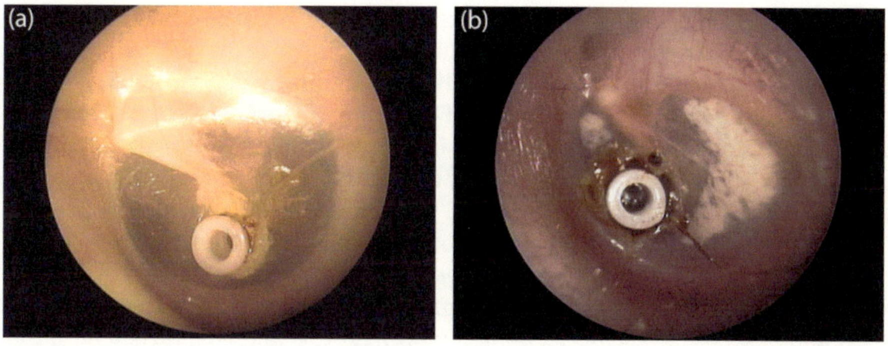

Figure 2.
(a–b) Tympanostomy tube insertion. Personal collection.

3.1.2 Synthetic sentence identification (SSI)

The SSI test consists of the presentation of 10 Brazilian Portuguese sentences at 40 dB HL, in the presence of a competing children's story in the same ear at a signal-to-noise ratio of 0, −10, or −15 dB. The task of the subject is to listen to the sentence and point to it in a frame. The ability analyzed in this test is figure-ground discrimination [21].

3.1.3 Frequency pattern test (FPT)

The FPT test is composed of three 150 ms tones presented at 50 dB HL and separated by 200 ms. The tones in each triplet are combinations of two sinusoids, 880 and 1122 Hz, which are designated as low frequency (L) and high frequency (H), respectively. Thus, there are six possible combinations of the three-tone sequence (LLH, LHL, LHH, HLH, HLL, and HHL). The subjects are instructed that they will hear sets of three consecutive tones that vary in pitch. Their task is to repeat the pattern by humming and verbalizing the frequency pattern (e.g., high–low–high). The FPT test checks temporal ordering ability [22].

3.1.4 Gaps-in-noise (GIN)

The GIN test consists of a series of 6-second segments of broad-band noise presented at 50 dB HL with 0–3 gaps embedded within each segment. The gaps vary in duration from 2 to 20 ms. The gap-detection threshold is defined as the shortest duration that is correctly identified at least four out of six times. The participants are instructed to indicate each time they perceive a gap. The GIN test measures temporal resolution ability [23].

To establish a difference between the right and left ears of subjects in the EG, it was necessary that there was a statistically significant difference in both the Dichotic Digits (p = 0.001) and GIN (p = 0.004) tests. No significant difference was found for gender in the behavioral tests. It was observed that the EG had lower mean responses than the CG for the DD test of approximately 5% in both ears; for the FPT 9.6% (humming) and 30% (naming); and 8% for the SSI test. For the GIN test, there was a statistically significant difference in the gap-detection threshold between the groups, with the highest threshold obtained in the EG compared to the CG (the higher the threshold, the worse the performance).

In summary, there was a negative effect of OM on the auditory skills of figure-background discrimination, resolution, and temporal ordering. The poorer results in CAP behavioral tests in the EG participants can be explained by the fact that OM, by generating a fluctuating auditory threshold and causing temporary auditory deprivation, hampers the maturation of auditory abilities (such as binaural integration, resolution, temporal ordering, and discrimination) which are fundamental for understanding speech. During this period of auditory deprivation due to episodic OM, the CANS received inconsistent and incomplete auditory information. That is, the period between clinical assessment and the decision to perform surgery may have been too long **Table 1**.

Recent research has demonstrated associations similar to those found in the present study. Borges et al. [11] studied the effect of OM in 69 children of different socioeconomic levels who underwent surgical intervention (insertion of ventilation tubes) and observed worse performance in both the DD and GIN tests. The authors concluded that a history of OM can lead to changes in central auditory functioning, regardless of socioeconomic status.

Khavarghazalani et al. [24] evaluated 12 children with a history of OM who had undergone surgical intervention for insertion of ventilation tubes and found worse performance in the DD and GIN responses than in normals.

Gravel and Wallace [25] also found a significant increase in signal-to-noise ratio in a prospective study of children with a history of OM. There was worse performance on the SSI test (responsible for the figure-ground ability) in the OM group.

Tomlin and Rance [26] recommend that children with a history of OM undergo an evaluation of spatial processing upon entering school. They studied 35 children with a history of chronic OM and found a statistically worse performance compared to the control group in the listening in spatialized noise-sentences test (LISN-S). They concluded that these children have altered spatial processing, difficulty in focusing attention on the relevant stimulus, and difficulty in simultaneously suppressing competing stimuli coming from other directions. It is hypothesized that fluctuating access to binaural cues, caused by OM, may negatively affect the development of spatial processing in the CANS.

3.2 Electrophysiological evaluation

Auditory evoked potentials are an extremely useful instrument for the study of auditory perception and its disorders, especially when a range of stimuli are used [27].

Procedure	Ear	Control group			Experimental group			
		N	Σ (%)	SD	N	Σ (%)	SD	*p*-value
DD	R	40	98.93	1.86	50	95.40	5.16	<0.001
	L	40	97.93	4.15	50	92.55	7.95	<0.001
FPT								
Humming	B	80*	73.50	21.2	100*	42.7	22.2	<0.001
Verbalizing	B	80*	73.50	21.2	100*	42.7	22.2	<0.001
SSI	B	80*	67.5	13.9	100*	59.8	16.9	0.020
GIN	R	40	4.65	1.00	50	6.22	1.40	<0.001
	L	40	4.72	1.06	50	6.56	1.52	<0.001

*n = number, * = number of ears, B = both, R = right, L = left; Σ = mean, SD = standard deviation, DD = dichotic digits, SSI = synthetic sentence identification, FPT = frequency pattern test, GIN = gaps-in-noise.*

Table 1.
Behavioral evaluation values of central auditory processing between control and experimental groups.

3.2.1 Click ABR

In the literature, there are contradictory results in Click ABR responses in individuals with a history of OM. Chambers et al. [28] and Folsom et al. [29] identified an increase in the latency of waves III and V in a group of children with a history of OM, whereas Shaffer [30] did not find a statistically significant difference in Click ABR responses in individuals with and without a history of OM. The majority of studies relating Click ABR results with OM history have investigated latency values; however, Maruthy and Mannarukrishnaiah [31] found a reduction in the amplitude of waves I and III. Sanfins et al. [32] observed statistically significant differences in the absolute latencies of waves I and V as well as in the amplitude of waves III and V from children with a history of bilateral OME compared to their healthy peers. Colella-Santos et al. [11] reported a significant increase in the absolute latency of wave III associated with a decrease in amplitude in children with bilateral OME. Finally, Sanfins [33] reported alterations in the values of waves III and V for both groups of children with a history of OME, seeing both bilateral and unilateral alterations **Figure 3**.

In animals, the effect of conductive hearing loss on CANS was studied by unilaterally removing the malleus and applying a fluid to simulate OM [34], finding a decrease in neuronal activity due to changes in various structures (wave III), upper olivary complex (wave IV), and lateral lemniscus (wave V). At the same time, based on the results of Maruthy and Mannarukrishnaiah [31], it has been suggested that the auditory nerve and cochlear nuclei are more susceptible to modifications after OM infection.

Sanfins et al. [32] suggest that different modifications may occur in CANS structures depending on the unilaterality or bilaterality of the infection. In episodes of bilateral OME, the latency values indicated that the auditory nerve (wave I, wave III) and the lateral lemniscus (wave V) were affected, whereas in unilateral OME, the cochlear nuclei (wave III) was affected. However, when the amplitudes were analyzed, the structures involved were the cochlear nuclei (wave III) and the lateral lemniscus (wave V), both for children with unilateral and bilateral involvement. It should be noted that when evaluating click ABR, the amplitude values show greater variability than the latencies. It is important to emphasize that a unilateral OM may not provide a better performance in the processing of auditory information than bilateral OM. The use of only one ear can lead to damage to the functionality of

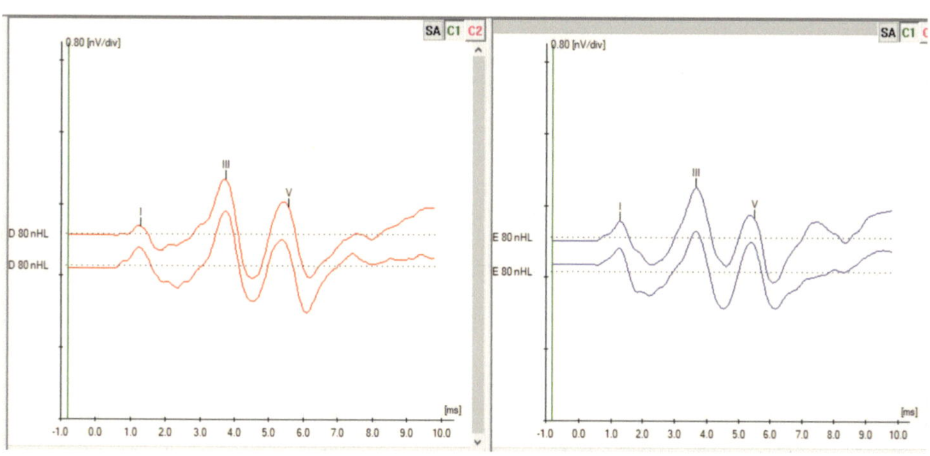

Figure 3.
Click ABR. Personal collection.

the CANS and, over time, activities that depend on binaural auditory processing (binaural interaction and binaural integration, among others) can be compromised due to the auditory imbalance arising from OM.

3.2.2 Frequency following response

Few studies have investigated the frequency following response (FFR) in cases of otitis media. A study of two groups of children with a history of bilateral OM (recent onset and long-term) showed that FFR responses were affected in a statistically significant way in the onset portions (waves V and A) and offset portion (wave O), along with reduced values of the VA complex (more specifically VA slope) when responses between the groups were compared. The findings suggest that long-term OM in children is associated with a reduced neural conduction velocity relative to the processing time of speech stimuli, either at the beginning (onset) or final portion (offset), resulting in a decrease in the coding of speech in the brainstem [35] **Figure 4**.

Sanfins et al. [32] reported that children with a history of SOM present an increase in the absolute latency of all FFR waves compared to children with no history of otological problems. In addition, children without hearing loss have more coherent responses in both ears, whereas the group of children with a history of OME has a greater dispersion of latencies in all FFR components (**Figure 5**). Colella-Santos et al. [11] also reported a decrease in VA slope in girls with OME.

3.2.3 Long latency auditory evoked potential

3.2.3.1 Tone burst

The literature reports alterations in the components of the LLAEP in children with language disorders and also in those with phonological disorders [36] changes that are frequently associated with problems arising from OM. Researchers note that OM can lead to changes in central auditory pathways [30, 37, 38]. However, there are few studies that have associated the LLAEP responses in children with a history of OM **Figure 6**.

Maruthy and Mannarukrishnaiah [31], Shaffer [30], Sanfins [33], and Colella-Santos [11] reported similar results, i.e., the presence of LLAEP changes in children with a history of OME. In the studies by Maruthy and Mannarukrishnaiah [32], all components of the LLAEP (P1, N1, P2, and N2) were significantly longer in

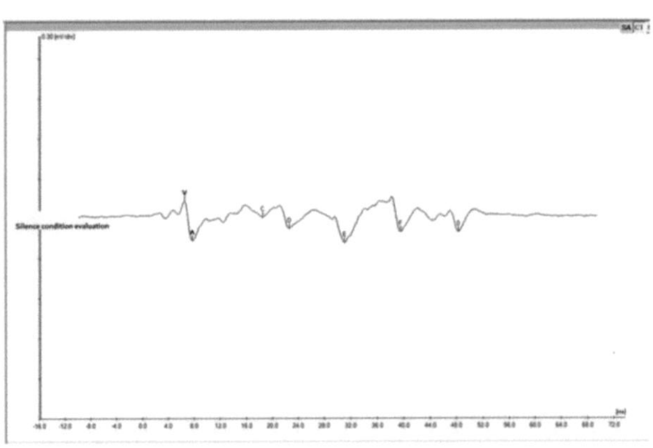

Figure 4.
FFR. Personal collection.

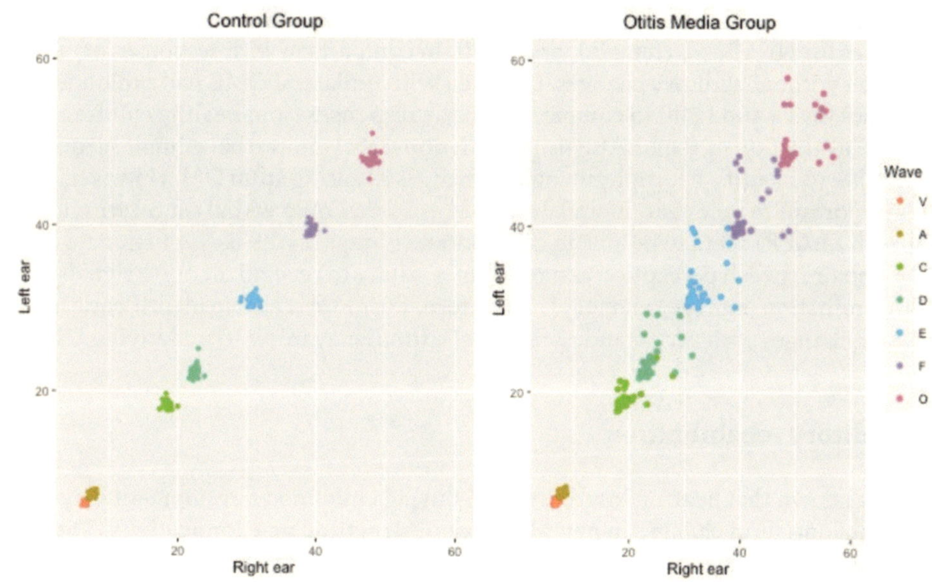

Figure 5.
Comparison (left vs. right ear) of absolute latency values of FFR components in children with a history of otitis media (right panel) and children with no history of otitis media (left panel), from Sanfins et al. [32].

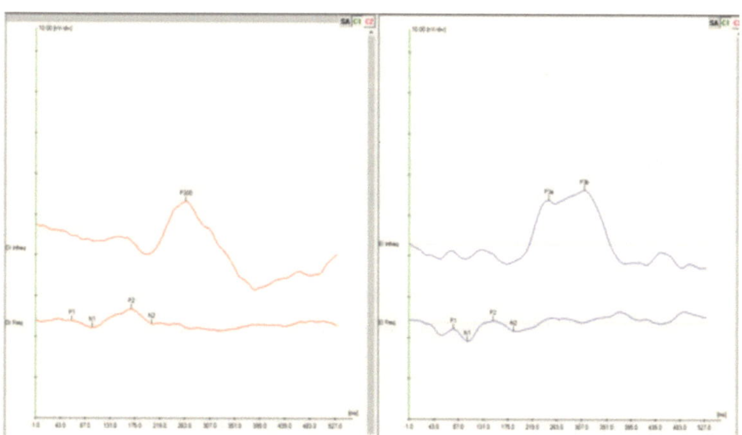

Figure 6.
LLAEP. Personal collection.

children with an SOM history. Shaffer [31] showed an increase in the latencies of N1 and P2, associated with the absence of the P300, in the majority of children evaluated. Sanfins [34] found prolongation of latencies only for P2 and N2 (for females), in comparison with the responses of children without otological alterations. Colella-Santos [11] observed an increase in P2, N2, and P300 latencies in children with a history of OME.

3.2.3.2 Speech

The LLAEP with verbal stimuli provides additional information about the biological processes involved in speech processing, enabling the collection of information complementary to those obtained by standard behavioral evaluations [30, 39, 40].

In the studies of Sanfins [33], children with bilateral OME presented prolonged latencies for N1, P2, N2 (female), and P300, in comparison with responses of children without auditory changes. Children with unilateral OME had prolonged latencies for P2 and P300 in comparison to the responses from healthy children.

The evaluation of the LLAEP using both nonverbal and verbal stimuli seems to be able to identify neurophysiological changes resulting from OM. However, it is important to note that, in unilateral OM episodes, only verbal sound stimuli (speech LLAEP) seem to be able to differentiate groups on the basis of latency. OM impairs speech perception as a result of a failure to recognize sound signals (discrimination, storage, memory). Therefore, the more accurate identification of LLAEP changes with verbal and non-verbal stimuli may relate to underlying OM.

4. Auditory rehabilitation

It is known that hearing loss due to OM during childhood development may result in long-term changes in neural function, structure, and connectivity. The changes are associated with a series of sensory, cognitive, and social difficulties suggestive of impaired brain function [41, 42] which may culminate in central auditory processing disorder (CAPD) [11].

Intervention for CAPD should be initiated as soon as the diagnosis, made through a series of behavioral and electrophysiological procedures, demonstrates the involvement of the CANS. Early identification, followed by intensive intervention, makes best use of the brain's inherent plasticity. Successful treatment outcomes depend on stimulation and repeated practice that induce cortical reorganization (and possibly reorganization of the brainstem), which is reflected in behavioral change [43–45].

Neuroplasticity is the key to the effectiveness of repeated auditory stimulation. Through experience and stimulation it induces reorganization of the cortex and brainstem, improving synaptic efficiency and neural density, giving rise to associated cognitive and behavioral changes [46–48]. The ability of the CANS to adapt to internal and external changes has important implications for learning [49].

Auditory training (AT) is defined as a set of (acoustic) conditions and/or tasks designed to activate the auditory system and related structures in such a way that their underlying neural processes and associated auditory behavior is altered in a positive way [8]. Both formal and informal AT procedures are conducted by audiologists in clinics; the difference between them is that formal training is acoustically controlled, meaning control over stimulus generation and presentation. Combined formal and informal AT offers an approach that provides more intensive practice and leads to better treatment efficacy [8]. AT performed in an individual with CAPD should include activities that aim to improve auditory skills such as sound localization and lateralization tasks, auditory discrimination, auditory pattern recognition, temporal aspects of audition, and auditory discrimination among competing acoustic signals [4].

Donadon and colleagues [50] have studied the efficacy of AT through behavioral CAP tests in children with a history of OM who had undergone bilateral tympanotomy for insertion of ventilation tubes. The sample consisted of 34 subjects who were divided into two groups: an auditory training group (ATG) formed by 20 children and adolescents, aged 8–13 years, diagnosed with CAPD, who were given an auditory training program; and a visual training group (VTG) formed of 14 children and adolescents, aged 9–13 years, diagnosed with CAPD who were given a visual training program. All subjects underwent peripheral auditory evaluation and behavioral evaluation of their CAP (using the dichotic digit test, sentence identification test with ipsilateral competing message, gaps-in-noise test, frequency pattern test, and dichotic vowel test). Auditory training was given through repeated verbal and non-verbal stimuli and associated tasks (available at the website www.afinandoocerebro.com.br) via headphones in an acoustic booth (the intensity was set at 50 dB HL). Each session lasted between 40 and 45 minutes and was performed once a week. The stimulation protocol was developed with the purpose of developing the auditory abilities of:

i. binaural integration-through dichotic listening exercises;

ii. temporal resolution-by means of minimum time interval perception exercises;

iii. temporal ordering-using nonverbal tasks related to frequency, intensity, and duration; and

iv. figure-background exercises with competing noise.

The visual stimulation protocol was elaborated using varied stimuli and tasks from the website via a 15″ notebook positioned in front of the subject on a table arranged in a sound booth. The stimulation protocol was designed with the purpose of stimulating the visual abilities of:

i. visual background;

ii. visual closure;

iii. perception and discrimination of sizes and formats; and

iv. visual memory.

All subjects were reevaluated after 8 weeks with the same battery of behavioral tests as performed at the initial evaluation. In the ATG the results showed a statistically significant difference in the abilities of binaural integration ($p = 0.001$), temporal ordering ($p < 0.0001$), temporal resolution ($p < 0.0001$), and bottom figure (<0.0001) in a comparison of before and after AT. These results suggest that the auditory stimulation performed during AT induced changes in the central auditory nervous system, as demonstrated by the better values recorded in the behavioral tests after intervention. Behavioral changes observed after AT in this population with a history of OM point to evidence of neuroplasticity, since auditory stimulation brought about improvements to the identified impaired hearing abilities.

For the visual training group, however, there was no significant difference in performance for any CAP behavioral tests when comparing pre and post interventions. Thus, auditory training appears to be effective as an intervention strategy for re-adjusting the auditory skills in subjects with a history of OM. Auditory stimulation brought about improvements in impaired hearing skills. AT was able to reorganize the neural substrate, providing appropriate experiences, shaping existing circuits in the CANS, and increasing neural density, reflected by an improvement in the behavioral evaluation **Figure 7**.

Modifications to a child's environment are also important aspects for teachers and parents to address in order to help individuals with CAPD improve access to

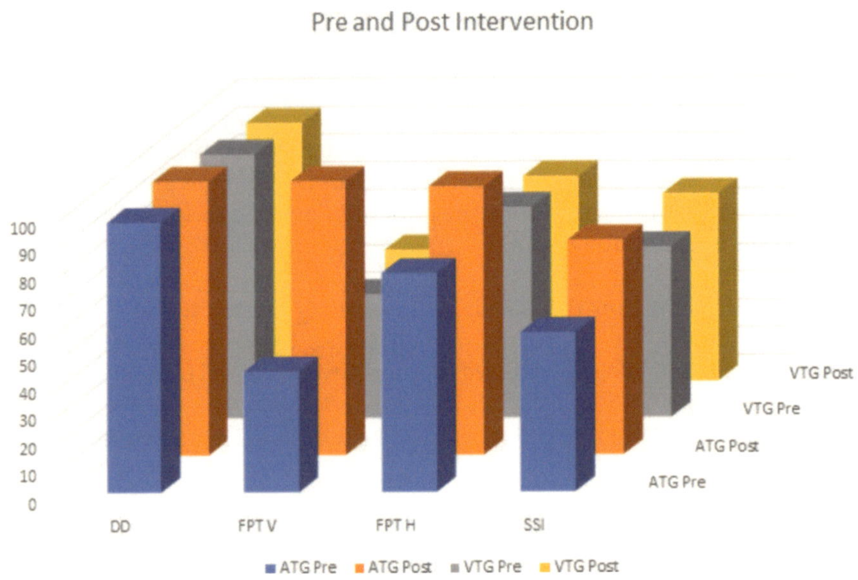

Figure 7.
Comparison of performance in behavioral evaluation pre and post intervention by groups. ATG Pre = auditory training pre intervention; ATG Post = auditory training post intervention; VTG Pre = visual training pre intervention; VTG Post = visual training post intervention; DD = dichotic digits; FPT V = frequency pattern test verbalizing; FPT H = frequency pattern test humming; SSI = synthetic sentence identification.

auditory information outside the therapy room. Some simple changes may bring many benefits to learning. Common recommendations for individuals with auditory disorders include the following:

- Preferred seating arrangements
- Addition of visual cues
- Clear language
- Making frequent checks for understanding
- Repetition or rephrasing
- Multimodality cues and hands-on demonstrations
- Preteaching of new information and new vocabulary
- Provision of a notetaker
- Recording information pictorially
- Gaining attention prior to speaking
- Positive reinforcement
- Reducing environmental noise
- FM systems

5. Summarize

- The negative effects of otitis media on the development of auditory abilities in children and the maturation of their central auditory pathways is undeniable;
- Early medical intervention in OM and family counseling is extremely important;
- The aim should be to avoid prolonged auditory fluctuation caused by OM, thereby minimizing the effects generated by fluid in the middle ear in the development of auditory abilities;
- The overall recommendation is that audiological diagnosis should include both behavioral evaluations and electrophysiological testing of auditory processing;
- In cases of auditory processing disorder, research shows that auditory training is the most effective procedure to re-adjust auditory skills.

Author details

Milaine Dominici Sanfins[1]*, Piotr Henryk Skarzynski[2,3,4] and Maria Francisca Colella-Santos[5]

1 CENA – Centro de Eletrofisiologia e Neuroaudiologia Avançada, São Paulo, Brazil

2 Department of Teleaudiology and Hearing Screening, World Hearing Center, Institute of Physiology and Pathology of Hearing, Warsaw, Poland

3 Department of Heart Failure and Cardiac Rehabilitation, Medical University of Warsaw, Poland

4 Institute of Sensory Organs, Poland

5 Faculty of Medical Science, State University of Campinas, Campinas, Brazil

*Address all correspondence to: msanfins@uol.com.br

References

[1] Baroch K. Universal newborn hearing screening: Fine-tuning process. Current Opinion in Otolaryngology & Head and Neck Surgery. 2003;**11**:003424-003427

[2] Yoshinaga-Itano C, Sedey A. Language, speech, and social emotional development of children who are deaf or hard of hearing: The early years. Volta Review. 2000;**100**(5):298

[3] AAA (American Academy of Audiology). Clinical practice guidelines for the diagnosis, treatment and management of children and adults with central auditory processing disorder. 2010. Available from: http://audiology.org/resources/documentlibrary/Documents/CAPD%20Guidelines%208-2010.pdf

[4] ASHA (American Speech-Language-Hearing Association). Central Auditory Processing Disorders [technical report]. Rockville, MD; 2005. Available from: http://www.asha.org/docs/html/TR2005-00043.html

[5] Bellis TJ. Assessment and Management of Central Auditory Processing Disorders in the Educational Setting: From Science to Practice. Clifton Park, NY: Thomson Learning; 2003. pp. 103-139

[6] Chermak GD, Musiek FE. Central Auditory Processing Disorders: New Perspectives. San Diego, CA: Singular; 1997

[7] Musiek FE, Chermak GD. Handbook of Central Auditory Processing Disorder: Auditory Neuroscience and Diagnosis. San Diego, CA: Plural Publishing; 2014

[8] Geffner D, Ross-Swain D. Auditory Processing Disorders: Assessment, Management and Treatment. San Diego, CA: Plural Publishing; 2013

[9] Billiet CR, Bellis TJ. The relationship between brainstem temporal processing and performance on tests of central auditory function in children with reading disorders. Journal of Speech, Language, and Hearing Research. 2011;**54**:228-242

[10] Borges L, Paschoal J, Colella-Santos MF. (Central) auditory processing: The impact of otitis media. Clinics. 2013;**68**(7):954-959

[11] Colella-Santos MF, Sanfins MD, Donadon C, Borges LR. Otitis media: Long-term effect on central auditory nervous system. BioMed Research International. 2019;**2019**:10. Article ID: 8930904. DOI: https://doi.org/10.1155/2019/8930904

[12] Rosenfeld RM. A Parent's Guide to Ear Tubes. Hamilton, Canada: BC Decker Inc; 2005

[13] Tos M. Epidemiology and natural history of secretory otitis. The American Journal of Otology. 1984;**5**:459-462

[14] Mandel EM, Doyle WJ, Winther B, Alper CM. The incidence, prevalence and burden of OM in unselected children aged 1-8 years followed by weekly otoscopy through the "common cold" season. International Journal of Pediatric Otorhinolaryngology. 2008;**72**:491-499

[15] Martines F, Bentivegna D, Di Piazza F, Martinciglio G, Sciacca V, Martines E. The point prevalence of otitis media with effusion among primary school children in Western Sicily. European Archives of Oto-Rhino-Laryngology. 2010;**267**:709-714

[16] Gravel JS. Hearing and auditory function. In: Rosenfeld RM, Bluestone CD, editors. Evidence-Based Otitis Media. 2nd ed. Hamilton, Canada: BC Decker Inc; 2003. pp. 342-359

[17] Sabo DL, Paradise JL, Kurs-Lasky M, Smith CG. Hearing levels in infants and young children in relation to testing technique, age group, and the presence or absence of middle-ear effusion. Ear and Hearing. 2003;**24**:38-47

[18] Katz J, Zalewski TR, Brenner MJ. Otitis media and central auditory processing disorder (CAPD). In: Geffner D, Ross-Swain D, editors. Auditory Processing Disorders. 3rd ed. San Diego, CA: Plural Publishing; 2019. pp. 307-326

[19] Rosenfeld RM, Shin JJ, Schwartz SR, Coggins R, Gagnon L, Hackell JM, et al. Clinical practice guideline: Otitis media with effusion (update). Otolaryngology and Head and Neck Surgery. 2016;**154**(1S):S1-S41

[20] Baran JA. Test battery principles and considerations. In: Musiek FE, Chermak GD, editors. Handbook of Central Auditory Processing Disorder: Auditory Neuroscience and Diagnosis. San Diego, CA: Plural Publishing; 2014. pp. 291-323

[21] Pereira LD, Schochat E. Testes Auditivos Comportamentais Para Avaliąˋcão Do Processamento Auditivo Central. Barueri, Brazil: Pró Fono; 2011

[22] Musiek FE, Baran JA, Pinheiro ML. Duration pattern recognition in normal subjects and in patients with cerebral and cochlear lesions. Audiology. 1990;**29**:304-313

[23] Musiek FE, Zaidan EP, Baran JA, Shinn JB, Jirsa RE. Assessing temporal processes in adults with LD: The GIN test. In: Convention of American Academy of Audiology. Vol. 203. Salt Lake City, Utah, USA: AAA; 2004

[24] Khavarghazalani B, Farahani F, Emadi M, Hosseni Dastgerdi Z. Auditory processing abilities in children with chronic otitis media with effusion. Acta Oto-Laryngologica. 2016;**136**(5):456-459

[25] Gravel JS, Wallace IF, Ruben RJ. Auditory consequences of early mild hearing loss associated with otitis media. Acta Oto-Laryngologica. 1996;**116**(2):219-221

[26] Tomlin D, Rance G. Long-term hearing deficits after childhood middle ear disease. Ear and Hearing. 2014;**35**(6):e233-e242

[27] Burkard R, Don M, Eggermont J. Auditory Evoked Potentials: Basic Principles and Clinical Application. Philadelphia: Lippincott Williams & Wilkins; 2007

[28] Chambers R, Rowan L, Mathics M, Novak L. Auditory brain-stem responses in children with previous otitis media. Archives of Otolaryngology – Head & Neck Surgery. 1989;**115**:452-456

[29] Folsom R, Weber B, Thompson G. Auditory brainstem responses and children with early recurrent middle ear disease. The Annals of Otology, Rhinology, and Laryngology. 1983;**92**:249-253

[30] Shaffer EK. Auditory evoked potentials in children with and without otitis medua. Tejas Journal of Audiology and speech pathology. 1999;**XXIII**:10-20

[31] Maruthy S, Mannarukrishnaiah J. Effect of early onset otitis media on brainstem and cortical auditory processing. Behavioral and Brain Functions. 2008;**4**:17

[32] Sanfins M, Borges L, Donadon C, Hatzopoulos S, Skarzynski P, Colella-Santos M. Electrophysiological responses to speech stimuli in children with otitis media. The Journal of Health Science. 2017;7(4):9-19

[33] Sanfins M. Electrophysiological evaluation with verbal and non-verbal sounds in children with a history of otitis media. State University of Campinas; 2017. Available from: http://www.repositorio.unicamp.br/handle/REPOSIP/330747

[34] Tucci DL, Cant NB, Durham D. Conductive hearing loss results in changes in cytochrome oxidase activity in gerbil central auditory system. JARO. 2001;**3**:89-106

[35] El-Kabarity RH, Abdel Rahman TT, Abdel Kader HA, Sanyelbhaa H. Effect of otitis media with effusion on brainstem timing in children. Hearing, Balance and Communication. 2016;**14**(1):20-24

[36] Tonnquist-Uhlén I. Topography of auditory evoked cortical potentials in children with severe language impairment. Scandinavian Audiology. Supplementum. 1996;**25**(44):1-40

[37] Hall JW, Grose JH, Buss E, Dev MB, Drake AF, Pillsbury HC. The effect of otitis media with effusion on perceptual masking. Archives of Otolaryngology – Head & Neck Surgery. 2003;**129**(10):1056-1062

[38] Gravel J, Roberts J, Roush J, Grose J, Besing J, Burchinal M, et al. Early otitis media with effusion, hearing loss, and auditory processes at school age. Ear and Hearing. 2006;**27**(4):353-368

[39] Oates P, Kurtzberg D, Stapells D. Effects of sensorineural hearing loss on cortical event-related potential and behavioral measures of speech-sound processing. Ear and Hearing. 2002;**23**(5):399-415

[40] Martin B, Tremblay K, Korczack P. Speech evoked potential: From the laboratory to the clinic. Ear and Hearing. 2002;**23**(5):399-415

[41] King AJ, Parsons CH, Moore DR. Plasticity in the neural coding of auditory space in the mammalian brain. Proceedings of the National Academy of Sciences of the United States of America. 2000;**97**(22):11821-11828

[42] Hogan SC, Moore DR. Impaired binaural hearing in children produced by a threshold level of middle ear disease. Journal of the Association for Research in Otolaryngology. 2003;**4**(2):123-129

[43] Kolb B. Brain Plasticity and Behavior. Mahwah, NJ: Lawrence Erlbaum; 1995

[44] Merzenich M, Jenkins W. Cortical plasticity, learning and learning dysfunction. In: Julesz B, Kovacs I, editors. Maturational Windows and Adult Cortical Plasticity: SFI Studies in the Sciences of Complexity. Vol. XXIII. Reading, PA: Addison-Wesley; 1995. pp. 247-272

[45] Russo NM, Nicol TG, Zecker SG, Hayes EA, Kraus N. Auditory training improves neural timing in the human brainstem. Behavioural Brain Research. 2005;**156**(1):95-103

[46] de Boer J, Thornton ARD. Neural correlates of perceptual learning in the auditory brainstem: Efferent activity predicts and reflects improvement at a speech-in-noise discrimination task. The Journal of Neuroscience. 2008;**28**:4929-4937

[47] Johnson KL, Nicol T, Zecker SG, Kraus N. Developmental plasticity in the human auditory brainstem. The Journal of Neuroscience. 2008;**28**:4000-4007

[48] Song JH, Skoe E, Wong PC, Kraus N. Plasticity in the adult human auditory brainstem following short-term linguistic training. Journal of Cognitive Neuroscience. 2008;**20**(10):1892-1902

[49] Bellis TJ. Assessment and Management of Central Auditory

Processing Disorders in the Educational Setting: From Science to Practice. USA: Plural Publishing; 2011

[50] Donadon C, Sanfins MD, Borges L, Colella-Santos M. Auditory training: Effects on auditory abilities in children with history of otitis media. International Journal of Pediatric Otorhinolaryngology. 2019;**118**:177-180

Section 2

Updates on Audiological Diagnosis and Therapy

Chapter 5

Electrical Stimulation of the Auditory System

Patrick J. Boyle

Abstract

In many healthcare systems electrical stimulation of the human auditory system, using cochlear implants, is a common treatment for severe to profound deafness. This chapter will describe how electrical stimulation manages to compensate for sensory-neural hearing loss by bypassing the damaged cochlea. The challenges involved in the design and application of cochlear implants will be outlined, including the programming of clinical systems to suit the needs of implanted patients. Today's variety of patient will be reviewed: unilaterally and bilaterally implanted, bimodal users of a cochlear implant as well as a contralateral hearing aid, CROS device users having either asymmetrical hearing loss or single-sided deafness. Alternative devices such as auditory brainstem implants will be described, and additionally the more experimental auditory mid-brain implants and intraneural stimulation approaches. Research that is likely to bring medium term benefits to the clinical application of cochlear implants will also be described.

Keywords: cochlear implant, electrical, stimulation, prosthesis

1. From the beginning to current practice

Electrical stimulation of the human auditory system is generally traced back to the pioneering experiments of Alessandro Volta, inventor of the battery. When Volta applied 50 volts to his own head, he reported hearing an unpleasant boiling sound [1]. However, the forerunner of a modern CI system is just over 60 years old: opportunistic stimulation of the auditory nerve [2] of a bilaterally deaf patient receiving a facial nerve graft. During the two decades following this work, various clinical studies [3–10] saw the implantation of single and then multi-channel cochlear implant (CI) systems in people suffering profound deafness. While many of these pioneers suffered ridicule at the hands of the mainstream scientific community, clinical considerations prevailed. The early devices that were produced in academic institutions were transferred to commercial organizations, these often building on prior medical device experience, for example experience gained in the pace maker field.

Today over half a million people, from babies under 6 months of age to adults in their late 90s, have been implanted with a CI. While it can be argued that the CI is the most successful medical device ever created, the outcomes are still highly variable (**Figure 1**). In the best of cases, CI users can make fluent use of a telephone, understand speech in adverse listening conditions where there is considerable competing noise and reverberation, hence enjoying independence spanning social lives and careers that would have been unimaginable without their CI device. Even where speech understanding is limited, a release from the isolation of deafness through access to environmental

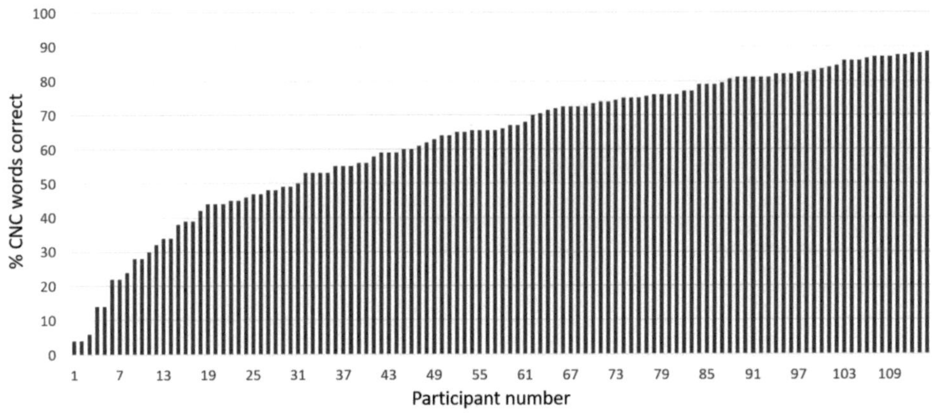

Figure 1.
Percent correct scores on the CNC word test ranked from poorest to best for 113 cochlear implant users showing a large variation in outcome reproduced from Holden et al. [79].

sounds, a reduction in the level of tinnitus and support of lip-reading with a reduction in the effort required for oral communication are all worthwhile benefits from use of a CI. It should also be noted that in many cases those most satisfied with their implant are not those who receive the highest scores on standardized tests of speech understanding.

The following sections will describe how electrical stimulation of the auditory system is achieved, with the main focus being on CI systems. The factors that influence outcome, so far as they are known, will be described, along with the challenges in delivering clinical service, both today and into the future. With the future in mind the major research topics that are currently being addressed will be outlined.

2. Overview of a cochlear implant system

Figure 2 shows the various components common to all of today's clinically applied cochlear implant systems. Sound is typically collected from microphones housed on a behind the ear (BTE) sound processor. The sound is first "cleaned" to remove noise and then processed to create the stimulation patterns destined for the implanted electrode array. Except in the case of one-piece processors, a lead connects the sound processor output to a radio frequency (RF) transmitter coil located above and behind the ear. The external coil is held in place over the implant's receiver coil through a pair of magnets: one external and one within the surgically implanted device under the skin. This arrangement supports reliable communication across the skin through the use of RF based telemetry. The RF signal provides both power for the implant's electronics and the information needed to produce electrical stimulation. Hence the implant consists of: its receiver coil, a hermetic package containing electronic circuits and an electrode lead assembly connecting to the electrode array that is placed inside the cochlea (**Figure 3**). In some of today's CI systems the sound processor and RF coil are a single component held in place by the magnet but having no wire or BTE part. This provides some esthetic advantage but may fall off more easily and compromises sound collection.

Additionally today's implants have the ability to make both physical and physiological measurements, using back-telemetry, to transmit these data to the sound processor. Through the use of wireless technology, information can be relayed to and from a host of devices: smartphones, tablets, laptops, remote microphones or other listening aids. Such connectivity leaves a CI user well placed to use many consumer devices to enhance their communication and support maintenance of their implant system.

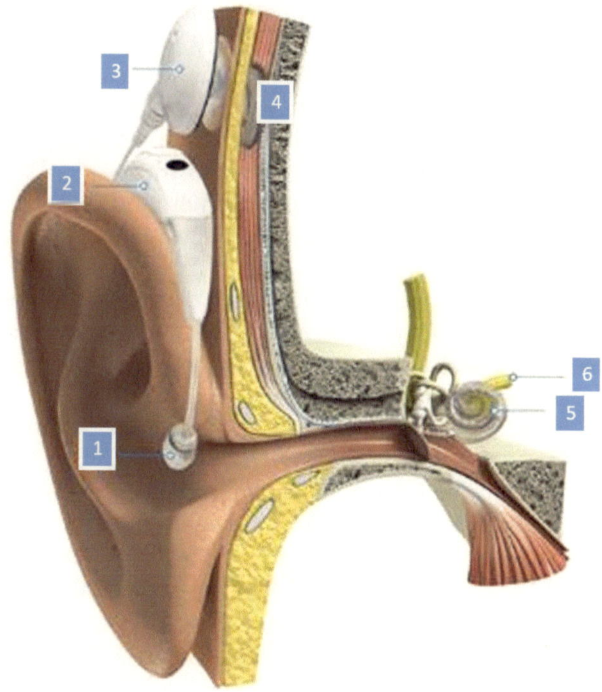

Figure 2.
The components of a behind the ear (BTE) model of cochlear implant showing (1) the T-mic placed in the external ear canal, (2) BTE sound processor, (3) radio frequency transmitting headpiece, (4) the implant body, (5) intra-cochlear electrode array and (6) the auditory nerve.

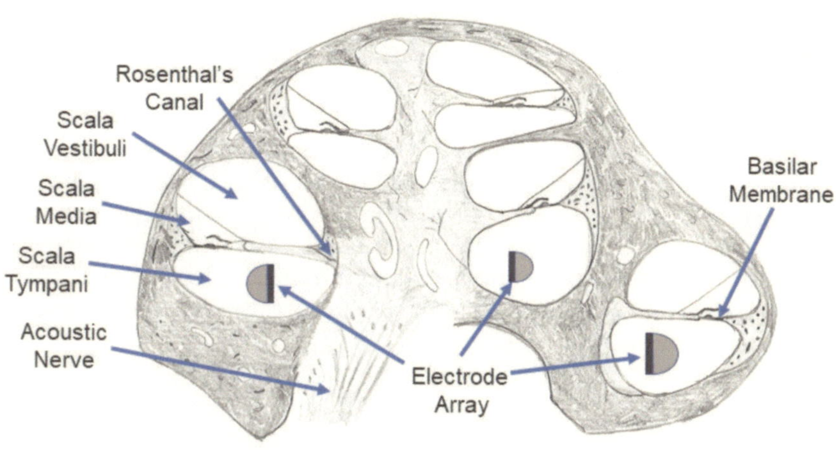

Figure 3.
A cross-section through the spirally-shaped cochlea showing the various compartments, including the scala tympani with a mid-scala located electrode array.

3. Electrical stimulation principles

In the earlier chapters of this book the auditory system has been described in some detail, including pathology that can result in the most debilitating degrees of hearing loss: severe to profound deafness. Fortunately, electrical stimulation can be delivered without external, middle or indeed even an inner ear. However, in the large majority of clinical cases the auditory nerve is intact and can be accessed via a very poorly or non-functioning cochlea: it being this damage that the CI bypasses. The operating principle of a CI is that small electrical currents are able to initiate activity on the auditory nerve that crudely mimics the activity produced by a normally functioning cochlea. Taking advantage of the cochlea's tonotopic organization, currents representing high frequency sounds are delivered to the base, while currents representing lower frequency sounds are delivered to more apical cochlear locations. This is achieved through the use of an electrode array containing multiple separate electrode contacts placed along the scala tympani (**Figure 4**). The number of intra-cochlear electrode contacts in clinical service varies by manufacturer between 12 and 24. In addition, either the shorting of adjacent contacts, or simultaneous delivery of synchronized stimulation patterns on multiple electrode contacts, seeks to increase the number of stimulation sites available [11–13].

In **Figure 4** an electrode array is shown placed in the scala tympani. Here the exposed electrode surface, from which stimulation current is delivered, faces towards the modiolar wall, behind which the auditory nerve's spiral ganglion cell bodies are located in Rosenthal's canal. The remainder of the array is composed of a soft silicone that supports the contacts and the insulated wires connecting the electrode contacts to the implant's electronics. A modiolar facing contact surface orients the electrical stimulation towards the primary stimulation targets, the spiral ganglion cell bodies. Since the perilymph fluid in the scala tympani is electrically conductive, it allows current to flow through the cochlea and achieve stimulation of neural elements. A downside is that current also tends to spread along the scala, rather than addressing only the area local to the electrode contact where we would like it to act. If peripheral processes still remain in the cochlea, extending from the cell bodies to the organ of Corti, then these may also be targets for electrical stimulation. Unfortunately it is not currently possible to accurately know the status of any individual's cochlea. It appears reasonable to assume that a great deal of the variability illustrated in **Figure 1** comes from variations in the health of those individuals' cochleae, this variation being part dependent on environment and disease and the individual's particular physiology, as well as how any one individual's immune system reacts to the insertion and presence of the electrode array itself.

The most common type of stimulation paradigm used in CIs today is monopolar stimulation (**Figure 5a**). Here the implant introduces current into the cochlea via a relatively small electrode contact. A typical surface area might be 0.2 mm^2. The density of current close to the electrode contact introduces a higher probability of activating elements close to the contact, the probability decreasing for elements

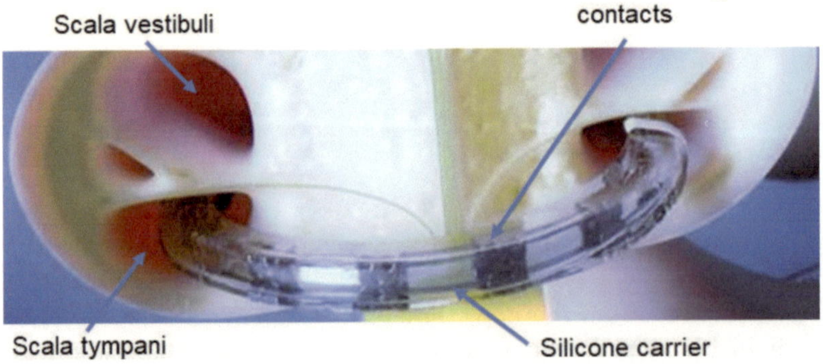

Figure 4.
View of how an electrode array will be positioned within the scala tympani of the cochlea, here with the electrode contacts facing towards the modiolar wall behind with the spiral ganglion cell bodies are located.

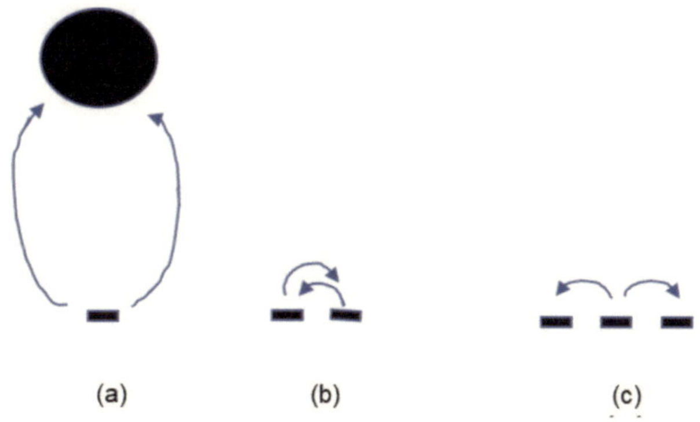

Figure 5.
Electrical stimulation configurations: (a) monopolar where current is returned to a large distant extra-cochlear return electrode, (b) bipolar where current flows between adjacent intra-cochlear electrodes and (c) tripolar where current is returned by two adjacent contacts.

at increasing distance. Monopolar stimulation current is returned to the implant using a distant extra-cochlear electrode that has at least 10 times the intra-cochlear electrode contact's surface area. This keeps the returning current density low, avoiding stimulation at this remote site. Typically there are two return electrodes on a cochlear implant, in case anything goes wrong with one of them. One is placed on the body of the implant while the other is on a separate flying lead, or on the electrode lead assembly but located outside of the cochlea. In **Figure 6** an implant can be seen with its various component parts indicated.

Alternative stimulation paradigms are sometimes used, but mainly for research. **Figure 5b** shows a bipolar stimulation paradigm. Here two intra-cochlear electrode contacts, separated by around 1 mm, operate as a pair. Stimulation is introduced by one contact and returned by the other. This in theory restricts stimulation to a small part of the cochlea, so should help with the spread of current mentioned above. However, in practice current introduced by one contact is returned to the other without activating enough neural tissue to create a loud enough hearing sensation. Hence, it is often necessary to form a bipolar pair using contacts that are not adjacent but for example are spaced 2 mm or more apart. In addition, the arrangement of contacts along the cochlea means that there will be a plane of zero field between the contacts, leading to a need to use higher currents and thus produce a wider spread in stimulation. In **Figure 5c** tripolar stimulation is shown. Now a group of three electrode contacts are used together. Stimulation is introduced via the middle contact and ideally half returned via the adjacent contact on either side. This avoids

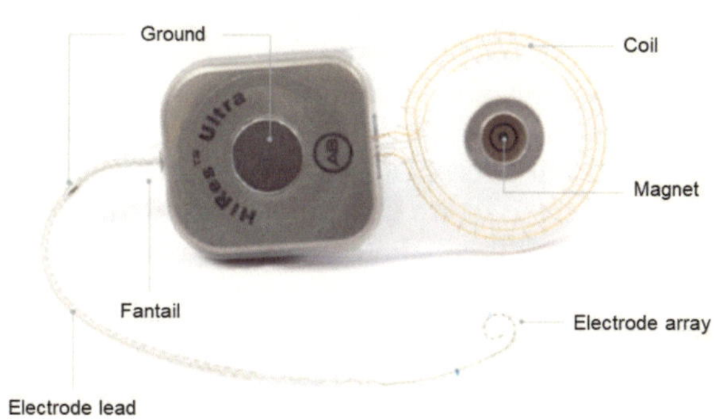

Figure 6.
A cochlear implant showing its various components. Note the two ground or return contacts, one on the case body and one on the electrode lead assembly.

the zero potential plane problem of bipolar stimulation and theoretically provides a tighter containment of stimulation. Again, in practice there is a need to recruit a given number of neurones to signal sufficient loudness and this means increasing the tripolar stimulation current on the middle contact. Eventually the current on the return electrodes will become high enough to cause stimulation, resulting in a wider spread of current than intended. In many cases it is not even possible to increase the current far enough to achieve sufficient loudness for a tripolar configuration. In such cases some of the current has to be returned to the remote extra-cochlear return electrode, a configuration referred to a partial tripolar. This even further increases the current spread. So, with these practical limitations in mind it is not difficult to see why monopolar is universally used as the default stimulation paradigm, despite the apparent large current spread that this entails.

3.1 Technical considerations

In the interests of simplicity some of the more technical aspects of electrical stimulation have been avoided in the text above, allowing focus on the broader application. In this section some of these more technical issues will be discussed. Where a reader is not interested in technical detail this section can be ignored.

Ohm's Law states that the electrical voltage difference required to drive a given current is directly proportional to the resistance through which the current has to flow. With changes in the resistance, or more generally the impedance if we consider frequency effects, a voltage source would lead to uncontrolled changes in the current being output. As described below this could lead to uncontrolled changes in loudness over time. Most of today's CI systems deliver electrical stimulation through one or more current sources. As the name implies, this circuit attempts to deliver the current requested of it regardless of how much electrical resistance is offered by the body. A current source is said to be in compliance when it is delivering the current requested of it. Given a finite amount of voltage being available within an implant, for example 8 volts, there will be a maximum impedance into which the implant's maximum output current can be delivered. With a typical stated maximum output current of 2000 μA, the maximum impedance for which this could be delivered will be 4000 Ohms (8 volts divided by 2000 μA, or $8/2 \times 10^{-3} = 4000$). For higher impedances the maximum stated output current will not be available. For lower impedances the implant will be limited to its maximum output current value, ensuring safe operation.

Since CI systems are designed to provide stimulation for essentially all waking hours, day after day over decades, their output must not damage the neural tissue that they are intended to stimulate. One obvious source of damage is the delivery of direct current (DC). If a DC current is applied to the body it will result a process called electrolysis. Here there will be a continuous transport of ions, charged atomic particles, leading to dissolution of the platinum electrode contact and destruction of the cochlear tissue: obviously a catastrophic situation. Research in animals indicates that even very small amounts of DC, 400 nA, is enough to cause tissue damage [14]. Several mechanisms are used in a CI to prevent the delivery of DC. Firstly, a balanced stimulus waveform is used, almost always a symmetrical pulse having two complementary phases (**Figure 7a**), although so long as the two phases contain an equal and opposite area they need not be symmetrical (**Figure 7b**). At the simplest interpretation, the first phase, referred to as cathodic, will depolarize neurones hence producing the electrical stimulation that we seek to achieve. The second, anodic, phase balances the stimulation resulting in no nett current being delivered to the body, thereby avoiding DC. Even with very careful design, there is likely to be some small imbalance between the two phases. To account for any in-balance,

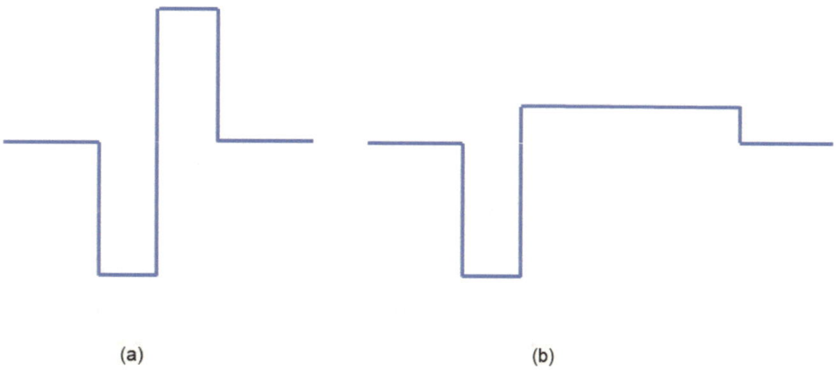

Figure 7.
Stimulation waveforms with balanced cathodic and anodic phases may have either symmetrical phases (a), or asymmetrical phases (b) where the area of each phases is identical.

following delivery of a stimulation pulse the electrode contact is connected to ground, ideally removing any residual DC. Finally, an electronic component called a capacitor is placed in series with the stimulating electrode contact. A capacitor does not allow DC to pass, so offers yet more protection in case some fault in the electronics interferes with either of the previous two protection mechanisms. Together these mechanisms appear successful in avoiding the delivery of DC. Devices do fail, particularly in the pediatric population, with reimplantation rates over tens of years being reported at 8% from the well-established Sydney clinic [15]. Typically half of these failures are medical issues and half device failures. However, in virtually all cases it is possible to re-implant the patient, with outcomes almost always being as good as those obtained when the original device was working well [16].

While we speak about stimulation current it is really the electrical charge that is at issue. Charge is simply the product of current times time and has units of Coulombs, C. Electro-chemical considerations mean that an electrode has a maximum charge injection capacity, such that a given size and material will only be able to handle a given charge limit in a reversible way, so that all of the charge injected in one phase can be removed in the second phase [17]. This is necessary to avoid the DC as discussed above. A conservative value for the maximum charge density, typically 30 $\mu C/cm^2$ [18], taking account of the electrode dimensions, is programmed into the implant's controlling software, ensuring that this limit is never exceeded. Animal experiments confirm that chronic stimulation with higher charge densities, for example 400 $\mu C/cm^2$, results in the dissolution of platinum but interestingly not to the loss of auditory neurones [19]. The loudness sensation produced by electrical stimulation is related to the amount of charge delivered in one phase of the stimulation waveform. There is also an effect of the rate at which stimulation pulses are delivered. However, for the stimulation rates used in clinical practice, typically 500–2000 pulses per second per channel (ppps/ch) the effect of rate is quite small and largely ignored.

Stimulation current flows through cochlear tissue as a result of voltage differences developed along the current's path. How these voltage differences are arranged along the length of a neurone's peripheral or medial process, or indeed across the cell body, determines which neurones depolarize, leading to action potentials being generated. The action potential may propagate to the medial synapse with the cochlear nucleus and hence initiate activity on the auditory pathway, leading to a sense of hearing being detected in the brain. Rather than stimulating a single neuron, typically hundreds, or even thousands, of neurones in a region of the cochlea will be addressed by a single electrode contact. These patterns of stimulation are interpreted as sound input by the higher levels of the auditory system, leading to the sense of electrical hearing. The next section will describe how sounds detected by the CI system's microphone will result in the generation of electrical stimulation patterns.

4. Operation of the cochlear implant system

The main cochlear implant system functions are shown schematically in **Figure 8**. Sound from the microphone is compressed by a single-channel automatic gain control (AGC) system. Compression ratios in CI systems tend to be substantially higher than those in acoustic hearing instruments: six to infinity, compared to two to three respectively. This reflects both the small electrical dynamic range of typically 10 dB [20] and the exponential like increase in loudness found for electrical hearing [21]. Both considerations require tight control of the stimulation current's amplitude to avoid discomfort. Research with different implant types shows a consistent advantage for slow-acting AGC, the benefit being a reduced compression of the information rich temporal modulations of speech [22–24], as well as a reduced co-modulation effect [25] associated with the single channel AGC.

Following AGC, the sound is broken into a number of frequency channels, this number varying between 12 and 24 channels, reflecting the number of intra-cochlear electrode contacts available in the implant model. In **Figure 8** only four channels are used to illustrate the principle. Today a Fast Fourier Transform (FFT) algorithm is often used to separate the incoming sound into discrete frequency channels. The amount of energy in each channel is then estimated by a rectification and low-pass filtering process. While the average energy is calculated over a period of perhaps 10 ms (milliseconds) or more, stimulation pulses will be delivered much more rapidly, typically once every millisecond. Hence calculations will be made that overlap in time in an attempt to follow the changes in speech energy over time. Next the acoustic energy in each channel is mapped to an electrical current amplitude that takes account of the CI user's sensitivity to electrical stimulation. The goal is to use smaller currents that barely produce a perception of electrical stimulation to represent low-intensity acoustic activity and larger currents that are perceived as loud to represent very intense acoustic events. This

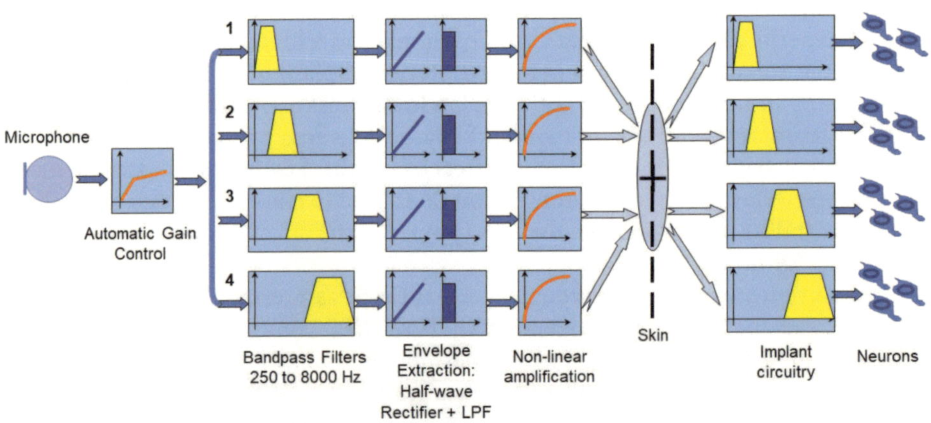

Figure 8.
A schematic of the sound processor system where sound is collected by the microphone, compressed by an automatic gain control, broken into discrete frequency channels, which have their energy assessed and mapped to the user's requirements. This information is then combined into a digital stream, transmitted by radio frequency to the implant where stimulation currents are generated.

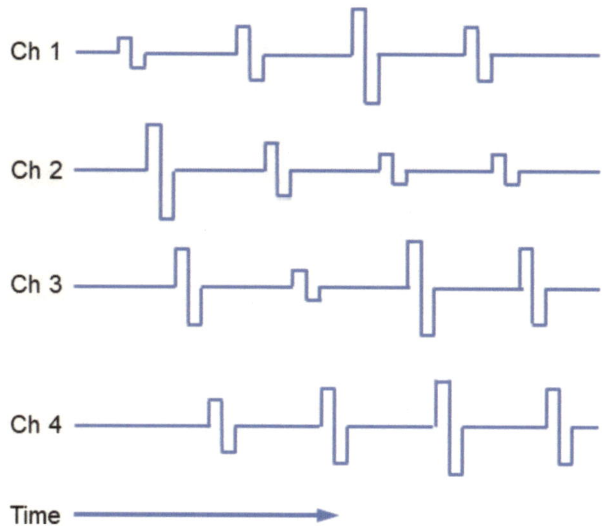

Figure 9.
An illustration of now non-simultaneous waveforms delivers information for each channel. Once a channel has been stimulated no more information may be delivered until the other channels have been updated.

needs to be managed separately for each channel, resulting in the continuous output of a stream of stimulation amplitudes for each channel. As shown schematically, these amplitudes are then combined together for transmission by the RF signal across the skin to the implant. Electronics inside the implant extract the digitally transmitted amplitudes, convert them to analogue values and then drive the implant's current source(s), resulting in stimulating currents being delivered by the intra-cochlear electrode contacts. For virtually all of today's clinical systems only one channel will be stimulated at a time. This approach avoids the channel interactions that would occur were channels presented simultaneously within the conductive scala tympani [26]. The disadvantage of this approach can be seen in **Figure 9** where a channel is only updated during its own time period, therefore, must wait until all the other channels have been updated until new information can be transmitted. Deliberately, very brief current pulses each of around 40–50 μs duration (20–25 μs/phase) are used, so that it is still possible to update each channel rapidly enough to keep up with the changes in acoustic energy over time. This often means stimulation at more than 1000 pps/ch. Such an approach generally leads to higher levels of speech understanding than where simultaneous stimulation is delivered [27, 28].

4.1 Sound coding strategies

Over the years the sound coding strategy, a software algorithm that relates audio from the sound processor microphones to the electrical patterns appearing at the electrode contacts, has changed. Initially it was believed that the damaged auditory system was not capable of transmitting much information, hence the most useful information was extracted from the speech and directly coded on sets of electrodes. For example an early feature extraction strategy F0F1F2 [29] extracted the first two formants of speech, F1 and F2, from which it is possible to estimate the vowel being articulated. Each formant range had a set of electrode contacts allocated, such that higher or lower frequencies for each formant lead to stimulation on more basal or more apical electrode contacts in that formant's electrode set. The rate at which pulses were delivered was related to the fundamental frequency (Fo) driving the vocal tract, leading to the strategies name. Such a strategy supported only very modest levels of speech understanding, around 8% correct for monosyllabic words presented in quiet [30]. The information extracted was limited to begin with and further reduced through errors generated in real life listening situations where background noise, reverberation and intensity and frequency response variations led to the algorithm making mistakes in both the extraction of formant frequencies and in the estimation of Fo.

It was eventually recognized that the brain was better at extracting information than the feature extraction algorithms and hence "whole-speech strategies" replaced feature extraction. Today's sound coding strategies simply average the energy in each channel's frequency range and generate levels of stimulation that represent this. In some cases a so-called n-or-m strategy will work out which subset (n) channels from the total (m) number available have the highest energy and then only stimulate this reduced set. Refinements to this may neglect adjacent channels on the basis that stimulating both will not add anything, so select a more distant lower amplitude electrode to transmit more information [31].

4.2 Neural population

Stimulation delivered by a CI system will result in the depolarization of neural elements, resulting in action potentials being generated that propagate to the next stage of the auditory system: the cochlear nucleus. With reference to the schematic of **Figure 8**, there is a population of spiral ganglion neurones associated with each electrode and hence each frequency channel of the CI system. As mentioned above, a channel's stimulation current will need to recruit a certain population of neurones whose firing indicates to the brain the amount of activity in a particular frequency range. Ideally, there will be a sufficient local neural population such that progressively increasing stimulation current initiates an appropriate number of action potentials, so that the brain correctly perceives the amount of acoustic activity in the channel's frequency range.

Unfortunately a discrete neural population for each channel as shown in **Figure 10a** is not always available. In **Figure 10b** only a reduced neural population is available for each channel. Hence, when there is a lot of activity in one channel, requiring recruitment of a full population of neurones, these are not available locally. It is still possible to increase the stimulation current, spread the electrical field further away from the electrode and depolarize neurones that should really be associated with another channel. While this will satisfy the perception of loudness, it generates channel interaction so that we are no longer able to deliver frequency specific information to a discrete part of the cochlea. The perception will be of a blurred or fuzzy sound, particularly a problem when trying to listen to speech in the presence of competing noise.

An alternative situation is shown in **Figure 10c** where most electrodes have a sufficient local neural population but one electrode is located in a so-called dead region [32]. When electrode 2 is stimulated it can only recruit neurones from the population belonging to electrodes either side of it, delivering information about channel 2's frequency region to other parts of the cochlea, spreading stimulation widely and interfering with the otherwise discrete frequency information being delivered by the neighboring channels.

Unfortunately, it is not currently possible to determine what the number and distribution of neural elements is for any individual. The literature is not always helpful in this area. As shown in **Figure 1**, there is great variability in outcome for the identification of monosyllabic words. Since this task involves little top-down processing, much of the variability in outcome must come from the electro-neural interface. Beyond speech understanding, examining the ability to discriminate adjacent electrodes, or intra-electrode stimulation sites [33], also showed both

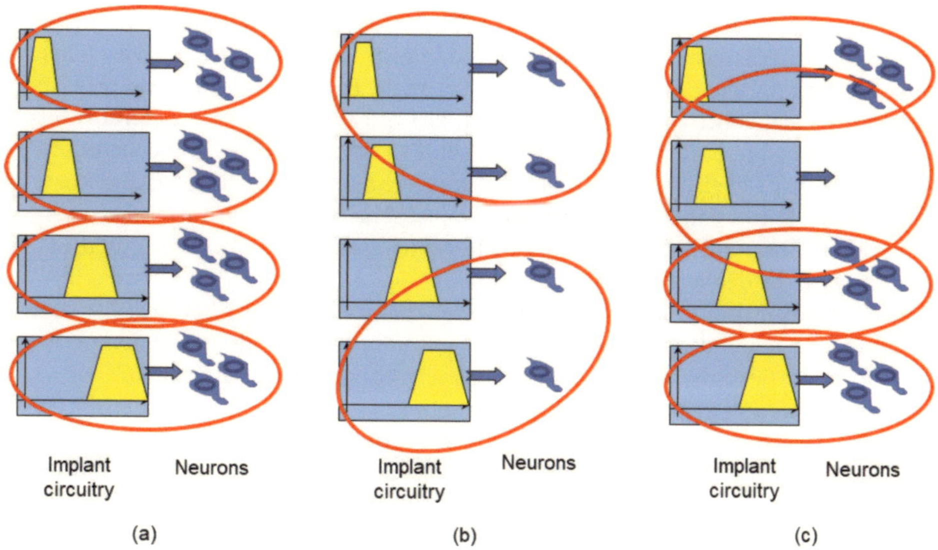

Figure 10.
A schematic representation of three different neural populations: (a) a full population exists for each channel, (b) a depleted population results in channel interaction and (c) a dead region requires recruitment from the populations rightly belonging to adjacent channels.

great variability between subjects and across the electrode array of individual subjects. This task, having no confound with cognitive processes related to speech understanding, further confirms the presence of peripheral variability and its likely contribution to variations in outcome.

It is unclear to what degree a loss of spiral ganglion cells (SGC) in humans will follow, even after years of severe to profound deafness. Histological studies of humans who had used a cochlear implant sometimes show a reasonable correlation between CNC word score and SGC count: for example R = 0.62 [34], R = 0.9 [35] but for a small group of only 6. However, the variability is such that the same SGC count can show variations of between 30% and 75% for CNC words, or the same CNC word score can be associated with 3000 or 18,000 SGCs. Examining the threshold current for detecting electrical stimulation in a group of 130 lateral wall electrode array users [36] showed significant differences between four groups: the increase in group mean threshold being associated with a reduction in monosyllabic word score. This works suggests that a higher SGC population (lower electrical threshold) is associated with better speech understanding.

The literature listed above indicates that there is a relationship between the number of spiral ganglion cells and the ability to identify monosyllabic words when using a cochlear implant. Contributions to speech understanding may also come from a large number of additional factors, some of which include: the distribution of SGCs, angular insertion of the electrode array, distance of the electrode contacts from the modiolar wall, presence or absence of peripheral processes, fibrous sheath formation and intrusion of new bone into the cochlea. How well a given implant user has had the parameters of their sound processor set, commonly referred to as their program, is another variable that we will examine next.

4.3 Programming the cochlear implant

As has been explained above, the small electrical dynamic range available to a CI user makes it necessary to carefully adjust the stimulation parameters to suit the requirements of each individual recipient. The most important adjustment is the amount of stimulation that will be delivered in response to acoustic activity. This must be done for each of the CI's separate channels. Each channel has two primary parameters that control its output. One will be typically called a most comfortable level, shortened to either M-Level or C-Level. The other is a threshold control, referred to as T-Level. The main CI manufacturers use these parameters slightly differently but to a good approximation T-Level sets the minimum stimulation level that the implant will deliver and M-Level will set the maximum stimulation level that can be delivered for an individual recipient. The sound processor will then arrange for the amount of acoustic range that it handles, somewhere between 40 and 80 dB depending on the user's setting and implant model, to be mapped to stimulation levels between T- and M-Level. In combination with the AGC of the system this will give the CI user access to their acoustic environment such that hearing levels of between 20 and 30 dB HL are achieved across the frequency range 250–8000 Hz. The combination of AGC and M-Level ensures that even high intensity sounds of 100 dB SPL do not produce uncomfortably loud sensations. Unlike acoustic hearing, it is generally possible to provide CI users with access to the full range of frequencies that are most important for speech understanding.

Which channels are activated is another important adjustment to make. Most audiologists are reluctant to deactivate channels, although sometimes a reduced set of channels can give a better outcome. In some cases an electrode array is not fully inserted into the cochlea, perhaps due to the cochlea being too small, or there being fibrosis tissue, or bone, that prevents a full insertion being obtained. Alternatively, electrode arrays can sometimes extrude from the cochlea [37, 38], either shortly after implantation or months to years later. In all these cases the more basal electrode contacts will need to be deactivated. Deleting electrodes from a program will lead to the frequency range being remapped across the remaining electrode contacts. There will be a coarser representation of frequency since fewer channels are now available. However, removing electrodes that are not inside the cochlea will produce a better outcome than simply leaving these electrodes active.

Beyond setting T- and M-Levels and defining an appropriate set of electrode contacts, there is sometimes adjustment made to the acoustic dynamic range mapped by the sound processor. This effectively controls the compression of acoustic sounds into the electrical dynamic range. It might seem logical to use as large an acoustic or input dynamic range (IDR) as possible, since this will maximize the range of sounds available to a CI user. However, it is the discrimination of different levels of sound in each channel that carries information. An excessively large IDR may squeeze these amplitude cues, reducing the ability of an implant recipient to understand speech. There are many parameters that can be adjusted in a CI system. However, it is common for the majority to remain at their default values. This may be through an inability to obtain user feedback, for example in young children, lack of time or knowledge on the part of the clinician, or a recommendation from the CI manufacturer.

How appropriate values are found for the T- and M-levels depends very much on the individual CI user. For a post-lingually deafened adult it is reasonably straightforward to find these. By presenting 200 ms bursts of stimulation and using a standard bracketing approach, the smallest detectable amount of stimulation for each channel can be found and this value set as the T-level. Similarly, progressively increasing the stimulation will allow an M-level to be found, the CI user often pointing to different categories on a loudness chart as the various levels of stimulation are presented. These measures can be made for each individual channel, channels can be programmed in groups of four, or only five or six channels across the electrode array measured with intermediate channels set to interpolated values.

For babies or young children and even for some adults, objective measures are often used to help set program levels. The most common measure used is the eCAP, the electrically elicited compound action potential [39]. The ability to record eCAPs is built into the fitting systems for all of today's major CI systems. Here masker-probe or alternate-subtraction techniques [40, 41] are used to reduce the large stimulus artifact. The amplitude of the remaining physiological signal, arising from synchronized activity on the auditory nerve, is then graphed against the stimulation level. A regression line extrapolates to intersect the stimulation axis which would correspond to a zero amplitude of eCAP. The stimulation value for which this occurs is then used as a guide for setting programming levels. Avoiding stimulus artifact and allowing sufficient neural synchronization, means that much lower stimulation rates are used when measuring eCAPs than for actual everyday stimulation. The means that the absolute eCAP values can fall at various parts of an individual's electrical range. Fortunately, it is the profile of values across the electrode array that it is important to determine. Once this is estimated a global change in level can be made to obtain appropriate loudness. In many cases the T-levels are set to 10% of the M-level since this is almost certainly not going to leave them set too high. Typically T-levels are measured at something like 25% of M-level [42]. When they can be measured and hence individually set, T-levels will tend to improve access to low intensity sounds. Often in clinical practice T-levels are set at a percentage of M-level even where they could be individually set: the additional benefit not being considered worth the additional effort needed for measurement.

Other objective measures are used to assist with programming, although less often due to these requiring additional equipment to be used in collaboration with the CI fitting system. There is a reasonable correlation between an electrically elicited stapedius reflex threshold (eSRT) and M-level [43]. Unlike eCAPs here the same stimulation rate can be used to measure eSRT as will be used in the everyday program. This simplifies the setting of levels and is partly behind why there is such a good correlation with M-level. Less commonly the electrically elicited auditory brainstem response (eABR) is used [44]. Again, eABR will require a lower stimulation rate to be used, so that the characteristic waveforms can be seen in up to 5 or 6 ms following stimulation. This tends to produce an extrapolated threshold for eABR quite high in an individual's electrical dynamic range. As with the eCAP and eSRT measures, it is the relative levels across channels that are important, the profile then being globally adjusted to determine the M-levels that will be used in the program.

Less frequently, some statistically based approaches are used for programming. Simple so-called "flat maps" are used where the T- and M-level is the same on each channel. These are justified by the spread of monopolar stimulation recruiting neurones form a larger section of the cochlea than associated with an individual electrode contact thus tending to produce a spatial averaging. Other approaches might use a template based on the statistical average of levels previously measured for earlier CI recipients. Approaches such as FOX [45] extend this technique, recommending a sequence of programs with progressively increasing levels that are used from the very beginning. For many CI users these techniques can work quite well, although numbers of outliers will require individually tailored programs to realize their potential outcome with comfortable stimulation and reasonable access to their acoustic environment.

Plasticity in the auditory system means that over time the M-levels will usually increase. The longer term M-levels might be typically double those that can be tolerated during the initial fitting. After the first 2 months of device use, neither T- or M-levels tend to change significantly over time [46]. Change in levels is highly individual requiring the initial program levels to be revised numbers of times during the first few months of implant use. Where a second (or third) fitting session is planned within around 2 weeks of the first fitting, most of the change can already be accommodated. Looking across large numbers of adult CI users, program levels will be stable by between 3 and 9 months following first fitting. Individual practice can result in pediatric levels being more slowly increased, leading to 6–12 months being needed to see stable levels.

5. Different patient groups

Cochlear implants were originally designed to help those suffering bilateral profound deafness who could not benefit from acoustic hearing aids. Traditionally candidacy would have required a loss of at least 90 dB HL across all of the audiometric frequencies from 125 to 8000 Hz. Over the past 30 years we have seen, improvements in outcome (speech understanding) through better sound coding strategies and electrode arrays, improvements in esthetics as the external equipment has moved from body worn to behind the ear or single piece processors, improvements in surgery with a skin to skin operating times of well under 1 hour, as well as much smaller incisions not requiring hair shaving and at least in some cases, the preservation of residual hearing. These developments have meant that a CI can now be considered for much more than the 0.2% of the population who suffer profound bilateral sensorineural deafness [47].

It is becoming more common for ears with useful low-frequency residual hearing to receive a CI. Candidacy can now include those with severe to profound levels of hearing loss above 1000–2000 Hz, but normal to moderate hearing loss for lower frequencies [48–50]: a group sometimes referred to as suffering partial deafness. Where the residual hearing can be preserved to within 10–20 dB of the pre-operative levels, many of these recipients use a combination of electrical and acoustic stimulation (EAS) in the same ear. Most CI manufacturers now make EAS processors so that a single instrument supports both modalities, offering comfort, convenience and allowing an EAS fitting to be made using a single piece of software.

Where there is some asymmetry in hearing, recent practice has seen only the poorer ear being implanted, while an acoustic hearing instrument is fitted to the contralateral ear. This is often referred to as bimodal hearing. Dedicated hearing instruments (HI) have been developed that match the compression characteristics and sound cleaning operations between the CI and HI, as well as offering wireless sharing of microphone and control signals. Such systems offer convenience for the user and can combine the natural acoustic low-frequency sound in the HI ear, with the high frequency information supporting speech understanding in the implanted ear.

Bilateral CI provision is now the standard of care in many healthcare systems, at least for children. Receiving two implants, either simultaneously or within a few months of each other, provides the best chance for the brain to have both sides work together. Redundancy, the countering of head shadow and a fuller sense of hearing are all advantages of bilateral implantation. It tends to be only considerations of cost that prevent bilateral CIs being offered universally to all those who could benefit from them. Again with wireless technology developing rapidly, the use of algorithms that combine microphones between the two CI sound processors can offer large improvements for listening in noise when beam formers are used to attenuate noise coming from directions other than directly ahead, particularly useful then the CI user is in a one-to-one conversation in a noisy location.

Where a second CI is not available and there is no aidable hearing on the contralateral ear a CROS, or strictly bi-CROS device can be used. Wireless CROS devices are available that essentially have their microphone pick up sound from the non-implanted side and wirelessly route it to the CI processor on the other side where it is mixed with the CI processor's microphone signal. This approach can reduce head shadow, although with stimulation only being delivered to one ear there is little ability to use the CROS device for localization. With the combination of HI and CI companies, for example Phonak and Advanced Bionics within the Sonova company, the migration of HI technology such as the ear-to-ear wireless technology has begun and will likely be more common in future.

In the German healthcare system, a CI is now available to those who suffer from single-sided deafness (SSD). Typically there may be also be some hearing loss on the better hearing side, making this a highly asymmetrical loss rather than a pure SSD. Those suffering with SSD would usually explore a CROS device and a bone conduction hearing aid before considering a CI. In the end around one third of SSD cases seen will elect to get used to hearing with only one ear, one third will use a bone conduction device and one third will receive a CI [51].

Tinnitus is another consideration that can influence treatment options, for SSD and beyond. Where the SSD is accompanied by intractable levels of tinnitus, a CI may provide relief [52]. The restoration of some input to the deafened ear can allow the tinnitus to either effectively disappear or at least be substantially reduced. In some cases, SSD in particular, the relief from tinnitus is found to be of much greater benefit than any hearing sensation arising from the implanted ear. The large majority of CI recipients report reduced amounts of tinnitus although in very rare cases tinnitus can be worsened through implantation.

5.1 Alternative stimulation sites

The cochlea is an attractive site for electrical stimulation, given that it presents tonotopic access to auditory nerve fibers with reasonably straightforward surgical access. However, where the cochlea has not formed properly or at all, due to some extreme malformation, a properly formed cochlea has been filled with bone or tissue, for example following bacterial meningitis, preventing all but minimal surgical access, or the auditory nerve is not available, either through malformation or following trauma, stimulation of the auditory system via the cochlea is not possible. In such cases alternative sites of stimulation may be used.

Auditory brainstem implants (ABIs) bypass the auditory nerve, targeting the next station of the auditory pathway: the cochlear nucleus located in the brainstem [53]. There is a tonotopic structure within the cochlear nucleus, although it is organized in the dimension of depth, so is not easy to access. Attempts to use a penetrating electrode array with a number of discrete needles has not been able to make better use of this tonotopic organization than a pad of flat electrodes placed on the surface of the cochlear nucleus [54]. Programming of an ABI device tends to be more difficult than that of a CI. Bone surrounding the cochlea usually keeps the CI's stimulation contained to auditory fibers. Only occasionally non-auditory stimulation of, for example, the facial nerve can be seen in muscle twitching around the mouth or eye. This is generally programmed around by deactivating electrode contacts or reducing stimulation levels. However, in the brain stem, functions such as respiration can be adversely effected by an ABI device. This calls for much more care when programming, leading to ABI devices being offered only by specialist centers. The surgery required to place an ABI is more invasive than that required for a CI, for example, requiring lifting of the cerebellum to gain sufficient surgical access. With ABIs being placed following removal of tumors there can be some distortion of brain structures. Some surgeons prefer to remove what are often sizable tumors, allow the brainstem and brain structures to settle into place again and then perform a second surgery during which the ABI is put into place. This two-stage approach is believed to provide less chance of the ABI's electrode pad moving out of position, risking substantial non-auditory stimulation. Outcomes with ABI devices are generally substantially poorer than with CIs. In some series there is essentially no open-set speech understanding possible [55], while in others the speech understanding is limited, with only occasional high levels of speech understanding [56]. The reasons for poor performance with ABIs are not fully understood. Beyond potential movements of the electrode pad, there are specialized auditory functions being carried out in the cochlear nucleus, meaning that simply assuming raw tonotopic stimulation patterns may not be sufficient. Additionally, those receiving ABI devices may have many other issues beyond deafness and these could also explain some of the difference in outcome.

Stimulation of even higher structures in the auditory system has been attempted through an auditory mid-brain implant (AMBI), where the electrode array is inserted into the inferior colliculus. Currently this is restricted to a pure research device [57]. Within the inferior colliculus it is possible to access a tonotopic organization, using a shortened version of a traditional CI electrode array, 10 mm long as opposed to 20–30 mm for most CI arrays. However, when accessing the auditory system at an even higher level than with an ABI device, the amount of pre-processing that should have already been done leaves a crude CI type coding strategy only able to support very limited outcomes. Already at this level higher stimulation rates are inappropriate, leaving limited sound coding strategy options [58].

While placement of an electrode array in the scala tympani, or where necessary in the scala vestibuli, leaves the electrode contacts quite close to their target neurones they are still some 1–3 mm away. This separation prevents discrete stimulation of local neural populations as discussed above. It has been proposed that an electrode array could be inserted directly into the auditory nerve, or failing this inside the modiolus. Promising results have been shown from acute experiments in cats [59]. Recording form electrodes placed in the inferior colliculus indicates that intra-neural simulation is more localized than stimulation using an electrode array placed in the scala tympani. There are considerable challenges to overcome before intra-neural stimulation could be considered for humans. The surgical access is not straightforward, risking losing all residual hearing. The human auditory nerve being in the order of 1 mm diameter would require very small stimulating contacts. While the stimulation currents required would be smaller than those needed for a traditional CI, charge density considerations require careful consideration. Also, how well an electrode array can be placed and be tolerated in the auditory nerve, without the destruction of auditory fibers or the formation of granulation tissue will need to be carefully studied. Finally, the structure of the auditory nerve is complex, with axons from different parts of the cochlea rolling into the tubular nerve, making tonotopic targeting an additional challenge.

6. Current research

With the CI field involving a wide range of professionals including, surgeons, nurses, audiologists, engineers, physicists, speech and language therapists, teachers of the deaf, hearing therapists, rehabilitation specialists, psychologists and health economists, research related to the field can cover a very wide range. Here some of the key topics that are most closely connected to extending current practice will be reviewed.

It is clear that when less trauma is inflicted during surgery that outcomes are better [60], this being the case whether or not there is residual hearing at risk [61, 62]. Hearing preservation is thus a key topic for surgeons. The design of less traumatic electrode arrays [48–50, 63] and the development of less traumatic surgical techniques [64], including the use of robot assisted insertion [65], are factors that can lead to reduced trauma. Providing real time feedback to the surgeon during electrode array insertion is a hotly researched area. Electrocochleography (ECochG), where acoustic stimulation of the ear produces a cochlear microphonic signal [66–68] that the surgeon can use to gauge proximity to structures, such as the basilar membrane, appears promising with clinical systems due to launch in 2019.

The use of drugs to reduce the body's reaction to implantation is also an area with some connection to minimizing trauma. Steroids such as Dexamethasone or antimitotic drugs [69] have been applied to suppress a fibrotic reaction during and immediately after surgery. Some longer term benefits have been shown but mainly in lower electrode contact impedances rather than significant outcome advantage [70]. Longer term deployment of steroids and other drugs has been proposed for some time [71] but has not yet seen clinical practice. Drugs such as neurotrophins [72] have been proposed to enhance the spiral ganglion but carry considerable risk of uncontrolled sprouting of new fibers that may not lead to any improvement in electrical stimulation [73]. Likewise the regrowth of hair cells or other cochlear structures [74] is an extremely challenging problem, although simply reconnecting peripheral processes that have been damaged while leaving intact hair cells [75] may be more manageable in the foreseeable future. While not a drug, near infrared light has been shown to promote tissue healing [76], helping reduce the extent of hearing loss following cochlear stress [77] and has been proposed as an approach that might also enhance the cochlea's ability to survive the traumas of electrode array insertion.

Deployment, development and assessment of sound coding strategies continues with a variety of goals. Optimizing compression parameters to maximize speech understanding [78], reviewing the effect of various parameters as well as individualized fitting approaches [79, 80] all promise improvements for strategies that are already available but could be fitted better. Likewise, tools to guide clinicians in fitting for performance, rather than simply for comfort [45] should also lead to substantial improvement in outcomes. Seeking improvement via limiting current spread, through tripolar stimulation [81, 82], phased array [83] or manipulation of field interactions [84] have so far not shown a general improvement, although benefits for some recipients have been demonstrated. New sound coding strategies that attempt to improve temporal information such as FSP [85], or reduce masking effects such as MP3000 [31] have been developed and introduced into clinical practice. There may be more benefit in reducing battery power from the likes of MP3000 or Optima than in any improvement in outcome. Connected to improving outcome, research into the listening effort required to understand speech has been increasing [86, 87].

The very nature of the speech tests used to evaluate CI systems is an active area of research. Standardization across languages has involved the use of matrix tests [88]. These tests involve a fixed syntax and a closed set of keywords so can be self-administered and used essentially indefinitely. However, the sentences are not fully representative of natural sentences. Avoiding fixed presentation levels, something necessary to evaluate the function of AGC has led to development of roving presentation level tests [89–91]. These tests provide insights into where a particular subject may have problems, so could be useful in supporting programming. As with the STARR tests, multiple speakers have been included in tests such as the AzBio [92] and taken much further with the coordinate response matrix test that can run on a multiple loudspeaker array and hence mimic a more realistic test environment [93].

Improved outcomes have been shown thorough use of the recently developed wireless technology supporting integrated bimodal [94, 95] and CROS [96]. EAS has also been shown to produce substantial improvements in outcome [97] although the test methodology used involves a questionable comparison of conditions.

With increasing numbers of CI recipients needing management by financially constrained health care systems, methods of improving patient management are being developed. These include the use of consumer devices, smart phones and tablets, to run Apps that can evaluate a CI user's speech understanding and qualitative condition as well as remotely analyzing the status of their implant and sound processor hardware [98] and delivering rehabilitation material usable by adult CI recipients and the families of young children. This whole area will necessarily see much development in the coming years.

The pressures on CI manufacturers to reduce size and improve comfort and ease of use continue, with cosmetic considerations playing a large role in the choice of which CI systems are selected. Reliability of both the implanted and external parts of the CI system needs to be continuously improved, underpinning consistent device use and reducing costs inherent in managing failures. Further pressures on cost are also critical to address so that the enormous unmet need for cochlear implants in developing countries can also be met.

Glossary of technical terms

Alternate-subtraction	a method of removing stimulus artifact that relies on delivering alternate phase stimulation pulses that are added to ideally cancel artifact and reinforce response
Anodic	refers to the positive going phase of a stimulation pulse that is generally assumed to provide change balance, hence avoiding the delivery of potentially harmful direct current
Automatic Gain Control (AGC)	a circuit that compresses the large acoustic dynamic range into a range that is more manageable for the restricted electrical dynamic range
Capacitor	an electrical component that does not allow direct current to pass so offers an additional level of protection to the body in the case of a fault occurring in an implant
Cathodic	refers to the negative going phase of a stimulation pulse that is generally assumed to depolarize neurones hence leading to electrical stimulation
Charge	an electrical measure formed by the product (strictly the integral) of current and time with units of Coulombs. The charge delivered determines the loudness perceived
Cochlear Implant (CI)	a surgically implantable prosthesis that bypasses a damaged cochlea providing hearing through direct electrical stimulation of the auditory nerve
Current source	an electrical circuit that delivers a programmed current, varying the amount of voltage necessary to achieve this depending on the electrical resistance offered by the body
Direct current	an electrical current that flows in one direction only that can be harmful to the body
Electrocochleography (ECochG)	the measurement of electrical signals produced by the cochlea in response to acoustic stimulation that indicates cochlear health

Electrolysis	an electro-chemical process whereby charged particles may be drawn towards an electrode with opposite charge, leading to electrode contact dissolution or tissue damage
Fast Fourier Transform (FFT)	an efficient software algorithm that evaluates the amount of sound energy at regularly spaced frequencies, so allows the channels to be calculated
Impedance	a measure of resistance to the flow of electrical current that takes account of frequency so is more general than resistance which strictly only applies to direct current
Masker-Probe	a method of stimulus artifact removal relying on introducing refraction to identify artifact and then using multiple stages of subtraction to isolate the neural response
Perilymph	a fluid contained within much of the cochlea that is electrically conductive
Resistance	a measure of how much the body will resist the flow of electrical current, strictly only considering direct current, impedance being the a more general factor
Spiral Ganglion Cell	the name given to hearing nerve neurones having a cell body in the cochlea's modiolus, axon in the auditory nerve and dendrites innervating the inner hair cells
Voltage source	an electrical circuit that delivers a programmed voltage but that will allow current to vary depending on the resistance (impedance) offered by the body

Author details

Patrick J. Boyle
Advanced Bionics GmbH, European Research Center, Hannover, Germany

*Address all correspondence to: patrick.boyle@advancedbionics.com

References

[1] Volta A. On the electricity excited by the mere contact of conducting substances of different kinds. Proceedings of the Royal Society. 1800;**1**:27-29

[2] Djourno A. Eyries, Prosthese auditive par excitation electrique a distance du nerf sensoriel a' l'aide d'un bobinage inclus a demeure. La Presse Médicale. 1957;**65**:1417

[3] Simmons FB. Electrical stimulation of the auditory nerve in man. Archives of Otolaryngology. 1966;**84**(1):2-54

[4] Michelson RP. Electrical stimulation of the human cochlea. A preliminary report. Archives of Otolaryngology. 1971;**93**(3):317-323

[5] Chouard CH, MacLeod P. Implantation of multiple intracochlear electrodes for rehabilitation of total deafness: Preliminary report. Laryngoscope. 1976;**86**(11):1743-1751

[6] House WF. Goals of the cochlear implant. Laryngoscope. 1974;**84**(11):1883-1887

[7] Graham JM, Hazell JW. Electrical stimulation of the human cochlea using a transtympanic electrode. British Journal of Audiology. 1977;**11**(2):59-62

[8] Hochmair ES, Hochmair-Desoyer IJ, Burian K. Experience with implanted auditory nerve stimulator. Transactions—American Society for Artificial Internal Organs. 1979;**25**:357-361

[9] Clark GM, Tong YC, Martin LF. A multiple-channel cochlear implant: An evaluation using open-set CID sentences. Laryngoscope. 1981;**91**(4):628-634

[10] Shannon RV. Multichannel electrical stimulation of the auditory nerve in man. I. Basic psychophysics. Hearing Research. 1983;**11**:157-189

[11] Frijns J, Kalkman R, Vanpoucke F, Bongers J, Briaire J. Simultaneous and non-simultaneous dual electrode stimulation in cochlear implants: Evidence for two neural response modalities. Acta Oto-Laryngologica. 2009;**129**(4):433-439

[12] Firszt JB, Koch DB, Downing M, Litvak L. Current steering creates additional pitch percepts in adult cochlear implant recipients. Otology & Neurotology. 2007;**28**(5):629-636

[13] McDermott HJ, McKay CM. Pitch ranking with nonsimultaneous dual-electrode electrical stimulation of the cochlea. The Journal of the Acoustical Society of America. 1994;**96**:155-162

[14] Shepherd RK, Linahan N, Xu J, Clark GM, Araki S. Chronic electrical stimulation of the auditory nerve using non-charge-balanced stimuli. Acta Oto-Laryngologica. 1999;**119**(6):674-684

[15] Wang JT, Wang AY, Psarros C, Da Cruz M. Rates of revision and device failure in cochlear implant surgery: A 30-year experience. Laryngoscope. 2014;**124**(10):2393-2399

[16] van der Marel KS, Briaire JJ, Verbist BM, Joemai RM, Boermans PP, Peek FA, et al. Cochlear reimplantation with same device: Surgical and audiologic results. Laryngoscope. 2011;**121**(7):1517-1524

[17] Leung RT, Shivdasani MN, Nayagam DA, Shepherd RK. In vivo and in vitro comparison of the charge injection capacity of platinum macroelectrodes. IEEE Transactions on Biomedical Engineering. 2015;**62**(3):849-857

[18] Stover T, Lenarz T. Biomaterials in cochlear implants. GMS Current Topics in Otorhinolaryngology—Head and Neck Surgery. 2009;**8**(10):10

[19] Shepherd RK, Carter P, Enke YL, Wise AK, Fallon JB. Chronic intracochlear electrical stimulation at high charge densities results in platinum dissolution but not neural loss or functional changes in vivo. Journal of Neural Engineering. 2018;**5**(10):1741-2552

[20] Zeng F-G. Compression and cochlear implants. In: Bacon SP, Popper AN, Fay RR, editors. Compression: From Cochlea to Cochlear Implants. New York: Springer; 2004. pp. 184-220

[21] Chatterjee M, Fu Q J, Shannon RV. Effects of phase duration and electrode separation on loudness growth in cochlear implant listeners. The Journal of the Acoustical Society of America. 2000;**107**(3):1637-1644

[22] Stöbich B, Zierhofer CM, Hochmair ES. Influence of automatic gain control parameter settings on speech understanding of cochlear implant users employing the continuous interleaved sampling strategy. Ear and Hearing. 1999;**20**:104-116

[23] Boyle P, Buchner A, Stone M, Lenarz T, Moore B. Comparison of dual-time-constant and fast-acting automatic gain control (AGC) systems in cochlear implants. International Journal of Audiology. 2009;**48**(4):211-221

[24] Khing PP, Swanson BA, Ambikairajah E. The effect of automatic gain control structure and release time on cochlear implant speech intelligibility. PLoS One. 2013;**8**(11):e82263. pp. 1-11

[25] Stone MA, Moore BCJ. Side effects of fast-acting dynamic range compression that affect intelligibility in a competing speech task. The Journal of the Acoustical Society of America. 2004;**116**:2311-2323

[26] Wilson BS, Finley CC, Lawson DT, Wolford RD, Eddington DK, Rabinowitz WM. Better speech recognition with cochlear implants. Nature. 1991;**352**:236-238

[27] Wilson BS, Finley CC, Lawson DT, Wolford RD, Zerbi M. Design and evaluation of a continuous interleaved sampling (CIS) processing strategy for multichannel cochlear implants. Journal of Rehabilitation Research and Development. 1993;**30**(1):110-116

[28] Buechner A, Frohne-Buechner C, Stoever T, Gaertner L, Battmer R, Lenarz T. Comparison of a paired or sequential stimulation paradigm with advanced bionics' high-resolution mode. Otology & Neurotology. 2005;**26**(5):941-947

[29] Patrick JF, Busby PA, Gibson PJ. The development of the nucleus(R) Freedom^{TM} cochlear implant system. Trends in Amplification. 2006;**10**(4):175-200

[30] Tye-Murray N, Lowder M, Tyler RS. Comparison of the F0F2 and F0F1F2 processing strategies for the Cochlear Corporation cochlear implant. Ear and Hearing. 1990;**11**(3):195-200

[31] Buchner A, Nogueira W, Edler B, Battmer RD, Lenarz T. Results from a psychoacoustic model-based strategy for the nucleus-24 and freedom cochlear implants. Otology & Neurotology. 2008;**29**(2):189-192

[32] Moore BCJ. Psychoacoustics of normal and impaired hearing. British Medical Bulletin. 2002;**63**:121-134

[33] Koch DB, Downing M, Osberger MJ, Litvak L. Using current steering to increase spectral resolution in CII and HiRes 90K users. Ear and Hearing. 2007;**28**(2 Suppl):38S-41S

[34] Kamakura T, Nadol JB Jr. Correlation between word recognition score and intracochlear new bone and fibrous tissue after cochlear implantation in the human. Hearing Research. 2016;**339**:132-141

[35] Seyyedi M, Viana LM, Nadol JB Jr. Within-subject comparison of word recognition and spiral ganglion cell count in bilateral cochlear implant recipients. Otology & Neurotology. 2014;**35**(8):1446-1450

[36] van der Beek FB, Briaire JJ, van der Marel KS, Verbist BM, Frijns JH. Intracochlear position of Cochlear implants determined using CT scanning versus fitting levels: Higher threshold levels at basal turn. Audiology & Neuro-Otology. 2016;**21**(1):54-67

[37] Dietz A, Wennstrom M, Lehtimaki A, Lopponen H, Valtonen H. Electrode migration after cochlear implant surgery: More common than expected? European Archives of Oto-Rhino-Laryngology. 2015;**12**:12

[38] van der Marel K, Verbist B, Briaire J, Joemai R, Frijns J. Electrode migration in cochlear implant patients: Not an exception. Audiology & Neuro-Otology. 2012;**17**(5):275-281

[39] Abbas PJ, Brown CJ. Electrically evoked auditory brainstem response: Growth of response with current level. Hearing Research. 1991;**51**(1):123-137

[40] Cohen LT, Richardson LM, Saunders E, Cowan RS. Spatial spread of neural excitation in cochlear implant recipients: comparison of improved ECAP method and psychophysical forward masking. Hearing Research. 2003;**179**(1-2):72-87

[41] Eisen MD, Franck KH. Electrically evoked compound action potential amplitude growth functions and HiResolution programming levels in pediatric CII implant subjects. Ear and Hearing. 2004;**25**(6):528-538

[42] Van der Beek FB, Frijns J. Population-based prediction of fitting levels for individual cochlear implant recipients. Audiology and Neurotology. 2014;**19**(20):1-16

[43] Hodges AV, Balkany TJ, Ruth RA, Lambert PR, Dolan-Ash S, Schloffman JJ. Electrical middle ear muscle reflex: Use in Cochlear implant programming. Otolaryngology—Head and Neck Surgery. 1997;**117**(3):255-261

[44] Allum JH, Shallop JK, Hotz M, Pfaltz CR. Characteristics of electrically evoked 'auditory' brainstem responses elicited with the nucleus 22-electrode intracochlear implant. Scandinavian Audiology. 1990;**19**(4):263-267

[45] Govaerts P, Vaerenberg B, De Ceulaer G, Daemers K, De Beukelaer C, Schauwers K. Development of a software tool using deterministic logic for the optimization of cochlear implant processor programming. Otology & Neurotology. 2010;**31**(6):908-918

[46] Gajadeera EA, Galvin KL, Dowell RC, Busby PA. The change in electrical stimulation levels during 24 months Postimplantation for a large cohort of adults using the nucleus(R) Cochlear implant. Ear and Hearing. 2017;**38**(3):357-367

[47] Davis A. Hearing in Adults. London: Whurr; 1995

[48] Skarzynski H, Lorens A, Matusiak M, Porowski M, Skarzynski P, James C. Partial deafness treatment with the nucleus straight research array cochlear implant. Audiology & Neuro-Otology. 2012;**17**(2):82-91

[49] Lenarz T, Stover T, Buechner A, Lesinski-Schiedat A, Patrick J, Pesch J. Hearing conservation surgery using the Hybrid-L electrode. Results from the first clinical trial at the Medical University of Hannover. Audiology & Neuro-Otology. 2009;**14**(Suppl 1):22-31

[50] Gantz BJ, Hansen MR, Turner CW, Oleson JJ, Reiss LA, Parkinson AJ. Hybrid 10 clinical trial: Preliminary results. Audiology & Neuro-Otology. 2009;**1**:32-38

[51] Arndt S, Laszig R, Aschendorff A, Hassepass F, Beck R, Wesarg T. Cochlear implant treatment of patients with single-sided deafness or asymmetric hearing loss. German version. Hals-Nasen-Ohrenheilkunde. 2017;**65**(7):586-598

[52] Mertens G, De Bodt M, Van de Heyning P. Cochlear implantation as a long-term treatment for ipsilateral incapacitating tinnitus in subjects with unilateral hearing loss up to 10 years. Hearing Research. 2016;**331**:1-6

[53] Shannon RV, Fayad J, Moore J, Lo WW, Otto S, Nelson RA, et al. Auditory brainstem implant: II. Postsurgical issues and performance. Otolaryngology and Head and Neck Surgery. 1993;**108**(6):634-642

[54] Otto SR, Shannon RV, Wilkinson EP, Hitselberger WE, McCreery DB, Moore JK, et al. Audiologic outcomes with the penetrating electrode auditory brainstem implant. Otology & Neurotology. 2008;**29**(8):1147-1154

[55] Shannon RV. Auditory implant research at the house ear institute 1989-2013. Hearing Research. 2015;**322**:57-66

[56] Colletti V, Shannon R, Carner M, Veronese S, Colletti L. Outcomes in nontumor adults fitted with the auditory brainstem implant: 10 years' experience. Otology & Neurotology. 2009;**30**(5):614-618

[57] Lenarz T, Lim HH, Reuter G, Patrick JF, Lenarz M. The auditory midbrain implant: A new auditory prosthesis for neural deafness-concept and device description. Otology & Neurotology. 2006;**27**(6):838-843

[58] Rode T, Hartmann T, Hubka P, Scheper V, Lenarz M, Lenarz T, et al. Neural representation in the auditory midbrain of the envelope of vocalizations based on a peripheral ear model. Frontiers in Neural Circuits. 2013;**7**(166):18

[59] Middlebrooks JC, Snyder RL. Intraneural stimulation for auditory prosthesis: Modiolar trunk and intracranial stimulation sites. Hearing Research. 2008;**242**(1-2):52-63

[60] Finley C, Holden T, Holden L, Whiting B, Chole R, Neely G, et al. Role of electrode placement as a contributor to variability in cochlear implant outcomes. Otology & Neurotology. 2008;**29**(7):920-928

[61] Aschendorff A, Kromeier J, Klenzner T, Laszig R. Quality control after insertion of the nucleus contour and contour advance electrode in adults. Ear and Hearing. 2007;**28**(2 Suppl):75S-79S

[62] Fitzpatrick DC, Campbell AP, Choudhury B, Dillon MT, Forgues M, Buchman CA, et al. Round window electrocochleography just before cochlear implantation: Relationship to word recognition outcomes in adults. Otology & Neurotology. 2014;**35**(1):64-71

[63] Dhanasingh A, Jolly C. An overview of cochlear implant electrode array designs. Hearing Research. 2017;**356**:93-103

[64] Todt I, Mittmann P, Ernst A. Hearing preservation with a Midscalar electrode comparison of a regular and steroid/pressure optimized surgical approach in patients with residual hearing. Otology & Neurotology. 2016;**37**:e349-e352

[65] Torres R, Jia H, Drouillard M, Bensimon JL, Sterkers O, Ferrary E, et al. An optimized robot-based technique for cochlear implantation to reduce array insertion trauma. Otolaryngology and Head and Neck Surgery. 2018;**159**(5):900-907

[66] Giardina CK, Brown KD, Adunka OF, Buchman CA, Hutson KA, Pillsbury HC, et al. Intracochlear electrocochleography: Response

patterns during cochlear implantation and hearing preservation. Ear and Hearing. 2019. (In Press)

[67] Bester CW, Campbell L, Dragovic A, Collins A, O'Leary SJ. Characterizing electrocochleography in cochlear implant recipients with residual low-frequency hearing. Frontiers in Neuroscience. 2017;**11**(141):1-8

[68] Harris MS, Riggs WJ, Koka K, Litvak LM, Malhotra P, Moberly AC, et al. Real-time intracochlear electrocochleography obtained directly through a cochlear implant. Otology & Neurotology. 2017;**38**(6):e107-e113

[69] Jia H, Francois F, Bourien J, Eybalin M, Lloyd RV, Van De Water TR, et al. Prevention of trauma-induced cochlear fibrosis using intracochlear application of anti-inflammatory and antiproliferative drugs. Neuroscience. 2016;**316**:261-278

[70] Wilk M, Hessler R, Mugridge K, Jolly C, Fehr M, Lenarz T, et al. Impedance changes and fibrous tissue growth after cochlear implantation are correlated and can be reduced using a dexamethasone eluting electrode. PLoS One. 2016;**11**(2):e0147552. p. 19

[71] Farahmand Ghavi F, Mirzadeh H, Imani M, Jolly C, Farhadi M. Corticosteroid-releasing cochlear implant: A novel hybrid of biomaterial and drug delivery system. Journal of Biomedical Materials Research. Part B, Applied Biomaterials. 2010;**94**(2):388-398

[72] Schindler JS, Schindler RA. Neurotrophin action in the cochlea: Implications for cochlear implants. Advances in Oto-Rhino-Laryngology. 1997;**52**:8-14

[73] Leake PA, Stakhovskaya O, Hetherington A, Rebscher SJ, Bonham B. Effects of brain-derived neurotrophic factor (BDNF) and electrical stimulation on survival and function of cochlear spiral ganglion neurons in deafened, developing cats. Journal of the Association for Research in Otolaryngology. 2013;**14**(2):187-211

[74] Taylor RR, Filia A, Paredes U, Asai Y, Holt JR, Lovett M, et al. Regenerating hair cells in vestibular sensory epithelia from humans. eLife. 2018;**18**(7):34817

[75] Wan G, Gomez-Casati ME, Gigliello AR, Liberman MC, Corfas G. Neurotrophin-3 regulates ribbon synapse density in the cochlea and induces synapse regeneration after acoustic trauma. eLife. 2014;**20**(3):03564

[76] Whelan HT, Smits RL Jr, Buchman EV, Whelan NT, Turner SG, Margolis DA, et al. Effect of NASA light-emitting diode irradiation on wound healing. Journal of Clinical Laser Medicine & Surgery. 2001;**19**(6):305-314

[77] Bartos A, Grondin Y, Bortoni ME, Ghelfi E, Sepulveda R, Carroll J, et al. Pre-conditioning with near infrared photobiomodulation reduces inflammatory cytokines and markers of oxidative stress in cochlear hair cells. Journal of Biophotonics. 2016;**9**(11-12):1125-1135

[78] Boyle PJ, Moore BCJ. Balancing cochlear implant AGC and near-instantaneous compression to improve perception of soft speech. Cochlear Implants International. 2015;**16**(S1):S9-S11

[79] Holden LK, Finley CC, Firszt JB, Holden TA, Brenner C, Potts LG, et al. Factors affecting open-set word recognition in adults with cochlear implants. Ear and Hearing. 2013;**34**(3):342-360

[80] Baudhuin J, Cadieux J, Firszt J, Reeder R, Maxson J. Optimization of programming parameters in children with the advanced bionics

cochlear implant. Journal of the American Academy of Audiology. 2012;**23**(5):302-312

[81] Fielden CA, Kluk K, Boyle PJ, McKay CM. The perception of complex pitch in cochlear implants: A comparison of monopolar and tripolar stimulation. The Journal of the Acoustical Society of America. 2015;**138**(4):2524-2536

[82] Srinivasan AG, Padilla M, Shannon RV, Landsberger DM. Improving speech perception in noise with current focusing in cochlear implant users. Hearing Research. 2013;**299**:29-36

[83] Smith ZM, Parkinson WS, Long CJ. Multipolar current focusing increases spectral resolution in cochlear implants. Conference Proceedings: Annual International Conference of the IEEE Engineering in Medicine and Biology Society. 2013;**9**(10):6610121

[84] Kalkman RK, Briaire JJ, Frijns JH. Stimulation strategies and electrode design in computational models of the electrically stimulated cochlea: An overview of existing literature. Network. 2016;**27**(2-3):107-134

[85] Muller J, Brill S, Hagen R, Moeltner A, Brockmeier SJ, Stark T, et al. Clinical trial results with the MED-EL fine structure processing coding strategy in experienced cochlear implant users. ORL: Journal for Otorhinolaryngology and Its Related Specialties. 2012;**74**(4):185-198

[86] Gagne JP, Besser J, Lemke U. Behavioral assessment of listening effort using a dual-task paradigm. Trends in Hearing. 2017;**21**:2331216516687287

[87] Firszt JB, Reeder RM, Holden LK, Dwyer NY. Results in adult cochlear implant recipients with varied asymmetric hearing: A prospective longitudinal study of speech recognition, localization, and participant report. Ear and Hearing. 2018;**39**(5):845-862

[88] Kollmeier B, Warzybok A, Hochmuth S, Zokoll MA, Uslar V, Brand T, et al. The multilingual matrix test: Principles, applications, and comparison across languages: A review. International Journal of Audiology. 2015;**2**:3-16

[89] Boyle P, Nunn T, O'Connor A, Moore B. STARR: A speech test for evaluation of the effectiveness of auditory prostheses under realistic conditions. Ear and Hearing. 2013;**34**(2):203-212

[90] Haumann S, Lenarz T, Buchner A. Speech perception with cochlear implants as measured using a roving-level adaptive test method. ORL: Journal for Otorhinolaryngology and Its Related Specialties. 2010;**72**(6):312-318

[91] Dincer D'Alessandro H, Ballantyne D, Boyle PJ, De Seta E, DeVincentiis M, Mancini P. Temporal fine structure processing, pitch, and speech perception in adult cochlear implant recipients. Ear and Hearing. 2018;**39**(4):679-686

[92] Spahr A, Dorman M, Litvak L, Van Wie S, Gifford R, Loizou P, et al. Development and validation of the AzBio sentence lists. Ear and Hearing. 2012;**33**(1):112-117

[93] Kitterick PT, Bailey PJ, Summerfield AQ. Benefits of knowing who, where, and when in multi-talker listening. The Journal of the Acoustical Society of America. 2010;**127**(4):2498-2508

[94] Veugen LC, Chalupper J, Snik AF, van Opstal AJ, Mens LH. Frequency-dependent loudness balancing in bimodal cochlear implant users. Acta Oto-Laryngologica. 2016;**136**(8):775-781

[95] Dorman MF, Cook Natale S, Agrawal S. The value of unilateral CIs, CI-CROS and bilateral CIs, with and without beamformer microphones, for

speech understanding in a simulation of a restaurant environment. Audiology & Neuro-Otology. 2018;**23**(5):270-276

[96] Snapp HA, Hoffer ME, Spahr A, Rajguru S. Application of wireless contralateral routing of signal technology in unilateral cochlear implant users with bilateral profound hearing loss. Journal of the American Academy of Audiology. 2018;**29**(10):17121

[97] Friedmann DR, Peng R, Fang Y, McMenomey SO, Roland JT, Waltzman SB. Effects of loss of residual hearing on speech performance with the CI422 and the hybrid-L electrode. Cochlear Implants International. 2015;**26**:26

[98] Cullington H, Kitterick P, Weal M, Margol-Gromada M. Feasibility of personalised remote long-term follow-up of people with cochlear implants: A randomised controlled trial. BMJ Open. 2018;**8**(4):2017-019640

Chapter 6

Endoscopic Ear Surgery in Children

Luca Oscar Redaelli de Zinis and Nader Nassif

Abstract

Endoscopic assistance is gradually gaining recognition in otology not only for office examinations but also during surgery. The first endoscopic surgical procedure that was started in our institution was endoscopic ventilation tube placement to manage children with stenotic and curved canals. Following this, endoscopy was used in all type I tympanoplasty and stage I cholesteatoma removals with the advantage of avoiding a postauricular or endaural approach. The last application of endoscopic assistance was to better visualize round window and scala tympani via posterior tympanotomy during cochlear implantation. There are several advantages in using endoscopes: the wide view obtained and the possibility to observe areas behind the angle with less invasiveness and its excellent resolution, in addition to its intense light and higher magnification that facilitates teaching and tutoring. The limits of endoscopic surgery are that one hand is always needed to hold the endoscope and the lack of a third dimension. Until miniaturization of 3D systems allow the possibility to work in the narrow external ear canal, in order to overcome the limitation that one hand is dedicated to the endoscope, we will describe the use of an endoscope holder in otologic procedures.

Keywords: endoscopy, ear surgery, pediatric otology, chronic otitis media, cholesteatoma, otitis media with effusion, cochlear implantation, myringoplasty, tympanic membrane perforation

1. Introduction

The impressive technological evolution of the last decades and the miniaturization of instruments have allowed to intervene with less invasiveness and better functional results in the field of surgical otology. The introduction of endoscopy with rigid endoscopes, coupled with video processors and high definition monitors, has opened new alternatives to conventional and validated otomicroscopic surgical approaches to the middle ear and avoid external access to various pathologies. Otoendoscopy is gradually taking hold on otology not only in diagnostic approaches but also in surgical aspects. It is well known that the endoscope has the advantage of offering a wider view compared to microscopes.

One of the first endoscopic surgical approaches to the middle ear was described by Thomassin [1]. Thereafter, many articles were published on otoendoscopic surgical procedures. Its limits are related to the lack of a third dimension, and in case of bloody fields, surgery becomes demanding and troublesome. In this chapter, we will discuss our experience developed over the years in endoscopic otology in children starting in 2004 as a diagnostic tool and then gradually as a surgical instrument

 IntechOpen

starting from ventilation tube placement, myringoplasty, cholesteatoma removal, ossiculoplasty, cochlear implantation, and, finally, the application of a holder for two-handed endoscopic surgery.

2. Endoscopy as a diagnostic tool

Endoscopy provides an easy and comfortable means of examining the external auditory canal (EAC) and tympanic membrane in children for diagnostic purposes and follow-up as data is recorded and archived on a hard disk. Most children are uncooperative during office sessions for pre- and post-operative evaluations when they are laid on a bed under the microscope. With endoscopy, the child can sit on the lap of his/her parent, feeling safe and being more compliant, but otherwise by him/herself (**Figure 1**).

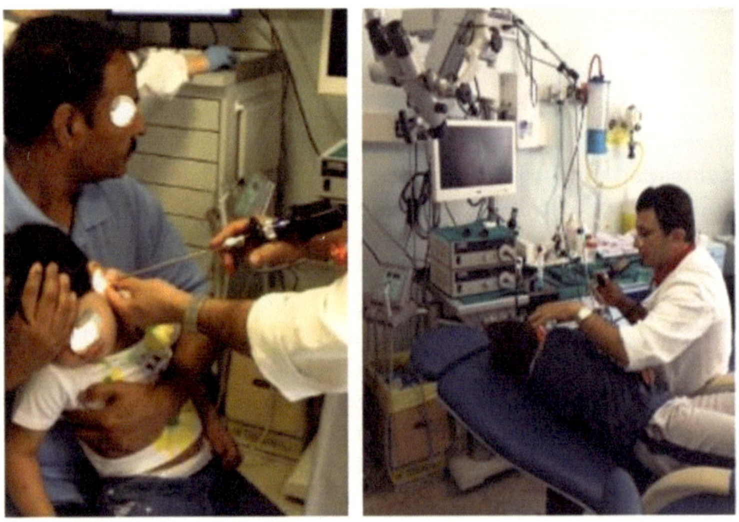

Figure 1.
On the left, a child sitting on the lap of his father during endoscopy; on the right, a boy lying on the examination bed by himself.

3. Ventilation tube placement

Myringotomy with placement of ventilation tubes (VT) is considered to be basic treatment for the ENT surgeon in certain pathologies of the middle ear such as otitis media with effusion (OME). The alternative to the use of a microscope in carrying out such a procedure is a rigid endoscope.

Endoscopic VT placement was started at our ENT Pediatric Department in 2008 by one colleague and was gradually followed by colleagues who shifted from a microscope to an endoscope until 2012, when all grommets were placed endoscopically. Nowadays, it is considered as standard technique, owing to the easy and straightforward nature of the procedure.

No special prior preparation is needed for endoscopic VT placement. In pediatric patients, the procedure is always performed under general anesthesia. The patient is placed in an otosurgical position, and cleansing of the surgical field is performed using an antiseptic product prior to removal of wax or debris when present. Wax is not a limitation in our hands, in contrast to what noted by other authors [2]; it is easily handled and cleaned endoscopically. Hair trimming was not needed in any patient. Antifogging liquid is needed to avoid blurred vision noted in another study [3].

The endoscopic procedure is stepwise starting with myringotomy in the anterior quadrants (**Figure 2A**) using a sickle knife; suctioning of secretions from the middle ear; positioning the rim of the grommet by forceps on the site of the incision (**Figure 2B**); and gentle slipping of the grommet inside the myringotomy with a needle or a pick (**Figure 2C**). We usually use two types of VTs: a Shepard grommet made of fluoroplastic and silicone Goode T-Tubes.

No age-related limitations were encountered in endoscopic VT placement and, in fact, the age of our patients at surgery ranged from 1 month to 15 years.

The **endoscope** to use in such procedures is 0° angled. At the beginning of our experience, surgical treatment was also performed with a 30° angled scope, but no significant benefit over the 0° endoscope was observed. Moreover, as described in other studies, different lengths of endoscopes, such as 6, 11, 14, 16, and 18 cm, can be used to carry out endoscopic management [2, 4]. In our hands, the use of 11 cm or longer scopes offers greater maneuvering space for other tools without limiting the excursion of the operative hand by touching the camera.

Endoscopes with different **diameters** (1.9, 2.7, 3, and 4 mm) are also available. Some authors have suggested the use of a small (2.7 mm) scope in pediatric patients [2]. In our experience, 3 or a 4 mm scopes can be used interchangeably except for stenotic, bending EAC or syndromic patients (i.e., Down S., Goldenhar S., Pallister-Killian S., etc.), and in that case, a 3.0 mm endoscope is recommended. A 2.7 mm, 18 cm long endoscope is avoided because in comparison with the 3.0 endoscope, it has less luminosity should slight pressure be exerted on the shaft and a black crescent appears laterally, reducing the operating field. As we all know, in narrow or oblique EACs, positioning of VTs with an operating microscope can be a troublesome task. The ear speculum itself occupies space in the membranous ear canal, further reducing its diameter. Secondly, in a curved EAC, it is often not possible to view the inferior quadrants of the eardrum where the tube should be placed. On the contrary, the use of an adequate endoscope in a narrow and curved canal offers a larger visual field. According to our experience, it is preferable to use an endoscope with a larger diameter that is compatible with the size of the EAC. There is no need for special **tools** for endoscopic placement of grommets. The same instrument set used in a standard microscopic approach is required. In our experience, sickle knives with the smallest tips, both straight and slightly curved, are the best choice. Additionally, delicate ear forceps having a working length of 8 cm are usually well handled. Suction tubes that fit well in the pediatric age are those with diameters from 3 to 5 French.

The standard **endoscopic technique** for grommet insertion has been described above, whereas T-tube placement is more articulated, especially in the presence of atelectasis. However, in these cases, the degree of difficulty is not different from conventional otomicroscopy procedures. After myringotomy, the T-tube is grasped

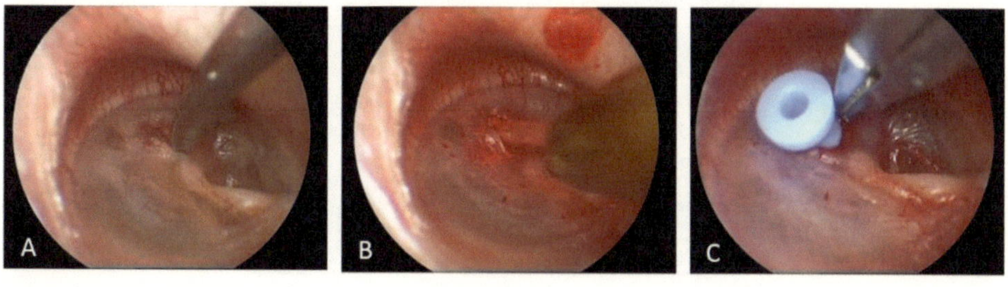

Figure 2.
(A) Myringotomy, (B) suction of secretions, and (C) positioning of the grommet.

either by both wings together or by one only and inserted inside the incision. The advantage of otoendoscopy compared to the operating microscope is that during placement of the T-tube, one can clearly evaluate the depth of insertion of the tube wings (**Figure 3**). The endoscope itself may be used to keep the T-tube in position while releasing and extracting the forceps.

Furthermore, an endoscopic surgical technique can be influenced by the size of EAC. Whenever the EAC diameter is stenotic, the surgical maneuver differs from standard cases. As performed in our patients, the tube is laid at the meatus and is then pushed medially and inserted in the myringotomy with a pick. The scope is not introduced deeply in the ear canal, but is stopped after its entrance. The impression of the endoscope on the membranous wall of the EAC can be noticed in **Figure 4A** and stenosis of the EAC with a diameter equal to that of the grommet is observed in **Figure 4B**.

In one of our patients, stenosis was not the only difficulty: the reduced space between the ear drum and the anterior wall of the EAC, that is, the acute anterior tympanomeatal angle, limited the surgical maneuvers was also problematic. Rotating the tube inside the incision was impossible due to the hindrance of the anterior wall. In this specific case, it was decided to switch to a T-tube to complete the procedure successfully (**Figure 5**). This was possible because in a very narrow space, holding the T-tube with a Hartmann forceps under endoscopic control allowed the flanges of the T to enter the myringotomy directly.

The type of **grommet** selected is another important issue in endoscopic myringotomies with VTs. Rigid materials such as fluoroplastic have some advantages over elastic ones (such as silicone or microgel). The reason lies in the need to push the grommet with a needle or a pick in smaller size EACs. The instruments would penetrate soft materials, thus not enabling correct insertion or getting stuck in it. In a couple of cases, the VT bounced back and we had to repeat the maneuver. Elastic materials might be preferred in wide EACs, where the grommet is applied directly by the ear forceps.

The **limitations** in handling instruments differ between left-handed and right-handed surgeons. For those right handed, tube placement in the right ear canal is more demanding, owing to the higher risk of excoriation of the inferior wall of the EAC and subsequent bleeding (**Figure 6**).

In left ears, the field of vision improves but there is a higher risk of skin lesions or hematoma of the anterior wall of the EAC. The opposite applies for those who are left handed; that is, excoriation of anterior canal wall in the right ear canal and inferior wall of the left ear canal (**Figure 6**).

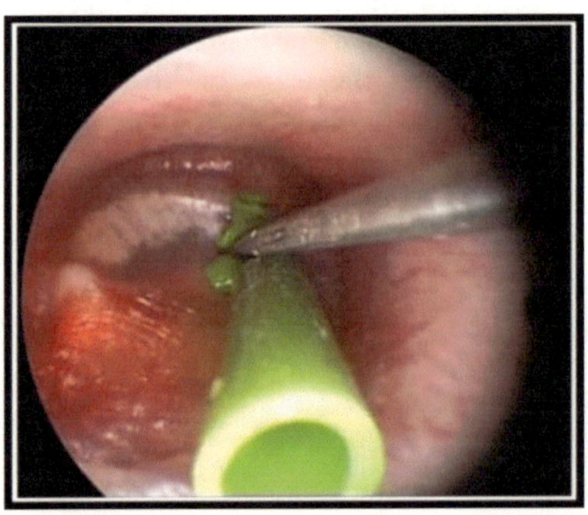

Figure 3.
Positioning of T-tube grabbed by its wings.

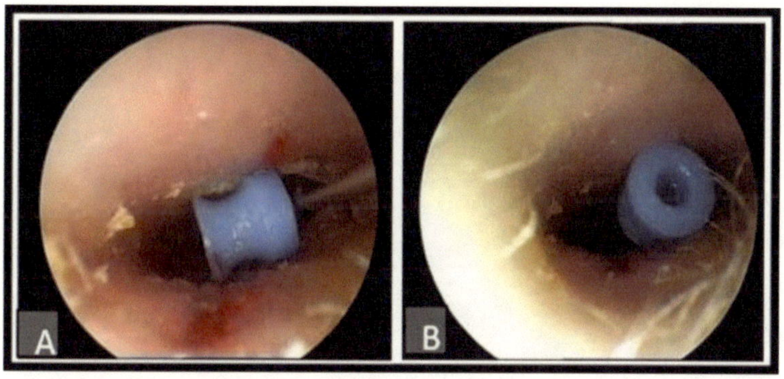

Figure 4.
Positioning of the grommet in stenotic EAC. (A) Notice the impression of the 2.7 mm endoscope greater in diameter w.r.t the EAC and (B) diameter of the grommet compared to that of the EAC.

According to the literature, a major limit of endoscopic ear surgery is bleeding. When it occurs, the surgeon is sometimes forced to interrupt or convert the procedure to an otomicroscopic, bi-manual technique [12, 13, 15]. Senior surgeons are accustomed to use the left hand for suction (if right-handed) and the right hand to maneuver the instruments within the EAC, due to traditional otosurgical training. This allows maintaining the tiny operative field within the ear canal clear from blood. In our series, bleeding was never a relevant issue: in no instance did the procedure become so demanding and troublesome that there was the need to convert it to the microscope. In case of bloody field due to inadvertent tearing of the canal skin, the simple application of a sponge soaked with a vasoconstricting agent such as epinephrine 1/1000 for a few minutes allowed adequate hemostasis. Following this, rinsing the ear canal with warm water or saline also helps to clear the field. It is clear that the surgical skills of the operator and his/her experience with the use of endoscopes play an important role in minimizing surgical trauma.

Single **handedness** in endoscopy is considered by different otologists as a limitation in ear surgery, even for simple procedures such as VT placement [2, 3]. According to our experience, this is not the case as long as the skin of the ear canal is respected by taking into consideration the above advice.

Unlike other studies in the literature [3], based on our experience, no special **training** for endoscopic grommet positioning is required. Special training is definitely required for more complex procedures such as tympanoplasties,

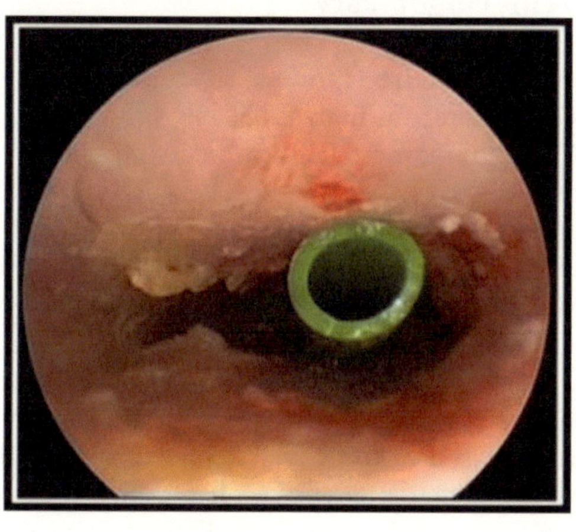

Figure 5.
T-tube placed in stenotic EAC.

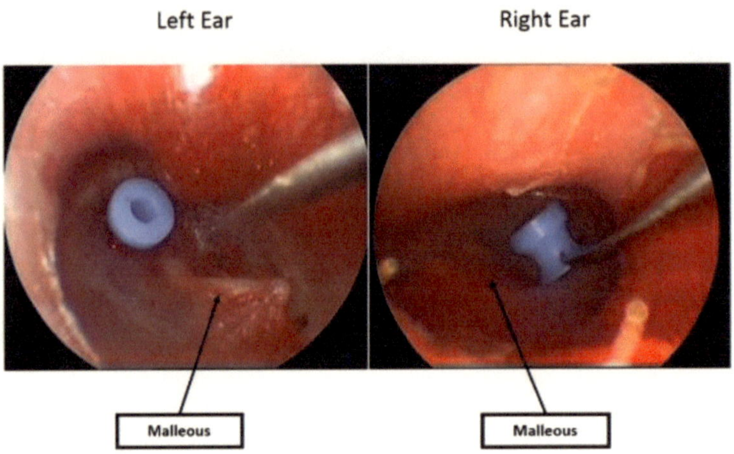

Figure 6.
Iatrogenic hematoma of the left ear anterior wall and excoriation of the right ear inferior wall for a right-handed surgeon.

ossiculoplasties, and stapedotomies. In fact, the learning curve for grommet insertion is very short, especially for colleagues who already perform endoscopic sinus surgery.

In conclusion, otoendoscopic VT placement is a valid and secure procedure. It is applicable in all patients independent of age, type of tube, Grommet tubes and T tubes, and anatomical conformation. The surgical approach is especially advantageous for grommet placement in narrow and curved EACs. In bloody fields, surgery may become more time consuming, but never compels the surgeon to abandon the technique. The learning curve is steep, especially for surgeons who are acquainted with endoscopic sinus surgery. Both cost-wise and in terms of logistic handling, the endoscope offers clear advantages. Nevertheless, the operating microscope must always be available in the operating room for rare cases of high jugular bulb or aberrant course of the internal carotid artery in the middle ear, which raises the hypothetical risk of bleeding that would not be controllable by an endoscopic approach alone. The endoscopic approach yields results that are comparable to traditional otomicroscopic techniques, but it is clearly superior in anatomically complex cases.

4. Tympanoplasty (myringoplasty)

Myringoplasty (MP) in children is one of the most common otologic procedures and can offer a success rate as high as 95% [5, 6]. It is considered a challenging procedure in children compared to adults due to narrowness of the EAC and generally smaller size of the ear [7, 8]. In a pediatric age, access to the tympanic membrane and elevation of tympanomeatal flap (TMF) to perform MP generally necessitates permeatal incision by employing an endaural or postauricular approach, especially in anterior and subtotal perforations, whereas a transmeatal approach is suitable only for small and posterior perforations [9–14]. In such cases, surgeons would not operate on children until the age of 10–14 years due to technical difficulties encountered in small anatomy, inability of the child to co-operate post-operatively, and increased risk of psychological trauma [10, 15–19]. In anterior perforation, surgery is more challenging as graft placement may be inaccurate [19], and the anterior aspect of the eardrum is more difficult to visualize, especially in children where the external ear canal dimensions are constraining [20, 21]. In our department, we have adopted an endoscopic technique since 2011 [22]. The use of endoscopy to perform myringoplasty may obviate such limiting factors. In agreement with our experience and that of others reported in the literature [23], an endoscopic approach can offer many advantages over a microscope approach in children. It provides the possibility to decrease morbidity by avoiding postauricular or endaural incision and applying a transcanal approach.

The **technique** consists of refreshing the margins of the perforation using a sickle knife or Rosen needle and grasping forceps. Two vertical incisions are performed at 12 and 6 o'clock on the skin, and at a horizontal one at about 0.2 cm from the annulus and the medial TMF is elevated. The graft is inserted under the anterior margin of the perforation, underlay fashion, and under or above the handle of the malleus depending of the extension of the perforation (**Figure 7A–L**); in the case that the perforation involves the anterior quadrants, it is applied over the malleolus (**Figure 8A, B**). Gelfoam is applied adequately in the middle ear, and then both the free part of the flap and the graft are repositioned.

According to literature reports, **bleeding** is considered as a limit of endoscopic ear surgery. When it does occur, the surgeon is sometimes forced to interrupt or convert to an otomicroscopic bi-manual procedure [4, 24, 25]. In our experience, bleeding was never a relevant issue: in no instance did the procedure become so demanding and troublesome that it had to be converted to a microscopic intervention.

Bleeding can be handled in several ways. Injection of the EAC with epinephrine 1:1000 solution and then waiting for a few moments before incising the skin. Another hint is that after the incision prominent bleeding will be noticed at the beginning, one may shift and start harvesting the tragal perichondrium in order to provide enough time for spontaneous hemostasis to take place. During elevation of the TMF, it could be controlled by irrigation with warm saline and local application of pledgets soaked with a vasoconstricting agent.

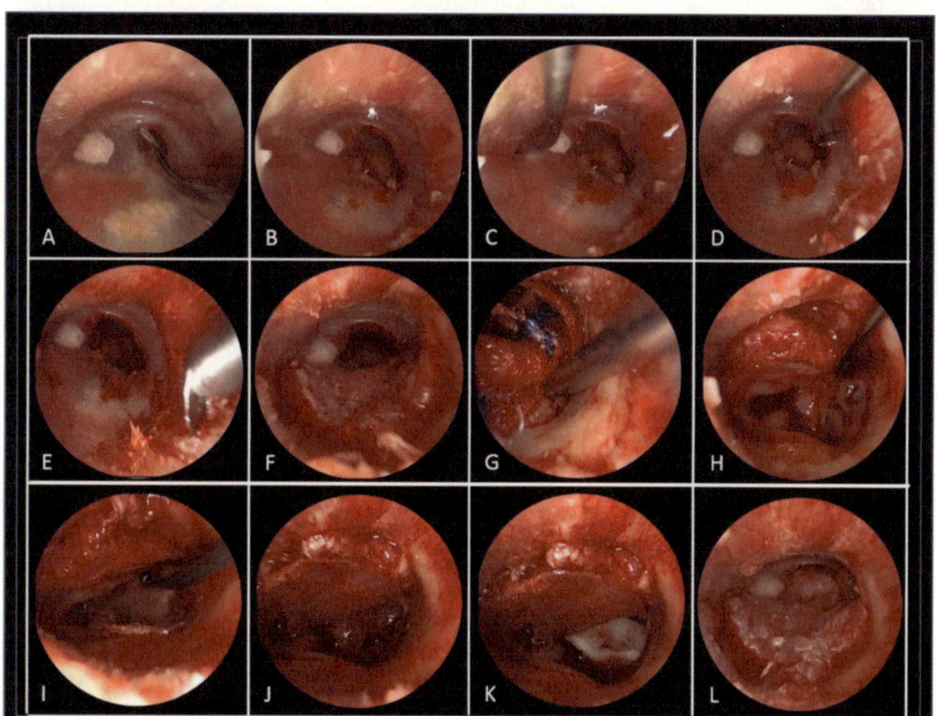

Figure 7.
Single-handed transcanal endoscopic myringoplasty, steps A through L.

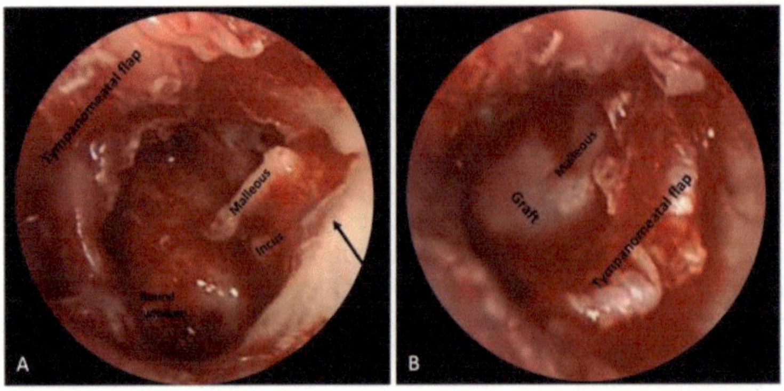

Figure 8.
(A) Tympanomeatal flap elevated and malleolus is denuded. (B) Graft positioned under the osseous annulus and over the malleolus.

A very helpful tool recently introduced as an otological instrument to overcome the presence of blood in the surgical field is the round knife with suction duct. During the dissection and elevation of the tympanomeatal, it offers the advantage of going through the dissection until reaching the annulus without the need to interrupt and use suction to aspirate blood from the surgical field (**Figure 9**).

The choice of the **graft** depends upon the preference of the surgeon. In our department, we either use the tragus (perichondrium ± cartilage) harvested and trimmed otherwise using biologic tissue. The advantage of biologic tissue is that one can reduce the surgical time by up to 15 min (**Figure 10**).

As described in other studies, endoscopes with different lengths, such as 6, 11, 16, and 18 cm, can be used to carry out endoscopic management [2, 4, 23]. According to our experience, the optimal length is between 11 and 16 cm in order to avoid impingement of instruments such as that noted on ventilation tube placement [22]. The rigid **endoscope** to use according to our experience is 3.0 mm and 16 cm long. This diameter and length fits all; both in a stenotic canal and a normal one, it provides good luminosity and adequate distance between the scope and otologic instruments. A 2.7 mm diameter, 18 cm long endoscope is not advisable due to its small field of vision, less luminosity, and inadequate rigidity of the shaft that will bend upon exertion of pressure, determining a crescent black spot field laterally (**Figure 11**).

In conclusion, endoscopic-assisted transcanal myringoplasty is feasible in all cases of tympanic perforation in children of any age and can be considered as a valid, alternative approach to the microscope. It is less invasive, especially in

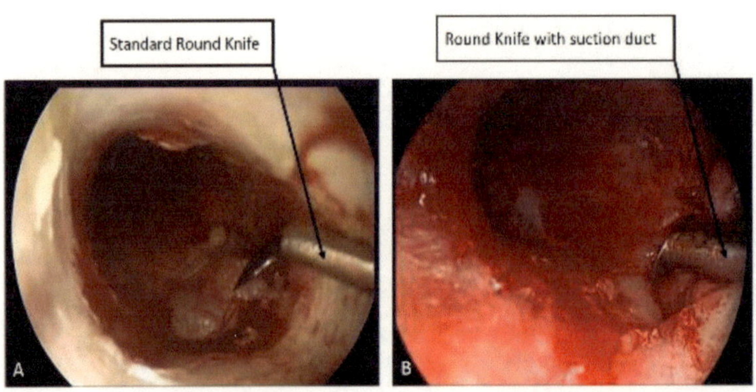

Figure 9.
(A) Standard round knife and (B) round knife with suction duct. Notice the difference of the surgical field in B, which is free of blood.

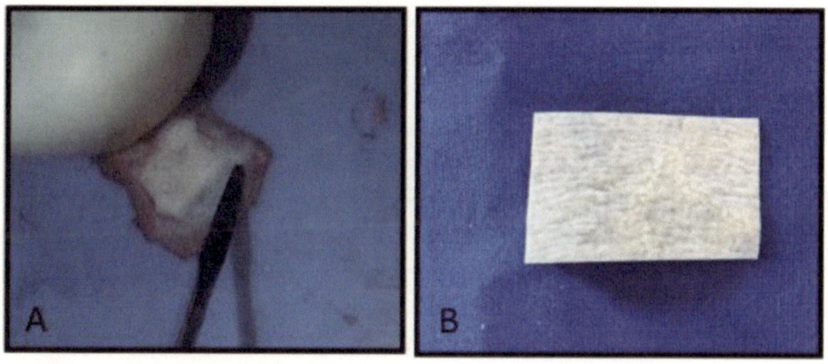

Figure 10.
(A) Autograft tragus perichondrium with cartilage isle and (B) xenograft acellular porcine small intestinal submucosa (Biodesign; Cook Medical Inc., Bloomington, IN).

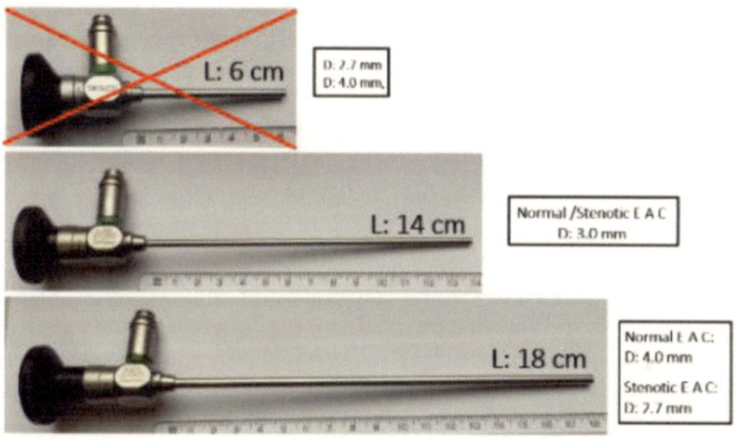

Figure 11.
Different lengths of endoscopes: 14 and 18 cm are indicated for ear surgery.

anterior perforations and in narrow and curved external canals where post-auricular or endaural approaches are otherwise required. An endoscopic approach grants better cosmetic outcomes and less psychological trauma with comparable anatomical and functional results vs. a traditional otomicroscopic technique.

5. Endoscopic approach to cholesteatoma

Endoscopy as a surgical tool in otosurgery is gradually evolving in daily practice when approaching middle and inner ear pathologies. Moreover, it is well known that cholesteatoma in a pediatric age, with an incidence that varies between 3 and 6 per 100,000 [26, 27], is an aggressive disease with respect to that in adults. The surgical approach is tailored according to the nature of cholesteatoma in being acquired or congenital, cystic or invasive. It often requires an extensive surgical approach, without undermining the preservation of hearing, as an attempt to eradicate the pathology due to its high rates of recidivism. The strategy to manage recidivism depends if it is residual, recurrent, or iatrogenic in nature. The standardized techniques at our disposition for such a destructive pathology, contrived upon the application of an operative microscope, are the transcanal (TC), canal wall-up (CWU), canal wall-down (CWD), and subtotal petrosectomy approaches. The surgeon should apply the indicated but least invasive and most effective approach, bearing in mind that the child will still have their entire life span ahead. Therefore, not only should eradication of the disease be considered, but functional outcomes such as hearing, balance, and stability of the cavity must be taken into consideration.

The application of otoendoscopy in recent years has allowed otosurgery to evolve and be less invasive in reconsidering certain standardized microscope operative techniques [28–30]. The major characteristic of endoscopy is to detect and dominate blind angles during surgery. Thus, in cholesteatoma surgery, endoscopy allows improved cholesteatoma removal while decreasing the rate of recidivism.

Cholesteatoma surgery is a step by step procedure that depends upon intraoperative findings. The approach cannot be decided a priori if it should be a TC, CWU, or CWD procedure. Radiologic work-up is another important element that is needed to plan the procedure. Nonetheless, during surgery, the extension of the disease may not be as expected due to the limits of radiology in defining the propagation of the pathology. Furthermore, the time elapsed between evaluating the result of radiology and the day of surgery may allow the disease to grow further and therefore invade and extend widely. For these reasons, a programmed, solely endoscopic surgical approach is not recommended, and an operative microscope is an indispensable tool to have next to the endoscope in the operating theater.

The advantages of endoscopy in ear surgery are to eradicate surgery and decrease morbidity, especially in children. According to our experience and that reported in the literature, the exclusive use of endoscopy for cholesteatoma removal should be limited when the disease involves only the tympanic cavity [31] and more precisely the mesotympanum and hypotympanum.

When cholesteatoma extends to the epitympanum, angled endoscopes offer a valid view to that area and for the aditus ad antrum. Therefore, the otosurgeon is at ease to decide the technical approach of the surgical procedure. Surgery may proceed as a TC approach, or extend to a combined endoscopic and microscopic one; therefore, mastoidectomy may become mandatory with a CWU or CWD approach in order to ensure that the pathology has been removed.

According to our experience, two major factors determine the choice of type or approach; the condition of the ossicular chain and nature and extension of the disease. In case of intact and healthy ossicles, and cystic cholesteatoma, such as congenital colesteatoma (**Figures 12–14**) being either lateral or medial to the ossicular chain, with an angled endoscope, there is a good chance of removing the cholesteatoma endoscopically.

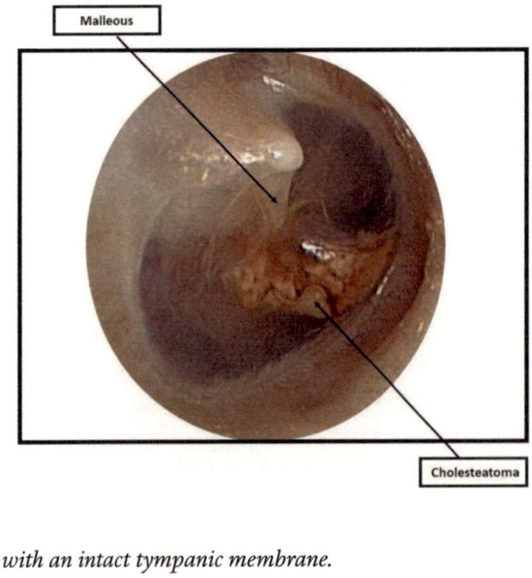

Figure 12.
Congenital cholesteatoma with an intact tympanic membrane.

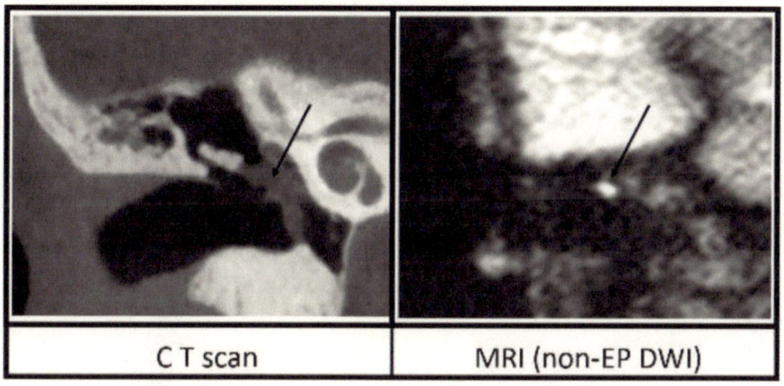

Figure 13.
Radiologic pre-operative work up: coronal CT and MRI showing the mass occupying the mesotympanum.

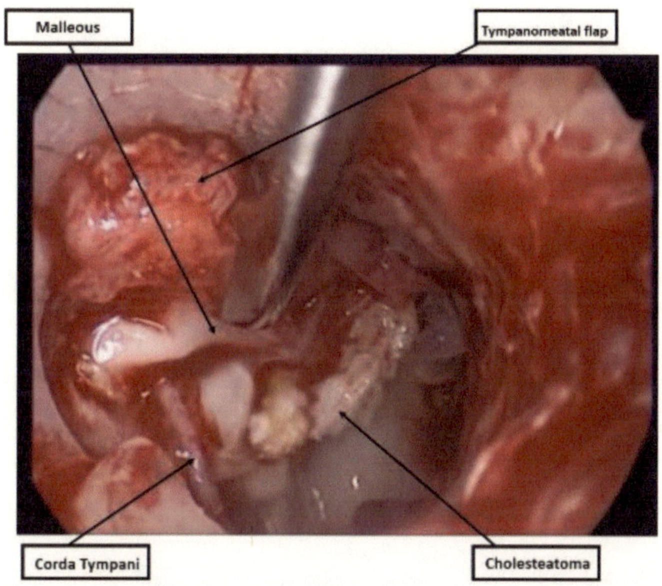

Figure 14.
Intraoperative endoscopic removal of cholesteatoma from the mesotympanum keeping the ossicles intact.

On the other hand, when the cholesteatoma is invasive and occupies the medial part of the head of the malleus and the body of the incus or extends to the antrum, it becomes a challenging task endoscopically. CWU mastoidectomy including antrostomy would be a plausible choice to remove the pathology and leave the ossicles intact. Conversely, in trying to approach such a pathology by an exclusive TC endoscopy, an extended atticotomy (reaching the antrum of the mastoid) and removal of the head of the malleus and the incus are required. This would determine the need to reconstruct both ossicles and the iatrogenic bony defect of the EAC with a piece of cartilage. In case of a small defect of the EAC, a small piece of cartilage may be adequate with good results in the future. If the defect is wide, a wedge of cartilage is needed to reconstruct the postero-superior part of the EAC. This closure would not be guaranteed in later years since the child is in growing phase. Therefore, an increase in the dimensions of the area of the EAC is inevitable with the consequence of a high probability of cholesteatoma recidivism, and CWD revision surgery would be necessary. As a consequence, otoendoscopy no longer has the advantages of being less invasive with reduced morbidity.

Children have an extremely active metabolism, and the risk of recidivism is higher than in adults. For this reason, children need closer follow-up and for a longer period, almost their entire lifespan. Non-EPI-DWI MRI is a very helpful tool in monitoring the disease and trying to avoid unnecessary surgical revisions. Most authors would consider MRI at 1 year postoperatively, otherwise explorative tympanoplasty in recommended [32].

In case of recurrence of disease, endoscopy plays an interesting role. If cholesteatoma is limited to the tympanic cavity such as the epitympanum, endoscopy grants fully transcanal removal of the disease without resorting to an post-aural approach (**Figures 15** and **16**).

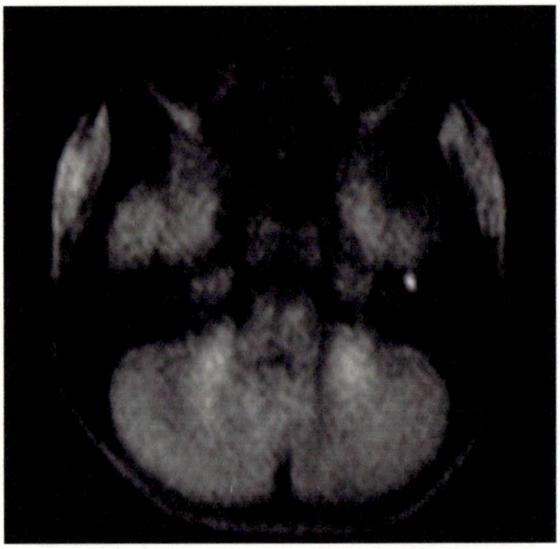

Figure 15.
Non-EPI-DWI MRI follow-up after 1 year of a 9-year-old child who underwent a CWU for cholesteatoma. Note the presence of the mass in the tympanic cavity.

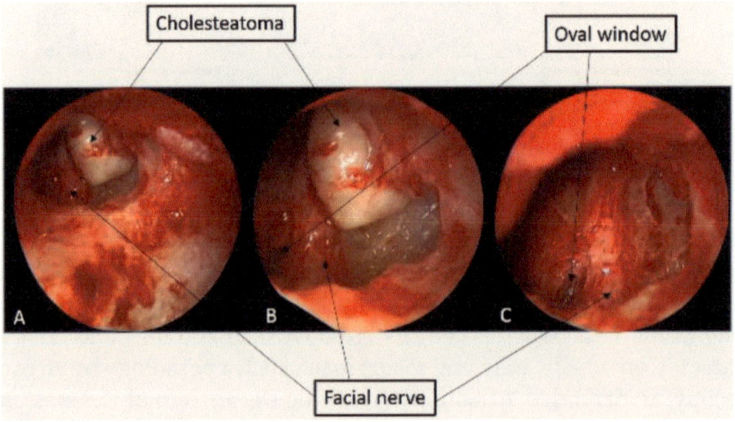

Figure 16.
Endoscopic transcanal of cholesteatoma removal for epitympanum.

In the case that cholesteatoma recurrence is in the mastoid cavity (**Figure 17**), endoscopy can also be an interesting tool for removal of cholesteatoma in a less invasive fashion, as shown in **Figure 18**.

Instrumentation for endoscopic cholesteatoma removal should include dedicated tools to better control its dissection and removal. The ones that really are helpful, besides being elongated and curved shafts, include a suction duct. They can be dissectors, knives, and suctions with sharp edges (**Figure 19**).

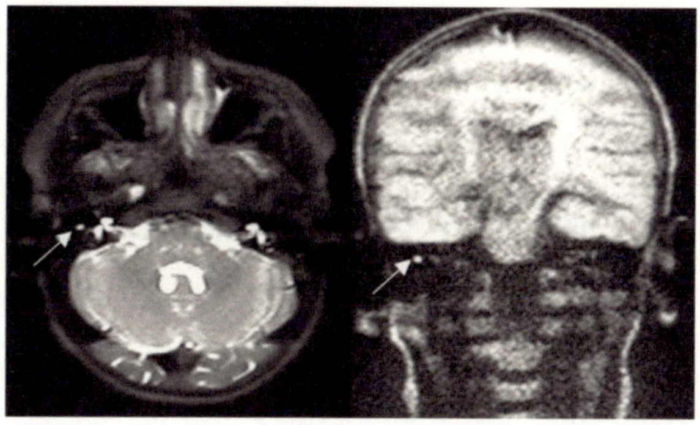

Figure 17.
MRI at 1 year follow-up in CWU in a 7-year-old child showing recurrence of disease in the mastoid cavity in proximity of the lateral semicircular canal.

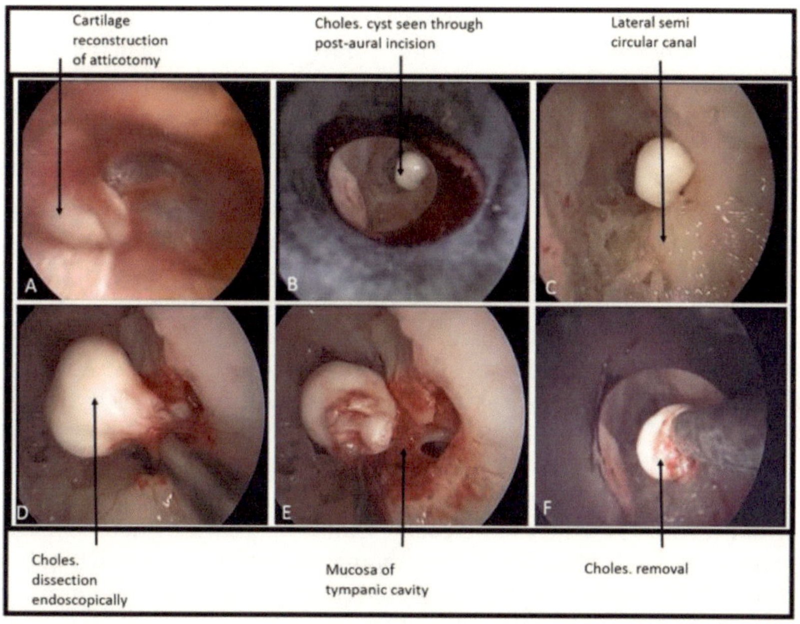

Figure 18.
Intraoperative removal of cholesteatoma through a small 1 cm post-aural incision to gain access to the mastoid after positioning the ear speculum.

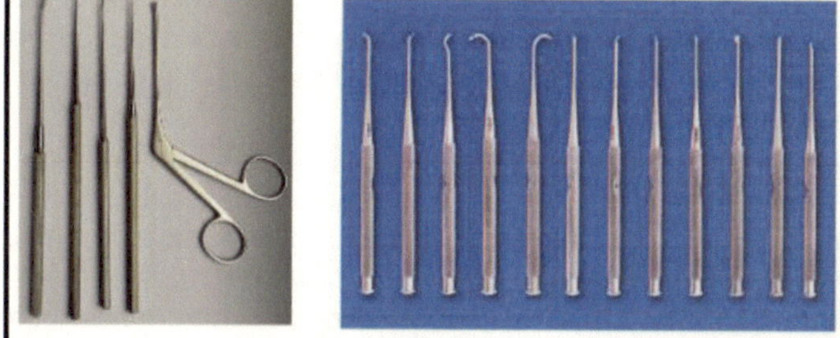

Figure 19.
On the left, standard surgical tools for otosurgery; on the right, Panetti Endoscopic Instrument Set (Spiggle and Theis, Overath, Germany) of dedicated tools containing curved sharp tips containing suction ducts.

6. Cochlear implantation

Nowadays, normal or near-normal low-frequency hearing threshold is present in many patients who are candidates for cochlear implantation, and combined electric acoustic stimulation can be proposed. Consequently, preservation of residual hearing is becoming a crucial issue. Avoiding trauma to inner ear structures when placing a multi-electrode array during the cochlear implant procedure is essential to preserve residual hearing, and the first important topic is the route of insertion of the multi-electrode array.

Scala tympani is the location for multi-electrode array cochlear implants when normal anatomy is present. The reasons for this are the greater sectional dimension compared to scala vestibuli, protection of cochlear duct by osseous spiral lamina and basilar membrane that are stronger than Reissner's membrane, close proximity to the spiral ganglion cell bodies and dendrites, and direct access through the round window (RW) [33, 34].

Access to the scala tympani to introduce the multi-electrode array during the cochlear implant procedure is obtained by opening the RWM, when the RW is sufficiently visible through a posterior tympanotomy or by performing a promontorial cochleostomy in the inferior-lateral wall of the scala tympani, when the RW exposure is not adequate. The advantages of RW access are: reduced risk of intracochlear trauma, wider stimulation surface, and facilitated perimodiolar position [35, 36].

Anatomical variations in the facial nerve, chorda tympani, and the RW niche may create obstacles to approach the RWM [37], and a classification of visualization to the RWM related to surgical approachability has been proposed. Following this classification, complete RWM exposure (considering the only one that guarantees a pure RW approach) was possible in 46% of children, while in 7% the RWM was not visible. [36].

Even if the RWM is completely visible, opening of the RWM does not provide straightforward access to the scala tympani because of the presence of a sharp bony crest in the anterior-inferior border of the niche called the "crista fenestrae," the morphology of which is highly variable [38].

When the RW is not visible and a promontorial approach is needed to consent opening of the scala tympani using a cochleostomy approach without damage to the osseous spiral lamina and basilar membrane, Adunka et al. [34] demonstrated, with an anatomical study, that drilling should proceed from the inferior to the RW annulus, with gradual progression toward the undersurface of the lumen. However, the most basal part of the scala tympani forms a fish hook-like curvature in three dimensions called the "hook" region of the cochlea [39, 40]. This portion contains the cul-de-sac of the endolymphatic space where the osseous spiral lamina, spiral ligament, and basilar membrane merge [40]. The lateral wall of the "hook" region shows large size variations with the consequence of the absence of a reliable landmark for safe access to the scala tympani when promontorial access is performed.

Finally, the angle formed by a line along the plane of the basal turn and the midline (a line between the nasal septum and the internal occipital protuberance) is variable, being inversely proportional with age, thus increasing the risk of damage of the cochlear duct in children. [41].

Applying rigid endoscopy through posterior tympanotomy consents the surgeon to gain both a view of the RW, in case it is invisible, and direction of the scala tympani in children, thereby allowing soft and less traumatic insertion of the multi-electrode array during cochlear implantation (**Figure 20**).

The use of an endoscope to assist cochlear implant surgery has been reported as a transcanal endoscopy by creating a tympanomeatal flap; this approach has been recently questioned because the basal turn of the cochlea align with a more posterior angle than that of the ear canal [42]. The idea to use a rigid endoscope

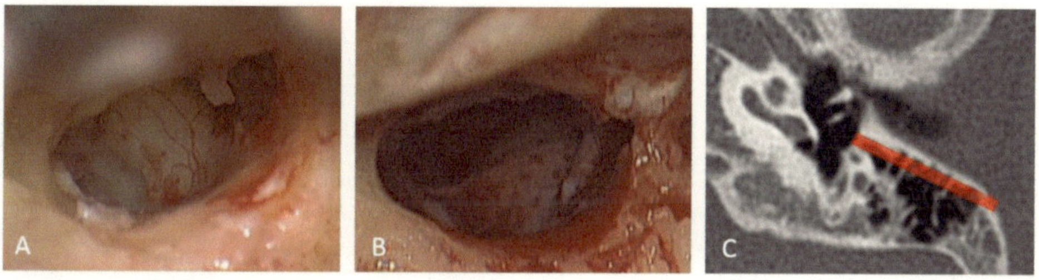

Figure 20.
(A) RW visible through posterior tympanostomy, (B) case where RW in invisible, and in (C) the advantage of the endoscope to visualize the RW.

to completely visualize the RW region from posterior tympanotomy has been recently reported: 0° (3 mm outer diameter, 14 cm length) and 0 and 30° (2.7 and 4 mm outer diameter, 11 and 6 cm in length) [43, 44]. Endoscopes of 2.7–4 mm diameter cannot likely enter a standard posterior tympanotomy and, that is, in case of posterior location of the RW, there is still suboptimal visualization. With a posterior tympanotomy of 2 mm and the introduction of a 1.9 mm outer diameter endoscope, the RW niche is always completely visible so that the thinning of the lips of the RW niche under endoscopic view until the projection of the lateral wall of the ST and the region of insertion of the basilar membrane are evident; this thus avoids damage to functionally relevant structures when detaching the annulus of the RWM.

The technique consists of a standard mini-invasive surgical approach under general anesthesia performed with a postauricolar access and transmastoid posterior tympanotomy of 2 mm (**Figure 21**).

The tip of a 0°, 1.9 mm diameter and 11 cm long endoscope is positioned in proximity of the upper part of the posterior tympanotomy to obtain a panoramic view of the inferior part of the medial wall of the tympanic cavity. The endoscope can be kept in place using a standard endoscope holder. The next steps are performed under direct endoscopic view. The bone overhanging the RW is lowered using a 1 mm microdrill under constant irrigation or irrigation/aspiration if an endoscope holder is used (**Figure 22**). At that point, exposure of the RW is now possible with a microscope through the posterior tympanotomy and the surgeon may proceed via the microscope approach.

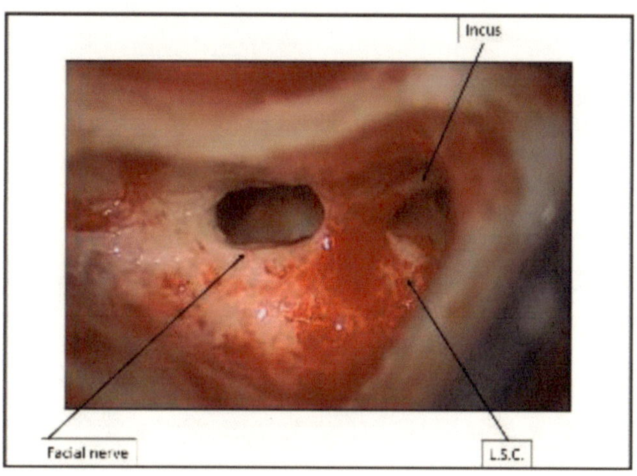

Figure 21.
RW not visible through posterior tympanotomy done by 2 mm burr.

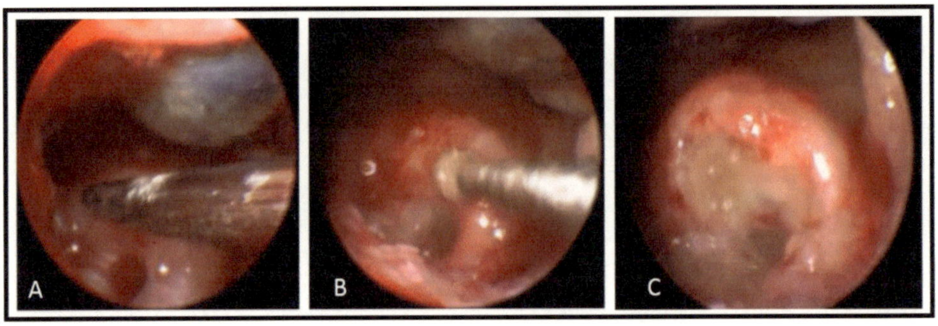

Figure 22.
(A) Visualization of the RW by 1.9 mm endoscope through the posterior tympanostomy, (B) drilling of overhanging bone, and (C) RWM exposed.

Otherwise, the surgeon may continue endoscopically and exploit the best use of endoscopy. The bone overhanging the RW in the anterior inferior aspect of promontory is thinned (**Figure 23A**) until a "blue line" inferiorly (corresponding to the ST) and a white line superiorly (corresponding to the insertion of the basilar membrane) are evident (**Figure 23B**). The anterior inferior part of the annulus of the RWM is detached from its bone insertion with a needle to open the ST (**Figure 23C**). The corresponding thinned bone, the crista fenestrae, is removed by a microcurette or a very low speed 0.5 mm microdrill until the direction of the canal is clearly evident. This allows gaining centered access to the ST and permits smooth linear introduction of the multi-electrode array into the basal turn (**Figure 23D**); attention must be given to avoid bone dust or fragments from entering into the ST. The diameter of the opening of the ST is not previously planned, provided that it is larger than the multi-electrode array diameter to leave perilymph coming out during insertion (**Figure 23**). Dexamethasone 4 mg/ml is gently flushed into the middle ear cavity and cochleostomy site. The electrode array is slowly inserted into the ST in a standard manner.

Endoscopic assistance through posterior tympanotomy has other advantages: in cases of particularly curved external auditory canal or small facial recess overthinning of the posterior wall with potential breakdown, extensive drilling of the fallopian canal with potential facial damage is no longer necessary; finally, in case of cochlear malformation, a panoramic view of the middle ear medial wall helps to identify the site for cochleostomy with no need of transcanal opening of the middle ear.

In conclusion, endoscope-assisted round window cochleostomy is a practicable alternative to the classical microscopic approach. It allows exposure of hidden RW niche, avoids over-thinning of the posterior external auditory canal by extensive drilling, and is useful to prevent luxation or removal of the external auditory canal when cochlear drillout is indicated. Thus, it optimizes array orientation and introduction into the scala tympani and leads to better results for residual hearing.

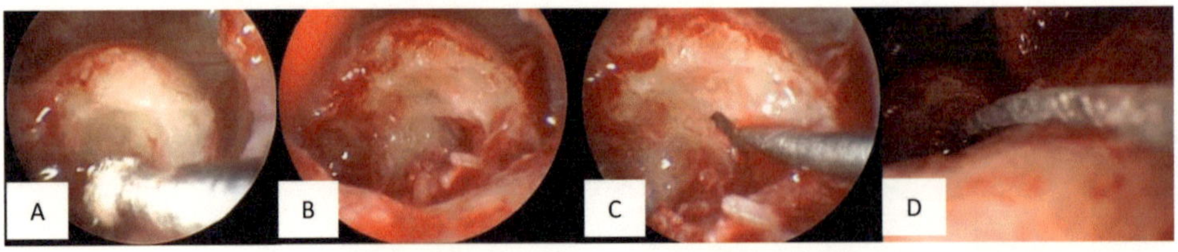

Figure 23.
Left ear. (A) Drilling of the RW lips, (B) exposure of the ST, (C) RW partially elevated, and (D) array insertion along the direction of the ST.

7. The holder, a two-handed technique

Otosurgeons are often skeptic and hesitant for endoscopic ear surgery for several reasons. First, single handedness in endoscopy is a limitation, especially in bleeding fields. When it occurs, bleeding is often a disturbing event, and frequent suction is needed so that the surgeon may be prone to interrupt the procedure and convert it to a traditional bimanual microscope technique [24, 45]. Second, otosurgeons are experienced with double-handed stereoscopic vision. Their teachings and therefore their maneuvers are based on two hands, whereas with endoscopy, otosurgeons have to manage maneuvers with one hand and lose the characteristic of the depth of vision. Differently, surgeons who practice sinus surgery are acquainted with a one hand procedure and for them approaching middle ear surgery is much preferred to operative microscope. This is why it seems to be more acceptable to nondedicated otosurgeons than to dedicated ones.

One of the drawbacks of the technique is being single handed. In case of a bloody field, especially in hyperplastic mucosa of the middle ear, surgery can become demanding and time consuming. Trying to overcome this limit, since January 2016, we started using the STORZ endoscope mechanical holding system followed a few months later by the Unitrack pneumatic holding system.

Different from other endoscopic procedures where a dynamic field is required, that is, cholesteatoma removal [46], during myringoplasty, the endoscope seldom needs to be moved to adjust the field of vision, so that the application of an endoscope holder is particularly favorable. The immediate advantage noticed is the rapidity of the procedure in elevating the tympanomeatal flap and fibrous annulus without frequently stopping to aspirate blood. Washing and suctioning simultaneously always guarantees optimal vision and cleaning of the endoscope. Another advantage is evident during introduction of the flap in case of liquid in the middle ear: suction by the second hand is promptly made. Positioning the graft underneath the anterior annulus with two hands is much easier by avoiding its wrinkling, and application of gelatin sponges under the graft itself is much easier. Finally, in one-handed surgery, the scope often has blurred vision due to blood clot or liquid left by hair in the EAC during the frequent introduction and extraction of the scope. The most important advantages of the use of a holding system are control of bleeding and shorter duration of surgery.

The endoscopic procedure consists of: (1) application of the endoscope holder on the operating table in front of the surgeon (**Figure 24**), (2) positioning of the endoscope at the mid level of the posterior part of the external auditory canal, (3) refreshing the margins of the perforation using a sickle knife and grasping forceps, (4) elevating a medial tympanomeatal flap with a semilunar incision at 12 and 6 o'clock, (5) inserting the graft under the malleus and the anterior margin of the perforation, and (6) applying gelatin sponges in the middle ear and, after repositioning the flap, in the ear canal. **Figure 25** shows the different steps of the endoscopic surgery and how it is handled bimanually, offering a clear advantage over a single-handed procedure.

The endoscopes used are 3 and 4 mm in diameter rigid 0° (Hopkins KARL STORZ GmbH & Co. Tuttlingen Germany), lengths 14 and 18 cm, respectively. The optic holder used is a mechanical articulating holding system (28,272 HC; 28,272 UGK; 28,172 HR: KARL STORZ GmbH & Co. Tuttlingen Germany) (**Figure 26**) or the Unitrack pneumatic holding system (Unitrac arm, RT040R, Aesculap AG, Tuttlingen Germany (**Figure 27**). All procedures are performed under general anesthesia.

The surgical maneuvers are managed better using a 3 mm vs. a 4 mm endoscope; according to our experience, we would recommend the 3 mm thanks to the greater space offered. The reason for using an 18 and 14 cm endoscope and not 6 or 11 cm

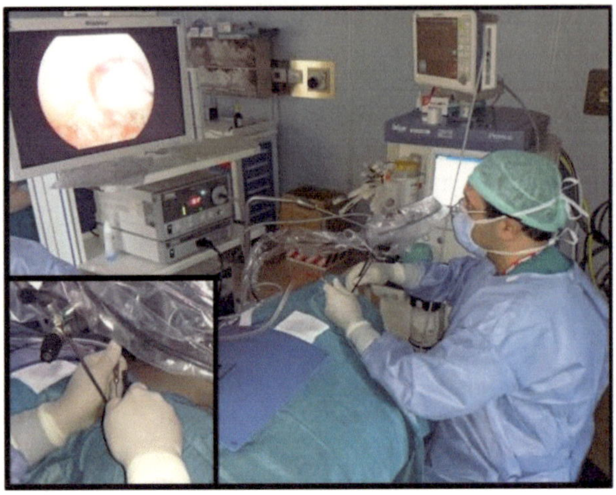

Figure 24.
Position of holder in front of the surgeon and screen applying the same concept of standard microscope surgery using both hands.

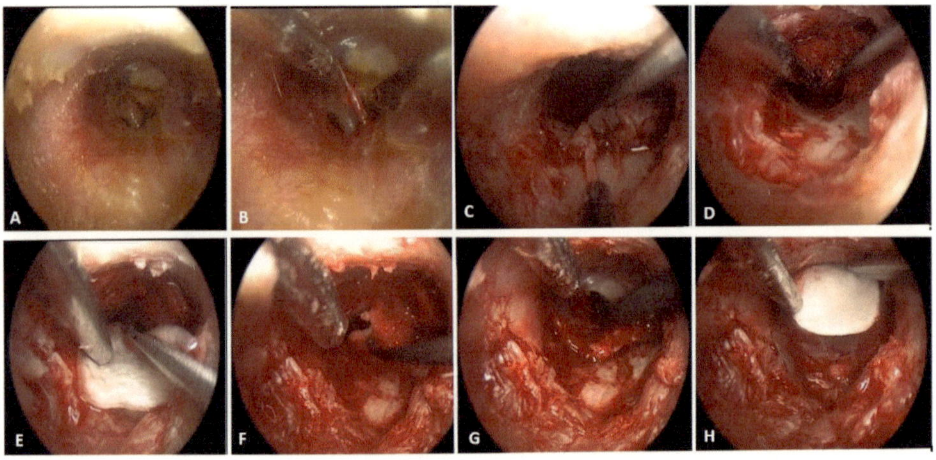

Figure 25.
(A–H) Steps of double-handed myringoplasty in an inferior perforation of the tympanic membrane.

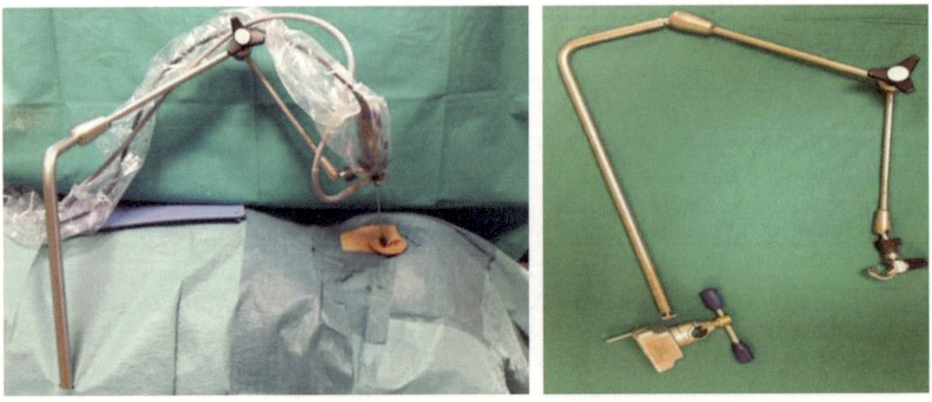

Figure 26.
Mechanical holding system.

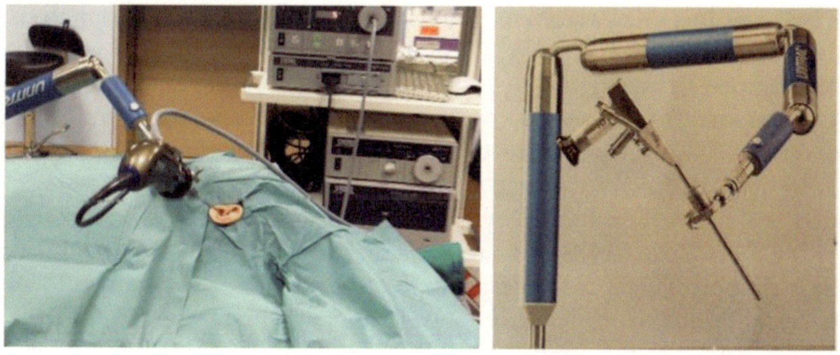

Figure 27.
Pneumatic holding system.

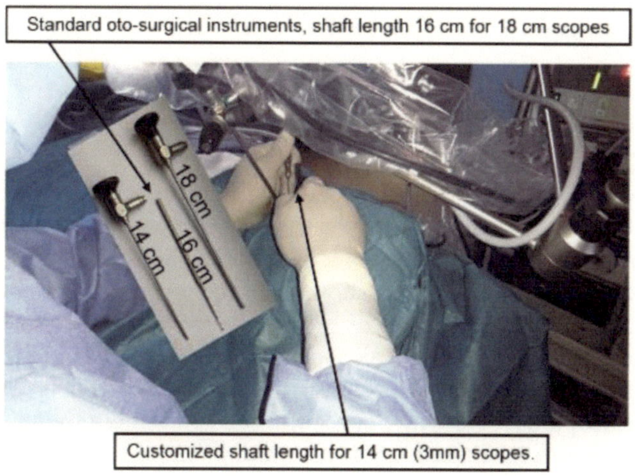

Figure 28.
Observe the working distance and instrument length by using 14 and 18 cm long endoscopes.

long is the possibility to maneuver both hands around the scope without encountering any obstacle by the camera and handle of the holder (**Figure 28**). The surgical instruments used for the microscope technique fit well with this technique.

There are some minor limitations of the technique: the endoscope is fixed in the canal, allowing a limited range of zooming and focusing. At the beginning of the procedure, the best view is to completely observe the ear canal, and then gradually magnify the middle ear throughout the surgery. Another limitation is that during the introduction of the graft in the EAC in some cases, the scope should be slightly pushed outward by the surgeon, which allows to see the graft lying completely on the posterior wall and therefore gliding it all the way through to the middle ear. It would be of a great help to have a camera with a foot pedal remote control in order to dynamically change the magnification and focus during the procedure and a motorized holder with fine movements to better manage the visual field.

8. Learning curve aspects

Learning curve varies upon both the complexity of otological procedure and prior experience of the otosurgeon with endoscopic sinus surgery. Furthermore, the profile of the learning curve is expressed through two factors.

The first factor is the variation of duration of the same surgery through time. According to our experience, surgical duration decreased with time in all types of endoscopic surgeries especially in ventilation tube positioning and tympanoplasty. For cholesteatoma endoscopic removal, since it was the last to introduce in our daily practice and therefore we were already acquainted with otoendoscopic surgery, duration of surgery did not undergo a tangible reduction.

The second factor is expressed by the anatomical results obtained through time. Concerning tympanoplasty, with time, we have experienced less reperforations in the reconstruction of the successive tympanic membranes. For cholesteatoma surgery, we become more confident in cholesteatoma removal from the hidden parts of the tympanic cavity, that is, sinus tympani and epitympanum. Such an experience permitted us to approach the middle ear less invasively such as creating less intracanal atticotomy or removing intact ossicles.

Concerning ossiculoplasties, stapedotomies, and cochlear implantation, it is preferable that they are advised after having obtained a certain confidence with otoendoscopic surgery.

9. Surgical training and young surgeons

Otosurgery belongs to the micro-surgery procedures. In order to progress through various techniques of otosurgery, the microscope is fundamental to adequately understand the anatomy in its three dimensions. For a beginner, the best choice is to proceed training with both techniques contemporary in order to understand the different perspective of the middle ear. Not all surgeries could be completely done by endoscope, and also in case of a complication or bleeding, a microscope surgery is necessary. For experienced otosurgeons, it is preferable to start their endoscopic surgery with ventilation tube placement then move to myringoplasty and therefore to cholesteatoma removal.

10. Conclusions

Endoscopy as a diagnostic and surgical tool in otology is now a valid instrument for different pathologies of the ear. Indications for solely endoscopic surgery should be well planned. Before starting endoscopic surgery, one should have good preparation in using the operating microscope and in certain surgeries should always be present in the operating theater. There are still some limits to endoscopy that in the future will be overcome thanks to technological advances such as miniaturization of 3D systems.

Conflict of interest

No conflict of interest is present.

DOI: http://dx.doi.org/10.5772/intechopen.84563

Author details

Luca Oscar Redaelli de Zinis[1]* and Nader Nassif[2]

1 Department of Medical and Surgical Specialties, Radiological Sciences and Public Health, Section of Head and Neck Surgery, University of Brescia and Pediatric Otolaryngology Head Neck Surgery, Children's Hospital "ASST Spedali Civili", Brescia, Italy

2 Pediatric Otolaryngology Head Neck Surgery, Children's Hospital "ASST Spedali Civili", Brescia, Italy

*Address all correspondence to: luca.redaellidezinis@unibs.it

References

[1] Thomassin JM. Endoscopically Assisted Diagnostics and Surgery in Otology, Technique, Equipment and Indications. Vol. 1. Tuttlingen: Endo–Press; 1988

[2] Abou-Elhamd KE. Telescopic myringotomy and tube application. The Journal of Laryngology and Otology. 2000;**114**:581-583

[3] Lee FP. An alternative use of video-telescopic guidance for insertion of myringotomy tube. The Journal of Laryngology and Otology. 2006;**120**:e10

[4] Ayache S. Cartilaginous myringoplasty: The endoscopic transcanal procedure. European Archives of Oto-Rhino-Laryngology. 2013;**270**:853-860

[5] Vrabec JT, Deskin RW, Grady JJ. Meta-analysis of pediatric tympanoplasty. Archives of Otolaryngology Head and Neck Surgery. 1999;**125**:530-534

[6] Yoon TH, Park SK, Kim JY, Pae KH, Ahn JH. Tympanoplasty, with or without mastoidectomy, is highly effective for treatment of chronic otitis media in children. Acta Oto-Laryngologica, Supplement. 2007;**558**:44-48

[7] Caylan R, Titiz A, Falcioni M, De Donato G, Russo A, Taibah A, et al. Myringoplasty in children: Factors influencing surgical outcome. Journal of Otolaryngology Head & Neck Surgery. 1998;**118**:709-713

[8] Singh GB, Sidhu TS, Sharma A, Singh N. Tympanoplasty type I in children-an evaluative study. International Journal of Pediatric Otorhinolaryngology. 2005;**69**:1071-1076

[9] Pignataro L, Grillo Della Berta L, Capaccio P, Zaghis A. Myringoplasty in children: Anatomical and functional results. The Journal of Laryngology and Otology. 2001;**115**:369-373

[10] Umapathy N, Dekker PJ. Myringoplasty: Is it worth performing in children? Archives of Otolaryngology Head and Neck Surgery. 2003;**129**:1053-1055

[11] Uyar Y, Keleş B, Koç S, Oztürk K, Arbağ H. Tympanoplasty in pediatric patients. International Journal of Pediatric Otorhinolaryngology. 2006;**70**:1805-1809

[12] Al-Khtoum N, Hiari MA. Myringoplasty in children: Retrospective analysis of 35 cases. Brazilian Journal of Otorhinolaryngology. 2009;**75**:371-374

[13] Ribeiro JC, Rui C, Natercia S, Jose R, Antonio P. Tympanoplasty in children: A review of 91 cases. Auris, Nasus, Larynx. 2011;**38**:21-25

[14] Knapik M, Saliba I. Pediatric myringoplasty: A study of factors affecting outcome. International Journal of Pediatric Otorhinolaryngology. 2011;**75**:818-823

[15] Saliba I. Hyaluronic acid fat graft myringoplasty: How we do it. Clincal Otolaryngology. 2008;**33**:610-614

[16] Friedberg J, Gillis T. Tympanoplasty in childhood. The Journal of Otolaryngology. 1980;**9**:165-168

[17] Carr MM, Poje CP, Nagy ML, Pizzuto MP, Brodsky LS. Success rates in paediatric tympanoplasty. The Journal of Otolaryngology. 2001;**30**:199-202

[18] Lau T, Tos M. Tympanoplasty in children. An analysis of late results. The American Journal of Otology. 1986;7:55-59

[19] Koch WM, Friedman EM, McGill TJ, Healy GB. Tympanoplasty in children.

The Boston Children's Hospital experience. Archives of Otolaryngology Head and Neck Surgery. 1990;**116**:35-40

[20] Gersdorff M, Garin P, Decat M, Juantegui M. Myringoplasty: Long term results in adults and children. The American Journal of Otology. 1995;**16**:532-535

[21] D'Eredità R, Lens MB. Anterior tab flap versus standard underlay myringoplasty in children. Otology and Neurotology. 2009;**30**:777-781

[22] Nassif N, Redaelli De Zinis LO, Berlucchi M, Zanetti D. Endoscopic ventilation tube placement in the pediatric age. Clincal Otolaryngology. 2014;**39**:50-53

[23] Dündar R, Kulduk E, Soy FK, Aslan M, Hanci D, Muluk NB, et al. Endoscopic versus microscopic approach to type 1 tympanoplasty in children. International Journal of Pediatric Otorhinolaryngology. 2014;**78**:1084-1089

[24] Tarabichi M. Endoscopic middle ear surgery. The Annals of Otology, Rhinology, and Laryngology. 1999;**108**:39-46

[25] Tarabichi M. Endoscopic management of limited attic cholesteatoma. The Laryngoscope. 2004;**114**:1157-1162

[26] Shohet JA, de Jong AL. The management of pediatric cholesteatoma. Otolaryngologic Clinics of North America. 2002;**35**(4):841-851

[27] Marchioni D, Soloperto D, Rubini A, Villari D, Genovese E, Artioli F, et al. Endoscopic exclusive transcanal approach to the tympanic cavity cholesteatoma in pediatric patients: Our experience. International Journal of Pediatric Otorhinolaryngology. 2015;**79**(3):316-322

[28] Badr-el-Dine M. Value of ear endoscopy in cholesteatoma surgery. Otology and Neurotology. 2002;**23**(5):631-635

[29] Ayache S, Tramier B, Strunski V. Otoendoscopy in cholesteatoma surgery of the middle ear: What benefits can be expected? Otology and Neurotology. 2008;**29**(8):1085-1090

[30] Barakate M, Bottrill I. Combined approach tympanoplasty for cholesteatoma: Impact of middle-ear endoscopy. The Journal of Laryngology and Otology. 2008;**122**(2):120-124

[31] Tarabichi M, Nogueira JF, Marchioni D, Presutti L, Pothier DD, Ayache S. Transcanal endoscopic management of cholesteatoma. Otolaryngologic Clinics of North America. 2013;**46**(2):107-130 (Review)

[32] Lecler A, Lenoir M, Peron J, Denoyelle F, Garabedian EN, Pointe HD, et al. Magnetic resonance imaging at one year for detection of postoperative residual cholesteatoma in children: Is it too early? International Journal of Pediatric Otorhinolaryngology. 2015;**79**(8):1268-1274

[33] Adunka OF, Buchman CA. Scala Tympani Cochleostomy I: Results of a survey. The Laryngoscope. 2007;**117**(12):2187-2194

[34] Adunka OF, Radeloff A, Gstoettner WK, Pillsbury HC, Buchman CA. Scala Tympani Cochleostomy II: Topography and histology. The Laryngoscope. 2007;**117**(12):2195-2200

[35] Paprocki A, Biskup B, Kozłowska K, Kuniszyk A, Bien D, Niemczyk K. The topographical anatomy of the round window and related structures for the purpose of cochlear implant surgery. Folia Morphologica. 2004;**63**(3):309-312

[36] Leong AC, Jiang D, Agger A, Fitzgerald-O'Connor A. Evaluation

of round window accessibility to cochlear implant insertion. European Archives of Oto-Rhino-Laryngology. 2013;**270**(4):1237-1242

[37] Hamamoto M, Murakami G, Kataura A. Topographical relationships among the facial nerve, chorda tympani nerve and round window with special reference to the approach route for cochlear implant surgery. Clinical Anatomy. 2000;**13**(4):251-256

[38] Angeli RD, Lavinsky J, Setogutti ET, Lavinsky L. The crista fenestra and its impact on the surgical approach to the scala tympani during cochlear implantation. Audiology and Neurotology. 2017;**22**(1):50-55

[39] Stidham KR, Roberson JB. Cochlear hook anatomy: Evaluation of the spatial relationship of the basal cochlear duct to middle ear landmarks. Acta Otolaryngologica. 1999;**119**(7):773-777

[40] Atturo F, Barbara M, Rask-Andersen H. On the anatomy of the "hook" region of the human cochlea and how it relates to cochlear implantation. Audiology and Neurotology. 2014;**19**(6):378-385

[41] Lloyd SK, Kasbekar AV, Kenway B, Prevost T, Hockman M, Beale T, et al. Cochlear rotation and its relevance to cochlear implantation. Cochlear Implants International. 2010;**11**(3):57-63

[42] Tarabichi M, Nazhat O, Kassouma J, Najmi M. Endoscopic cochlear implantation: Call for caution. The Laryngoscope. 2016;**126**:689-692

[43] Chen YH, Liu TC, Yang TH, Lin KN, Wu CC, Hsu CJ. Using endoscopy to locate the round window membrane during cochlear implantation: Our experience with 25 patients. Clincal Otolaryngology. 2018;**43**:357-362

[44] Güneri EA, Olgun Y. Endoscope-assisted cochlear implantation. Clinical and Experimental Otorhinolaryngology. 2018;**11**:89-95

[45] James AL. Endoscopic middle ear surgery in children. Otolaryngologic Clinics of North America. 2013;**46**:233-244

[46] Presutti L, Marchioni D, Mattioli F Villari D, Alicandri-Ciufelli M. Endoscopic management of acquired cholesteatoma: Our experience. Journal of Otolaryngology Head & Neck Surgery. 2008;**37**:481-487

Chapter 7

The Frequency Following Response: Evaluations in Different Age Groups

Milaine Dominici Sanfins, Michele Vargas Garcia, Eliara Pinto Vieira Biaggio and Piotr Henryk Skarzynski

Abstract

In this chapter, recent data on the clinical application of the frequency following response (FFR) in different age groups will be presented. The chapter begins with the importance of using speech sounds in electrophysiological assessments. Then the FFR methodology is presented, giving normative data and the expected responses in different age groups: infants and young children, children and adolescents, and adults and the elderly. Finally, the unique responses of each age group are presented in order to show how this new technology can be an extremely useful tool for diagnosing hearing dysfunction.

Keywords: electrophysiology, frequency following response, speech perception, hearing, auditory evoked potential

1. Introduction

Until recently, electrophysiological evaluations were performed exclusively with nonverbal stimuli such as clicks and tone bursts which allow rapid and synchronous stimulation of neurons. However, the use of verbal stimuli, such as speech sounds, allows a more accurate analysis of the auditory system, especially if the aim is to investigate how the system decodes speech sounds involved in daily communication. Verbal and nonverbal stimuli are decoded in different ways and follow different trajectories through the central auditory nervous system.

Human communication consists predominantly of verbal stimuli, and it is important to understand how verbal sounds are coded at various levels of the auditory system. The need to develop research methods that are objective and accurately represent daily listening led to the development early this century of electrophysiological tests for measuring how speech sounds are perceived [1, 2]. Subsequently, a number of research groups have focused their efforts on using complex stimuli such as speech for diagnostic purposes [3–10].

The initial studies were performed in animal models [11] aiming to evaluate how the temporal and spectral properties of verbal stimuli were coded, and later human responses were also analyzed [12]. Among the electrophysiological procedures for investigating the processing and coding of verbal sounds, we highlight the frequency following response (FFR).

2. Frequency following response

Acquisition of an FFR is very similar to collecting an ABR with a click stimulus. However, interpretation of an FFR requires that the audiologist has a more sophisticated knowledge base. Speech stimuli allow a more complex analysis of the responses, such as their:

- timing;
- magnitude;
- frequency content and magnitude;
- frequency tracking;
- phase consistency;
- intrinsic factors; and
- difference between individual responses.

An FFR evaluation can be performed on different clinical populations and age groups, and below we give details of how the procedure varies depending on the patient's age. Because FFR is a relatively new procedure, initial work was done on adult subjects. Afterward, researchers turned their interest to the study of responses in infants and young children, children and adolescents, and the elderly.

In order for an FFR assessment to be useful in identifying auditory disorders at an early stage, normative values using different equipment and recording parameters need to be established and compared with language acquisition markers.

The distinctive features of FFRs in different age groups will be presented in three parts:

- evaluation in infants;
- evaluation in children and adolescents;
- evaluation in adults and the elderly.

3. Frequency following response: evaluation in infants

In clinical practice, a comprehensive hearing evaluation for infants and young children is essential, since the integrity of their auditory system is the basis for acquiring oral language. In this context, if one measures only the functioning of the peripheral auditory pathway, perhaps by recording and analyzing otoacoustic emissions and/or auditory brainstem evoked potentials, it significantly constrains one's knowledge of the patient's hearing status. Moreover, behavioral assessments of hearing in very young children are often inconclusive, considering the diversity of neuropsychomotor development in this age group.

The perception of speech is important for the development of receptive and expressive language [13]. Through auditory experiences, infants and toddlers acquire and master the linguistic elements necessary for effective communication. The experiences are associated with information from the other senses, and together they allow the acquisition and development of oral language. Through listening, the subject understands oral language and creates concepts, finally inter-relating them and expressing them through speech [14]. Thus, the importance of hearing for the acquisition and development of language is vital, and any disturbance to the auditory pathway has implications for oral communication as a whole [14].

FFR testing can be used with infants and young children as a predictor of the extent of future language appropriation—in other words as a way of identifying children who are at risk of deficits in oral language acquisition [2, 15]. Assessment by FFR of infants and young children is relatively recent, and published studies of its potential have only been done over the last decade. Before discussing what is known about FFR in this population, it is first necessary to clarify an important factor: maturation of the auditory pathway.

It is known that peripheral hearing is functional even before birth, whereas myelination and the organization of neural connections keep developing after birth [16, 17]. Indeed, the central structures, such as the subcortex and cortex, develop throughout the early years of human life. There is an ascending myelination of the auditory pathway, evidenced by magnetic resonance imaging. Up to the 13th week of life, there is an increase in myelination density of the cochlear nucleus, the superior olivary complex, and the lateral lemniscus, with the inferior colliculus demonstrating an increase in density around the 39th week of life [18]. This continuous process of myelination of the higher structures of the auditory pathway during the first year of life must be considered when evaluating the FFR, for it means that the lower the age of the evaluated subject, the greater the latency of the FFR waves [19, 20]. This increase in latency can also be seen in other auditory evoked potentials [21]. An FFR can be recorded from a neonate, but the responses only become readily apparent from the third month of age [15]. The existence of a series of FFR waves—V, A, C, D, E, F, and O—in neonates has been pointed out by several researchers [15, 19, 22–26]. FFR evaluations have been performed with the vowel /i/ [15, 24], the syllables /ba/ and /ga/ [26], and the syllable /da/ [23].

The FFR has been studied in neonates of different nationalities (Chinese, American) during the first days after birth, and the FFRs were nearly the same. This finding makes it possible to infer that, independent of the mother tongue, there is an innate capacity for speech coding in neonates at the subcortical level [22].

The evaluation of subcortical representation of speech coding was studied by evaluating FFRs in 28 healthy North American infants, 3–10 months of age. The study focused on the fundamental frequency (F0), the response time of the FFR, and the representation of harmonics. To analyze the data in the frequency domain, spectral amplitudes were calculated by fast Fourier transform (FFT) and divided into three frequency ranges: F0, 103–125 Hz; first formant (F1), 220–720 Hz; and high harmonics (HH), 720–1120 Hz. The F0 responses were more robust in infants 3 months of age and the amplitude of F0 did not show significant changes over the entire 6 months. For the F1 and HH frequencies, there was a rapid and systematic increase of amplitude from 3 to 6 months of age.

To analyze the data in the time domain, the peaks were identified manually and confirmed by a second observer. Waves I, III, and V were first identified in response to a click, and then, in the FFR, the same peak and following valley (V and A), the peaks (D, E, and F), and the displacement peak (O). Non-detectable peaks were marked as missing data points and were excluded from analysis. The latencies and amplitudes (baseline to peak) were extracted from the identified waves. The time domain analysis demonstrated a decrease in neural conduction time and an improvement in amplitude with increasing age. The latencies of A and O, the time interval between A and O, and the slope between V and A were shown to have a negative correlation between latency and age. In addition, there was an improvement in the morphology of all waves as age increased. It was also observed that infants 3–5 months of age had longer latencies, smaller intervals between A and O, and a lower V/A slope compared to those 6–10 months of age. This negative correlation between the latencies and the age of the infants, as well as the decrease of slope in the smaller children, is due to a maturational process occurring in the subcortical auditory system and shows that there is less neural synchrony in younger infants [23]. The authors also note that these findings indicate that at approximately 6 months of age, the coding of speech characteristics, both spectrally and temporally, becomes more like those of an adult, although the changes continue through to school age. These findings indicate that FFR evaluation can detect early disorders in the perception of speech sounds.

The researchers also investigated the development of subcortical speech processing in Chinese infants born in households in which the mother tongue was Mandarin. They recorded FFRs at two ages: 1–3 days of life and at 3 months. This prospective-longitudinal design study included only infants who had undergone auditory screening at birth, who had no obvious neurological disorders, and did not have any risk indicator for hearing loss. Initially, 44 newborns were tested by FFR during natural sleep. After that, the sample was divided into groups. For each group, the researchers selected different speech stimuli for the evaluation of FFR (monosyllables contrasting with Mandarin). Only 13 infants completed the follow-up protocol at the third month. The processing and tracking of the fundamental frequencies of human speech at the subcortical level, evidenced by the FFR, showed more robust responses when the babies were 3 months old. Researchers acknowledged the limitations of the study, including statistical analysis and data interpretation. A research weakness was the relatively low completion rate (i.e., 17/44 infants or 38.64%). This factor undermined the power of the conclusions and prevented the possibility of performing statistical analyses for each Mandarin tone used. Despite the limitations of the study, the findings fill a gap in understanding the developmental trajectory of subcortical processing during the first 3 months of life [25].

From the theoretical assumptions highlighted in the previous reference, it should be noted that the linguistic environment of a newborn has a substantial effect on the development of its speech perception. Even at birth, children are able to detect subtle differences in verbal sounds. Newborns can effectively differentiate all the features of human speech and most infants who participated in an FFR follow-up showed improvement in pitch tracking and response amplitudes at 3 months of age [25]. Such neural refinements observed by FFR are often highlighted in the literature for both infants [22, 24] and young infants [15, 23]. For example, in a longitudinal case report of one infant, the researchers obtained FFR records when the infant was 1, 3, 5, 7, and 10 months old. The results showed an evolving trajectory of development with a transition point of about 3 months [15].

Using FFR evaluation in preterm infants may also be an alternative for the early diagnosis of auditory disorders in this population related to the perception of speech sounds. Premature babies are at high risk of developing language disorders, so using FFR may be a way of measuring immature neural activity and predicting possible changes in the processing of verbal sounds. In order to do so, one study evaluated 12 premature Indian infants through FFR with the aim of exploring how an immature auditory system responds to complex acoustic stimuli such as speech [27]. Peaks V, A, C, D, E, and F were detected in almost all babies and with latencies and amplitudes similar to those reported in the literature. The waves could be replicated. The authors conclude that FFR may be a way of understanding how the human brain-stem receives speech signals and that such an assessment might be important for all high-risk babies. Although the findings of this study cannot be generalized, mainly due to the limited data (small sample and absence of a controls, among others), they point out the potential of FFR in evaluating infants from neonatal intensive care units.

More recently, studies that record FFRs in the presence of background noise have been published. It is known that competing noise can make speech comprehension more difficult in people of all ages. Speech-in-noise tests are clinically available but cannot be given to infants. Thus, the use of FFRs in noise may be an alternative for evaluating impaired speech perception in young children who are unable to respond to behavioral tests.

In this context, with the objective of examining the electrophysiological responses in the presence of noise, researchers have evaluated the FFR in 30 children with typical development under conditions with and without noise (a signal-to-noise ratio of +10 dB in the former) [28]. Babies were divided into two age groups: 7–12 and 18–24 months. For all infants, frequency analysis of the FFR with a Fourier transform was performed, analyzing the latency and amplitude of waves V, A, D, E, and F, and correlation tests were carried out. In both groups, the mean latency of all recorded waves was higher in the presence of noise. According to the authors, this suggests that, at least for infants up to 24 months, the presence of noise causes a delay in the appearance of FFR waves independent of age. In addition, they observed a greater amplitude of F0 in the noise condition in the group of older babies; this difference was not seen in the silent condition. Thus, the authors point out that, at 2 years of age, infants are less vulnerable to the degrading effects of noise compared to children younger than 12 months.

The development of phase lock and frequency representation has also been evaluated in infants. This was the focus of a study that included an initial sample of 56 typical babies, aged between 2 and 12 months, and evaluated the FFR with /ba/ and /ga/ stimuli presented in the right ear using the SmartEP equipment from Intelligent Hearing Systems [26]. These responses were also obtained in young adults to provide a reference for the course of development of neural synchrony (represented by phase lock) and response amplitude (represented by spectral magnitude). The results obtained in this study demonstrate that the strength of phase-lock in the fine structure at CV transition is higher in young adults compared to infants. However, phase lock for F0 was equivalent between adults and infants. The frequency of F0 was found to be higher in older infants compared to younger infants and adults. Thus, these data demonstrate that speech coding can be evaluated in infants from 2 months of age and that such data are of value in a clinical setting, since it is known that performing electrophysiological evaluation of hearing in young children is difficult because they are less able to remain still during a test. The data indicate that the FFR may be a way of testing babies who are at risk of developing a language disorder, examining the auditory coding mainly of the midbrain, but also reflecting contributions from the auditory nerve, brain stem, and cortex.

The most commonly used parameters in FFR evaluations are: monoaural stimulus, right ear stimulation, intensity of 80 dB SPL, syllable /da/ speech stimulus, alternating polarity, presentation rate of 10.9 stimuli per second, vertical placement of electrodes, insert headphones, and the subject sitting distracted or awake during recording [29].

Regarding the latency parameters, when FFR is done with the Navigator Pro AEP System (Natus Medical, Inc.) and a syllable stimulus, one group of researchers [19] pointed out that in 23 normal-hearing babies (0–12 months) the wave latencies were on average: V = 7.22 ms, A = 8.22 ms, D = 23.14 ms, E = 31.5 ms, F = 39.91 ms, and O = 49.64 ms. FFR wave latencies were also investigated in 53 children aged 3–5 years (**Tables 1** and **2**).

Parameters of FFR evaluation in infants and young children used in the Hearing Electrophysiology Service of the Federal University of Santa Maria, Brazil, are presented in **Table 3**.

	Waves					
	V	A	D	E	F	O
	Lat	Lat	Lat	Lat	Lat	Lat
Σ	7.22	8.22	23.14	31.51	39.91	49.64
SD	0.42	0.43	0.66	0.49	0.45	1.32
Detect (%)	86.9	86.96	91.30	91.30	82.61	65.22

Σ: average (ms), SD: standard deviation, Detect: the percent detectability for each peak.
Sample: 23 babies (0–1 years old).

Table 1.
FFR latency values using syllable /da/of 40-ms duration performed on babies with normal hearing (silent background) [19].

	Waves					
	V	A	D	E	F	O
	Lat	Lat	Lat	Lat	Lat	Lat
Σ	6.59	7.56	22.36	30.90	39.34	48.14
SD	0.26	0.35	0.38	0.37	0.32	0.42
Detect	100	100	88.67	98.11	100	90.57

Σ: average (ms), SD: standard deviation, Detect: the percent detectability for each peak.
Sample: 53 children (3–5 years old).

Table 2.
FFR latency values using syllable /da/ of 40-ms duration performed in children with normal hearing (in silence) [19].

Presentation parameters	Setting
Equipment	SmartEP, Intelligent Hearing Systems (IHS)
Transducer	Insert phones
Electrodes	Fz; Fpz; M1; M2 or Cz, M1, M2
Stimulation	Right ear
Stimulus	Syllable /da/
Duration of stimulus	40 ms
Presentation rate	10.9/s
Window	80–100 ms
Filter	Low pass of 100 Hz and high pass of 2000 Hz Low pass of 100 Hz and high pass of 3000 Hz
Polarity	Alternating
Intensity	80 dBnHL
Number of stimuli	6000
Reproducibility	2 × 3000 stimuli
Condition of evaluation	Awake and quiet
Impedance	3k Ohms
Artifact rejection	Acceptance if <10%

ms, millisecond; s, second; Hz, hertz; dB, decibel; HL, hearing level.

Table 3.
Parameters of FFR in infants and young children.

Source	Latency (ms)		Amplitude (μV)		VA measures	
	Σ	SD	Σ	SD	Σ	SD
V	6.61	0.25	0.31	0.15		
A	7.51	0.34	0.65	0.19		
C	17.69	0.48	0.36	0.09		
F	39.73	0.61	0.43	0.19		
Slope VA (μV/ms)					0.13	0.05
Area VA (μV × ms)					1.70	1.23

Σ: average, SD: standard deviation.
Sample: 36 and 38 children and adolescents (8–12 years old) with normal hearing.

Table 4.
FFR latency and amplitude values using the syllable/da/of 40-ms duration, performed in children with normal hearing on the right ear (silent conditions) [12].

		Waves													
		V		A		C		D		E		F		O	
	Sex	Lat	Amp	Lat	Amp	Lat	Amp	Lat	Amp	Lat	Amp	Lat	Amp	Lat	Amp
Σ	M	6.53	0.10	7.53	0.19	18.43	0.08	22.29	0.17	30.86	0.21	39.31	0.17	48.02	0.13
	F	6.49	0.13	7.43	0.23	18.33	0.12	22.28	0.15	30.81	0.29	39.27	0.24	47.95	0.21
Med	M	6.49	0.10	7.53	0.18	18.28	0.07	22.24	0.09	30.86	0.21	39.28	0.7	48.11	0.13
	F	6.49	0.12	7.37	0.22	18.37	0.09	22.11	0.13	30.78	0.22	39.11	0.24	47.86	0.21
SD	M	0.19	0.05	0.32	0.04	0.44	0.05	0.32	0.07	0.53	0.07	0.44	0.08	0.45	0.07
	F	0.22	0.07	0.35	0.90	0.44	0.11	0.67	0.09	0.58	0.35	0.56	0.26	0.75	0.28

Σ: average, Med: median, SD: standard deviation, M: male, F: female.
Sample: 40 children and adolescents (8–16 years old).

Table 5.
FFR latency and amplitude values for males and females using syllable /da/ of 40-ms duration performed in children with normal hearing (silent conditions) [30].

		Complex VA	
	Sex	Slope VA (ms/μV)	Area VA (ms × μV)
Σ	M	0.31	0.29
	F	0.39	0.34
Med	M	0.29	0.31
	F	0.36	0.31
SD	M	0.11	0.09
	F	0.14	0.14

Σ: average, Med: median, SD: standard deviation, M: male, F: female.
Sample: 40 children and adolescents (8–16 years old).

Table 6.
Complex VA (slope and area) values for males and females using syllable/da/of 40-ms duration performed in children with normal hearing (silent conditions) [30].

The early identification of hearing disorders through FFR evaluation allows a speech-language pathologist to intervene, lessening the damage that this disorder can have on the development of speech skills in early childhood [2, 20, 22, 31]. This

	Waves														
		V		A		C		D		E		F		O	
	Ear	Lat	Amp	Lat	Amp	Lat	Amp	Lat	Amp	Lat	Amp	Lat	Amp	Lat	Amp
Σ	R	6.50	0.12	7.46	0.22	18.33	0.10	22.21	0.14	30.89	0.30	39.37	0.24	48.00	0.21
	L	6.51	0.11	7.48	0.21	18.41	0.11	22.36	0.13	30.78	0.23	39.20	0.19	47.95	0.16
Med	R	6.45	0.12	7.45	0.21	18.33	0.08	22.12	0.14	30.86	0.23	39.24	0.19	47.99	0.15
	L	6.53	0.11	7.41	0.21	18.33	0.09	22.28	0.11	30.78	0.21	39.07	0.18	48.03	0.15
SD	R	0.21	0.06	0.33	0.09	0.42	0.08	0.66	0.09	0.50	0.39	0.55	0.29	0.75	0.30
	L	0.21	0.06	0.36	0.07	0.46	0.10	0.44	0.08	0.61	0.09	0.47	0.09	0.54	0.12

Σ: average, Med: median, SD: standard deviation, R: right, L: left.
Sample: 40 children and adolescents (8–16 years old).

Table 7.
FFR latency and amplitude values for right and left ears using syllable/da/of 40-ms duration performed on children with normal hearing (silent conditions) [30].

	Complex VA		
	Ear	Slope VA (ms/μV)	Area VA (ms × μV)
Σ	R	0.37	0.33
	L	0.34	0.31
Med	R	0.32	0.31
	L	0.32	0.31
SD	R	0.14	0.13
	L	0.13	0.13

Σ: average, Med: median, SD: standard deviation, R: right, L: left.
Sample: 40 children and adolescents (8–16 years old).

Table 8.
Complex VA (slope and area) values for right and left ears using syllable/da/of 40 ms duration performed on children with normal hearing (silent conditions) [30].

	Waves														
		V		A		C		D		E		F		O	
	Age range	Lat	Amp	Lat	Amp	Lat	Amp	Lat	Amp	Lat	Amp	Lat	Amp	Lat	Amp
Σ	8–11	6.53	0.12	7.44	0.22	18.37	0.11	22.26	0.15	30.80	0.25	39.34	0.21	47.95	0.17
	12–16	6.46	0.11	7.51	0.21	18.36	0.10	22.32	0.10	30.89	0.28	39.19	0.21	48.02	0.21
Med	8–11	6.53	0.11	7.45	0.21	18.37	0.09	22.20	0.14	30.78	0.23	39.28	0.20	47.95	0.15
	12–16	6.45	0.12	7.45	0.17	18.28	0.08	22.20	0.09	30.86	0.20	39.11	0.15	48.03	0.13
SD	8–11	0.23	0.06	0.32	0.10	0.46	0.09	0.53	0.08	0.62	0.19	0.56	0.11	0.75	0.14
	12–16	0.17	0.06	0.37	0.07	0.41	0.08	0.63	0.45	0.43	0.22	0.42	0.32	0.46	0.33

Σ: average, Med: median, SD: standard deviation, R: right, L: left.
Sample: 40 children and adolescents (8–16 years old).

Table 9.
FFR latency and amplitude values for various age ranges using syllable/da/of 40-ms duration performed on children with normal hearing (silent conditions) [30].

assertion can be understood by appreciating the relationship between language development and the presence of stimulating auditory experiences in the first few months of life.

Future studies evaluating FFRs in infants will no doubt benefit from interdisciplinary collaboration which seeks to deepen understanding of the underlying mechanisms involved in the typical and atypical development of the auditory system during early childhood.

4. Frequency following response: evaluation in children and adolescents

Auditory impairment is almost invariably associated with language and communication deficits. Learning a spoken language depends on assimilating the acoustic and phonetic elements of a language [32]. The development of the central auditory nervous system begins in intrauterine life and continues until adolescence, over which time hearing abilities become more complex and elaborate.

Because of the close relationship between hearing, language, and learning, it is extremely important to monitor hearing over the course of life. Especially in children, be it pre-school or school age, the aim should be to monitor auditory function, either through behavioral or electrophysiological assessments. The ideal would be a combination of both behavioral and electrophysiological methods, so that with numerous evaluations there are crosschecks which allow a more accurate diagnosis to be made.

The electrophysiological procedure traditionally used in clinical practice is the click ABR. However, in evaluating children with language deficits, this type of sound stimulus is not ideal for making diagnoses. Assessments using verbal sound stimuli, such as used in FFR, appear to be more effective and reliable in cases of learning problems or school difficulties [6]. Evaluation via an FFR allows a detailed analysis of how verbal stimuli are encoded in the central auditory nervous system to be done.

The FFR allows fine-grained auditory processing deficits associated with real-world communication skills to be identified. As well as being used for the early identification of auditory processing, it can also be used to assess hearing across different clinical populations [33, 34]. This electrophysiological procedure can provide reliable and objective information about acoustic patterns such as timing, pitch, and timbre [35]. These three elements can be evaluated using different parts of the FFR, as follows:

- timing—via analysis of the onset and offset portions;
- pitch—by analysis of the fundamental frequency (F_0);
- timbre—from analysis of the harmonics of F_0.

Simplistically, it can be said that the FFR helps in understanding which speech sounds were spoken (their timing and harmonic cues) and who said it (pitch cues) [36]. In addition, an FFR test can be performed under two conditions: (i) in silence (presentation of verbal stimuli only), and (ii) in noise (presentation of verbal stimuli plus background noise).

In children and adolescents, studies have shown that FFRs change in latency as age increases. FFRs of children aged around 5 years appear to be very similar to the responses of children aged 8–12. However, the FFR pattern of children under 5 years has a somewhat different morphology and latency. According to Johnson et al. [33], the differences in children younger than 3 years are more evident in the initial portion of the responses (the onset), while in older children the change is more evident in the final portion (the offset) [3, 37].

Initial studies have focused on understanding the FFRs in children and adolescents under silent conditions and in subjects who have normal hearing and typical development. For the benefit of clinical audiologists, some of these studies are summarized below (**Tables 4–9**).

Table 10 shows the parameters used in children and adolescents at the Electrophysiology Department of the State University of Campinas using Biologic equipment and BioMARK software.

Because FFR is a new procedure, unstudied pathologies are gradually being added and, little by little, we are gaining new information about what effects the pathologies have on the responses of affected children and adolescents.

The FFRs of children diagnosed as poor readers frequently present as alterations in the timing and magnitude of timbre components [38]. The perception of the duration of a sound stimulus is essential for proficient reading, and the FFR can evaluate or monitor a decline in temporal and spectral precision. Children and adolescents with dyslexia commonly have difficulty perceiving speech sounds either in silence or in competing noise backgrounds. If a child has difficulty in perceiving speech sounds, their reading can be severely impaired [39]. Recently, Sanfins et al. [6] highlighted the importance of FFR as a biological marker in scholastic difficulties.

FFR evaluation in children who have suffered from secretory otitis media in the first 6 years of life, and who have undergone myringotomy for bilateral ventilation tube placement, exhibit changes in their FFR compared to normal children [5]. This study found that evaluating the FFR seems to be a promising method of identifying

Parameter	Settings
Equipment	Biologic Navigator Pro
Software	BioMARK
Electrode montage	Cz, M1, and M2
Stimulated ear	Right ear
Stimulus	Speech
Stimulus type	Syllable /da/
Stimulus duration	40 ms
Stimulus polarity	Alternating
Stimulus intensity	80 dB SPL
Stimulus rate	10.9/s
Number of sweeps	6000
Replicability	Twice for 3000 sweeps
Transducer	Insert
Assessment condition	Watching a movie
Impedance	1k Ohms
Window 85.33 ms	85.33 ms
Filter	100–2000 Hz
Artifact rejection	>10%

Cz: vertex, M1: left mastoid, M2: right mastoid, ms: millisecond, dB: decibel, SPL: sound pressure level, s: second, Hz: hertz.

Table 10.
Parameters of FFR in children and adolescents.

changes in the coding of speech stimuli in these children which might be undetected using traditional electrophysiological evaluation. The changes in their electrophysiological responses might serve as an alert to parents and educators, who can then adopt strategies to minimize the negative consequences on language development and academic achievement.

Another possibility for using FFR assessment may be in monitoring an auditory training program or even tracking the effect of therapeutic interventions. Studies have shown that children with learning disabilities can benefit from an auditory remediation program, and it might therefore be usefully accompanied by FFR examinations (because FFRs have good repeatability in test and retest) [40, 41]. In addition, bilingual children can also be monitored through FFR assessment. Researchers have confirmed that neural perception of speech seems to be more consistent in bilinguals than in monolinguals [42, 43]. Bilingual experience during childhood may favor plasticity in the neuronal coding of sound and improve fundamental frequency perception (F0).

Recently, the neurophysiological aspects of speech perception have been investigated in cases of autism spectrum disorder (ASD). The results showed that children with ASD tend to have changes in the sensation of pitch (frequency), which might explain a withdrawal from speech reception. The fundamental frequency (F0) and its harmonics contain speech information which is essential in conveying affect [44], so changes in FFRs are consistent with a defect in perceiving prosody. The inference is that prosody deficits in some ASD patients may derive from an inability to encode and transmit auditory information in the brainstem [45].

Traditionally, FFR testing is done by presenting verbal stimuli through an insert earphone with a silent background. However, the perception of speech in a noisy background is a much discussed topic. In the presence of noise, normally hearing individuals need to make constant adjustments in their central auditory nervous system to satisfactorily understand and process speech information. Of course, there are others who, in the presence of competing noise, experience great difficulty in understanding speech [46].

The evaluation of FFR in the presence of noise can be effectively used to diagnose children with learning disabilities [47]. Thus, identification of such children could lead to improvements in their reading and writing skills and in daily communication.

5. Frequency following response: evaluation in adults and the elderly

In the adult and elderly population, the need for detailed audiological investigation increases when the patient complains of hearing difficulties, even if auditory thresholds appear normal.

The evaluation of the FFR first involves time and prosody recordings, which provide important information about consonant and vowel discrimination and also aid in the perception of intonation [48]. For adults, but especially in the elderly, participation in these sorts of tests can assist in rehabilitation, either using a hearing aid or auditory training (or both).

The clinical usefulness of the FFR in gauging how well auditory information is being processed is unquestionable. In adults and the elderly, many studies have already been done to identify how the FFR can help in diagnosing complaints related to central auditory processing, thereby allowing better rehabilitation.

The latencies (mean and standard deviation) for adults and the elderly are presented in **Table 11**. The values come from Skoe et al. [19] who used Biologic and

Complex VA			
	Age range	Slope VA (ms/μV)	Area VA (ms × μV)
Σ	8–11	0.38	0.31
	12–16	0.33	0.34
Med	8–11	0.37	0.31
	12–16	0.28	0.31
SD	8–11	0.12	0.11
	12–16	0.16	0.16

Σ: average, Med: median, SD: standard deviation, R: right, L: left.
Sample: 40 children and adolescents (8–16 years old).

Table 11.
Complex VA (slope and area) values for age range using syllable/da/of 40-ms duration performed in children with normal hearing (silent conditions) [30].

Age (years)	Number	Latencies (maximum in milliseconds + 2 SD)					
		V	A	D	E	F	O
17–21	54	7.04	8.15	23.21	31.9	39.50	48.94
21–30	143	7.17	8.28	23.4	32.54	40.84	49.79
30–40	32	7.27	8.39	23.64	32.09	40.38	49.13
40–50	11	7.05	8.22	24.26	31.86	39.93	49.6
50–60	26	7.5	8.77	24.5	32.97	41.46	50.72
60–73	24	7.68	8.81	24.27	32.47	40.60	50.02

Data from [19].
SD: standard deviation.

Table 12.
*FFR latency values based on mean values in **Table 11** plus two standard deviations.*

Navigator Pro equipment. In this study, subjects aged between 18 and 72 years and distributed in 6 age brackets were used. In the case of adults, the authors list values for subjects aged 21–30 years (n = 143) and found that latency values tended to increase with age. Thus, the researchers emphasized the importance of conducting research on FFRs in different age groups, since normative values can be modified with the aging process.

In **Table 12** the maximum values of each wave are listed by adding two standard deviations to those in **Table 13**. Assuming the distribution is Gaussian means that this measure will cover 95% of the population.

Undoubtedly, the largest number of FFR studies have been performed using the Navigator Pro model from Biologic. Researchers tend to use this equipment together with the Intelligent Hearing Systems and SmartEP software [7, 49, 50].

One study aimed to assess the processing of auditory information in those with hearing loss through an evaluation of eight individuals, aged 46–58 years, with hearing loss [7]. FFRs (collected by SmartEP) were correlated with results from two auditory processing behavioral tests—the masking level difference test and the random gap detection test. No correlation was found between FFR and these tests. The researchers found that the generation of this potential is extremely complex and could encompass several functions and does not depend on just temporal resolution

Age	Number	Latency ∑ (mean in milliseconds)						Standard deviation											
		V	A	D	E	F	O	V (SD)	%	A (SD)	%	D (SD)	%	E (SD)	%	F (SD)	%	O (SD)	%
117–21	54	6.58	7.53	22.41	31.02	39.50	48.26	0.23	100	0.31	96.30	0.40	92.6	0.44	94.44	0.46	98.15	0.34	98.15
221–30	143	6.65	7.60	22.60	31.12	39.61	48.33	0.26	100	0.34	100	0.67	95.8	0.71	100	0.62	99.30	0.73	97.90
330–40	32	6.61	7.53	22.52	31.09	39.54	48.21	0.33	100	0.43	100	0.56	96.88	0.50	96.88	0.42	96.88	0.46	93.75
440–50	11	6.67	7.64	22.84	31.26	39.49	48.30	0.19	100	0.29	100	0.71	90.90	0.30	100	0.22	100	0.65	90.90
550–60	26	6.86	7.89	23.08	31.57	39.92	48.72	0.32	92.31	0.44	92.31	0.71	76.92	0.70	96.15	0.77	92.31	1.00	88.46
660–73	24	6.92	7.89	23.05	31.37	39.68	48.84	0.38	91.67	0.46	91.67	0.61	83.33	0.55	83.33	0.46	83.33	0.59	100

∑: Average (ms), SD: standard deviation, %: percent detectability for each peak.

Table 13.
FFR latency values for syllable /da/ of 40-ms duration, (silence) performed in adults and the elderly with normal hearing [19].

or selective attention [7]. Also seeking to correlate FFRs with hearing loss, Peixe et al. [49] evaluated 11 individuals, aged 23–59 years, with moderately severe hearing loss. They concluded that hearing loss may cause an increase in the FFR wave latency, but the waves are still present so long as the stimulus intensity is adjusted. In other words, the presence of FFR waves is related to the audibility of the signal.

Another interesting study was conducted with 30 young Indian adults aged 18–25 years [50]. The evaluation was carried out with the SmartEP equipment, and FFRs were present in all subjects evaluated. The latency and amplitude values of the analyzed elements were: wave V (lat = 6.81 ms and amp = 0.19 μV), wave C (lat = 16.82 ms and amp = 0.24 μV), wave D (lat = 24.75 ms and amp = 0.32 μV), wave E (lat = 31.36 ms and amp = 0.37 μV), and wave F (lat = 40.04 ms and amp = 0.29 μV).

Worldwide, there is a large increase in the number of elderly people. This entails providing better care for the elderly in all aspects of their health. With aging, there are structural changes in the peripheral and central auditory system which can lead to a decline in hearing. This, in turn, causes complaints of difficulty in understanding speech, especially in unfavorable environments [51, 52]. These impairments have a great impact on the life of the elderly, since in addition to causing social isolation, it can also lead to a depression and reduce cognitive function [53].

Only a few studies have focused on FFR in the elderly, with the most reported population being young adults [54]. Some researchers have pointed to the clinical applicability of FFR in different populations and with different pathologies [7, 19, 37, 55].

The effects of presbycusis on FFRs have been investigated in 18 individuals aged 61–78 years with hearing loss at frequencies of 2, 4, and 8 kHz (and compared with the responses of a control group of 19 young adults aged 20–26 years with normal hearing) [37]. The elderly group had lower amplitudes and increased latencies compared to the control group, demonstrating that the FFR can be affected by aging as well as hearing loss, but in different ways.

The effects of hearing loss on FFRs were described in a sample of 30 elderly individuals aged 60–71 years who were divided into two groups matched by gender and intelligence quotient: (i) normal hearing, and (ii) mild to moderate hearing loss [35]. With ABR clicks, all subjects had normal responses. FFR testing indicated that individuals with hearing loss could be assessed with this procedure, but there were changes in the frequency responses. In the elderly with hearing loss, there was a breakdown in the perception of the speech signal, which resulted in differences in signal parameters compared to the group with normal thresholds. This breakdown in neural synchrony may explain the greater difficulty subjects with hearing loss have in speech perception.

The evaluation of FFR in noisy environments is becoming more widespread, Thus, one study was carried out with 111 individuals between 45 and 78 years of age (mean 61.1 years) with normal to moderate hearing loss [56]. All subjects presented values within normal limits for the Montreal Cognitive Assessment (MoCA) and click ABR. In addition, they were tested on the SSQ (Speech, Spatial, and Qualities of Hearing Scale) which relates to auditory quality, as well as to the Quick Speech-in-Noise test (QuickSIN), in which phrases are presented binaurally with a verbal background babble. The FFR assessment demonstrated an increase in O-wave latency associated with speech comprehension difficulty in competing noise environments.

Supporting the observation that FFR traces are affected by increasing age, research on 34 individuals aged 22–77 years with normal hearing [57] found a decrease of the amplitude was associated with an increase in latency (**Figures 1** and **2**).

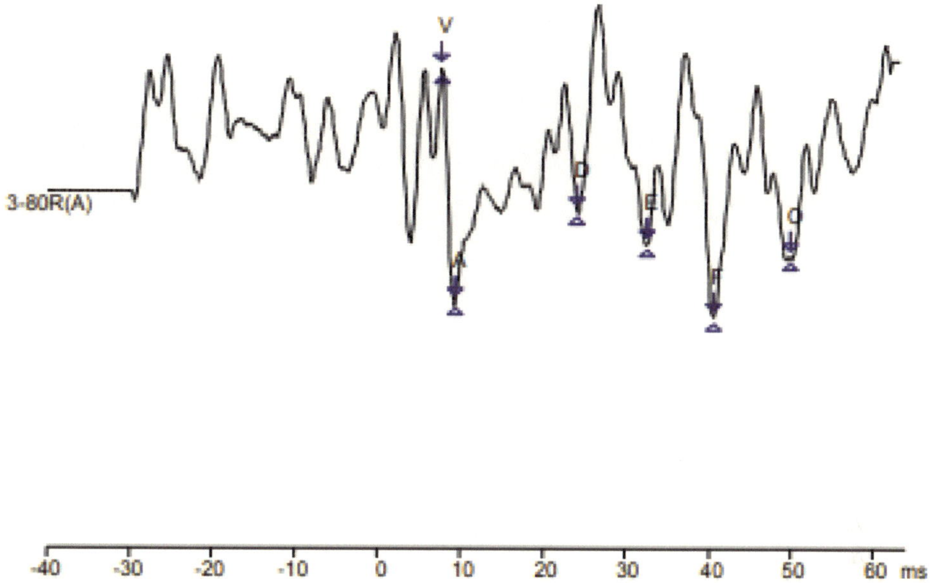

Figure 1.
FFRs of an infant 13 days old. Authors' data with FFR performed using SmartEP.

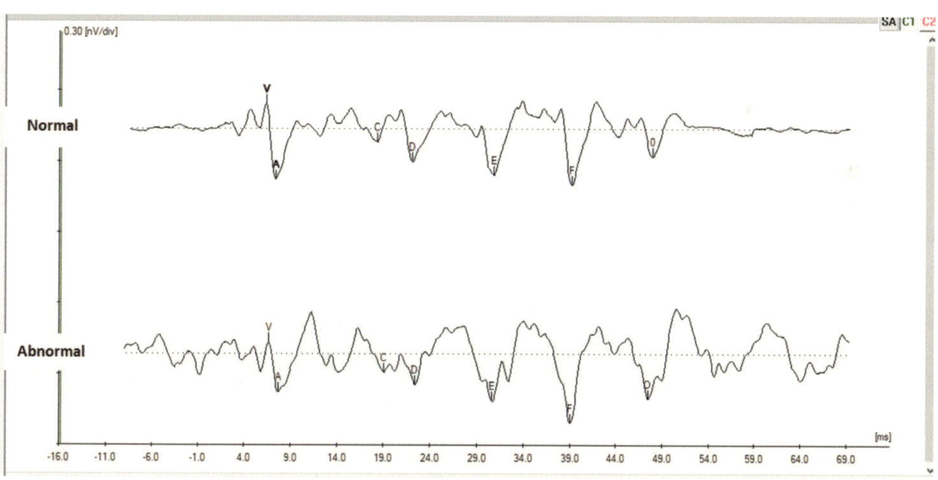

Figure 2.
FFRs of two 9-year-old-children. The top trace represents a normal response and the second represents an abnormal response. Authors' data using BioMARK software and Biologic equipment.

Figure 3 shows an FFR done on an adult aged 25 and on one aged 70. The shape of the FFR is similar in both, but there is an increase in latencies and some waves appear to be absent.

In these FFR tracings, it can be seen that the elderly subject had an increase in latency of all waves compared to the younger adult. Aging causes a progressive loss of structure or functioning of neurons, which can be seen as decreased auditory evoked potentials. Through the FFR, it is seen that there is also a reduction in the speed of neural activation from brainstem to cortical structures.

Our FFR evaluation in adults and the elderly used IHS equipment and the parameters are shown in **Table 14**.

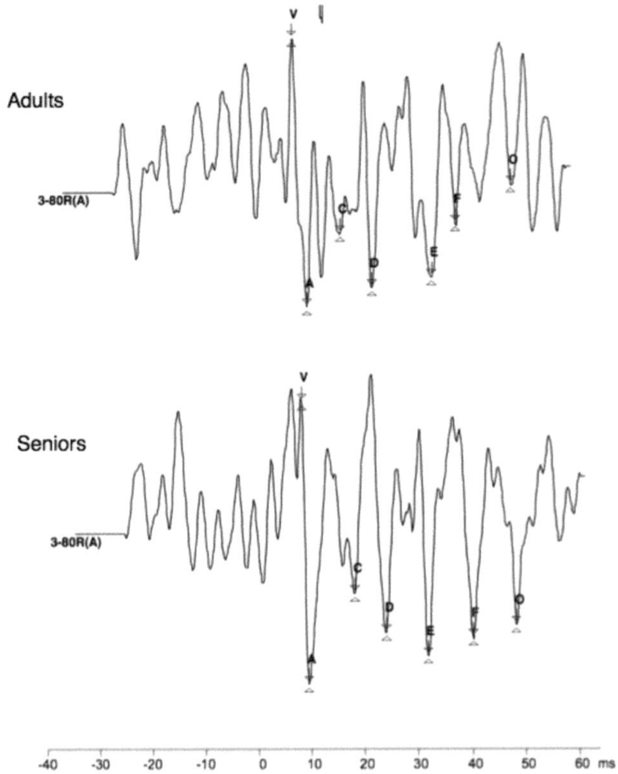

Figure 3.
FFRs of an adult aged 25 years (top) and another aged 70 (bottom). Note the increase in latency of the waves. Authors' data using SmartEP equipment.

Presentation parameters	Setting
Equipment	SmartEP Intelligent Hearing Systems (IHS)
Transducer	Insert phones
Electrodes	Fz, Fpz, M1, M2 or Cz, M1, M2
Stimulation	Right ear
Stimulus	Syllable /da/
Stimulus duration	40 ms
Presentation rate	10.9/s
Window	80–100 ms
Filter	Low pass of 100 Hz and high pass of 2000 Hz Low pass of 100 Hz and high pass of 3000 Hz
Polarity	Alternating
Intensity	80 dBnHL
Number of stimuli	6000
Reproducibility	2 × 3000 stimuli
Condition of evaluation	awake and quiet
Impedance	3k Ohms
Artifact rejection	>10%

Table 14.
Parameters of FFR in adults and the elderly.

6. Conclusion

FFR evaluations can be included as an extra examination in diagnostic testing and have an important role in crosschecking the results. It can also greatly assist making differential diagnoses in different clinical populations. However, each age group has FFRs with specific characteristics, so it is important that the audiologist has access to good normative values for the different age groups (infants and toddlers, young children, children and adolescents, adults and the elderly).

Terminology

10–20 International System	a standard system for electrode location
ABR	auditory brainstem response
AEP	auditory evoked potential. Evoked potential when using an auditory stimulus
BioMARK	Biological Marker of Auditory Processing is software that compares responses from a click to those from a synthetic syllable (usually /da/)
CANS	central auditory nervous system
CAP	central auditory processing
CAPD	central auditory processing disorder
CNS	central nervous system
CV syllable	a phoneme produced by a consonant and a vowel
FFR	frequency following response
Onset portion	the first part of an FFR that reflects the consonant
SAB	Scale of Auditory Behavior, a questionnaire for monitoring auditory processing skills
Sustained portion	the second part of an FFR that reflects the vowel Synthesized speech

Artificial human speech produced by a computer

Author details

Milaine Dominici Sanfins[1,2*], Michele Vargas Garcia[3], Eliara Pinto Vieira Biaggio[3] and Piotr Henryk Skarzynski[4,5,6]

1 Faculdade de Ciências Médicas, Universidade Estadual de Campinas, Campinas, Brazil

2 Centro de Eletrofisiologia e Neuroaudiologia Avançada, São Paulo, Brazil

3 Departamento de Fonoaudiologia, Universidade Federal de Santa Maria, Santa Maria, Brazil

4 Department of Teleaudiology and Hearing Screening, World Hearing Center, Institute of Physiology and Pathology of Hearing, Warsaw/Kajetany, Poland

5 Department of Heart Failure and Cardiac Rehabilitation, Medical University of Warsaw, Poland

6 Institute of Sensory Organs, Kajetany, Poland

*Address all correspondence to: msanfins@uol.com.br

References

[1] Johnson KL, Nicol TG, Kraus N. Brainstem response to speech: A biological marker of auditory processing. Ear and Hearing. 2005;**26**(5):424-434

[2] Skoe E, Kraus N. Auditory brainstem response to complex sounds: A tutorial. Ear and Hearing. 2010;**31**:320-324

[3] Anderson S, Parbery-Clark A, White-Schwoch T, Kraus N. Aging affects neural precision of speech encoding. Journal of Neuroscience. 2012;**32**(41):14156-14164

[4] Bidelman GM. Multichannel recordings of the human brainstem frequency-following response: Scalp topography, source generators, and distinctions from the transient ABR. Hearing Research. 2015;**323**:68-80

[5] Sanfins M, Borges L, Donadon C, Hatzopoulos S, Skarzynski P, Colella-Santos M. Electrophysiological responses to speech stimuli in children with otitis media. Journal of Hearing Science. 2017;**7**(4):9-19

[6] Sanfins M, Borges L, Ubiali T, Colella-Santos M. Speech auditory brainstem response (speech ABR) in the differential diagnosis of scholastic difficulties. Brazilian Journal of Otorhinolaryngology. 2017;**83**(1):112-116

[7] Sanguebuche TR, Peixe BP, Bruno RS, Biaggio EPV, Garcia MV. Speech-evoked brainstem auditory responses and auditory processing skills: A correlation in adults with hearing loss. International Archives of Otorhinolaryngology. 2018;**22**(1):38-44

[8] Sanju HK, Mohanan A, Kumar P. Speech-evoked auditory brainstem response in individuals with diabetes mellitus type 2. The Journal of International Advanced Otology. 2017;**13**(1):77-82

[9] Tahaei AA, Ashayeri H, Pourbakht A, Kamali M. Speech evoked auditory brainstem response in stuttering. Scientifica (Cairo). 2014;**2014**:328646

[10] Elkabariti RH, Khalil LH, Husein R, Talaat HS. Speech evoked auditory brainstem response findings in children with epilepsy. International Journal of Pediatric Otorhinolaryngology. 2014;**78**(8):1277-1280

[11] Young E, Sachs M. Representation of steady-state vowels in the temporal aspects of the discharge patterns of populations of auditory-nerve fibers. The Journal of the Acoustical Society of America. 1979;**66**:1381-1403

[12] Russo N, Nicol T, Musacchia G, Kraus N. Brainstem responses to speech syllables. Clinical Neurophysiology. 2004;**115**:2021-2030

[13] Holt L, Lotto A. Speech perception as categorization. Attention, Perception, & Psychophysics. 2012;**72**:1218-1227

[14] McLaughlin S. Introduction to Language Development. Edmond: University of Central Oklahoma; 2007. 508 p

[15] Jeng F, Schnabel E, Dickman B, Ju J, Li X, Lin C, et al. Early maturation of frequency-following response to voice pitch in infants with normal hearing. Perceptual and Motor Skills. 2010;**111**:765-784

[16] Neville H, Bavalier D. Specificity and plasticity in neurocognitive development in humans. In: Gazzaniga M, editor. The New Cognitive Neurosciences. London: Cambridge Mass; 2000

[17] Bellis T. Assessment and Management of Central Auditory Processing Disorders in the Educational Setting. Thomson Delmar Learning: California; 2003

[18] Sano M, Kaga K, Kuan C, Ino K, Mima K. Early myelination patterns in the brainstem auditory nuclei and pathway: MRI evaluation study. International Journal of Pediatric Otorhinolaryngology. 2007;**71**:1105-1115

[19] Skoe E, Krizman J, Anderson S, Kraus N. Stability and plasticity of auditory brainstem function across the lifespan. Cerebral Cortex. 2015;**25**(6):1415-1426

[20] Jeng FC. Infant and childhood development: Intersections between development and language experience. In: Kraus N, Anderson S, White-Schwoch T, Fay R, Popper A, editors. The Frequency-Following Response: A Window into Human Communication. Switzerland: Springer; 2017. pp. 17-43

[21] Picton T, Woods D, Baribeau-Braun J, Healey T. Evoked potential audiometry. The Journal of Otolaryngology. 1977;**6**:90-119

[22] Jeng F, Hu J, Dickman B, Montgomery-Reagan K, Tong M, Wu G, et al. Cross-linguistic comparison of frequency-following responses to voice pitch in American and Chinese neonates and adults. Ear and Hearing. 2011;**32**:699-707

[23] Anderson S, Parbery-Clark A, White-Schwoch T, Kraus N. Development of subcortical speech representation in human infants. The Journal of Acoustical Society of America. 2015;**137**:3346-3355

[24] Jeng F, Peris K, Hu J, Lin C. Evaluation of an automated procedure for detecting frequency-following responses in American and Chinese neonates. Perceptual and Motor Skills. 2013;**116**(2):456-465

[25] Jeng F, Lin C, Chou M, Hollister G, Sabol J, Mayhugh G, et al. Development of subcortical pitch representation in three-month-old chinese infants. Perceptual and Motor Skills. 2016;**122**:123-135

[26] Van Dyke K, Lieberman R, Presacco A, Anderson S. Development of phase locking and frequency representation in the infant frequency-following response. JSLHR. 2017;**60**:2740-2751

[27] Ayas M, Yaseen H, Rajashekhar B. Auditory brainstem processing of complex speech sounds in high risk infants-a preliminary study. International Journal of Current Research and Review. 2015;**7**:22-27

[28] Musacchia G, Ortiz-Mantilla S, Roesler C, Rajendra S, Morgan-Byrne J, Benasich A. Effects of noise and age on the infant brainstem response to speech. Clinical Neurophysiology. 2018;**129**:2623-2634

[29] Gabriel L, Vernier L, Ferreira M, Silveira A, Machado M. Parameters for applying the brainstem auditory evoked potential with speech stimulus: Systematic review. International Archives of Otorhinolaryngology. 2018;**22**(4):460-468

[30] Sanfins M. Electrophysiological Evaluation with Verbal and Non-verbal Sounds in Children with a History of Otitis Media. State University of Campinas; 2017. Available from: http://www.repositorio.unicamp.br/handle/REPOSIP/330747

[31] Hornickel J, Lin D, Kraus N. Speech-evoked auditory brainstem responses reflect familial and cognitive influences. Developmental Science. 2013;**16**(1):101-110

[32] Cacace AT, McFarland DJ. Central auditory processing disorder in school-aged children: A critical review. Journal of Speech, Language, and Hearing Research. 1998;**41**(2):355-373

[33] Johnson K, Nicol T, Zecker S, Kraus N. Developmental plasticity

in the human auditory brainstem. The Journal of Neuroscience. 2008;**28**(15):4000-4007

[34] Kraus N, Hornickel J. cABR: A biological probe of auditory processing. In: Geffner DS, Ross-Swain D, editors. Auditory Processing Disorders: Assessment, Management, and Treatment. 2a ed. San Diego: Plural Publishing; 2013. pp. 159-183

[35] Anderson S, Parbery-Clark A, White-Schwoch T, Drehobl S, Kraus N. Effects of hearing loss on the subcortical representation of speech cues. The Journal of the Acoustical Society of America. 2013;**133**(5):3030-3038

[36] Hornickel J, Kraus N. cABR can predict auditory-based communication skills. The Hearing Journal. 2012;**65**:28-30

[37] Vander Werff KR, Burns KS. Brain stem responses to speech in younger and older adults. Ear and Hearing. 2011;**32**(2):168-180

[38] Benasich A, Fitch R. Developmental Dyslexia: Early Precursors, Neuro-behavioral Markers and Biological Substrates. Baltimore: Brookes Publishing; 2012

[39] Bogliotti C, Serniclaes W, Messaoud-Galusi SLSC. Discrimination of speech sounds by children with dyslexia: Comparisons with chronological age and reading level controls. Journal of Experimental Child Psychology. 2008;**101**:137-155

[40] Kumar P, Singh NK. BioMARK as electrophysiological tool for assessing children at risk for (central) auditory processing disorders without reading deficits. Hearing Research. 2015;**324**:54-58

[41] Song JH, Nicol T, Kraus N. Test-retest reliability of the speech-evoked auditory brainstem response. Clinical Neurophysiology. 2011;**122**(2):346-355

[42] Krizman J, Marian V, Shook A, Skoe E, Kraus N. Subcortical encoding of sound is enhanced in bilinguals and relates to executive function advantages. Proceedings of the National Academy of Sciences. 2012;**109**(20):7877-7881

[43] Krizman J, Skoe E, Kraus N. Bilingual enhancements have no socioeconomic boundaries. Developmental Science. 2016;**19**(6):881-891

[44] Patel A, Peretz I, Tramo M, Labrecque R. Processing prosodic and music pattern: A neurophsychological investigation. Brain and Language. 1998;**61**:123-144

[45] Russo N, Skoe E, Trommer B, Nicol T, Zecker S, Bradlow A, et al. Deficient brainstem encoding of pitch in children with autism spectrum disorders. Clinical Neurophysiology. 2008;**119**(8):1720-1731

[46] Wible B, Nicol T, Kraus N. Abnormal neural encoding of repeated speech stimuli in noise in children with learning problems. Clinical Neurophysiology. 2002;**113**:484-494

[47] Cunningham J, Nicol T, Zecker S. Neurobiologic responses to speech in noise in children with learning problems: Deficits and strategies for improvement. Clinical Neurophysiology. 2001;**112**:758-767

[48] Ladefoged P. A Course in Phonetics. Boston: Thomson Higher Learning; 2006

[49] Peixe BP, Silva DD, Biaggio EPV, Bruno RS, Sanguebuche TR, Garcia MV. Applicability of evoked auditory brainstem responses with complex stimuli in adults with hearing loss. International Archives of Otorhinolaryngology. 2018;**22**(3):239-244

[50] Sinha S, Basavaraj V. Speech evoked auditory brainstem responses: A new

tool to study brainstem encoding of speech sounds. Indian Journal of Otolaryngology and Head & Neck Surgery. 2010;**62**(4):395-399

[51] Schoof T, Rosen S. The role of auditory and cognitive factors in understanding speech in noise by normal-hearing older listeners. Frontiers in Aging Neuroscience. 2014;**6**:307

[52] Palmer SB, Musiek FE. Electrophysiological gap detection thresholds: Effects of age and comparison with a behavioral measure. Journal of the American Academy of Audiology. 2014;**25**(10):999-1007

[53] Lin F, Ferrucci L, Metter E, Um Y, Zonderman A, Resnick S. Hearing loss and cognition in the Baltimore longitudinal study of aging. Neuropsychology. 2011;**25**:763-770

[54] Sanfins M, Colella-Santos M. A review of the clinical applicability of speech-evoked auditory brainstem responses. The Journal of Health Science. 2016;**6**(1):9-16

[55] Ahadi M, Pourbakht A, Jafari AH, Jalaie S. Effects of stimulus presentation mode and subcortical laterality in speech-evoked auditory brainstem responses. International Journal of Audiology. 2014;**53**(4):243-249

[56] Anderson S, Parbery-Clark A, White-Schwoch T, Kraus N. Auditory brainstem response to complex sounds predicts self-reported speech-in-noise performance. Journal of Speech Language and Hearing Research. 2013;**56**(1):31-43

[57] Clinard C, Tremblay K. Aging degrades the neural encoding of simple and complex sounds in the human brainstem. Journal of the American Academy of Audiology. 2013;**24**(7):590-599

Chapter 8

Role of Glucocorticoids in Hearing Preservation in Partial Deafness Treatment

Magdalena B. Skarżyńska

Abstract

During the last 15 years, cochlear implantation became available as a method of treatment for different types of hearing impairment. Leading, specialized centers have now introduced the analysis of the nonsurgical factors that could contribute to improve rates of hearing preservation in preoperative period, during surgery, or in postoperative period in patients who suffer from partial deafness. One of the approaches is using pharmacotherapy (glucocorticoids) as a factor that may improve hearing functions. Preservation of hearing in patients who suffered from partial deafness and underwent cochlear implantation by using two different regimes of corticosteroid therapy was the aim of the study carried out by the *World Hearing Center* (WHC). Forty-six patients were enrolled in the trial and divided into three subgroups. Hearing preservation (HP) was evaluated using pure tone audiometry (PTA) (11 frequencies ranging from 125 to 8000 Hz). The impact of administrated substances was evaluated by pure tone audiometry during six different periods: before cochlear implant surgery, during activation of audio processor, and 1, 6, 9, and 12 months after activation of audio processor in comparison with control group. According to *hearing preservation (HP)* classification, patients from the second group, to whom combined glucocorticoid therapy was administrated, achieved the best HP results. The complete hearing preservation index was observed in the highest percentage of patients from the second subgroup. The dispersion of measured values was lesser than in other subgroups. According to the results, administration of glucocorticoids (dexamethasone and prednisone or dexamethasone only) to the patients, who suffered from partial deafness and underwent cochlear implantation surgery, may be important in stabilization of hearing thresholds and in protection of hearing.

Keywords: glucocorticoids, dexamethasone, prednisone, partial deafness treatment, cochlear implant

1. Introduction

In the past cochlear implantation was a "gold standard" treatment method for patients who suffer from hearing impairment and dedicated only for patients totally deaf. During the last 15 years, cochlear implantation became available also as a treatment for different types of hearing impairment. Regardless of surgical technique for cochlear implantation (cochleostomy or an approach through the round window), specialized centers have now introduced the comprehensive analysis of the nonsurgical factors that could contribute to improve rates of hearing preservation in preoperative, during surgery, or in postoperative period in patients who suffer from partial deafness. One of the approaches of many leading research centers is using pharmacotherapy (glucocorticoids) as a factor which may improve hearing functions following cochlear implantation [1, 2]. Glucocorticoids play an important role in pharmacotherapy of many different otorhinolaryngological diseases, such as Meniere's disease, sudden sensorineural hearing loss (SSNHL), and tinnitus, and as a part of otorhinolaryngological procedures in surgeries (e.g., cochlear implantation) [3, 4]. The effects of treatment listed diseases are different and mainly depend on treatment results, adverse effects of used medications, and additional pharmacological treatment that was used during treatment. Unfortunately, the side effects of glucocorticoids are serious, and as the result, sometimes pharmacotherapy has to be stopped, and discontinuation of therapy is the only solution.

On the one hand, insertion of specific electrode of cochlear implant requires perfection in surgical techniques, but on the other it is difficult to do perfectly. Clinically approved algorithm of corticosteroid therapy (local or systematic) is discussed as one factor in reducing oxidative stress, an inflammatory reaction, and as a result apoptosis of hearing cells. A major challenge in effective administration and delivery medicines is the *blood-labyrinth barrier* (BLB) and physical inaccessibility of the inner ear, especially the apical part of the cochlea. It seems to be crucial especially for patients who suffer from partial deafness (PD) (apical hair cells are responsible for receiving low frequencies).

1.1 Characteristic: glucocorticoids

The adrenal cortex synthesizes two classes of steroids: corticosteroids (mineralocorticoids and glucocorticoids) and androgens. One of the differences is the number of carbon atoms. Corticosteroids have 21 carbon atoms, and androgens have 19 carbon atoms. In the human body, the main glucocorticoid is cortisol, and the main mineralocorticoid is aldosterone [5].

Glucocorticoid receptor (GR) is located in the cytoplasm in inactive form until it binds with the molecule of glucocorticoid. This action results in activation of receptor and translocation complex: *glucocorticoid receptor for glucocorticoid* to the nucleus. Activation of receptor is based on dissociation from the associated proteins. After translocation to the nucleus, a complex of glucocorticoid receptor interacts with specific, short DNA sequences with the regulatory regions. The regions are termed *glucocorticoid responsive elements* (GREs) and allow induction of the gene transcription by glucocorticoids. This process is very complicated because of interaction with specific cofactors and proteins and still not well and completely understood by scientists and researchers [5]. Not only positive response to glucocorticoid is possible. According to Webster and Cidlowski, genes negatively regulated by glucocorticoids were also identified [6]. An example of downregulation (negative regulation) is to repress the expression of gene responsible for encoding cytokines or enzymes (e.g., collagenase). Both play an important role in inflammatory and immune reactions. According to the information provided, this negative expression appears to play a key function in anti-inflammatory and immunosuppressive effects of the glucocorticoids. Anti-inflammatory activity of representative glucocorticoids is presented below (**Table 1**). Dexamethasone and betamethasone are two glucocorticoids with the highest anti-inflammatory activity. If cortisol has anti-inflammatory activity defined as 1, then prednisone, prednisolone, triamcinolone, and 6α-methylprednisolone have 4–5 times stronger anti-inflammatory properties, with longer half-life than cortisol. Examples of representative glucocorticoids and their properties are shown in **Table 1**.

	Anti-inflammatory activity	Biological half-life $t_{1/2}$ (h)
Cortisol	1	Short: $t_{1/2}$ = 8–12
Cortisone	0.8	Short: $t_{1/2}$ = 8–12
Fludrocortisone	10	Intermediate: $t_{1/2}$ = 12–36
Prednisone	4	Intermediate: $t_{1/2}$ = 12–36
Prednisolone	4	Intermediate: $t_{1/2}$ = 12–36
6α-methylprednisone	5	Intermediate: $t_{1/2}$ = 12–36
Triamcinolone	5	Intermediate: $t_{1/2}$ = 12–36
Betamethasone	25	Long: $t_{1/2}$ = 36–72
Dexamethasone	25	Long: $t_{1/2}$ = 36–72

Table 1.
Characteristics of representative corticosteroids [5].

1.2 Functions and activity of glucocorticoids

The two key roles glucocorticoids play as biological and pharmaceutical compounds are anti-inflammatory and immunosuppressive roles. Glucocorticoids also affect:

1. Carbohydrate and protein metabolism. Glucocorticoids stimulate the liver to form glucose in biochemical reaction (gluconeogenesis) from amino acids and glycerol or/and stimulate the liver to release glucose from glycogen. At the same time, the diminishing of glucose is reduced, reaction of lipolysis and protein breakdown increases, and as a result the blood glucose level rises. Patients suffering from diabetes or other forms of hyperglycemia during glucocorticoid therapy should be under special control. Glucocorticoids induce increased protein metabolism and deliver compounds such as amino acids for further reactions.

2. Lipid metabolism. One effect of therapy with corticosteroids is redistribution of fat tissue known as Cushing's syndrome.

3. Water and electrolyte balance. Glucocorticoids exert negatively on metabolism of Ca^{2+} due to reduction of absorption from the digestive system and increased excretion via kidneys. Prophylaxis of osteoporosis requires supplementation of Ca^{2+} ions and physical activity adequate to possibilities of patient. Additionally, glucocorticoids reduce activity of osteoclasts and stimulate the activity of osteoblasts. Glucocorticoid therapy sometimes may cause increased retention of Na^{+} ions and decrease in concentration of K^{+} ions because of interaction with receptor for mineralocorticoids. The deterioration in ion level may affect the cardiovascular system.

4. Impaired wound healing. Due to reduction of synthesis of collagen, glycosaminoglycans and disturbance in fibroblast function problem with healing wound may occur.

5. Anti-inflammatory and immunosuppressive activity. Glucocorticoids can suppress or prevent inflammatory reactions in different ways: reduction in diapedesis of granulocytes and proliferation of lymphocytes Th, inhibition/

reduction of activation of macrophages, neutrophils, mast cells and cytokines (interleukins 1, 2, 3, 4, 5, 6, 8), and tumor necrosis factor alpha (TNF-α); reduction in the expression of cyclooxygenase 2 (COX-2) resulting in dropping of production of a few prostanoids; intensification of activity of catechol amines; and reduction in production of histamine by basophils [7, 8].

Sometimes glucocorticoid treatment in the otorhinolaryngological diseases requires high doses or long time of therapy. It may cause adverse effects. Some of them are listed below:

1. Repression in responding on infections and injuries

2. Propensity to opportunistic infections

3. Propensity to hyperglycemia

4. Muscular dystrophy

5. Cushing's syndrome

6. Glaucoma (mostly in patients genetically predisposed)

7. Osteoporosis

Acute discontinuation of treatment may cause adrenocortical insufficiency, especially when therapy was long-term. It is important to reduce the dose of glucocorticoids slowly, not suddenly [5].

1.3 Pharmacokinetics of glucocorticoids and studies on animal model

Many factors have an impact on pharmacokinetics of drugs. Pharmacokinetics is described by acronym *LADME (liberation, absorption, distribution, metabolism, and elimination)*. Firstly, the therapeutic (or its carrier) must be water-soluble, because of distribution in the blood. The protein binding of drugs is one of the key factors in initial parts in pharmacokinetic process. The greater the protein binding of drug is, the longer the activity of therapeutic due to its function as a stock of drug in the organism. Absorption depends on lipophilicity and solubility of drugs. According to the data and publications, only a few medical substances can effectively be used in otorhinolaryngological practice due to achieving sufficient concentration in the inner ear [9]. Two main groups of drugs are used in clinical practice: aminoglycoside (mainly gentamicin) in pharmacotherapy of Meniere's disease [10] and corticosteroids (dexamethasone, triamcinolone and dexamethasone) in pharmacotherapy of *idiopathic sudden sensorineural hearing loss* (ISSHL) and other cases of acute hearing loss [11]. The inner ear from a pharmacokinetic point of view is a multicompartment model [12, 13] with stable fluids and balance between them (due to the presence of *blood-labyrinth barrier* (BLB)). Distribution process depends on many different factors such as route of administration, model of administration (single or repeated administration), dose of medicine, ionic composition, and pH or osmolarity of solution. The same factors of drug chemical and physical properties influence the elimination of drug from the organism (clearance, rate of removal).

In a study published in 2017, authors in animal model (guinea pig) compared dexamethasone with saline. Both substances were administrated intravenously 60 min before implantation. As a final conclusion, authors stressed that dexamethasone could reduce scarring process as the electrode negotiated the hook region or near the electrode tip, but they did not observe the relation between dexamethasone and reduction of fibrosis relating to cochleostomy [13]. In vitro studies showed the correlation between reduction (loss) of auditory cells after exposure to *tumor necrosis factor* alpha (TNF-alpha) and dexamethasone-releasing polymer used to coat electrode of cochlear implant carries.

Research carried out on animal model proved that prolonged steroid therapy could significantly improve hearing preservation rate (including pharmacokinetic and morphological analysis) when the electrode of cochlear implants was covered with dexamethasone (special formulation with controlled drug release) [14]. However, Honeder et al. did not confirm that steroids could have a positive impact on residual hearing in a guinea pig model. One reason why both authors gain different results may lie in different types of steroid therapy. In the first study, dexamethasone was used on the contrary to the second study where triamcinolone was administrated [15]. Douchement et al. investigated the effects of steroids using a gerbil animal model. Animals were implanted with an electrode with controlled dexamethasone delivery (1 and 10% concentration of dexamethasone) on one side and a conventional electrode on the contralateral side. Hearing levels were established based on the tone bursts on auditory brain stem responses at 4–6-week postimplantation and at 1-year postimplantation period for older gerbils. A 1-year observation period showed significantly improved results obtained for the high auditory frequencies, but the results for the low frequencies were ambiguous [16].

Cho et al. analyzed the efficacy of preoperative and intraoperative schemes of administration of steroids for hearing preservation. Dexamethasone was administrated systematically at the dose of 5 mg/ml in the preoperative period and then topically (off-label) during cochlear implantation surgery. Pure tone audiometry (PTA) was measured in four frequencies: 250, 500, 1000, and 2000 Hz. Statistically significant differences were observed between the steroid group and the control group, supporting the observation and beneficial impact of administration steroid treatment [17].

1.4 Glucocorticoid administration: the possibilities

During treatment of otologic diseases, two routes of administration of glucocorticoids are possible: local and systemic. Local administration (e.g., transtympanic injection) allows to achieve high concentration of glucocorticoid in the middle ear, but due to presence of Eustachian tube, the medication may be partly evacuated. Local drug administration to the middle and inner ear avoids "the first pass effect." The main advantages of local drug delivery are as follows:

1. Reduction dose of medicine
2. Achieving high concentration
3. Better effects of treatment
4. Reducing possibility of adverse effects
5. Bypassing of the blood-labyrinth barrier (BLB)

Local drug administration may involve intracochlear administration (e.g., stem cell, gene therapy) or extracochlear administration (e.g., intratympanic injection). A combination of both routes of drug delivery to the ear is also possible [13]. According to publications, systemic administration in treatment of otorhinolaryngological diseases is known as noninvasive route of drug delivery, due to lack of damaging of tympanic membrane. Adverse effects of systemic delivery may be one of the purposes of discontinuation of the therapy. The presence of *blood-labyrinth barrier* (BLB) in the inner ear is one of the causes of problems with reaching high concentration of drug.

1.5 The experience in using glucocorticoids in PDT patients

This study was the first study which was carried out in the Institute of Physiology and Pathology of Hearing World Hearing Center and will be continued with different groups of patients and different implants and algorithms of glucocorticoid administration. The aim of the study was to evaluate different regimes of administration of glucocorticoids: dexamethasone and dexamethasone/prednisone to *partial deafness* treatment (PDT) patients who underwent cochlear implantation on the hearing preservation. Implant used in the study was the MED-EL implant with an electrode length of 28 mm (Flex 28). The impact of administrated glucocorticoids on hearing was measured in six different periods:

1. Preoperatively
2. During activation of audio processor
3. One month after activation of audio processor
4. Six months after activation of audio processor
5. Nine months after activation of audio processor
6. One year (12 months) after activation of audio processor

Forty-six patients were enrolled to the trial and then divided randomly into three subgroups. Patients from the first subgroup underwent intravenous steroid therapy (**Figure 1**). According to the scheme, 30 min before implantation, dexamethasone at the dose of 0.1 mg/kg body mass was administrated intravenously to patients from the first subgroup. For the next 3 consecutive days in every 12 h, dexamethasone was administrated intravenously at the same dose to each patient.

Patients from the second subgroup underwent prolonged (combined) steroid therapy: oral and intravenous. Three days prior to the surgery, prednisone was administrated orally at a dose of 1 mg/kg body mass/day. Then, 30 min before the cochlear implantation surgery, dexamethasone was administrated intravenously at a dose of 0.1 mg/kg body mass. For the next 3 consecutive days in every 12 h, dexamethasone was administrated intravenously (the same as in the first subgroup). For the next 3 days, prednisone was administrated orally at the dose of 1 mg/kg body mass/day. After this period, the dose of prednisone was reduced (10 mg per every

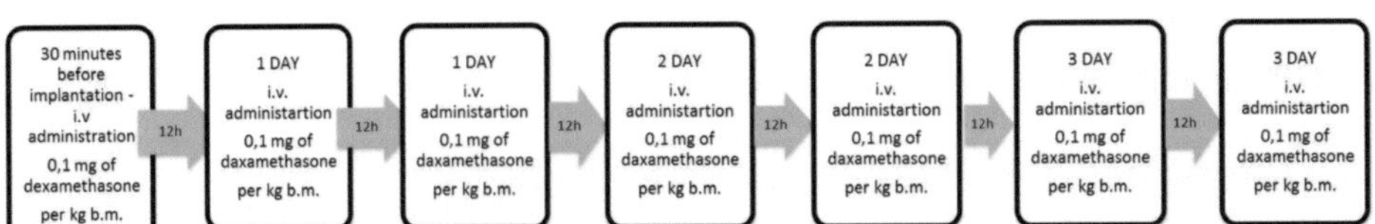

Figure 1.
Scheme of steroid administration in the first subgroup of patients.

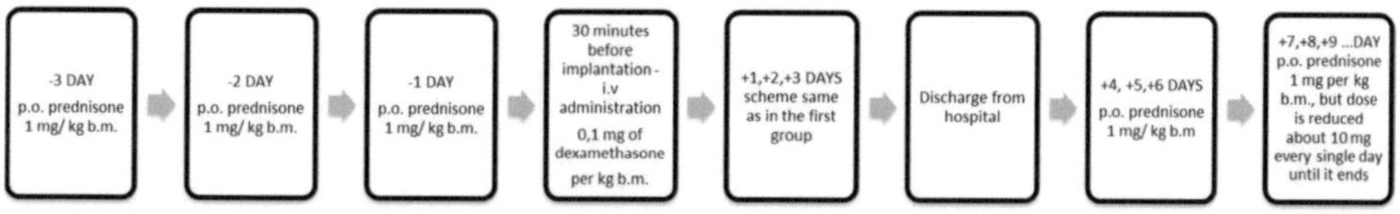

Figure 2.
Scheme of steroid administration in the second subgroup of patients.

day) till complete reduction of dose, due to reducing the risk of adverse effects. The algorithm of administration of glucocorticoids to patients form the second subgroup is presented below (**Figure 2**).

The third subgroup was a control group and underwent standard cochlear implantation procedure [18].

1.6 Administrated glucocorticoids

According to the protocol of this study, two different algorithms of administration with two different glucocorticoids were proposed in the study. Although both substances belong to the same pharmacological group, both of them have different pharmacokinetics and pharmacodynamic properties. Dexamethasone is a synthetic glucocorticoid (*molecular weight 392.46 g/mol*) with anti-inflammatory, anti-allergic, and immunomodulating activity. In common practice dexamethasone is administrated *intravenously* or off-label, e.g., *transtympanic injections*. After *intravenous* administration the mean time to peak concentration (C_{max}) is between 10 and 30 min, and the half-life ($t_{1/2}$) is from 2.2 to 3.8 h. Transport proteins are responsible for transport and distribution of dexamethasone in blood. Dexamethasone is mainly metabolized by the liver and eliminated with the bile. Only 2.6% of the chemically unchanged dose is eliminated via kidneys.

Prednisone is a synthetic glucocorticoid (derivative of cortisone) and classified according to the *Anatomical Therapeutic Chemical (ATC) Classification System* as H02 AB 07. Prednisone is *prodrug* which converts into *active metabolite—prednisolone* (higher anti-inflammatory activity). According to characteristic of medical product and literature data, bioavailability of prednisone administrated orally is between 70 and 90%. The mean time to peak concentration (C_{max}) is between 1 and 2 h. Half-life ($t_{1/2}$) is between 3.4 and 3.8 h in plasma and 18–36 h in tissue. Binding prednisone with plasma proteins is between 70 and 73% (binding prednisolone (active metabolite) to the plasma proteins is higher (90–95%)). Similar to dexamethasone, prednisone is metabolized mainly by the liver and eliminated with the bile. Pharmacodynamic and pharmacokinetic data were based on characteristics of medical products: dexamethasone and prednisone.

1.7 Methodology of the study: primary and secondary outcomes and inclusion and exclusion criteria

The primary outcome variables were mean values of hearing thresholds averaged across all 11 frequencies (125–8000 Hz). The secondary outcome variable was hearing preservation (HP). Hearing preservation (HP) was calculated by comparing hearing thresholds in the 1-year postoperative period with the preoperative hearing thresholds, according to the hearing preservation (HP) formula (below) and converted to three levels: minimal, partial, and complete hearing preservation.

$$\mathrm{HP} = \left(1 - \frac{PTApost - PTApre}{PTAmax - PTApre}\right) * 100\ (\%) \quad (1)$$

In this equation, *PTApre* is the pure tone average measured preoperatively, *PTApost* is the pure tone average measured postoperatively, and *PTAmax* is the maximal sound intensity generated by a standard audiometer, usually 120 dB hearing level (HL), and *HP* is the rate of hearing preservation in percentage [19].

The protocol of this prospective clinical trial was approved by the Bioethics Commission. Patients enrolled to the study suffered from severe-to-profound hearing loss and were classified according to Prof. H. Skarżyński *partial deafness treatment (PDT) classification* into two groups: *partial deafness treatment-electrical* stimulation (PDT-EC) and *partial deafness treatment-electroacoustic* stimulation (PDT-EAS) (**Figure 3**) [20, 21].

Inclusion and exclusion criteria were in accordance with the consensus of the international *HEARRING* group on hearing preservation in cochlear implant. Study eligibility criteria included participants ≥18 years of age with a cochlear duct length of ≥27.1 (measured by computer tomography), with:

1. Hearing sound levels in the range of 10–120 decibels (dB) and sound frequencies of 125–250 hertz (Hz)

2. Hearing sound levels of 35–120 dB and frequencies of 500–1000 Hz

3. Hearing sound levels of 75–120 dB and frequencies of 2000–8000 Hz [18]

Exclusion criteria included suffering from severe diseases when the steroid treatment could worsen the patient's condition or when there would be a possibility of interaction between medication intake by patients and steroids. Nonparametric tests were used in the study due to discrepancies in the number of participants between all subgroups, small number of participants in the study, and violation of normal distribution of pure tone audiometry results [18].

1.8 Main results and observations

Preoperative hearing threshold levels of patients from the first, the second, and the control subgroup were similar. The difference between patients from the three subgroups was not statistically significant, which means that hearing

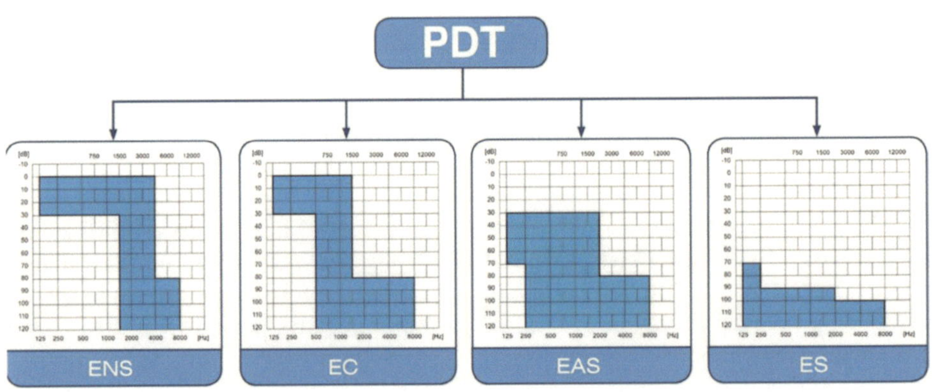

Figure 3.
Partial deafness treatment groups for cochlear implantation. ENS, electro-natural stimulation; EC, electrical complement; EAS, electrical-acoustic stimulation; ES, electrical stimulation.

thresholds in preoperative period of all participants, who were enrolled to the study, were similar.

The deterioration of average hearing thresholds (measured by pure tone audiometry) was observed from the first point of observation—the activation period. A significant difference was observed between two groups: patients from the second subgroup (combined steroid therapy, prednisone + dexamethasone) and the control subgroup. Patients from the second subgroup had better pure tone audiometry (PTA) results considering low frequencies in comparison with the results of patients from the control group. Similar observation was done in 1, 6, 9, and 12 months after activation follow-up periods. The results of the study may be even more a promise and beneficial for patients. The hearing of participants of the study, to whom combined (prolonged) glucocorticoid therapy was administrated, remained stable during all observed follow-up periods (*activation, 1-month, 6-month, 9-month, and 12-month post-activation follow-ups*), and they did not vary significantly (**Figure 4**).

The hearing preservation (HP) rate is calculated using hearing preservation formula by comparing hearing threshold in the 12-month postoperative period with the preoperative hearing thresholds. Then the results were divided, according to the hearing preservation (HP) formula, into minimal HP, partial HP, and complete HP according to **Table 2**. The smallest variability of results was observed in the second subgroup (patients to whom prednisone and dexamethasone were administrated) as well as the highest overall HP rate. Patients from the second subgroup (prolonged steroid therapy) and nearly 69% of the patients from the first subgroup had partial or complete hearing preservation. The majority of patients from the control group had minimal hearing preservation (**Table 2** and **Figure 5**).

1.9 Final conclusion

This study is the first study to report the results of two different regimes of steroid administration in human subjects who underwent cochlear implantation in comparison with the control group. As it was said in the previous paragraph, the findings of this study have shown that glucocorticoid therapy not only stabilizes hearing thresholds but also preserves hearing ability in adult patients. The combination of *intravenously* administrated dexamethasone and *orally* administrated prednisone in one scheme of administration seems to be the optimal treatment regimen.

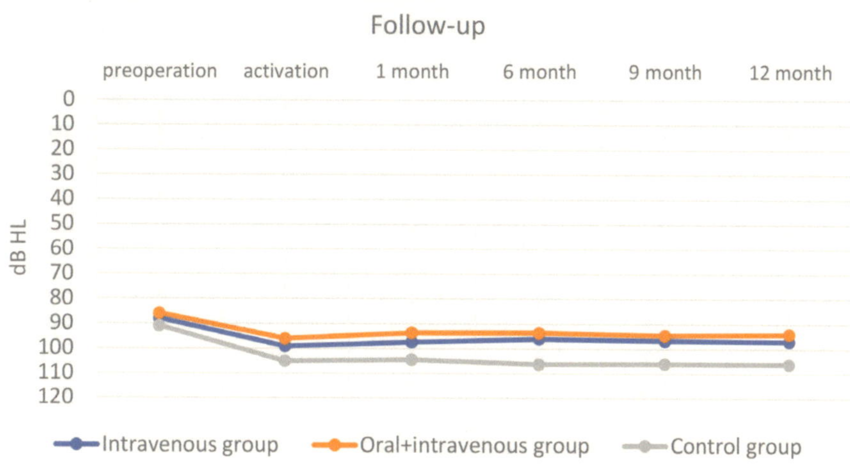

Figure 4.
Mean hearing thresholds of patients with standard steroid therapy (subgroup No 1), patients with prolonged steroid therapy (subgroup No 2), and control patients (subgroup No 3) in the preoperative period, upon activation, and at 1-month, 6-month, 9-month, and 12-month follow-up after CI implantation.

	Minimal (0–25%)	Partial (26–75%)	Complete (≤75%)
Subgroup No 1	5 (31.2)	7 (43.8)	4 (25.0)
Subgroup No 2	0 (0.0)	8 (61.5)	5 (38.5)
Control group	12 (70.6)	3 (17.6)	2 (11.8)

Table 2.
Hearing preservation 12 months after CI implantation, according to the type of treatment (data are given as the number of patients (percentage in brackets).

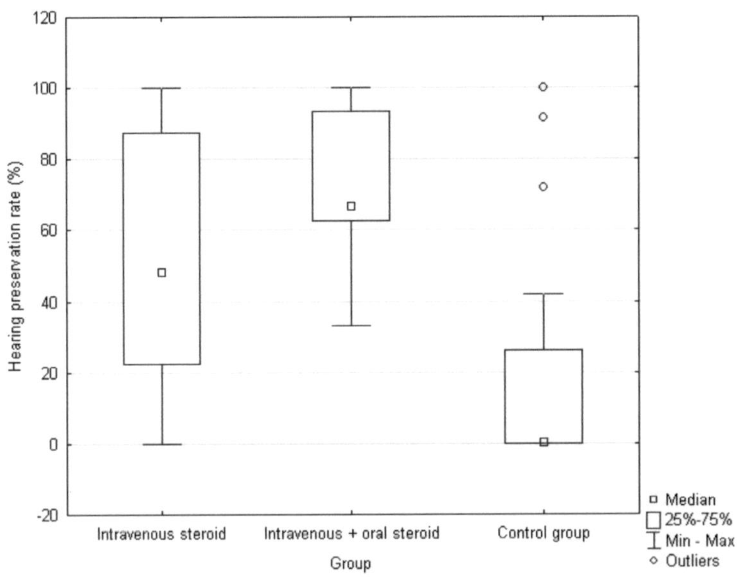

Figure 5.
Hearing preservation (HP) rate in three subgroups.

Previously published studies have shown that there have been new directions in the development and use of electrodes and cochlear implant surgery in recent years. Currently, researchers, clinicians, and commercial companies are working on developing modern steroid-eluting electrodes or electrodes with controlled drug delivery. The results of the preliminary study described in this chapter suggested that combined glucocorticoid administration (according to scheme of administration in the second subgroup) is beneficial in preserving and stabilizing hearing thresholds in patients undergoing cochlear implantation surgery. The findings of this study are supported by the results of similar studies [1, 17]. However, the present study adds to the findings of previous studies by having a relatively long follow-up period, of 12 months after activation, with study analysis conducted during six different follow-up periods. According to the results, administration of glucocorticoids (dexamethasone and prednisone or dexamethasone only) to the patients, who suffered from partial deafness and underwent cochlear implantation surgery, may be important in stabilization of hearing thresholds and in protection of hearing. The dispersion of measured values in the second group (the second subgroup) was lesser than in the first and the control group.

Author details

Magdalena B. Skarżyńska[1,2,3]

1 Institute of Sensory Organs, Kajetany, Polnad

2 Center of Hearing and Speech Medincus, Kajetany, Poland

3 World Hearing Center, Institute of Physiology and Pathology of Hearing, Warsaw/Kajetany, Poland

*Address all correspondence to: m.skarzynska@csim.home.pl

References

[1] Sweeney AD, Carlson ML, Zuniga MG, Bennett ML, Wanna GB, Haynes DS, et al. Impact of perioperative oral steroid use on low-frequency hearing preservation after cochlear implantation. Otology & Neurotology. 2015;**36**(9):1480-1485

[2] Rah YC, Lee MY, Kim SH, Kim DH, Eastwood H, O'Leary SJ, et al. Extended use of systemic steroid Is beneficial in preserving hearing in guinea pigs after cochlear implant. Acta Oto-Laryngologica. 2016;**136**(12):1213-1219

[3] Chrousos GP. Adrenocorticosteroids and adrenocortical antagonists. In: Basic and Clinical Pharmacology. The McGraw-Hill Companies Inc.; Chapter 39, edition 13. 2015

[4] Alles MJRC, der Gaag MA, Stokroos RJ. Intratympanic steroid therapy for inner ear diseases, a review of the literature. European Archives of Oto-Rhino-Laryngology. 2006;**263**(9):791-797

[5] Brunton L, Knollman B, Hilal-Dandan R. Goodman and Gilman's The Pharmacological Basis of Therapeutics. T. 1. 13. wyd. McGraw-Hill Education; 2017

[6] WebsterJC, CidlowskiJA. Mechanisms of glucocorticoid-receptor-mediated repression of gene expression. Trends in Endocrinology and Metabolism. 1999;**10**(10):396-402

[7] Munck A, Guyre PM, Holbrook NJ. Physiological functions of glucocorticoids in stress and their relation to pharmacological actions. Endocrine Reviews. 1984;**5**(1):25-44

[8] Lo J, Campbell L, Sale P, Chambers S, Hampson A, Eastwood H, et al. The role of preoperative steroids in atraumatic cochlear implantation surgery. Otology & Neurotology. 2017;**38**(8):1118-1124

[9] Salt AN, Plontke SK. Pharmacokinetic principles in the inner ear: Influence of drug properties on intratympanic applications. Hearing Research. 2018;**368**:28-40

[10] Lange G. Gentamicin and other ototoxic antibiotics for the transtympanic treatment of Menière's disease. Archives of Oto-Rhino-Laryngology. 1989;**246**(5):269-270

[11] Hamid M, Trune D. Issues, indications, and controversies regarding intratympanic steroid perfusion. Current Opinion in Otolaryngology & Head and Neck Surgery. 2008;**16**(5):434-440

[12] Liebau A, Plontke SK. Lokale Medikamententherapie bei Innenohrschwerhörigkeit. HNO. 2015;**63**(6):396-401

[13] Plontke SK, Götze G, Rahne T, Liebau A. Intracochlear drug delivery in combination with Cochlear implants. HNO. 2017;**65**(1):19-28

[14] Liu Y, Jolly C, Braun S, Stark T, Scherer E, Plontke SK, et al. In vitro and in vivo pharmacokinetic study of a dexamethasone-releasing silicone for Cochlear implants. European Archives of Oto-Rhino-Laryngology. 2016;**273**(7):1745-1753

[15] Honeder C, Zhu C, Schöpper H, Gausterer JC, Walter M, Landegger LD, et al. Effects of sustained release dexamethasone hydrogels in hearing preservation cochlear implantation. Hearing Research. 2016;**341**:43-49

[16] Douchement D, Terranti A, Lamblin J, Salleron J, Siepmann F, Siepmann J, et al. Dexamethasone eluting electrodes for Cochlear implantation: Effect on residual hearing. Cochlear Implants International. 2015;**16**(4):195-200

[17] Cho HS, Lee K-Y, Choi H, Jang JH, Lee SH. Dexamethasone is one of the factors minimizing the inner ear damage from electrode insertion in cochlear implantation. Audiology & Neuro-Otology. 2016;**21**(3):178-186

[18] Skarżyńska MB, Skarżyński PH, Król B, Kozieł M, Osińska K, Gos E, et al. Preservation of hearing following cochlear implantation using different steroid therapy regimens: A prospective clinical study. Medical Science Monitor: International Medical Journal of Experimental and Clinical Research. 2018;**24**:2437

[19] Skarżyński H, Lorens A, Skarżyński PH. Electro-natural stimulation (ENS) in partial deafness treatment: A case study. The Journal of Health Science. 2014;**4**(4):67-71

[20] Skarzynski H, van de Heyning P, Agrawal S, Arauz SL, Atlas M, Baumgartner W, et al. Towards a consensus on a hearing preservation classification system. Acta Oto-Laryngologica. Supplementum. 2013;(564):3-13

[21] Skarzynski H, Lorens A, Dziendziel B, Skarzynski PH. Expanding pediatric cochlear implant candidacy: A case study of electro-natural stimulation (ENS) in partial deafness treatment. International Journal of Pediatric Otorhinolaryngology. 2015;**79**(11):1896-1900